Substance and Method

Studies in Philosophy of Science

Substance and Method

Studies in Philosophy of Science

Chuang Liu

World Scientific

NEW JERSEY • LONDON • SINGAPORE • BEIJING • SHANGHAI • HONG KONG • TAIPEI • CHENNAI • TOKYO

Published by

World Scientific Publishing Co. Pte. Ltd.
5 Toh Tuck Link, Singapore 596224
USA office: 27 Warren Street, Suite 401-402, Hackensack, NJ 07601
UK office: 57 Shelton Street, Covent Garden, London WC2H 9HE

Library of Congress Cataloging-in-Publication Data
Liu, Chuang, 1958– author.
Substance and method : studies in philosophy of science / Chuang Liu.
pages cm
Includes bibliographical references.
ISBN 978-9814632188 (hardcover : alk. paper) -- ISBN 981463218X (hardcover : alk. paper)
1. Science--Philosophy. 2. Science--Methodology. I. Title.
Q175.L588 2015
501--dc23
2015031775

British Library Cataloguing-in-Publication Data
A catalogue record for this book is available from the British Library.

In-house Editor: Christopher Teo

Printed in Singapore

Preface

This volume contains a selection of my written works between 1992 and present. Most of the included papers have been published previously or are forthcoming. The collection is divided into parts that represent the areas I have worked in during this period, and the selection is aimed at including the major results of my effort in those areas. The papers are not arranged chronologically but grouped into parts with titles that fit them loosely so that introductory notes can be written for each part.

No attempt has been made to revise or update any of the included papers from their original published forms, but besides brief summaries, the introductory notes contain accounts of the new developments and their relevant references. Readers who find the papers out-of-date but still wish to know more will find useful information to further pursue the issues.

Unlike one of the stars in the philosophical constellation in my intellectual heaven, David Lewis, who once said that he "should have liked to be a piecemeal philosopher" (Lewis 1983, ix), working to solve one philosophical problem or paradox and then another, without an overarching program, but eventually "succumbed to" (*ibid.*) the gospels that include Humean supervenience and realism of possibilia, I have included in this volume of a body of work that is piecemeal in its appearance and piecemeal by nature.

To those I have acknowledged in the included papers on the occasions of their publication, I here again register my gratitude. And I leave my thanks to those who helped me in writing the few unpublished pieces included in this volume to the footnotes of those pieces. I thank Dr. K. K. Phua for inviting me to publish my works at the World Scientific, and Mr. Christopher Teo for patiently editing the volume and helping me in every step through the process.

Acknowledgement

I thank the editors and publishers for granting me the permissions to reprint the papers that appear in this volume of collected papers. Below I list the included papers and the places they are first published.

2015. "Re-inflating the Conception of Scientific Representation," *International Studies in the Philosophy of Science* (forthcoming)

2005. (with G. Emch) "Explaining Quantum Spontaneous Symmetry Breaking," *Studies in History and Philosophy of Modern Physics* **36**: 137-163.

2005. "Review: *Symmetries in Physics: Philosophical Reflections* (CUP, 2003)" *International Studies in the Philosophy of Science* **19**: 85-88.

2004. "Laws and Models in a Theory of Idealization," *Synthese* **138**: 363-385.

2004. "The Possibility of Unifying Gravity and Quantum Gauge Fields," *International Studies in the Philosophy of Science **17**: 143-159.*

2004. "Approximations, Idealizations, and Models in Statistical Mechanics," *Erkenntnis* **60**: 235-263.

2003. "Classical Spontaneous Symmetry Breaking," *Philosophy of Science* **70**(S): S1219-S1232.

2003. "Spontaneous Symmetry Breaking and Chance in Classical World," *Philosophy of Science* **70**: 590-608.

2001. "Infinite Systems in SM Explanations: Thermodynamic Limit, Renormalization (semi-) Groups, and Irreversibility," *Philosophy of*

Science **68**(S): S325-344.

1999. "Explaining the Emergence of Cooperative Phenomena," *Philosophy of Science* **66**(S): S92-S106.

1999. "Approximation, Idealization, and the Laws of Nature," *Synthese* **118**: 229-256.

1998. "Theories and Models II: Issues and Applications," *International Studies in the Philosophy of Science* **12**: 111-128.

1997. "Theories and Models I: the Semantic View Revisited," *International Studies in the Philosophy of Science* **11**: 147-164.

1996. "Potential, Propensity and Categorical Realism," *Erkenntnis* **45**: 45-68.

1996. "Realism and Spacetime: Of Arguments Against Metaphysical Realism and Manifold Realism," *Philosophia Naturalis* **33**: 243-263.

1996. "Gauge Invariance, Cauchy Problem, Indeterminism, and Symmetry Breaking," *Philosophy of Science* **63**(S): S71-S80.

1994. "The Aharonov-Bohm Effect and the Reality of Wave Packets," *British Journal for the Philosophy of Science* **45**: 977-1000.

1994. "Is There a Relativistic Thermodynamics: A Case Study of the Meaning of Special Relativity," *Studies in History and Philosophy of Science* **25**: 983-1004.

Contents

Part 1
Models and Fiction

Three of the four papers included in this part appear for the first time in this volume of collected papers. The papers deal with issues with respect to scientific representation, especially with scientific models. Two themes run through the papers: one against deflationism, and the other against fictionalism, on scientific models. Scientific models are shown to be fundamentally different from symbols or symbolic representational vehicles in that they function mainly as vehicles of showing or demonstration rather than as vehicles of detonation or reference (these latter functions being often fulfilled by the symbols that are embedded in models). Scientific models are also fundamentally different from fictional representations of reality, despite sharing many features, such as the use of imagination and pretending. Apart from the differences of aims and methodologies (such as the presence or absence of the demand for the checking of factual details), model building and fiction writing also differ in the way in which their "characters" are introduced in the works (cf. Paper 2).

Since the papers are of recent origin, the references given in their bibliographies are current.

Fictional Models in Science

Ah! sachez-le: ce drame n'est ni une fiction, ni un roman. *All is true*, il est si véritable, que chacun peut en reconnaître les éléments chez soi, dans son coeur peut-être.

– Honoré de Balzac, *Pére Goriot*

All the world's a stage,
And all the men and women merely players;

– Shakespeare, *As You Like It*, II. vii

1. Introduction

I shall investigate the question of whether models in science can in general be properly understood as similar to works of fiction. And since the answer, as I shall argue, is "no, not in general," I further inquire whether or not there are some models in science that are indeed fictional, or better which ones are obviously so and which ones may well turn out to be so despite their initial appearance to the contrary.

Before I go into the main argument, a brief summary of what has so far been ascertained in the recent studies of models and modeling should help set the ground for our discussion. This is especially important because to find out whether or not models are works of fiction without having a firm and clear idea of what models are may be all but impossible.

Whether scientific models are part of scientific theories (à la van Fraassen (1980); Giere (1988); and their supporters) or "mediators" that are external to theories, as some philosophers of science would call them (à la Cartwright (1999) and her supporters), they are distinguished from theories by being non-truth-apt, i.e., they are not the sort of things that by themselves can be said to be true or false (cf. Godfrey-Smith 2009; Frigg 2010; Contessa

2009, 2010; Levy 2010, 2012). Statements can be made about them or their parts that are truth-apt, and therefore, models are more like things and events in reality than statements or equations in theories or explanations. Whichever approach one adopts, our representation of the world around us is no longer, as traditionally conceived, essentially a "word-world" relationship, whose main component comprises meaning and reference. A different and apparently more opaque relationship between the models and the world appears on the scene that generates a new set of problems about representation that are different from the traditional challenges in semantics (cf. Godfrey-Smith 2009; Frigg 2010).

Models are of very diverse kinds: some are physical objects, such as a scale model of an airplane, and others are apparently non-physical, such as the mechanical model of luminiferous aether; some are used to represent systems that are of different kinds, such as using a plastic structure to represent an organism, but others are of the same kind as the represented, such as a model organism to represent the type of that organism (cf. Keller 2002; Rowbottom 2009; see also Goodman 1976 for a more general point). Some models are constructed to represent the whole target system, such as an elaborate scale model of a bridge, while others may only represent one or a few selected aspects, such as a mechanical model of an economy or a map of a metropolitan area. Some models are abstract, while others are concrete. The abstract-concrete pair can be distinguished at two different levels. Among the models that are physical objects, some may be abstract and some concrete. A fruit fly *qua* organism is a concrete system, and yet as a biological model, an exemplar for some salient aspects of biological organisms, it is an abstract model, while a wood model for the earth is concrete as a model, for it is not meant to represent a type of objects, but a particular one, the earth. Among the non-physical models, most of them are of course abstract; Mendel's model for heredity in sexual reproduction is an abstract model and so is Maxwell's model for electromagnetism. They may be realized in different physical copies with different initial and boundary conditions, and yet the copies let us visualize the model but are not the model itself, just as John Gielgud's Hamlet is a physical realization of Hamlet but not Hamlet himself. But there are also concrete non-physical models in science, the Ptolemaic geocentric model of the universe being an example.

In spite of their diversity and ubiquity, it would, however, be a mistake, as Godfrey-Smith (2006, 2009) rightly pointed out, to regard all enterprises of science as essentially involving models and model building. Still, one couldn't go wrong by saying that the "model-based" (Godfrey-Smith's phrase) approach is one of the most effective and widely used strategies in science today, which cuts across all major disciplines from physics to economics. Now, are there any common characteristics of models that can help us to begin our study of them in general?

One common feature of all models is that they are vehicles of representation, a functional characteristic, if you will. However, this feature is so general that many more things, such as linguistic or symbolic vehicles of representation, share this feature with models. Some (cf. Callender and Cohen 2006) find this fact remarkable and argue for a deflationary view on scientific modeling or representation. According to this view, models are no different from other types of representational vehicles, such as symbolic ones, and while the suitability of different types of models for different jobs in science must be separately determined by pragmatic concerns (specifiable in their job descriptions), all models represent their targets in virtue of being conventionally chosen as representational vehicles (see also Teller 2001; van Fraassen 2008). Arguments against deflationism also surfaced recently (cf. Toon 2010b).

Another common feature of models is the tripartite nature of model representation. Instead of the traditional picture of scientific representation, which regards all representation, model-based or not, as a binary relation between symbolic systems and phenomena, we now have, according to Giere (1988), which is later modified by Godfrey-Smith (2009) (see also Frigg 2010), a three term structure of "description", "models," and "phenomena," where the relation between the first two parts is a specification of models, and the relation between models and phenomena is non-semantic; and this latter relation remains mysterious until one gets a good grip on what models are ontologically and functionally. Note that the models in this picture do not have to be regarded as abstract entities; concrete physical systems can serve the role just as well (cf. Goodman 1976; Rowbottom 2009).

Adopting such an account raises the question concerning the ontology of models. If they are non-truth-apt and bear a relationship with reality or phenomena that is not part of logic and semantics, then what are they and how do they represent (cf. Contessa 2009, 2010; Levy 2010, 2012)? Partly because of difficulties in answering such questions and partly because of the temptation from other quarters, a sizable literature has been generated in recent years in which mostly positive explorations have been made into the possibility that "scientific models are works of fiction" (cf. papers in Suárez 2009; Frigg 2010, 2011a, b; Godfrey-Smith 2006, 2009; Toon 2010a, b). Representations via models, according to the fictionalist view, are very much like the representations in art and literature: they "make fiction" of the target systems or phenomena, which yield fictional objects, i.e. models, on which scientists operate and learn. Idealization or something like it is often taken to be the means by which models are made (or reality is made fictional); and how models represent reality may be gleaned from how fictional characters and events relate to actual people and events, such as the case in the relationship between *War and Peace* and the Napoleonic War or between *Middle Earth* and Europe in the Middle Ages (cf. Godfrey-Smith 2006).

In this paper, I begin with a discussion of Giere's recent work arguing against taking models to be works of fiction. I then move on to explore a spectrum of scientific models that scientists themselves regard as obviously fictional or fictional but not obviously so. And then I discuss the modeling of the unobservable and make a case for the idea that despite difficulties of defining them, unobservable systems are modeled in a fundamentally different way than observable systems. While idealization and approximation is key to the making of models for the observable systems, they are in fact not operable, at least not straightforwardly so, regarding models for the unobservable. And because of this point, which is so far neglected in the literature, I argue that fictionalism may have a better chance with models for the unobservable. However, this fictionalism is not the one defended in the above-cited literature; it belongs to an old tradition of fictionalism in philosophy of science that is incompatible with scientific realism (cf. Fine 1993 and references therein).

2. No Works of Fiction

In a recent article, Giere (2009) explained "Why Scientific Models Should Not Be Regarded as Works of Fiction." As a conclusion for all models in science, I disagree with Giere, as I shall explain later; and yet I agree with him that many types should not be so regarded. Some of Giere's arguments for his conclusion I cannot concur with, yet I found more and stronger reasons to support his conclusion (see also, Giere 2002, 2011).

Giere begins the article by claiming that models and fictional objects are ontologically the same but functionally different. "Theoretical models and works of fiction are, I will argue, ontologically on a par. It is their differing functions in practice that makes it inappropriate to regard scientific models as works of fiction." (Giere 2009, 249) Because Giere has always regarded models as abstract entities (Giere 1988), this "ontologically on a par" might be interpreted as saying that both models and fictional objects are abstract entities of one sort or another. An immediate observation about this claim, given what I have said above, is that it does not apply to those models that are physical objects. If the fruit fly can be regarded as a model organism, it cannot be said to be ontologically fictional, at least not obviously so. So for those of us who believe that "model" is properly used in science to refer to abstract as well as concrete objects that are used to serve as representational vehicles, this claim of ontological identity between models and fiction is, *prima facie*, too strong.

Even with regard to abstract models, I still find ample reasons for doubting that models in general are ontologically fictional, because it does not fit into what we typically regard as the same type of things that do different jobs. Typical examples of things that are ontologically on a par but functioning differently are nouns and verbs in English, pistols and cannons, women and cats; and typical examples of those that are ontologically not the same are numbers and atoms, names and the things they name, and humans and angels (if they exist). When we think about our ontological commitment to a model of an ideal pendulum or gas, is it not more like our commitment to mathematical objects, such as numbers and equations, than our commitment to fictional characters such as Sherlock Holmes? Unless one thinks the ontology of fictional characters is also "on a par" with number and sets, etc., ontologically identifying models with fiction seems wrongheaded.

But wait, isn't there a fictionalist approach to the ontology of mathematical and other abstract entities (Field 1989; Yablo 1998, 2010)? The idea of factionalism on abstract entities is roughly that, instead of being a Platonist or a nominalist one should treat such entities as fictional. An in-depth discussion on how such a fictionalist view on abstract entities would bear on the model-fiction literature has to wait for another occasion (but see

Godfrey-Smith 2009). It suffices to mention here that complications begin almost on the first step into that discussion. It is not clear, first of all, that models in science are in the same league as numbers and propositions, and secondly, the grounds for a fictionalist approach to mathematics and such are far from being secure.

Be that as it may, there are *prima facie* differences between models and fictional characters/events that do not appear to be a difference in function alone. First off, the imagination of a simple pendulum or an ideal gas, for example, can be regarded as a natural extension of a method we use daily, namely coming up with an idea or image of something by omitting non-essential details. On the contrary, the imagining of a fictional character or event does not seem to be the result of such a method. While idealization is indispensible for building *good* models in science, it produces only *bad* characters or events in fiction making. True, bad fiction is still fiction and if model-building in science is analogous to bad fiction making in literature, the fictionalists can still bite the bullet and say, so what? Secondly, when we have a scale model for a simple pendulum, what it represents is no doubt real pendulums, not the abstract model for them, just as a scale model of Boeing 787 Dreamliners should be taken to represent the airplanes, not the model of the airplanes. However, there is no other way of conceiving any physical realizations of a fictional character, such as Sean Connery's portrayal of James Bond, as anything other than representing the character (even if there was a real person on whom James Bond, the character, is based)[1]. But the fictionalists would argue that what the scale models represent are the abstract models, which are fictional objects. Yes, these are only *prima facie* differences that suggest but cannot prove the inadequacy of ontological identity; but the suggestions have force.

Arnon Levy (2010, 2012) suggests that we distinguish two different readings for acts of modeling: *de novo* modeling vs. *de re* modeling, and defends to some extent the latter. Very roughly (since Levy's view will not be

[1] This statement is only true when we are comparing the original or near original acts of model-making and fiction-making. If we are thinking about a secondary act of descriptively reporting on what one read in a textbook in which a pendulum is given as an example as oppose to reporting on reading the "Adventure of Sherlock Holmes," then there is indeed little difference. However, we can rarely give any evaluation of ontological commitments by looking at such secondary acts.

critically discussed here), a *de novo* modeling is an act that intends to make an imaginary object out of a particular system or phenomenon and treats the result as such, while a *de re* modeling is one that operates directly on the system and generates idealized or simplified statements about its relevant aspects. Although inclined to the *de re* reading, Levy did not use it to argue against the idea that scientific models are works of fiction. I think we could and should use it to do so.

Literary genres such as historical novels, in which entirely fictional details are attributed to actual individuals, appear to suggest that *de re* modeling could still be regarded as fiction making, if additional reasons convince us that model building is somewhat analogous to the writing of historical novels. But given the simple fact that scientific models, however idealized or abstract, are still subject to empirical testing, while for historical novels, it makes no sense to check their main characters' sayings and acts with historical facts (although there may be factual constraints on the description of historical background). Some (cf. Frigg 2010, 2011a) may disagree with this point, for there must be enough verifiable features about the fictionalized figure, e.g., a Washington or a Lincoln. However, if the work is indeed a historical *fiction* of, say, Lincoln, the important stuff, the stuff that readers care the most, the stuff that is going to make or break it as a fictional account of Lincoln, cannot be something that should be held accountable by historical facts. If it is, then it is simply no fiction and needs no pretense when we read it (cf. Walton 1990).

To conclude, there is really no obviously good reason to declare defeat to the fictionalists on the ontological front; no good reason to think that when talking about diluted gas as ideal gas (as in Boyle's ideal gas model), we are not talking about *real, albeit very diluted, gas* and *its idealized behavior* but rather some *abstract object* (and therefore a fictional one) and *its true and actual behavior*. In other words, there is good reason *not* to say that "scientific models should not be regarded as works of fiction" *ontologically.*

Giere's main argument comprises three components: scientific models should be compared to works of non-fiction rather than fiction; model-making involves "[i]dealizations, abstractions, and approximations, yes. Fictions, no[;]" (Giere 2009, 254) and conceiving models as fictional helps the enemies of science, such as creationists, which is not something a philosopher of science should do with good conscience. I shall not comment on the last argument in this essay but will focus on the first two.

It is a bit of a mystery to me that invoking the fiction vs. non-fiction distinction, vague though it is and borderline cases abound notwithstanding, has not stopped the fictionalists on scientific modeling in their tracks. The comparison is so stark that unless the fictionalists are actually touting a line that regards all creative representations, verbal or otherwise, as fictionalizing reality, and there is no real difference between the fictional and non-fictional depiction of events (i.e. history and historical fiction are fundamentally the same enterprise), it is difficult for me to see how anyone could argue for fictionalism for *all acts* of scientific modeling. The reason that fiction and non-fiction are fundamentally distinct categories of writing is clearly given by Giere: while non-fictional writings are judged, *inter alia*, by empirical evidence for details of what is written, it makes no sense to demand the same level of verifiable factual details for fictional writings.[2] This is true even when we take into account the fact that fictional writings do in a sense represent reality; but factual details, such as a particular meeting between Newton and Huygens, may make or break a biography of Newton or of Huygens but can be entirely disregarded when judging a fictional life of one or of the other or of both. If the meeting is crucial in making the fictional story of Newton and Huygens more attractive, all things being equal, it should "happen."

It is equally false, as shall be seen, to say that all acts of legitimate or good scientific modeling are squarely analogous to non-fiction rather than fiction. This is so even when we notice that on many occasions scientists would call with good reason a model "a piece of fiction" that fails to qualify as a legitimate model. For we shall see in the next section, there are straightforwardly fictional models in science, which are often regarded as such by the scientists who create and use them, and there are borderline cases (but never dismissively called a piece of fiction by scientists). All of them are perfectly good models and some disciplines often cannot do without them. Those are the models that it would make no or not a whole lot sense to regard in the *de re* mode. However, even if these are fictional models in science, we may still agree with Giere that most models are not, especially not those models that are intended for representing corresponding target systems in reality.

[2] Essentially identical reasons can be found in Walton (1990, 70-73), where scientific textbooks are simply included in the category of non-fiction. Note, "[w]orks of nonfiction do not, in general, qualify as representation in our special sense. (70)"

To Giere's second point about models being results of idealization, approximation, and/or abstraction, but not of fictionalization, I want to add a further observation. While idealization when constrained by good approximation is justly regarded as the essential ingredient in model building, it is not at all so regarded in fiction making, as I mentioned in an earlier point. In fact, although highly idealized characters or events may be valued in advertising and political propaganda, reputable fiction makers, which include writers, painters/sculptors, and moviemakers, inevitably shy away from idealization. It is a virtue for them to make their imaginary creations as concrete or particular, as unidealized, as possible. Models, because they represent real things, need to be appropriately "unreal;" while fictional characters or events, because they don't represent anything real, need to be as "real" as possible (so as to be effective in acts of make-believe).[3]

For example, when a model population is conceived in epidemiology for the purpose of studying an outbreak of influenza, the actual population is stripped of, in the researchers' mind, the inessential and irrelevant aspects and added standardized features such as "medium health" for every member and the population is conceived as infinitely large. The purpose of such a maneuver is to allow the researches to concentrate on salient properties of a population, which usually helps to simplify and therefore expedite the study. Such idealizations are often necessary to the extent that without them no study is possible. The creation of fictional characters, including populations of them, is usually a very different process. Because the purpose of such a creation is to generate imaginary people whom the consumers of the fiction are able to imagine as actual in their acts of make-believe, the creator, if she is any good, is supposed to create individual as concrete, vivid, and real as possible (cf. Walton 1990). It is true that whatever description a writer could possibly give to her creations, the result is always far from being detailed

[3] Regarding idealization in science or in fiction, there may well be a difference between the practitioners' attitudes towards it and their actual usage of it. It might be argued that with respect to what they actually do in their works, fiction writers use just as much idealization as modelers in science. However, there is no question that writers condemn idealization while modelers extol it. Try to imagine asking Philip Roth to tell you how much he cleaned up one of his characters, David Kepesh, to please his readers or critics. Wouldn't you be kicked out of his house? While Stephen Hawking might take great pleasure to tell you how cleverly he idealized in getting his blackhole model so that it yields critically acclaimed results.

enough; but one should not mistake the sketchiness of literary or artistic creations for the sketchiness of idealized models: in the former enough characteristics are added by imagination to create particular characters and/or events in the mind of a reader or viewer to realize acts of make-believe, while in the latter enough characteristics of the target systems or events are removed[4] by a principled scientific method — idealization — so that something simple and abstract is created for the purpose of theoretical investigations. Even if both come to the same thing at the level of depiction, what is suggested (as an implicature?) are quite different between them: even if the descriptions of both Sherlock Holmes and "the average man" do not include the mentioning of their toes, toes of a particular sizes and shape are part of the character of Holmes, while there is no need to imagine such things for the average man who frequents works of economics.

Godfrey-Smith (2006, 2009) has argued for regarding scientific models as imaginary concrete systems that might have existed, and used the point to support the idea that models are fictional systems. There is an ontological view of fiction that takes fictional characters as non-actual possible people, people who don't actually exist but exist in some possible world (cf. Kripke 1971). Godfrey-Smith's view appears to go with such a view about fictional characters and events. This view does not seem plausible without a lot of heavy padding (cf. Kripke 1972; Thomasson 1999) and besides, if Sherlock Holmes is supposed to be a possible person, existing in some non-actual possible world, then there is no need for make-believe when we read Conan Doyle's works or any other fictional works. If we couple this view with Lewis's realist view on possible worlds, we should *actually believe, not make ourselves believe,* that Holmes is a real person, period. This does not seem to agree with our normal practice of reading fictional works. A fictional realm is not a possibly real realm but a non-real, non-actual one that readers may pretend to be real and actual; and what is required in reading fiction is to suspend our disbelief that Holmes is not an actual person and did not actually live in 221B Baker Street. There is no sense of make-believe of this sort when we

[4] It may seem that we often add things to effect idealization, such as when we add points to make an otherwise discontinuous object a continuous one in a model; however, one can easily conceive of that as "taking away" the *gaps* between the parts of the object. For the sake of conceptual uniformity, we should always regard acts of idealization as removing characteristics from actual systems so as to create uniformity and abstraction.

are dealing with scientific models. It would not make good sense in which we could regard scientists, when studying Rutherford's model of hydrogen atoms and the Crick-Watson model of DNA molecules, as imagining them as concrete objects *existing alongside* hydrogen atoms and DNA molecules.

Moreover, fictionalists along the Godfrey-Smith and Frigg line (see also Toon 2010b) usually borrow from Walton (1990) the pretense theory for fiction as their basis for viewing models as fictional. A *prima facie* problem of such a move is the following. There is no good evidence to indicate that modelers in science, when creating or using their models, pretend that the models rather than what the models represent are real. It might be more reasonable to say that they pretend that the real systems their models represent are the same or nearly the same as their models. But this is an entirely different sort of pretense from pretending that fictional characters or events are real. There is a perfectly good *de re* reading for the former, while there is none for the latter; and the reason is fairly obvious. It does not make good sense to pretend that your model of DNA molecules exists *alongside* the real DNA molecules, while nothing in reality is preventing fictional characters or events from being imagined as real and actual in acts of make-believe.[5]

3. Typically Fictional Models

By now we should agree that there are strong arguments against taking scientific models in general to be works of fiction. The strongest reason or the best argument comes down to the simple fact that for many models in science, their details often must match corresponding facts with appropriate levels of approximation, while it makes no sense to require that for fictional characters and/or events.

It is equally obvious that science does allow for models or elements of

[5] One may object to this point by saying that modeling in science is more like historical fiction, where real people are fictionalized, just as in science real things are fictionalized in models. This is indeed a valid point, but then the further point about being held accountable against facts, as I have pointed out earlier, would prevent models from being compared or identified with works of historical fiction.

models that are purely fictional.[6] Let me now give some detailed evidence for this point. Giere (2009) discussed a model that Teller (2009) used to illustrate an aspect of model building. It is a case in electrostatics, where in order to calculate the induced charge distribution on a large metal plate (modeled as of infinite size) by a unit charge above it at a distance, *d*, one is asked to replace in one's imagination the metal plate with a negative charge on the opposite side of, and at distance *d* to, the surface where the plate is.[7] This imaginary charge can best be seen as *ontologically* as well as *functionally* fictional. It is a prop, in the sense of an imaginary replacement of one object by another non-existing object. It is also a case of make-believe in that one can pretend that instead of an infinite metal plate, a negative charge exists in its place and proceeds to calculate the effect. Very much like in the cases of fiction, especially in staged fiction as in films and plays, we proceed *as if* there is a negative charge in the supposed location. Following through this line of argument, we may very reasonably conclude that whatever ontological and functional status a fictional character has, the same is true for this and other similar imaginary models that are used for convenience or for equivalent effects.

To add a further point for this conclusion, we may again use Levy's *de novo* vs. *de re* reading of modeling building as a test. If the case can admit a *de re* reading, it is not analogous to works of fiction, and if it does not, if it *only* admits *de novo* reading, then it is very much analogous. Does it make sense to think of a particular piece of metal plate (with a positive charge hovering above it) as a negative point charge and proceed to seriously talk about a charge distribution on the plate in terms of an imagined negative charge at an imagined location? If that is possible *de re*, then it is equally possible to talk about an able (actual) detective as Sherlock Holmes *de re*, if the description of Holmes better illustrates her. I assume the latter is not appropriate, and therefore our test is negative: there is no way of giving a *de*

[6] Frigg (2011a) distinguished "Two Kinds of Fiction" in science, and what is discussed in this section may in some sense be seen as belonging to the first kind. Since my overall conception of fictional models in science is different from Frigg's, a simple identification is not appropriate.

[7] The fictional negative charge is supposed to be where the "lines" of the electric field from the positive charge above the metal plate would "end up." By imagining its presence at that location simplifies the finding of the distribution of the "field-lines" on the surface of the metal plate.

re reading of Teller's case and many other similar cases. These models are just like works of fiction proper.

However, Giere doesn't think this model involves any work of fiction, his reason being that there are two models involved here. An actual (or concrete) system of a charge above a metal plate is first *idealized* into a model of a charge above an *infinite* model plate, and then this model is replaced by a effectively equivalent second model, and that is the model with the imaginary negative charge on the opposite side of the infinite plate. This analysis is certainly correct and yet I do not see how the second model is not a work of fiction, unless Giere's point is that the first model, which only involves idealization, is the real and the only significant model and it does not involve fiction. However, without the second entirely imaginary model, the idealization in the first model is in vain; in fact, the idealization taken to make the metal plate infinite makes sense only if one is going to introduce the imaginary negative unit charge.[8] So, the two models taken as a whole involve idealization and fiction.

There are similar cases in geometric optics, such as the virtual object in the two-lens system (see Virtualobject). Scientists who invent and teach such models have the good sense of correctly calling them "virtual" objects or models, which in our sense, the sense used in the model-fiction literature, means "fictional." (Also see Rowbottom 2011 for another interesting case: the introduction of virtual particles to explain the Casimir Effect.)

However, there are examples of scientific models that obviously do not belong to the type of imaginary or "virtual" objects, but are they like the common sort of models, such as the simple pendulum and ideal gas, which should not be regarded as fictional, as we have argued along with Giere in the previous section? Are they *sui generis* and if so, are they fictional in any proper sense?

Harmonic oscillators as a type of models might be thought of as posing some difficulty. One might think on behalf of Giere that no difficulty exists because "harmonic oscillator" names a class of different systems, such as the pendulum system and the electric circuit, each of which is a model that represents its target in a normal (i.e. nonfictional) way. This may appear to

---

[8] If the metal plate is conceived as having edges, the "field-lines" from the positive charge will bundle up at the edges, and therefore replacing the plate with the negative below its surface would no longer be a good model.

be the right diagnosis if the matter stops at that. However, when computer simulation technology becomes available, scientists can easily construct and study a "harmonic oscillator" in a simulation that is neither a pendulum nor an electric circuit nor anything that exists in reality but simply a harmonic oscillator. There is the concrete model system that the term refers to that does not exist in reality, nor is it an obvious idealization of something that does exist in reality. The computer-simulated harmonic oscillator is obviously neither an idealized pendulum nor an idealized electric circuit nor any sort of ordinary (or recognizable) physical oscillators. And there are plenty of simulated or simulatable models in science, especially in today's science, that are like the harmonic oscillator. They are conjured up by scientists to study a wide variety of systems whose behavior exhibit sufficient similarities to one another such that a general/generic description is given and a virtual object as a particular thing is generated by a computational device.

The second sort of models that may create problem is a grab bag of items. We couldn't help but notice that in Teller's example, an infinite metal plate is first introduced to set up the model in which the charge distribution of an actually large but finite metal plate is represented and calculated. The size is obviously an idealization but it is clearly introduced for convenience. In other words, one could just as well use a very large but finite metal plate for the calculation, even though the task is more difficult and the result so close to the ideal case that introducing the infinite plate, though not strictly necessary, makes a lot more sense. There is a type of model in science in which taking the limit of infinity — call them infinity models — is *necessary*, at least necessary for the theoretical framework in which such models are invented and used. For the Teller example, the charge distribution won't be "ill-defined," or simply indefinable, even if the plate is not imagined to be infinitely large; and yet there are phenomena such as phrase transitions in condensed matter physics where without an infinity model, the target phenomenon, such as the boiling of water through which the liquid turns into vapor wouldn't even "occur" in its theoretical framework. The problem with not having such phenomena occurring in the framework is that the framework cannot provide any explanation for why such simple events as watering boiling and turning into steam occur in nature. And if the framework is deemed with good reason the only plausible option, then we've got a problem.[9]

However, suppose we ask why this type of models should be thought of as special. Some sense of "infinity" seems to be implicit in all idealization cases; for example, isn't making a surface frictionless the same as making the fiction infinitely small? Or isn't making the earth a point and its orbit a curve the same as reducing the size of the planet and its orbit to infinitely small? Here are the important differences. First, relaxing the ordinary "infinities" — restoring small sizes and thin tracks — results in messier but in principle manageable and similar outcomes. That's why we say the results of such idealizations are approximate. Some such idealizations are indeed later relaxed theoretically, as with the model of van der Waals's "real" gas. This maneuver is not possible for the infinity models for phase transitions and critical phenomena (PT&CP henceforth). No finite model that provides non-rigorous explanation, such as the "mean-field" model, can be regarded as the result of relaxing the infinity idealization (cf. Stanley 1971; Liu 1999; Callender 2001). Second, no model of the ordinary infinity sort needs to introduce the infinity in order to even "have" the actual phenomenon, such as boiling water, it represents. When friction is completely removed, the rolling ball will move on forever in the same velocity, which is no actual phenomenon that Galileo needs to account for with his theory; it is rather a principle that may be displayed in the ideal situation.

To see these points clearly, let us look at one of the simplest infinity models for PT&CP in some details. To some extent this model resembles the harmonic oscillator I mentioned earlier. Different systems or phenomena, such as gases and magnets or evaporation and ferro-magnetization, have different models and equations, and yet the idea and structural features of this model applies to all such systems that undergo PT&CP. And very much like in the case of the harmonic oscillator, a computer-simulated system can be created that is independent of the details of the actual systems involved. What is different and distinguishes this model (and those of its type) from other abstract models is the necessity of taking the limit of infinity for

[9] There is a controversy over whether the infinity model is necessary for understanding PT&CP (cf. Liu 1999; Callender 2001). But there is no doubt that within the mathematical physicists community, the infinity models are currently the best rigorous models. So the philosophical question is not what these infinity models are and signify but whether modeling finite system with infinite models is on sound foundations.

the *possibility* of the actual phenomena (i.e., PT&CP) and for supplying explanations for them.

The model is known as the Ising model (cf. Stanley 1971), and the same mathematical structure can be applied to magnets as well as to fluids (e.g. lattice gas). One may construct such a model of any dimension, but the least complex Ising model for studying PT&CP is the two dimensional one. Imagine a large two-dimensional spatial region in which N "sites" (positions in space) are spread out at equal distance, where N is a large integer. Each site is endowed with one simple binary property, s_i, and between any two sites a coefficient, J, such that for any distribution, d, of s_i ($i \in N$), we can define a "moment" function,

$$M_d = (1/N)\sum_{i=1}^{N} s_i(d), \tag{1}$$

and a near-neighbor "connection" function

$$C_d = \sum_{\langle i,j \rangle} -Js_i(d)s_j(d), \tag{2}$$

for all the adjacent pairs $\langle i, j \rangle$ of sites in the lattice. This function has the same dimension as the Hamiltonian of the system.

Imagine again a global parameter T that is associated with the sites' ability to switch from one to the other value of s_i, such that it and J are the only two global parameters of the lattice that determine the probability of any given distribution of the values. With such a model, it is known that the Boltzmann distribution of the values of s_i applies, and therefore, the probability of a given distribution, d, should be,

$$P_d = Z^{-1}\exp[-C_d/kT], \tag{3}$$

where k is the Boltzmann constant, and Z "the partition function," a special function that provides the weight to the probability such that it adds up to unity. This model can be used to study a magnet (e.g. for its ferromagnetism) if the binary values of the sites have + or –; or we can use it to study fluids if the values are 1 or 0 (occupied or not occupied). And the model is only good for thermal systems in equilibrium *near the critical temperature*, for it does not hold for anything far from that temperature regime. It is a remarkable fact that near the critical temperature, systems that are otherwise radically different in physical composition and structure are remarkably similar or essentially the same thing, endorsing the cliché: "it's nothing but a bunch of molecules."

Here is in a nutshell what the "problem of PT&CP" is and how it is resolved in mathematical rigor. And it also shows why the "infinity models" are rather perculiar.[10] Thermodynamic quantities of a system are represented in thermo-statistical physics as the weighted (e.g., with a probability measure as in Eq. (3)) averages of the system's molecular quantities. Mathematically, we can express them as various derivatives of the partition function Z for the system. (For instance, the pressure p is $p = -\left[\partial F/\partial V\right]_T$ (i.e. partial derivative under constant temperature T), where $F = -kT(\ln Z)$ is the free energy and the main connection via Z to the molecular quantities.) Phase transitions happen at a critical temperature and are expressed as singularities (in terms of divergent derivatives) in the thermodynamic quantities. Translating this into statistical mechanics, it turns out that phase transitions are possible only if there are positive solutions to the equation, $Z_{gran} = 0$, (which will make, e.g., dp/dV, at the critical temperature divergent).[11] However, it is proven that for any finite models of the Ising type, no such solutions are possible (known as the "no-go" theorem of PT&CP). And such a solution appears when we take N to infinity (while keeping the model's density finite). This becomes known as taking the thermodynamic limit so as to make sense of PT&CP in statistical mechanics. What is special about this and other infinity models, what distinguish them from the other typical idealized models, such as the frictionless plane and nonviscous fluids, is shown by the "no-go" theorems attached to them: *without going to infinity, one cannot even "find" the phenomena in question much less to begin studying them and look for explanations.*[12]

Another crucial area in physics that needs infinity models is that of

---

[10] For details of how a thermodynamic function is related to a function in statistical mechanics, see any standard textbooks on the subject or see Liu 1999.

[11] This is not strictly correct, but it conveys the idea. Z_{gran} is the partition function of the grand canonical ensemble; for details see Liu 1999.

[12] A qualification is here in order. There are finite models (albeit still idealized) that exhibit PT&CP (and other phenomena that require the infinity models), for instance, the

quantum measurement. When a quantum system in an entangled state of some observable, such as its momentum or spin, is measured for the values of that observable, the state "collapses" into a particular eigenstate of the observable in which a definite value is obtained (e.g. spin up or down). This so called collapse that destroys the coherence of a quantum entanglement, which constitutes the quantum measurement "process," is no ordinary physical or causal process a quantum theory can account for, and has generated over the years competing interpretations of quantum theory in which the process is either explained away or made more realistic. Among the many alternative interpretations, the decoherence solution to the quantum measurement (QM) problem has, for what it worth, garnered the most support from working physicists (cf. Joos *et al.* 2003). The best version of the decoherence model of QM is to conceive the environment of any QM experiment, which according to the model produces a complete dissipation possible of the coherent state, as an infinite system. The disentanglement as expressed mathematically is an asymptotic process that approaches its completion only as the size of the environment approaches infinity. Just as no transition in PT&CP is possible without taking the infinite limit, no definite values can emerge from QM unless a similar sort of limit of infinity is taken.[13]

These *infinity models* (models that carry "no-go" theorems for any of their finite versions)[14] are certainly not like the virtual models simply because they are impossible systems, while the latter, though obviously unreal, are all ordinary systems that are imagined to be where they are not. We *pretend* that the charge is there for some equivalent effects, and that's why such models are so much like the fictional characters. The infinity models seem to be the opposite: we know that PT&CP is real and QM actually happens regularly, and yet what makes them happen, if the models are literally believed, is impossible.

---

mean-field models, but interestingly, all such finite models require the introduction of some essentially non-statistical-mechanical elements in order to reproduce the phenomena in question. All exact microscopic models require going to infinity. For a different view on this issue, see Callender 2001

[13] There is also a "no-go" theorem for QM cases, which is known as the von Neumann-Stone theorem and which roughly says that for quantum systems represented in a Hilbert space, no physical processes can reduce an entangled state to a mixture of pure states.

[14] For many excellent examples of such models, see Ruetsche 2011.

Again, when we are dealing with the sort of idealization that amounts essentially to *neglecting the negligible*, such as ignoring the small effect of friction on a smooth surface or through thin air and ignoring the sizes of, and interactions among, the molecules in a diluted gas, the resulting "unrealistic" models can be modified towards more realistic models with specifiable rules. In most cases, what needs to be done mathematically is to introduce modification factors in the equations of motion (or of state) such that their adjustable values may be obtained empirical (i.e. through matching with the experimental data). The prime example of this can be clearly seen in the transition from Boyle's equation of state for ideal gas to van de Waals's equation for "real" gas: $pV = kT \Rightarrow \left(p + a/v^2\right)\left(v - b\right) = kT$, where the "$a$" and "$b$" are the factors for interaction and volume, respectively.

To generalize a bit, here seems to be two necessary conditions for this and other similar types of idealized models.

1. The modification factors must be introducible at the same level of description that the original idealized models are on.

2. If the models are mathematically given, the factors must be introducible into the same equations we get from the original models.

These are necessary simply to ensure that what is "idealized away" is indeed negligible and the resulting models of such idealizations are close to being correct.

None of the infinity models I discussed above satisfy either of these conditions. The only alternative to the Ising model and its derivatives, as I mentioned above, is the mean-field type models for PT&CP. It begins with a quantum mechanical model and its Hamiltonian, but then going beyond the quantum level, a "mean," i.e. average, field is added to mimic the intermolecular interaction that is of long-range. This field is *macroscopic* (i.e. not quantum-mechanical) and the equations that show CP below the critical temperature are not the same type of equations as Ising type models give. The situation is even worse for the QM problem. Apart from the Bohmian Mechanical Interpretation, which adds non-quantum mechanical elements to quantum mechanics, no solution to the problem exists without an infinity model.[15]

Moreover, we can think of describing real but diluted gases as ideal gas in the *de re* mode, but it does not seem that we can describe phase transition or quantum measurement in the same way.

4. Microscopic Models: Idealization or Fictionalization?

When philosophers of science talk about modeling and idealization they often assume that one cannot do the former without the latter. Degrees of idealization may vary greatly from one job of modeling to another, but some amount must be involved, for otherwise there wouldn't be a point of model building. However, if one seriously reflects on such models as the Crick-Watson model of the DNA molecule or Rutherford's model of the hydrogen atom, it is not clear that specific idealizations, as I explained earlier, are taken there at all. We can probably safely assume that the models in these and other similar cases are simpler than their target systems, but what we don't know and should really not assume that we know is whether the simplification is gotten through *idealization* (cf. Lakhtakia 1996; Watson 2001; and also Karp 2013, ch. 13).

People who are not convinced that anything can be said intelligently, if intelligibly at all, about the division of the observable and the unobservable, people who tout a hardline version of scientific realism, may object by saying that we can and do observe the structure of a DNA molecule or a hydrogen atom, and it is of such observable structures that the models are made as a result of precise idealization. The objection is flawed in that the Crick-Watson model of DNA is a mechanical/combinatorial model while the target system is a biological/living thing. We may be able to "observe" the combinatorial patterns of the bases in a DNA molecule, but how the bases are chemically and biologically connected and what makes such connections work is not represented in the model. There is no sense in which neglecting these aspects in the model is the result of idealization, anymore than my daughter's use of a piece of Paz candy dispenser as her dad in a play, which

[15] The solution attempted in the algebraic approach to quantum theory in terms of quantum spontaneous symmetry breaking (cf. Liu & Emch 2005) also requires taking the infinite limit.

very much simplifies the interactions among family members, may be counted as an act of idealization. If we are supposed to also regard these modeling practices as no more than the results of idealization, then I think it is fairly obvious that we couldn't simply agree with Giere, as I quoted him earlier, that we should say, "Idealizations, abstractions, and approximations, yes. Fictions, no." (Giere 2009, 254)

The point may be more obvious for Rutherford's model of hydrogen atoms. The "solar-system" model that Rutherford advocated sometime in early 20th century can hardly be viewed as a model of idealization or abstraction. True, we have some experimental observations regarding the interior of atoms from Rutherford and his research group. When the alpha particles from high-speed bombardments of gold film are scattered afterwards in a certain pattern, the interior of the atoms that creates such a pattern must be of a certain sort and the "solar-system" simile fits quite well. A piece of brilliant analogical reasoning, perhaps, but obviously not a carefully constructed idealization or abstraction.

One may object at this point by asking: what is the difference between analogy in modeling and idealization or abstraction? Are they not more or less the same sort of practice? No, they should not be taken as the same sort of scientific method (broadly construed), and the models we obtain by using them do not appear to be of the same type. To see this point, let us begin with Rutherford's solar-system model for hydrogen atoms (cf. Lakhtakia 1996). In a rough and ready manner, let us reflect for a moment on how that model might have been conceived.[16] There was evidence for the existence of electrons as small particulars that are freed from atoms; and since atoms are neutral in charge, there must be positively charged elements in them. Rutherford's scattering experiments seemed to show that most of the space in an atom is "empty," which implies that the positive charge must also be concentrated in small particles. The nature of electric charges, namely, how opposite charges

[16] The argument here is necessarily speculative for two main reasons. First, it is very unlikely that Rutherford or any of his collaborators would have left any remarks on whether it was "idealization" or "analogy" or any other method that he actually used in coming with the model. Second, that is probably because he and his collaborators didn't even think about such a question when they were constructing the model. They might be aware that some sort of analogous reasoning was involved, but what sort exactly and whether is might be regarded as a species of idealization couldn't possibly be on their mind at any time of their career.

attract and same charges repel, was also well known. Given the size and the empty space of an atom, the electrons must be moving rapidly around the nucleus of positive charges in order to maintain that space. From these pieces of evidence and perhaps the unconscious awareness that the same patterns are often repeated at different scales in Nature, Rutherford's model for atoms was born. Nowhere in this process is idealization or abstraction involved. The fact is that there couldn't be observable data that directly show what how exactly are the charges distributed inside the hydrogen atom. Therefore, the conjecture that the distribution must be similar to the mass distribution in a one-planet solar system has no base to be idealized and abstracted from.[17] The Rutherford model is clearly a brilliant piece of Analogical Reasoning, and this is modeling in the conception of Mary Hesse (1966).

According to Hesse, the quintessential character of analogical modeling is setting up a system that resembles something we are already familiar, or at least know better, and that share properties with the modeled system that appear to be essential to it. That is how electromagnetic waves in the luminiferous aether (according to the 19th century conception of aether) are modeled by waves in fluid (cf. Hesse 1966, 11; also see Whittaker 1910). The problem of what the medium is that carries light through space emerged after unequivocal experimental evidence appeared to support the wave theory of light. However, like space itself, the optic aether is not observable and therefore it is not possible to find out what light waves are like in the same way we can find out what waves in fluids are like. Because aether transmits a stuff that produces comparable effects as produced, e.g., by waves in fluid, analogical reasoning suggests that a model of aether be constructed from our knowledge of fluid. If we bring in the Maxwell equation for light waves and compare it to the mechanical equation of waves in fluid; and couple that with the measureable effects of both types of waves, we have what Hesse defined as the positive analogy. Assuming the general causal connection between a wave and its medium, we can set up a model, as Lord Kevin did, for the optic aether that goes beyond those aspects that it shares with the

[17] True, the electron and the nucleus might be idealized as point particles and there are some other such idealization being taken into consideration, but they are subsequent to an already constructed model, the solar-system-like model, and they are not essential to the model because nobody can use the point-and-curve-orbit structure to calculate, e.g. how fast the electron revolves around the nucleus.

fluid waves.

It is obvious that idealization (and approximation) is not the method by which this type of modeling is done: what small bumps must be taken away from the aether or small gaps or holes be filled or viscosity among parts that renders it less than perfectly rigid along the direction perpendicular to light beams, what anything of this sort could be carried out in order to come up with an aether model for electromagnetic waves?

The same can be said and must be true in equal measure regarding the modeling of Rutherford's atoms. Rutherford and his colleagues (Hans Geiger and Ernest Marsden) did get glimpses of what the interior of atoms is like from how particles passing through them behave, and yet his model can only be viewed as the result of analogical reasoning. And it had to be so primarily because it was not feasible to come up with such a model by, among other things, acts of idealization.

Again, this is radically different from how models for pendulums or the solar system are made. With Boyle's model of ideal gas, different options are possible, because it may be a borderline case. It all depends on whether it is plausible to regard the idea of diluted gas as comprising mostly free molecules as deriving from "observation" or from speculation by means of analogy. I regard the former as most reasonable and therefore with good reason we should take the "ideal gas" model for what its name signifies: a product of idealization. The mechanical differences among different molecules are neglected, their sizes are made to vanish, any interactions among them are also removed, and so on. Voila, out drops the model, a perfect product of idealization.

However, one should not take the above point to be that there are no *actual* acts of idealization/abstraction in modeling the unobservable and there are no *actual* analogy used in modeling the observable. The point is the following.

1. Whether or not a modeler thought she has got her model of an unobservable system directly from idealization/abstraction, she couldn't possibly have succeeded if she did.

2. Whatever reason by analogy a modeler used in coming up with a model of an observable system, she could have gotten an equally legitimate model through idealization/abstraction.

Hence, my point is about possibility not about what is actually done, which often includes misidentifications. It is not my intention to argue that idealization and analogy are the only methods by which models in science are built. But one thing seems true enough, namely, if idealization can properly be used for the modeling of the unobservables, there must be a separate hypothesis about what the objects are like from which the idealization is taken, which is independently well confirmed. To put the point dramatically, Boyle's model would not have been a good model had there been good evidence that gas is a continuous fluid.

Let me now return to our central question about which models in science are fictional. After the foregoing analysis should we say that models of the apparently unobservable systems, which cannot be gotten from idealization, should be regarded as works of fiction? To answer this question in any clarity, let us first reflect on the senses in which we can attribute fictionality to a model without getting into obvious trouble with ontology. To begin with, there is no such trouble with the obviously fictional models of imagery charge and virtual objects. With such cases, "fictional" simply means ontologically nonexistent: the charge does not actually exist nor does it represent anything that actually exists; it is there to stand in for an equivalent physical effect. This point is similar to what we regard fictional things ontologically: Orwell's Animal Farm (to use one of Godfrey-Smith's examples 2009) neither actually exists nor represents any actual places, or anything else, for that matter.[18]

The infinity models are very different, for though physically impossible, they can be understood as representing something real, namely, the *actual finite* systems in which the critical phenomenon or the quantum measurement takes place. But they are marked off by a subtle and yet

[18] Godfrey-Smith (2009) believes that fiction represents actual people or places in a more straightforward manner: 'Tolkein's "Middle Earth" is fairly similar to the world of Malory's King Arthur tales (Morte D'Arthur)'; and 'the events in Orwell's Animal Farm are similar to those in Russia in the first part of the 20th century (2009, 5-6).' But these are not the sense in which models in analogy represents their target systems. Fictional characters or events may have some original "models," but they are by no means what fictional creations represent or even about. If the Animal Farm is, like Rutherford's model being about hydrogen atoms, only about early 20th century Russia, Orwell's fiction would have been in the dustbin of literature and few people would have known its existence.

nontrivial feature. There is little doubt (i.e. for independent reasons) that the actual systems they are supposed to represent are finite, and yet according to the model, no PT&CP or QM can take place in such systems (because of the "no-go" theorems). If the theories which contain the infinity models are true or approximately true, it cannot be about the actual systems; it can only be about the infinity models, which in turn represent what happens in the actual PT&CP or quantum measurements in some indirect and unspecifiable manner. Again, it is similar to the fact that the Middle Earth neither actually exists nor represents any actual place, and yet the fictional story tell us about Medieval Europe in a way not easily specifiable exactly. (Think about the inscrutability of the many alternative literary theories about what and why we learn from fiction.)

Models for the unobservable, even if fictional, are not fictional in the same sense as the virtual models, nor do they appear to be similar to the infinity models. They are in general by no means physically impossible. Were it not for the generic feature of representing the unobservable, they are just like regular models. Imagine that there are, or we can create, intelligent beings that can change their sizes at will and therefore could on demand shrink themselves to the size of electrons and get into an atom and obtain a sense impression of its interior. Afterwards, they grow back to our sizes and show us verbally or by pictorial depiction what the interior of an atom is like. A model based on such a report would indeed be an ordinary model from idealization and have nothing to do with fiction, as Giere has so convincingly argued — they would have to answer to the facts about the atom's interior that the creatures provide us. The same thing could be imagined for any other models for the unobservables, and such models would not be ones from analogy but from standard idealization/abstraction procedures.[19] In other words, models for the unobservable could have been regular models in a world of make-believe. As is, such models are more akin to the infinity models than to the regular ones.

[19] For the aether model of electromagnetic waves, the imagination needs to be more fanciful than most science fiction writings. The creature would have to be able to change its characters from normal material made of atoms, say, to extraordinary material made of aether such that it could perceive the medium in the way we perceive regular material. And then it has to be able to change back to tell its story. Unlike changing sizes, this may not be physically possible.

Acknowledgements

The author thanks Darrell Rowbottom for carefully reading a draft of the paper and providing detailed constructive criticisms.

References

Callender, Craig (2001) Taking Thermodynamics Too Seriously *Studies in History and Philosophy of Science Part B* **32**:539-553

Callender, Craig & Jonathan Cohen (2006) There Is No Special Problem About Scientific Representation *Theoria* **55**: 7-25

Cartwright, Nancy (1999) *The Dappled World: A Study of the Boundaries of Science* Cambridge: Cambridge University Press

Contessa, Gabriele (2009) Scientific Models and Representation, in Steven French and J. Saatsi eds. *The Continuum Companion to the Philosophy of Science*

Contessa, Gabriele (2010) Scientific models and fictional objects *Synthese* **172**:215–229

Field, Hartry (1989) *Realism, Mathematics and Modality* Oxford: Blackwell

Fine, Arthur (1993), "Fictionalism", Midwest Studies in Philosophy 18:1-18.

Frigg, Roman. (2010a). "Models and Fiction," *Synthese.* **172** 251-268.

Frigg, Roman. (2010b). "Fiction and Scientific Representation," in R. Frigg, M.C. Hunter (eds.), Beyond Mimesis and Convention, Boston Studies 97 in the Philosophy of Science 262, Springer Science+Business Media B.V., 97-138.

Frigg, R. (2011). Fiction in Science, in Woods, J. ed. *Fictions and Models: New Essays.* Munich: Philosophia Verlag (forthcoming).

Giere, Ron (1988). *Explaining Science: A Cognitive Approach.* Chicago: Chicago University Press

Giere, Ron (2002) How Models Are Used to Represent Reality *Philosophy of Science* S742-S752

Giere, Ron (2009) Why Scientific Models Should not be Regarded as Works of Fiction, In Mauricio Suárez ed. *Fictions in Science: Philosophical Essays on Modeling and Idealization* 248-258 (Routledge)

Giere, Ron (2011) Representing with Physical Models, in Paul Humphreys &

Cyrille Imbert eds. *Models, Simulations, and Representations* (Routledge)

Godfrey-Smith, Peter (2006) The Strategy of Model-Based Science *Philosophy of Biology* **21**:725–740

Godfrey-Smith, Peter (2009) Models and Fiction in Science *Philosophical Studies* **134**:101-116

Goodman, Nelson (1976) *Languages of Art: An Approach to a Theory of Symbols* Indianapolis: Hackett Publishing Company, Inc.

Hesse, Mary G. (1966) *Models and Analogies in Science* Notre Dame: University of Notre Dame Press.

Joos, E. *et al.* (2003) *Decoherence and the Appearance of a Classical World in Quantum Theory* 2nd ed. Berlin: Springer-Verlag

Karp, Gerald (2013) *Cell and Molecular Biology: Concepts and Experiments* 7ed. Wiley and Sons

Keller, Evelyn Fox (2002) *Making Sense of Life: Explaining Biological Development with Models, Metaphors, and Machines* Cambridge, MA: Harvard University Press

Kripke, Saul (1971) Semantical Considerations on Modal Logic, in Linsky, L ed. *Reference and Modality* Oxford: Oxford University Press

Kripke, Saul (1972) *Naming and Necessit,* Cambridge: Cambridge University Press

Lakhtakia, A. ed. (1996) *Models and Modelers of Hydrogen* Singapore: the World Scientific

Levy, Arnon (2012) Models, Fictions, and Realism: Two Packages *Philosophy of Science* **79**: 738-748

Levy, Arnon (2010) Fictional Models *De Novo* and *De Re* (preprint)

Liu, Chuang (1999) Explaining the Emergence of Cooperative Phenomena Philosophy of Science 66: S92-S106.

Liu, Chuang and Gerard Emch (2005) Explaining Quantum Spontaneous Symmetry Breaking Studies in History and Philosophy of Modern Physics 36: 137-163.

Rowbottom, Darrell P. (2009) Models in Biology and Physics: What's the Difference? *Foundations of Science,* **14**: 281-294.

Rowbottom, Darrell P. (2011) The Instrumentalist's New Clothes *Philosophy of Science* **78**: 1200-1211

Ruetsche, Laura (2011) *Interpreting Quantum Theories* Oxford: Oxford University Press.

Stanley, H. Eugene (1971) *Introduction to Phase Transitions and Critical*

Phenomena Oxford: Clarendon Press

Suárez, Mauricio (2004) An Inferential Conception of Scientific Representation *Philosophy of Science* **71**: 767-779

Suárez, Mauricio (2009a) Fictions in Scientific Practice, in Mauricio Suárez ed. *Fictions in Science: Philosophical Essays on Modeling and Idealisation* Routledge: 1-15

Suárez, Mauricio (2009b) Scientific Fictions as Rules of Inference, in Mauricio Suárez ed. *Fictions in Science: Philosophical Essays on Modeling and Idealization* Routledge:158-178

Thomasson, Amie (1999) *Fiction and Metaphysics* Cambridge: Cambridge University Press

Teller, Paul (2001) Twilight of the Perfect Model Model *Erkenntnis* **55**: 393-415

Teller, Paul (2009) Fictions, Fictionalization, and Truth in Science, in Mauricio Suárez ed. *Fictions in Science: Philosophical Essays on Modeling and Idealization* Routledge: 253-247

Toon, Adam (2010a) The Ontology of Theoretical Modelling: Models as Make-Believe *Synthese* **172**: 301-315

Toon, Adam (2010b) Models as Make-Believe, in Roman Frigg and M. Hunter eds. *Beyond Mimesis and Convention, Boston Studies in the Philosophy of Science* 71-96, Dordrecht: Springer

van Fraassen, Bas (1980) *The Scientific Image* Oxford: Clarendon Press

van Fraassen, Bas (2008) *Scientific Representation: Paradoxes of Perspective* Oxford: Clarendon Press

Virtualobject: http://science.sbcc.edu/~physics/flash/optics/virtualobject.html

Walton, Kendall L. (1990) *Mimesis as Make-Believe* Cambridge, Mass: Harvard University Press

Watson, James D. (2001) *The Double Helix: A Personal Account of the Discovery of the Structure of DNA* Touchstone

Whittaker, E. T. (1910) *A History of the Theories of Aether and Electricity: From the Age of Descartes to the Close of the Nineteenth Century* London: Longmans, Green, and Co.

Yablo, Stephen (1998) Does Ontology Rest on a Mistake? *Supplement to Proceedings of the Aristotelian Society* **72**/1: 119-37.

Yablo, Stephen (2010) Go Figure: A Path through Fictionalism, in *Things: Philosophical Papers 2* Oxford University Press, 177-199.

The Hypothetical versus the Fictional

In this essay I argue against the idea that modeling in science is analogous to fiction making in literary works by pointing out that a typical move in the former, which is widely acknowledged in philosophy literature as a signal for fictionalization, is never present in works of fiction (not even in science fiction). I further argue that the reason for such a disparity is profound and profoundly against conceiving modeling as fictionalization. I then explain the difference between the hypothetical and the fictional, and argue that modeling in science belongs to the former.

I

Scientific models are traditionally regarded in philosophy of science as a major component of scientific theories (cf. van Fraassen 1980; Giere 1988; Suppe 1989). In recent years, a sizable literature has grown up in which models are treated apart from the theories; and this is mostly because they are found to be non-truth-apt. And this non-truth-aptness seems to further suggest that they need a separate ontology (Contessa 2010; Levy 2010, 2012). Perhaps inspired by fictionalism in philosophy of mathematics or perhaps not, a fictionalist view of scientific models has recently emerged as an attractive approach (Godfrey-Smith 2006, 2009; Frigg 2010 and references therein). The view takes models, when regarded as abstract entities, to be works of fiction. They are said to be the result of imagination or invention; they are objects that inhabit a realm that is analogous to the realm inhabited by fictional characters and events (and therefore, sharing the same ontology); and they serve as the middle part of a tripartite scheme of representation in science (cf. Figg 2010).

More accurately, in the tripartite account there is the description of a model that may be given verbally or mathematically; and this is the part whose elements can be true or false. The model that belongs to the second part may either be an abstract entity or a physical system. Some people believe that all scientific models are abstract; the concrete/physical things we sometimes also call models are rather the physical realizations of the abstract models. But others think concrete things can be used as models without the mediation of any abstract entities. A scale model for a Boeing 747 is not a physical realization of any abstract airplane but simply an object created and used for representing all the actual Boeing 747's. Either way, it does seem right to regard models as the middle "thing" in the scheme of scientific representation. The third part comprises the real systems that are represented in the practice of modeling. It is argued (cf. Godfrey-Smith 2006) that in the "model-based" approach to scientific representation, which is one of the major approaches, the verbal or mathematical part of a theory does not speak to the part of real systems directly; it first speaks to the models and then the models fulfill their representational job in a rather circuitous manner (cf. Frigg 2010).

Reasons for taking scientific models to be abstract entities are several, and the main reason seems to be that models are almost inevitably the result of idealization. To say that a physical artifact — a physical copy of an ideal pendulum or a DNA molecule, for instance — is the model may therefore be problematic. However well constructed, it cannot be the idealized thing. But only the appropriately idealized thing is accurately accounted for by the descriptive account of pendulums or DNA molecules. Since our theory never account for the flaws and irrelevant aspects of the artifacts, it seems more reasonable to take the artifacts only as rough displays of the models. Model cannot be ideas or images in people's heads either for similar reasons. For physicalists (in philosophy of mind), those are just physical objects, anyway; and therefore they cannot be models. Even if one is not a physicalist about ideas and images in people's heads, those ideas and images are nevertheless particulars that do not necessarily satisfy the uniform ideal conditions. Hence, our theories cannot be said to be accounting for such particulars in people's heads.

I am not committed in this paper to this view of scientific models, for I do not believe in fictionalism. But as far as I can see, fictionalism on scientific models only appears to be viable if one can defend the abstract-entity view. Fictionalists would more naturally regard physical artifacts that we sometimes call "models" as physical realizations of models. And this point appears to have a parallel in fictional characters (and events). *Prima facie*, fictional characters may be invented by verbal descriptions and then they can also be physically realized by actors in a dramatization, in a movie or on stage, of the fiction; but presumably neither the set of all verbal descriptions for a character nor the actors playing it can be identified as the character itself.

Given what has been said so far about scientific models, it should come as no surprise that the traditional debate in metaphysics about the ontology of universals and abstract entities also applies to models. Similar to the well-known positions on the ontology of, e.g., mathematical entities, such as numbers and geometrical figures, we could have Platonism or Nominalism regarding models. The Platonists regard models as abstract entities and exist as such. The Nominalists deny that; for them there are no such abstract things as models; what we call "models" are all concrete objects that serve as vehicles for representing other concrete objects (including self-representations). When confronted with such matters as idealization and abstraction in modeling, the Nominalists would argue that those are only ways in which to think about the models, having nothing to do with models themselves, which are always concrete objects. Or a more practical way to be a Nominalist on models is to think of a model representation of a target system as *de re*, rather than *de novo* (cf. Levy 2010). For instance, rather than thinking of the double-helix model of DNA molecules as an imagined entity that bears some special relationship to the actual DNA molecules, we think of it as a mere label, such that whenever we say something about the model, we are saying it of the real molecules rather than the model. Constructing the model allows us to talk about the actual molecules in a special manner, and nothing more than that.

A third position, now well entrenched in philosophy of mathematics, is fictionalism (Field 1989; Yablo 1998, 2010). According to this view, the best way to accommodate our intuition about numbers and such is to think of them as fictional objects. We have no trouble feeling as if the fictional characters, such as Sherlock Holmes and Harry Porter, were real persons, and yet we know those names do not refer to any such. Similarly, we feel on the one hand that numbers, such as one, two, and three, are real, but on the other hand, we do have reason to believe that they do not refer to anything. We find rules and laws that apply to numbers and geometrical figures that do not strictly apply to any concrete objects, such as occupying spatio-temporal regions and having causal efficacy of some sort. And this is analogous to our finding stories among fictional characters that do not obtain to any actual people. If there is a problem of ontology for mathematical objects, the same (or similar) problem also besets fictional characters and events; and hence, ontologically, according to this view, the mathematical is on a par with the fictional. And now, we join models to mathematical objects and there we have fictionalism for models.

Partly because of the apparent parallel between models and fictional characters and events as imagined and invented objects and partly because of the fictionalist option for abstract entities, fictionalism on scientific models has gained a sizable following in recent years (cf. Contessa 2010; Frigg 2010; Giere 2009; Godfrey-Smith 2009; Levy 2010, 2012; Suárez 2009a, b; Toon 2010). The question I shall investigate in the next two sections is whether there are serious problems with such an approach.

II

Let me now point out the little noticed disparity that threatens to upset fictionalism for scientific models. In Frigg (2010), he writes,

> This framework [i.e. one for fiction] has the resources to explain the nature of model-systems. Typically, model-systems are presented to us by way of descriptions, and these descriptions should be understood as props in games of make-believe. These descriptions usually begin with expressions like "consider" or "assume" and thereby make it clear that they are not descriptions of fact, but an invitation to ponder — in the present idiom, imagine — a particular situation.

We have another similar remark in Levy (2012), where we read,

> There are further discursive markers that point in the direction of fiction. Like Nowak, modelers often introduce their work with locutions such as "consider…" or "imagine…" And in their more reflective moments, many tend to speak of their work as depicting "artificial systems", "simplified scenarios" or "stylized versions" of the phenomena under study.

These quotes are fairly representative of how philosophers regard model building in science as similar to fiction making. And there is little question that they are right about how models are introduced in science. Leafing through scientific journals and textbooks, models are indeed typically introduced in this manner; and there is equally little question that the purpose of introducing models this way is to alert the readers that the objects or situations that they are asked to consider or imagine are only intended for *studying*, not yet for *believing*. By such an act, a realm of assumed objects and situations is invented in which many statements will be made for the readers to hold up and consider. Is this not analogous to works of fiction, in which realms of similar nature are invented for their readers to entertain rather than to actually believe?

But the question is: does this parallelism actually exist? Do writers or producers of literary fiction introduce their fictional world in like manners? Let us see some examples of how fictional realms, populated with characters and events, are in fact introduced. James Joyce's *Ulysses*, a depiction of an entirely fictional but ordinary world, begins thus,

> Stately, plump Buck Mulligan came from the stairhead, bearing a bowl of lather on which a mirror and a razor lay crossed.

And here is how J. R. R. Tolkien's *The Fellowship of the Ring* as the first part of *The Lord of the Rings*, a depiction of a fictional and mythical world, begins.

> When Mr. Bilbo Baggins of Bag End announced that he would shortly be celebrating his eleventy-first birthday with a party of special magnificence, there was much talk and excitement in Hobbiton.

More interestingly, after Honoré de Balzac introduced Mme. Vaugque and her boardinghouse on rue Neuve-St-Geneviève between the Latin Quarter and the faubourg St-Marcel in exactly the same, matter-of-fact, manner as the above, and reflected on how his readers might regard his account of what unfolds in that house, he writes,

> Ah! sachez-le: ce drame n'est ni une fiction, ni un roman. *All is true*, il est si véritable, que chacun peut en reconnaître les éléments chez soi, dans son coeur peut-être.

What about pure science fiction? In the opening sections of Asimov's *Foundation*, we read, "His name is Gaal Domick and he was just a country boy who had never seen Trantor. ... He had steeled himself just a little for the Jump through hyper-space, a phenomenon one did not experience in simple interplanetary trips. The Jump remained, and would probably remain forever, the only practical method of traveling between stars."

There is no difference, again, even in plays or screenplays. Open Shakespeare's *The Tragedy of Julius Caesar*, what do we find there? A list of people's names and brief descriptions of who they are, and then:

> [Scene] *Rome; the neighbourhood of Sardis; the neighbourhood of Philippi.*
> [Act I. Scene I] *Rome. A street.*
> *Enter* Flavius, Marullus, *and certain* Commoners.
> Flavius: Hence! home, you idle creatures, get you home. Is this a holiday?
> …… ……

In all these and many other examples, there is no trace in the written words that anything is imagined or invented, much less than "assumed" or taken up for "consideration." No, they are straightforward account of facts about real people and events. As far as the texts are concerned, there is no difference between a fictional work and a historical or biographical work. The difference is of course what is understood beyond what is written, and we are supposed to understand (we are in trouble if we don't) that the stories and dramas the above quoted begin are entirely fictitious, non-existent, made-up. (And for this fact, a disclaimer is sometimes, but not often, given before the body of the work.) But why won't the authors tell us explicitly about this assumption within their works, as the scientists do about their models? Why do they always tell their fictitious stories straightforwardly as if they were factual? And another thing: for good and serious fictional works, it is almost inevitable that, all things being equal, the stories could well have been accounts of what actually exist and has happened. And modal scope of the "could have been" should include what is barely physically possible to cover such genres as science fiction.

III

A fictionalist on scientific models, when confronted with the difference of how the "models" in science and in fiction are introduced, may easily respond by saying that it is simply a superficial difference, a difference of style of writing only; there isn't anything substantial enough in this to raise objections to the fictionalist original claims. Even though writers introduce their fictional tales as if they were accounts of facts, what they introduce are in fact results of imagination that do not correspond to facts, and most consumers of fiction know or should know it.

Let us not be too hasty and reflect a bit further on the cause or reason why two groups of "inventors" invent what are allegedly the same sort of things, models and fictional characters/events, in such markedly different manners. In other words, let us inquire separately why the scientists who propose a model for a type of systems, such as the ideal gas model, do not straightforwardly speak about the gas and its molecules but rather use the preamble of "assume or suppose there is a gas of such-and-such properties." And why do writers of literary fiction do the opposite, knowing full well what they tell is mainly shear fabrication? (Historical facts are often included, but they usually only serve as the background of the fictional story.) I have not done any exhaustive search on this matter, but I am confident that the writing of literary fiction follows the above convention regardless of what language or culture in which the works are found. If that is right, it does not seem reasonable to think that such a universal feature could be the result of a mere historical accident. Scientific literature is much less diverse worldwide than literary works, and so it might not be a surprise to find that the way in which models are introduced is uniform as well.

A simple and also largely correct answer to why literary fiction is always introduced as if it were about facts may well be this: the very purpose of inventing a fictional realm populated with fictional characters and events (often mixed with real but less important ones) is to get the consumers to suspend their disbelief of the account and "consume" it as if it were an account of something real. Or on the flip side, if the creator of the account fails to get her readers/consumers to suspend their disbelief, the account is usually regarded as a failure. Imagine what would happen if a reader of a novel feels that she is constantly reminded of the fakeness of the tale and cannot make herself pretend to believe anything written in there; wouldn't she soon abandon the book? Now, no matter how excellent the novel is as a literary fiction, if the author had inserted in her narrative frequent reminders, as scientists do when discussing their models, that the tale is an invention of imagination or only to be entertained but not believed, wouldn't she have simply spoiled her work?

The above is only a naïve and intuitive answer. What do philosophers who investigate the nature of fiction say about this? Concerning the nature of fiction or, more generally, of representational arts, there are different philosophical approaches. Older views of fiction as part of philosophy of language, which treat fictional discourse as an extraordinary species of discourse, may take it as using assertions of "pretense." When a writer writes about her characters, she is pretending to be making assertions about them but in fact made none. Formally, there is little difference between an assertion and a pretended one, and yet the former, if true, must be about something factual, while the latter, even if true in some sense, is about none; it is pretended to be about some facts; it is asserted *as if* it was about something (cf. Searle 1975; Lewis 1983).

Kenneth Walton (Walton 1990) rejects such a "linguistic" approach, partly because he sees fictional representation as an integral part of artistic representation that includes fine and plastic arts. According to his theory, works of fiction, as any other forms of artistic representation, are acts of make-believe towards props, both on the part of producers and on the part of consumers. Writers of fiction and their readers engage in a game of make-believe in that both pretend that the created world is real, which comprises imagined characters and events, some of which being purely fictional and others not entirely so. The game of make-believe is to suspend one's belief about the fictionality of the work and pretend that it is straightforwardly about a portion of the actual world we inhabit. Treating artistic representation that includes fiction as games of make-believe has the benefit of not having to postulate and believe in the existence of fictional characters and events. When a painter pains a portrait of Sherlock Holmes, she and her audience do not have to believe that the portrait is a portrait of somebody or anybody at all. The portrait and all the other artistic representations are props in games of make-believe.

Jamie Thomasson (1999) espouses the opposite, in terms of ontology, of Walton's theory, i.e., a realist theory of fiction. According to Thomasson, fictional characters and events exist as created artifacts in the culture in which they are made. Very much as the Chrysler Building and the Forbidden City are actual existents in reality, Sherlock Holmes and Harry Potter are as well; the only difference being that the former are *physical* things while the latter are *abstract* entities. As abstract artifacts, the fictional are born and exist in the medium in which they are created. For instance, Sherlock Holmes and Harry Potter exist in the books that tell their stories, not as physical things such as ink marks that form sentences on the books' pages but as abstract entities embodied in the sentences. In other words, they are entities that readers may encounter when they open the books and read. With such real artifacts populating the fictional world, all the sentences about them within or without the fictional contexts genuinely refer and have truth-values in accordance with a straightforward semantics. The statement "Sherlock Holmes lives in 221b Baker Street, London" is true of Holmes in the fiction, and "Sherlock Holmes is more famous than any detectives in history" may also be true; and if it is true, it is true outside the fictional world and true of the character as a real thing (it says that the abstract thing that is this detective by the name of Sherlock Holmes is more widely known and esteemed than any actual detectives, real people, in history; and this statement may very well be true.)

Much more needs to be said about these and other approaches to fiction, and yet I hope this briefly canvassed summary of the major ones is sufficient for us to observe that what they have in common regarding what constitutes the fictional is the idea that fictional characters and events are made up and yet they are made up to be real in a world of make-believe. The authors of fictional works are suppose to fabricate things that their readers or audiences are able to suspend their disbelief and pretend, at least while enjoying the products, that they are for real. And what other means could be better suited for helping such pretending or make-believe than by beginning the narrative as if it is straightforwardly telling a true story of what has happened to real people?

If make-believe is an essential feature of fictionalization, and hence it is necessary to open one's narrative for a work of make-believe as straightforward factual account, then any other enterprises that are similar to works of fiction and essentially engage in fictionalization in their practice should exhibit the same feature. If scientific models are similar to works of fiction and modeling is essentially fictionalization, why do scientists not introduce their models straightforwardly as if they were talking about real systems?

The reason why scientists in making a model cannot introduce it as writers introduce their realms is subtle but profound. It has to with how models are taken to represent their targets in reality. The fictional also represents reality, but the way in which it represents reality is fundamentally different from how models do it; and there lies the reason we now seek. Imagine a case where Boyle is trying to introduce his model of ideal gas, where diluted gases are regarded as comprising sizeless particles that do not interact with one another. Is it reasonable for us to imagine Boyle introducing his model in the same way a fiction writer introduces her fictional world? Could he open his account of ideal gas by directly talking about how sizeless molecules populates an otherwise empty space of a certain volume and move completely free? Among the results of his account so introduced are the familiar equations for the temperature and pressure of the gas as functions of the molecules' average movement (I do not mean the "ideal gas law," which only contains macroscopic variables).

The problem of imagining Boyle doing such a thing is obvious. Introducing the model of ideal gas directly would have the same effect as introducing a fictional world in which such gases exist, and would tantamount to asking us to suspend our disbelief and pretend that such gases exist in reality. But that cannot work because the readers are also supposed to know that there are diluted gases in the world that Boyle's model is really *about*. Boyle's introducing his model straightforwardly as if making factual claims would, in analogy to fictionalization, have the undesirable result of forcing his readers to believe that there exist ideal gases in reality on top of the actual gases (some of which animals and plants couldn't survive without), *and yet — here is the most unreasonable demand — we are supposed to believe that the only reason for the ideal gases to exist is for them to represent the actual diluted gases.*

The problem would not occur to writers or artists of fiction precisely because what is made up is not supposed to be about anything in reality that correspond to it. There should be no problem (make) believing that Sherlock Holmes exist *alongside* with all the actual British detectives in history because he is not *about* any of them. If he is supposed to be about one of them, the matter then becomes totally different. I shall discuss the case for historical fiction separately in the next section.

Therefore, modelers introduce their creations with a preamble that alerts the readers of their *direct representational* role as well as *hypothetical* nature. The preamble does a double duty: (1) alerting the readers that what is introduced is different from the actual systems it is introduced to represent; and (2) asking the readers to suspend their disbelief of the introduced artificial system and consider it *as if it were the actual thing*. By putting their account of the models in a suppositional or hypothetical mode of presentation, scientists are not in any particular sense trying to be "honest" as if novelists are constantly trying to "cheat;" no, this is obviously the best policy given the constraints that modelers in science work under, which are very different from the novelists.

Let us look at this difference between the HYPOTHETICAL and the FICTIONAL in more details. Writers that fabricate literary stories are not intended on deceiving, but rather they are playing, so to speak, a socially accepted game of make-believe. Their profession is to fabricate "worlds" in which consumers as players can play, in imagination or in physically realized settings, while suspending their disbelief (or suspending the belief that what they experience is no more than an account of a game). By the same principle, simulation games that create what we now call "virtual reality" are introduced to us in the same way. Once we are strapped up with the gears, what we experience in the virtual reality games are supposed to be no different from our experience from reality (at least that is the aim of such games). Implicit in this is a sort of lack of responsibility on the part of the creators of such games, literary or virtual. Precisely because what they invent are understood as fictionalization, the creators of works of fiction need not worry about the factual relationship between their works and reality. The assumed relationship between a fictional world and the actual one is a two-dimensional relationship. The *external* relationship is commonly understood (minimally at least) by the consumers as a complete detachment — there is no connection between what is said literally in the fiction and whatever actually exists or happens; while the *internal* relationship is commonly understood as complete faithfulness to the fabricated facts in the fictional world. Because of this feature in works of fiction, the technique of introducing the story straightforwardly appears essential and indispensible to the genre, which explains the ubiquity (or universality) of such a technique across cultures and histories.

Modeling in science — the task of scientific fabrication — is quite different. When a scientist asks her reader to consider or imagine some aspect of a natural system, such as melanoma, she invents an artificial system that is different from a real piece of skin and yet it has all the relevant properties of skin to fully show the different stages and manifestations of skin cancer (cf. Nowak 2006). The invented system — in imagination or with a physical realization — that she asks her reader to consider or assume is what we call "her model" for melanoma, and it is usually a fabrication through either idealization or analogy; it is usually not and should not be regarded as a factual representation of human skin in any shape or form. It is in this sense that modeling is compared with fictionalization. But note, the properties of this model system that scientists propose for us to study and understand are supposed to closely resemble the real or actual aspects of melanoma. This statement of course must be taken with a grain of salt because not all modeling jobs in science share this feature. Acts of modeling in science are ubiquitous and also diverse, and it would not be wise for me to even attempt a brief enumeration of all the types. But what I am discussing here, the model for the study of melanoma, or of cancer in general, is a type of modeling that represents a large category. What the model is given to represent are usually systems that modelers can have direct sensory access to; here we have organs of human or animal bodies. Selected aspects are to be singled out for investigation and, because of the complexity of the actual systems or the nonseparability of their parts, much simpler artificial systems are imagined via the method of analogy to accentuate the aspects. Rather than figuring out how carcinogens operate in real pieces of skin, we have a mechanical model in which carcinogens can be clearly seen to produce melanoma.

The clear division between the internal statements (i.e. statements about what is going on inside the invented realm) and the external statements (i.e. statements about the representational relationship between the invented and the real), a division adhere to rather strictly in works of fiction, is totally untenable in most and the most common acts of modeling in science. While the name Sherlock Holmes only refers to a person internally (no external reference), most names or kind terms in a model do the opposite. They do not refer internally to the components inside the model that they are attached to. For instance, in a picture of Boyle's ideal gas, the label "molecule" attached to the dots in the picture does not refer to those dots or sizeless points that never interact with one another. No, the word actually refers to, as a type, molecules in real diluted gases. The same is true for most representational vehicles, such as in maps and scale models. In fact, the representational link between a model and its target is not secured by the structure of the model, which can appear very differently from the target system's. The link is provided by terms attached to the parts in a model. Hence, when the ideal gas model is introduced, it can and should not be introduced as "look at these sizeless points that move without interaction," but "try to imagine that molecules in diluted gases are without size and interactions." The external connections are buried in the model with those referring terms or labels;

and that is something that never happens in real fictionalization.[1]

We can now see that the relationship between a model and what it represents is very different from the one between fiction and reality. First of all, the clean separation between the internal and the external connection in the fictional case is guaranteed by the fact that fictional characters and events are not in any direct representational relationship with any real persons or events. Scientific models cannot achieve such a separation from facts. Note, what I said above about works of fiction by no means imply that they are necessarily mere fantasies totally separate from real life; no, it only implies that factual correspondence with reality of characters and events is not assumed in any degree of approximation. Although the mechanical model of human skin is very different from the skin, and yet it cannot be thought of as an invention that does not correspondent to it, as in the case of a Sherlock Holmes or a Harry Potter. The essential details of the model melanoma must be compared with facts from experiments with real skin, whereas no such essential details of either character, or of any other fictional characters, can be made responsible for matching, however remotely approximate, any factual details of a corresponding actual object.

Furthermore, the relevant properties, such as the specific ways in which carcinogens causes melanoma in some graphically clear way in the model should, if the model is adequate, be regarded as closely corresponding to the actual causal processes that happen in real skin; for otherwise (as a realist would strongly argue) it would be a miracle that our story from the model actually explains what happens to actual melanoma cases in life.

Because of this tight but also complex epistemic relationship between scientific models and their target systems, it would be awkward, if not downright deceptive, if scientists introduce they model-based theory of melanoma (as sketched in our above example) straightforwardly as if they were simply talking about the actual cases in real skin. This point also

[1] There are models in science that contain names and labels that do not refer to objects in reality in any literal sense. For instance, there is a famous model in statistics that is called "dog-flea" model. It represents a situation in which fleas jump between two dogs in precisely stipulated ways. It has little to do with the behavior of fleas and dogs; it is rather a possible dynamic structure that happens to be "fleshed out" by a particularly fertile imagination with fleas and dogs. This model and models like this, which widely populate areas such as economics, may truly be called "fictional models." This point does not invalidate my argument because the fictionalist view on scientific models is a view about all acts of modeling.

explains an important difference between character (or event) fabrication in fiction and model making in science. When the mechanical model of melanoma is imagined, it is not intended as an imaginary "piece of skin" that is different from actual skin but has as many, albeit different, details. It is rather a simplified system that shares with actual skin (within our theoretical framework) the relevant properties for melanoma. Fictional characters and events, at least in good works of fiction, are very different in this respect. They are imagined as concrete and detailed as any faithful account of real people and events. And this is why fictional objects cannot be used to represent any real objects in the literal sense, while objects in models are supposed to literally represent actual systems in different degrees of approximation. The double-helix DNA molecules in the Crick-Watson model are supposed to "stand-in" for the actual DNA molecules, and so are the sizeless molecules in Boyle's ideal gas model for actual gas molecules. But it makes no sense to regard James Bond as a stand-in for any particular actual British special agent. As a corollary, this is also why it is perfectly legitimate to have a plastic or steel model of a piece of cancerous skin, while it has to take a real person, whether in a movie studio or on a theatre stage, to physically realize any fictional character.[2] This is one and perhaps the main reason that I regard scientific models, although very much the result of highly developed methods of fabrication, as hypothetical, while I regard fictional characters and events as properly fictional.

There is another reason why models are hypothetical. Model-based scientific research should not be written off so easily towards the anti-realist direction. Even though there is reasonable ground in thinking that models are "not real" or "not true," there is also good reason to believe that they are close to the real in some suitable sense of "being close to." But what does "being reasonably close to" mean? To take our mechanical example

[2] Cartoons and puppet shows are also used as physical realizations of fictional characters, but the difference is still obvious between model realization and character realization. While it makes no sense to try to use a biologically "life" material to physically realize a mechanical model of melanoma, it is by no means senseless to employ a real person rather than drawings and stick figures to represent a fictional character. Special circumstances or genres of fictional representation may favor cartoons and puppets, but while no genre of modeling melanoma favors the use of real skin, no genre of character representation of any works of fiction precludes the use of actors.

of melanoma as an example; we know that real skin as a whole is not as simple as our mechanical model, but it is still a scientific/empirical question whether or not the real skin is not a composite system, some of whose basic components are indeed mechanical subsystems rather similar to our model. The current "fashion" in science, especially in physics, leans towards the anti-mechanistic, and yet nothing in our empirical tradition of scientific research conclusively rejects the idea that what is modeled, often in terms of simple mechanisms, is not actually what is or very close to being the case. The idea that *nature is fundamentally modular* is not yet falsified by any scientific discoveries nor is it completely rejected by philosophy of science; and model making under a strong assumption of nature's modularity may be neither fictionalization nor fabrication. It is not entirely implausible that apart from the material, which the abstract models usually do not care to postulate any way, what is imagined in a highly idealized model may well be, in great approximation, true of the corresponding component of a system, whose complexity may initially make this idea incredible.

And this leads to our third reason of why scientists and fiction writers introduce their "worlds" so differently. Sherlock Holmes and his world are introduced straightforwardly as if they were real because they are not by default, while the mechanical model of melanoma is introduced tentatively as a hypothetical object because it is never regarded as simply "false" or unreal. It is more reasonable to see it as the first step of an approach to a scientific representation of melanoma that may eventually be transformed into one is that is *true of* the phenomenon. The case of Boyle's ideal gas model later being enriched (and replaced) by van de Waal's "real" gas model and then even later by the quantum mechanical version of that model, etc., is a good case in support of such a view. One cannot imagine such an improvement scheme being applied to works of fiction, can one? We can see that the different styles with which the narratives are introduced tell us profoundly about the fundamental differences of two seemingly analogous enterprises.

Therefore, we should say that models are hypothetical objects that typically appear in science. It is not a good idea to look for further illustration or understanding of their nature, such as in the realm of fiction.

IV

Historical fiction is a literary genre that is generally perceived as lying between fiction and history. The characters and events in a historical novel are supposed to be real people and actual events in history (at least the major ones are such), while most of the details pertaining to the characters and events, such as what the characters say and how the events unfold, are made up to serve ends that such details in pure fiction do.

Shall we say that scientific models are rather similar to characters and events in historical fiction? Couldn't we regard our mechanical model of melanoma a fabricated "story" about something real, namely, actual cancerous lesions in real skin? The minimal sense in which modeling may be said to be similar to historical fictionalization is perhaps this: the account is not true and is not aimed at literal truth while the entities so accounted for do actually exist. Here both scientific modeling and historical fiction appear to fit into what Levy (2010) called *de re* modeling (i.e. for x, rather than fabricating a new object from x and say things true of it, we say things about x that are not literally true of it). But apart from this minimal point, the two enterprises have little in common; they have especially little or nothing in common in their respective functions. To use the terminology we have established in the previous sections, the difference of function is simply this: models are hypothetical while characters and events in historical fiction are first and foremost fictional. A Lincoln or a Mao Zedong in works of historical fiction is no more than a fictional character; the fact the name refers to an actual person in history has no *factual* significance. What is said in a historical novel about Lincoln is neither intended nor objectively about Lincoln, the actual person; and so the reference of "Lincoln," even if held, does not do the work it does in normal semantics. And what is said about this character, therefore, does not tell us anything factually significant about Lincoln. This is emphatically not the case with scientific models. None of this can be said with justification to, say, the mechanical model of melanoma or Boyle's ideal gas. The details of the models, though they are equally non-factual, are nevertheless important to our factual understanding of real cases of melanoma or real gases; and they have to be tested through the normal procedures of hypothesis-testing.

Whether a fictional account of Lincoln does matter to our understanding of the real Lincoln will depend on what the account is intended for. It is quite possible that the fictional account is intended for "pure entertainment," in the sense that it gets readers to imagine a possible, and perhaps less tragic, alternative life of Lincoln. The parallel in science of such fabrications may result in what we regard as "bad models," ones that by default cannot withstand empirical tests. But the entertaining fictional Lincoln is not necessarily bad fiction; at least it cannot be so simply because what is said of Lincoln there does not match historical facts in any senses of matching. However, a fictional account of Lincoln may well be intended for something much more significant or profound; it may be aimed at discovering some intrinsic or law-like characteristics in Lincoln's personality that the actual life of Lincoln, because of all the accidental factors in history, failed to display. The result of such an account would be a better understanding of Lincoln, the person, or even more ambitiously, Lincoln as an exemplar of a type of persons. If this were the aim, then I would think that there is little difference between this case of fictionalization and the typical cases of scientific modeling; but I doubt that this is often the aim of historical fiction. In fact, to aim at such a historical fiction of Lincoln seriously, wouldn't the writer have to be well equipped with scientific findings in, say, psychology and social sciences, where scientific models abound?

References

Contessa, Gabriele (2010) Scientific models and fictional objects *Synthese* **172**:215–229

Field, Hartry (1989) *Realism, Mathematics and Modality* Oxford: Blackwell

Frigg, Roman (2010) "Models and Fiction," *Synthese*. **172** 251-268.

Giere, Ron (1988) *Explaining Science: A Cognitive Approach*. Chicago: Chicago University Press

Giere, Ron (2009) Why Scientific Models Should not be Regarded as Works of Fiction In Mauricio Suárez ed. Fictions in Science: *Philosophical Essays on Modeling and Idealization* (Routledge) 248-258

Godfrey-Smith, Peter (2006) The Strategy of Model-Based Science *Philosophy of Biology* **21**:725–740

Godfrey-Smith, Peter (2009) Models and Fiction in Science *Philosophical*

Studies **134**:101-116

Levy, Arnon (2010) Fictional Models *De Novo* and *De Re* (preprint)

Levy, Arnon (2012) Models, Fictions, and Realism: Two Packages *Philosophy of Science* **79**: 738-748

Lewis, David (1983) Truth in Fiction. In *Philosophical Papers vol. 1.* Oxford: Oxford University Press.

Nowak, Martin A. (2006) *Evolutionary Dynamics: Exploring the Equations of Life* Cambridge, Mass: The Belknap Press of Harvard University Press

Searle, John (1975) The Logical Status of Fictional Discourse. *New Literary History* **6**: 319-332

Suárez, Mauricio (2009a) Fictions in Scientific Practice, in Mauricio Suárez ed. *Fictions in Science: Philosophical Essays on Modeling and Idealisation* Routledge: 1-15

Suárez, Mauricio (2009b) Scientific Fictions as Rules of Inference, in Mauricio Suárez ed. *Fictions in Science: Philosophical Essays on Modeling and Idealization* (London: Routledge)158-178

Suppe, Frederick (1989) *The Semantic Conception of Theories and Scientific Realism* Chicago: University of Illinois Press.

Toon, Adam (2010) Models as Make-Believe, in Roman Frigg and M. Hunter eds. *Beyond Mimesis and Convention, Boston Studies in the Philosophy of Science* (Dordrecht: Springer) 71-96

van Fraassen, Bas (1980) *The Scientific Image* Oxford: Clarendon Press

Yablo, Stephen (1998) Does Ontology Rest on a Mistake? *Supplement to Proceedings of the Aristotelian Society* **72**/1: 119-37.

Yablo, Stephen (2010) Go Figure: A Path through Fictionalism, in *Things: Philosophical Papers 2* Oxford University Press, 177-199.

What Is Wrong with the New Fictionalism of Scientific Models

This paper defends an approach to modeling and models in science that is against model fictionalism of a recent stripe (the "new fictionalism" that takes models to be abstract entities that are analogous to works of fiction). It further argues that there is a version of fictionalism of models to which my approach is neutral and which only makes sense if one adopts a special sort of antirealism (e.g. constructive empiricism). Otherwise, my approach strongly suggests that one stays away from fictionalism and embraces realism directly.

Keywords: models, representation, realism, empiricism, fictionalism, maps, and reference

1. Introduction: the new fictionalism

Philosophical discussions on scientific modeling and models have concentrated for some time on the role they play in scientific theorizing and their relationships with theories. One approach, lead by Cartwright (cf. Cartwright 1983, 1999; Morgan & Morrison 1999), takes models to be "mediators" between phenomena and theories. Models differ from theories in the roles they play in the praxis of science, and also because they are non-truth-apt entities. Unlike statements or propositions, they are not the type of entities that can be straightforwardly said to be true or false. (For those who take models to be physical objects, models are obviously non-truth-apt and not part of any scientific theories.) Another approach comes from the deflationists (cf. Teller 2001; Suárez 2003, 2004, 2010; Callender and Cohen 2006) who take models to be part of scientific theories and yet their uses and meanings are entirely pragmatic, such as tools for inferences (a la Suárez). Yet another approach, à la van Fraassen and Giere (cf. van Fraassen 1980; Giere 1988), is to regard models as the core components of theories (model + empirical hypothesis = theory) or theories as classes of structures/models (the semantic approach to theories). With inquiries into the functions of modeling and the structures of theories approaching exhaustion, attentions have recently shifted towards the metaphysical or ontological issues about models: what sort of entities are scientific models? (cf. Contessa 2009, 2010; Frigg 2010a, b, 2011; Godfrey-Smith 2006, 2009; Levy 2010, 2012; Rowbottom 2009; Suárez 2009a, b; Toon 2010a, b).

Godfrey-Smith (2006) is undoubtedly correct in pointing out that even though not all disciplines of science resort to modeling and use models, a large number of them do, and so at least with respect to such "model-based" disciplines, it is worth finding out what models are metaphysically and how they represent. One of the results of this intense effort is what I shall call the new fictionalism on models (cf. Contessa 2009, 2010; Frigg 2010a, b, 2011; Godfrey-Smith 2006, 2009; Levy 2010, 2012; Toon 2010a, b), of which I give a sketch below (more details appear later in the discussion).

The model-based scientific representation, according Frigg (2010a, b, 2011), is an enterprise of a *tripartite structure* (descriptive elements $\Rightarrow$ models $\Rightarrow$ real systems), and two salient features stand out among the variety of models and modeling jobs in science: models are *non-truth-apt* (or *non-discursive*) entities and modeling involves an invention of *imaginary systems.* As non-linguistic and a "mediator" between the linguistic and the target system, a model is best seen as either *an abstract artifact* or *a physical artifact* (cf. Keller 2002; Rowbottom 2009 for how even natural objects, such as a fruit fly, can be models). If one chooses the former, as the new fictionalists do, one has the following options for models: Plationism (realism), nominalism, and fictionalism, as they are options for other abstract entities such as numbers and propositions. And given the second salient feature on models, fictionalism becomes an attractive choice. Besides, adopting Platonism for models appears awkward, very much unlike the situation with, say, numbers and propositions, because while adding natural numbers, etc. to reality on top of physical things does not seem too problematic, adding representational vehicles (including models) of physical things to that reality does seem odd. In that case, we would have in reality the physical things, the physical realizations of the models (of some) of those things (such as scale models) and, in addition, the models. Although not an impossible stance, it does not seem a natural or reasonable position to take. And nominalism is not viable if Platonism is not a sensible option. On the other hand, ample examples have been paraded to show similarities between modeling and fiction writing (cf. Frigg 2010a, b; Levy 2012; Toon 2010a, b; Godfrey-Smith 2009). Godfrey-Smith reminds us that *War and Peace* is intended to and does represent Napoleon's campaign against Russia and *Lord of the Rings* Europe in the Middle Ages. Frigg also argued that the "t-representation" in his account between model systems and their targets is analogous to how fiction teaches us about real life (see Frigg 2010a, pp. 103, 121, will discuss this point later). To further illustrate and dispel misunderstandings about the matter, let us use a very simple example to show the parallel between modeling in science and fictionalizing in literature that the new fictionalism holds, we can think of a model of simple pendulum with a "string" holding a "bob" that swings in a regular manner describing a "trajectory." And this is analogous to a story about a "murderer" walking down a "street" that results in a set of "footprints." Both are not in the actual world but exist in their respective fictional realms, and yet in some indirect way both represent actual objects and their aspects.

The next question is whether such a fictionalist turn is compatible with scientific realism. While some of the new fictionalists (e.g. Frigg 2010a, b, 2011) suggest that we shelve the issue, others (e.g. Levy 2012) confront and explore it, short of making a commitment. Contessa (2010, pp. 219-220) rightly warns us against a confusion between the new fictionalism and the more traditional fictionalist view theoretical entities in scientific theories á la Fine (1993) and Cartwight (1983, 1999). While the traditional fictionalism is about theoretical/unobservable objects being unreal or fictional, being necessary only to save phenomena, and it is therefore only an option for antirealism, the new fictionalism is said to be compatible with realism if not conducive to it.

> In fact, one can be a scientific realist and still believe that the models that we use to represent atoms or quarks are fictional entities even if the atoms and quarks themselves are actual concrete entities just like tables and chairs. (Contessa, 2010, p. 220)

Levy (2010, 2012) fully confronts the compatibility issue between fictionalism and realism and explores two packages of fictionalism towards reconciliation: the 'whole-cloth'-fiction (Levy 2012) package and the worldly-fiction package. In the former approach, whole systems are imagined or fabricated and treated as targets (or as if they were real systems), while in the latter approach modeling is seen as abstracting aspects of real systems, forming statements about the abstracted aspects, and forming an idealized or abstracted theory of the real system. In the earlier paper (Levy 2010) a version of the former is called *de novo* modeling and a version of the latter *de re* modeling. *Prima face*, the worldly-fiction package is more amenable to realism than the world-cloth version.

Whichever version of new ficionalism one is inclined to embrace, whether it is compatible with scientific realism is undoubtedly treated as a serious issue in the recent literature.

This paper defends an approach to modeling and models in science that is against the new fictionalism. It further argues that there is a version of fictionalism on models to which my approach is neutral, and that it only makes sense to adopt it if one adopts a special sort of antirealism (e.g. constructive empiricism). The other option is to stay away from fictionalism altogether and embrace realism directly.

2. A hybrid approach to modeling

I shall begin my approach to modeling with a look at how maps work. The map metaphor for modeling is quite common and goes a long way back, and Frigg (2010b) gives perhaps its latest detailed study (he uses maps to show how his t-representation relation may work (p. 125)). My account differs from his, as I shall explain later.

When we look at, for instance, a map of Beijing (in English) and see a thick straight line-segment in the middle bearing the name "E. Chang-an Ave," we know that the line-segment represents in the map the East Chang-an Avenue in Beijing. Other streets and buildings in Beijing and their geological relations are represented in the map the same way: a line segment or a dot and a name or a numeral attached to it (with numerals, a key is provided at a corner). The names denote the objects they refer to, and then the geometrical figures the line segments refer to tell us how those objects in the city of Beijing are represented, or *what they are like, in the map.*[1] The

[1] Throughout this paper, I assume that geometrical figures are abstract entities that we use, e.g., marks on surfaces, to represent. However, if one rejects all abstract entities as an ontological commitment, then one can easily dispense the reference talk such as line segments

geometric figures (picked out by the line segments) are abstract objects and they appear in other places and fulfill other purposes just like any other geometric figures do. They serve a special purpose in this map because the line segments and the dots are *attached with* the appropriate names. However, the names do not refer to or denote the geometric figures, even though it has the appearance of doing that, because, for instance, the name "University of Florida" on a plague at the entrance of the university refers to the university (and its campus). The thick line-segment in the middle of the map of Beijing does not by itself represent East Chang-an Avenue. It represents what that avenue should look in relation to other figures represented in the map. The whole map is usually called "The Map of Beijing," which is the name of the map, not of the city; but all the names, including "Beijing" in "the Map of Beijing," in the map are names of parts of the city, not of the figures in the map. By the direct reference of the names and symbols, the map is made to represent a complex city as Beijing in a simple geometrical image. And the image is *shown* by the lines and dots that refer to pure geometrical figures of difference types and sizes. Is the map like (= good representation) the city? Yes, if it represents the *geometrical features* of the city correctly; otherwise, it is not like the city (= bad representation). Does the map resemble the city? No, not by a long stretch; but that is beside the point for representations of this sort. Our map of Beijing necessarily distorts the actual features of the city for the sake of accenting those features that map users want, and there are good distortions and bad distortions, but good or bad, once the names are attached correctly or nearly correctly, it is a map that represents Beijing.

I suggest that the sort of scientific models the recent literature has been covering (the models understood in the new fictionalism) can and should be understood in essentially the same way.

Figure 1 shows a model for atoms in an experiment (Rutherford's), and "The Rutherford model" is a name of the model, not the actual experiment, but of in model, we have "atoms," "electron," "α particle," "nucleus," etc.

---

and dots representing geometric figures and just talk directly about line segments and dots on paper and their geometric relations.

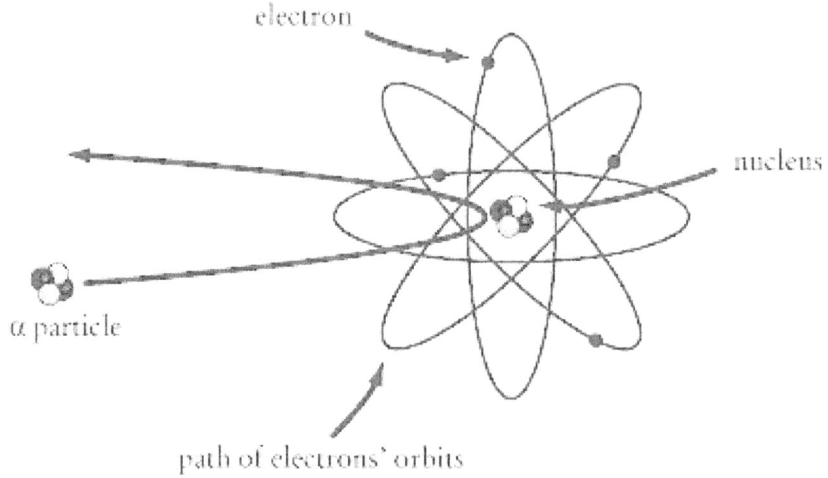

Figure 1: The Rutherford Model of atoms (and experiment)

These are terms referring to the objects in reality, not the components in the model. It is not difficult to see that most other models, differences in details notwithstanding, can and should be viewed essentially in this manner[2]. The names, (natural) kind terms, or any other sorts of symbols used in such a model all refer to the specific components of the target systems that the model represents. They do not refer to or denote components in the model. The components are in most cases abstract objects that may serve in a great variety of models and bear a great variety of terms (or symbols). If we strip all the *referring devices* (RD) from a model, what is left is an abstract object of a certain structure,[3] which may or may not resemble something real.

I shall henceforth call such things "models sans RD," (where RD refers to all symbolic devices, not just words and labels, though they are the most often used,) and they are generic abstract entities (or physical objects if you will);

---

[2] I say "most," for there are exceptions to this claim, as we shall discuss separately later. There are "toy" models or fictional models in science that do not fit this claim here.

[3] NB: this point remains valid if one rejects the abstract-entity conception of modeling and takes all models to be concrete physical objects (such as a scale model of the Rutherford atoms or diagrams on paper as the one shown above). Striping the labels from the parts and removing the name of the model would make the "thing" just another physical object or set of ink marks. This is consistent with what is said in footnote 3.

and they are better understood as proto-models — entities that are ready to be used as models but not yet such. Note: the point is not that abstract model components, such as geometric figures, cannot serve as notational vehicles. They can, of course, just as a realistically drawn portrait of Obama can serve (without any labels) the double duty of denoting and showing the likeness of Obama (but what if Obama had an identical twin brother? cf. Goodman 1976)[4]. The point is that in most cases, and more importantly by the nature of scientific modeling, the duties of denoting and showing likeness in selected aspects are distinct and separate. What modeling does is to show us what the denoted components are like and how they are related. The referring job is done by devices that already perform the same duty in other occasions in our language, and the showing job is done by the abstract objects as model components (again, if you reject abstract entities, the job is done by the physical components or drawn geometric figures). This conception of modeling as a combination of abstract (or concrete) objects as modelistic elements and RDs as symbolic elements is, I argue, the right conception; and it provides a simple explanation for a well known fact in the literature on models: why resemblance, isomorphism, or any other intrinsic relationship between the structure of a model and its target is neither necessary nor sufficient for representation *per se* (cf. Goodman 1976; Suárez 2003, 2004; Callender and Cohen 2006, among others). Such a relationship between the modelistic elements and the target in terms of structural resemblances of some sort is not needed for the sake of denotation/reference. The job is done essentially by RDs or their equivalence. In my approach, modelistic elements (elements without RDs) are not supposed to *also* tell the inventors and users of models what they represent because whenever necessary, there are always attached RDs to rely on.

Let me now provide more arguments for this approach to modeling, which one may call a *hybrid view* of modeling, because it says that a model involves symbolic as well as modelistic devices, where the former denote and the latter show. And one essential feature of most modeling jobs is a sort of division of labor.

First, the quote above from Contessa (see p. 4) suggests that there are atoms and quarks in models — let us call them "schmatoms" and "schwarks,"

[4] Much more liberty may be taken in arts in how and what things and pictures can represent, we are here only concerned with scientific representation.

and there are atoms and quarks in reality (like tables and chairs). Schwarks, etc. are fictional objects, but Contessa claims that one can still be a realist in the sense of believing that quarks, etc. exist and are represented, through some magical ways, by schwarks, etc. ("magical" because it connects quarks, the real, with schwarks, the fictional). Contessa and the other new fictionalists do not talk about this arrangement in this transparent manner; they think the same terms such as "atoms" and "quarks" can be used to denote modelistic elements in some contexts and the real objects in other contexts. Therefore, in models, we have idealized or abstracted objects that are also called "atoms" and "quarks." This is obviously the wrong way of looking at modeling and models, because terms, such as "atom" and "quark" do not shed their normal meaning when combined with modelistic elements (or more general with pictorial elements) and become "schmatom" and "schwark," that denote those elements.

Why can't terms function differently in different situations? They do and can but not in this case because we also use the same terms, such as "quark," in sentences that talk about quarks. Don't tell me that scientists have to constantly disambiguate the term as meaning quark or meaning schwark when they read high-energy physics papers. (There are special terms referring to model components that do not include the word "model", such as "Bohr's atoms" and "lattice gas," but they are just like "schmatom" and "schwark" and they are never ambiguous.)

Second, this hybrid approach bears some resemblance to Frigg's view (2010a, b, 2011), but it is different. I have no problem agreeing with the framework of the tripartite conception of modeling, where abstract systems are imagined even "whole-cloth" (to borrow a term from Levy (2012)). In Frigg's version (2010a, b, 2011), we see a structure that may be simply rendered this way: Model Descriptions =(p-represent)=> Model Systems =(t-represent)=> Real Systems, where the first two parts are related by the "p-representation" and the last two parts by a different and separate representational relation he calls "t-representation." Given our analysis above regarding the nature and functions of RDs, it is simply not correct to think of the model-description part of the enterprise as only about the models systems in the middle part. The RDs in the sentences and equations in that first part refers to the objects in the third part where real systems reside that are the targets of modeling. Frigg (2010b) has tried to deal with the problem of how his t-representation may work, and for delivering a "First-Stab" he enlisted the map metaphor (p. 125) as I mentioned earlier. No doubt his effort in making this connection between map and modeling continues and enriches a long tradition of exploring that metaphor for representation, but by separating the p-representation and the t-representation, he fails to see that RDs function independently or free from these two separate stages of modeling. RDs in most cases of modeling do not refer to modelistic elements in the middle part, but are transferred from the first part to these elements so that the original referential relation is kept intact. In this way, when we consider the t-representation (if we opt to follow Frigg's structure), the referential relation that is transferred from the p-representation (always pointing to the components of real systems) is used to secure the connection between the modelistic and the real systems.

The RDs do the job I interpret them to do in a model the same way they do in other contexts; and this is a merit of my hybrid view of modeling. Symbolic vehicles of representation, which include terms and labels, function, and should function, the same way whether they appear in a sentence, a picture, or a more elaborate scientific model. They refer to the objects they are used to refer to in accordance with the convention that connects them to their referents and secure that connection. Unless special circumstances, as the ones we shall soon turn to, require the terms to work differently (but also by agreement among users for the special usage), there is no good reason to believe that when combined with modelistic elements, they change their function and refer to those elements instead. (The special cases include, e.g. that the terms "hawk" and "dove" don't really refer to hawks and doves in such models as the Hawk-Dove Model in evolutionary game theory.)

Not only is this the right way to resolve Frigg's "t-representation" problem, it also explains why modeling in science can be so imaginatively creative and not be bound by any apparent constraints on whether or how much the model (sans RDs) resembles the target. When the target of a model is securely anchored by its RDs, we can create whatever we want to show what it and its parts are like in the aspects we choose as long as there is a collective agreement. We may be bad modelers, but no one can accuse us of modeling the wrong thing. On the other hand, we can be said bad modelers precisely because the RDs in our models hold us accountable. Otherwise, we can acquit ourselves by saying we are talking about schmatoms and schwarks, and they are what we say they are; and unless we are inconsistent, we are therefore good modelers of atoms and quarks. This consequence, which may be reasonably attributed to new fictionalism, is clearly unacceptable.

Third, what about the point in favor of new fictionalism that models are often studied for their own sake, apart from what and how they represent? This happens only for models sans RDs and they are inevitably studied not as models of x and y, but rather as *proto-models*, something generic and can be used to model x and y and many other things when RDs are introduced. They may be mathematical objects or more substantial but still abstract artifacts (or they are just physical artifacts), and they may have special terms for their parts (such as "lattice gas" or "auxiliary loops"); but a study of them is not to be confused with studying the full-fledged model. That is a study of the target (through the model), which is chiefly the aim of science. We may study how non-interacting points operate in a confined space with rigid walls as a study of what modeling job this combination of abstract objects can do. But when we study ideal gas in which gas molecules behave like the abstract system, we are studying the behavior of molecules, not that of schwolecules! This is not so different from how differential equations are studied for their own sake. It is quite different from how Maxwell's equations are studied to understand electromagnetism. In parallel, the RDs for the variables in Maxwell's equations are exactly analogous to the RDs in, say, Rutherford's model. For instance, the vectors **E** and **B** are variable (symbols) for the intensity of electric field and magnetic field, not for their stand-in's in any model of electromagnetic field.

Models sans RDs are not models for they do not represent anything (yet). Metaphysically, there is an option for taking all such objects as fictional objects, just as there is such an option for numbers and propositions (cf. Field 1989; Yablo 1998, 2010). When attached with RDs, they become models, and the devices refer as they do in other appropriate contexts, and the referred real systems are represented and displayed by the models. In the next section I shall discuss in a more systematic manner (though mentioned earlier in passing as in the "Hawk-Dove Model") how there are fictional models in science (which the modelers themselves recognize as such), and they are not such models as Rutherford's model or the model for phlogiston or for luminiferous aether.

What is not plausible at all is the conception of new fictionalism that there is a realm that is populated with objects that are different, i.e. idealized or distorted or abstracted, from the real systems that they are supposed to represent; nevertheless, they wear the same "clothes" as the real systems do. So, a sizeless thing in an ideal gas model should still be referred to as a molecule although it only stands in for the real thing. And this realm somehow resembles the imaginary world of fiction, where similar things appear to be happening. There we have people and buildings that bear names real people and buildings have or might well have, and such "models" of society somehow can be used to represent our lives, but they don't represent them in any straightforward manner because the names in those stories do not refer to real people. From what we have argued so far, we can see that is emphatically never how models in science or in daily life work to represent reality.

3. Realism, empiricism, and fictionalism

There are no special challenges to scientific realism when we understand models according to the hybrid view. Realists can argue for the position as they have always done, namely, to show that science can and does aim at a truthful representation of the referred systems, such as electrons and economies, via model systems (be they abstract entities or scaled physical objects). I take it that to the extent that scientific realists have any success in arguing for their position, the fact that real systems are often represented in scientific theories as systems of selected aspects with idealization (and approximation) is no special challenge to their position. Fictionalism only becomes an attractive option when one is convinced of antirealism but needs to deal with the ontological problem for theoretical entities.[5] However, it may not be an attractive option, if one is also an instrumentalist. For instrumentalism, theoretical models, such as the Rutherford Model, contain non-referring terms, and therefore they are not *literally about* anything beyond itself (not fully representational). Claims about electrons, etc. in a model are neither true nor false in the standard way (when they are treated

[5] One must however be clear again that the problem for the antirealists is not whether it makes sense to regard the model sans RDs as existing. If it makes sense to be a Plationist on abstract entities while remaining an antirealist on hypothesized unobservable systems, then a model of quarks exists for such an empiricist while quarks are not regarded as real as desks and chairs.

as external claims). They may be trivially/analytically true or false if they are taken to be internal claims about model components or, more plausibly, they are empirically testable if they are taken to be reducible to claims of observational results. Historically, naïve instrumentalism has never been an attractive philosophical position.

Not all empiricists are instrumentalists, and van Fraassen is clearly a non-instrumentalist empiricist. It turns out that fictionalism makes the most sense for van Fraassen's constructive empiricism (CE) (van Fraassen 1980). In a subtle and illuminating paper, Rosen (1994) argues that the only sense one could make of van Fraassen's CE is to see it as a piece of fiction in social science, namely, the aim of science as accepting empirically adequate theories despite the demand for their literal construal is best seen as a piece of fiction among fictional communities of scientists. As a side point, but not an unimportant one, he argues that it is also best to think of van Fraassen as taking a fictionalist stance on theories that must be literally construed (although empirical adequacy is the only criterion for accepting them) (see especially, pp. 148-154). Let me flesh out this point of Rosen's and show why fictionalism can help us make good sense of van Fraassen's demand on theories for the unobservable.

Van Fraassen takes scientific realism to be committed to two claims regarding theories (or models): the terms in them genuinely refer and to accept them is to believe that they are true with good reasons. In coming up with CE, van Fraassen explicitly keeps the first claim but rejects the second, replacing it with the criterion of empirical adequacy. But what does it mean for the terms in a theoretical model to "genuinely refer"(cf. van Fraassen 1980, chapter 2) while suspending beliefs about claims made by using them? Let's see how a map works in a fictional setting. Figure 2 is a map of Thomas Hardy's Wessex.

Figure 2: Map of Wessex

What do terms in it such as "Upper Wessex" and "Wintoncester" mean? As in the case of the map of Beijing, neither the former refers to the enclosed space (containing the words) nor the latter refers to the dot beside "Wintoncester." No, they refer to the territory, Upper Wessex, and the town of Wintoncester. The only difference here is while "E. Chang-an Ave" refers to the Avenue in Beijing, an actual avenue, "Upper Wessex," etc. refers to *nothing* or at least not to any actual regions of the earth's surface bearing those names, because Wessex is a fictional region made up by Thomas Hardy for his novels and poems.

There are strong reasons for us to understand in the same way Rosen's fictionalized van Fraassen's CE. Unlike the new fictionalists who regard models as self-referential and claims as mostly internal, the CE'st account of models, with Rosen's interpretation, is a truly fictionalist approach on the unobservable. The terms in the models or theories (according to van Fraassen's semantic approach to theories, theories are classes of models!) do refer to things beyond the models (as "Upper Wessex," etc. refer to regions beyond the map) but the things simply do not exist; they could be fictional. It is not the model that is fictional, it is what the model represents (or genuinely refer to) that is non-existent, and therefore, fictional. As I cautioned in footnote 5, one can still be a Platonist (i.e. a realist) about the abstract entities that constitute models sans RDs, even if one does not believe what such models (when equipped with referring devices) represent can be true or one may be agnostic about that. This is the only version of fictionalism on scientific models that is analogous to the case of literary fiction and that makes good sense, but this is obviously not a fictionalism that can be made compatible with scientific realism.

I do not think van Fraassen would have objected to this fictionalist interpretation of his view on literally construed models/theories (in a prompt response to Rosen's critique, van Fraassen (1994) does not raise any objections to this particular point of Rosen's). However, in its original form, a CE'st needs only to hold an agnostic attitude towards unobservable objects, while a fictionalist has to at least *prima face* deny their reality.[6]

But aren't literary fictional worlds model-worlds, and if so why can't we analogously take *all* scientific models to be fictional? Are not works of fiction, such as Hardy's *The Mayor of Casterbridge* or *Tess of the d'Urbervilles*, themselves models rather than representing anything beyond? This is a challenge the new fictionalists would, I think, raise against my approach and analysis of the fictional. Let us look at the map case again. The question is: does it make sense to say that since there is nothing that "Upper Wessex" refers to, it must denote the space in the map? But that is tantamount to saying because "Tess" does not refer to any actual person, it must refer to the appropriate set of sentences or propositions in *Tess of the d'Urbervilles*. But how could that even make good sense? It does not seem to make sense to say that "Tess," clearly a name for a person, may name a person if there is actually one for it to name; otherwise, it may name a set of sentences or pictures. Similarly, it does not seem to make sense either to say that "Tess" refers to Nastassja Kinski in the film version of *Tess of the d'Urbervilles* directed by Roman Polanski, does it? Therefore, there is no good reason to believe in the new fictionalism in this respect.

What is the upshot here? Again, it is the distinction between the symbolic and the modelistic. If all the new fictionalists are claiming is that models sans RDs (= modelistic elements) are fictional entities, then I have no objection at all, as I do not have any objection to fictionalism about numbers and other

[6] Some violence is done to van Fraassen's view on models (the semantic view of theory), but hopefully no confusion arises from it.

mathematical entities. But that does not entitle one to say that models with RDs are also fictional unless one is committed to taking what the RDs refer to as objects of fiction. That means one takes airplanes, galaxies, and quarks, etc. to be fictional because that's what those corresponding symbols refer to. This is not what the new fictionalists are ready to hold, and it only makes sense for CE with respect to unobservable objects.

A work of fiction may be taken as a model of a fictional world, because it is more likely than not an abstraction (or idealization) of that world. For example, Thomas Hardy's fictional world of Wessex appears in his many novels and poems. Each of those can only show us some aspects of Wessex and its inhabitants (= abstractions), and yet all of them refer to the same world, which might have been modeled further had Hardy lived longer and produced more works about it. This must be true if we can, and I think we should, believe that, for instance, there is a definite number of houses in Wintonchester even though the number was never mentioned anywhere in Hardy's works. So models, either scientific or literary, are not fictional objects (other than being fictional in the trivial sense of being abstract artifacts if one rejects both Platonism and nominalism regarding abstract entities); they are vehicles that represent fictional as well as real objects.

Besides, whether one agrees with Rosen's fictionalist CE on the unobservable, one must admit that there are purely fictional models in science. Models that scientists themselves call "virtual models" are an example. This category of fictional models in science is a grab bag of items, and it is a conception that should clearly be distinguished from taking all scientific models to be fictional (as the new fictionalism does). In another paper I have enumerated and categorized such models in a more systematic way, but here the mentioning of some examples should suffice. In physics, there is a well-tested method of introducing imaginary objects at chosen places in a system to create identical effects that divergent or near-infinite substances cause. The physicists know that the objects are not actually there in the system but all relevant causal effects can be conveniently obtained by pretending that they are there. There are also models that exemplify a large variety of types of systems, such as the "harmonic oscillator" or the "rational man" or the "hawk-dove game." These terms do not refer to any particular types of systems in reality. They are rather like terms for a trope of properties or functions; in other words, they are analogous to terms referring to geometric figures that can be used as modelistic elements. The study of a harmonic oscillator or a rational man or the hawk-dove game is not a study of a natural system, but rather a study of a model. Terms for the components of such systems are clearly terms referring to the parts of the models. The hawks and the doves in the evolutionary game of hawk-dove do not refer to hawks or doves, nor do they refer to any groups of people in our society, not even idealized groups of people. They refer to strategies that idealized people adopt in idealized situations. Are there hawk-dove games in reality? No, not directly. There are situations in real life that we can use this model to represent, and most likely, when the hawk-dove game models is used to represent some real situations, additional names will be given to them to denote the situations as hawk-dove games. This is no different from using a geometric figure known as the "vertical line segment" to represent the free fall of an object in the Galileo case, when it is attached with a label "free-falling body."

Space limit prevents me from doing justice to Rosen's fictional interpretation of CE, but one of his critical points of van Fraassen needs to be addressed, for it further illustrates what a proper fictionalism should look like. After pointing out that van Fraassen could not at the same time be a Platonist on unobservable entities, and nominalism and fictionalism shares the belief that those entities do not exist, Rosen raised the question: does *T* exist in the claim "*T* is empirically adequate" (p. 165)? (Here, *T* is supposed to be a literally construed theory.) Rosen argues that "*T*" must refer to an abstract entity, given van Fraassen's semantic approach to theories; and it has to exist in order for us to believe the claim, and so van Fraassen has to commit to the existence of at least one abstract entity in order for CE to work, but he can't (pp. 167-169).

I think we have the resources in this paper to defend van Fraassen. First, we note the distinction between models sans RDs and what they represent. The former are no different from, e.g. mathematical entities, while the latter include observable and unobservable objects. A CE'st does not have to deny existence to all abstract entities; but she has to deny it to those that would be physical if they existed. Rosen rightly points out that van Fraassen cannot avail himself of this move (p. 165), but it ought to be available to CE'sts, because (1) even scientific realists may consistently adopt nominalism on numbers and propositions and (2) only those unobservables that may make a difference for empirical adequacy matter to CE (must be denied or held agnostic of its existence); and abstract entities such as numbers and propositions clearly don't, even if they *are real*. Hence, *T* could well be real for a CE'st; and that avoids Rosen's challenge, and yet what is genuinely referred to inside *T*, when they are unobservable, is still fictional. This version of CE actually makes more sense and here is why. What does it mean to be fictional in its plainest or original sense? That Tess is fictional means that if we suspend our disbelief of her non-existence, we can pretend that we are reading a history about an actual person named "Tess," among others, entitled *Tess of the d'Urbervilles*. We do the same for the unobservable in *T* if we are CE'st, but it makes no sense for us to do the same for numbers, sets, and models. For instance, it would not make sense to ask whether *Principia Mathematica* is empirically adequate, whether or not one believes that numbers and sets are real, would it?

Therefore, there is a defendable version of a fictionalist CE that does not treat all unobservable entities (which include mathematical ones by default) equally. To hold this version of CE, one may take, if one wants, all those abstract entities that make no difference to the empirical adequacy of scientific theories to be real (be a Platonist for them); but one has to be agnostic about those abstract entities that may make a difference for empirical adequacy and take them to be fictional.

References

Cartwright, Nancy (1983) *How the Laws of Physics Lie* Oxford: Clarendon Press

Cartwright, Nancy (1999) *The Dappled World: A Study of the Boundaries of Science* Cambridge: Cambridge University Press

Contessa, Gabriele (2009) Scientific Models and Representation, in Steven French and J. Saatsi eds. *The Continuum Companion to the Philosophy of Science* (forthcoming)

Contessa, Gabriele (2010) Scientific models and fictional objects *Synthese* **172**:215–229

Field, Hartry (1989) *Realism, Mathematics and Modality* Oxford: Blackwell

Frigg, Roman (2010a) Models and Fiction, *Synthese*. **172** 251-268.

Frigg, Roman (2010b) Fiction and Scientific Representation, in Roman Frigg & M. Hunter (Eds.), *Beyond Mimesis and Convention, Boston Studies in the Philosophy of Science* Dordrecht: Springer 97-138.

Frigg, Roman (2011) Fiction in Science, in Woods, J. ed. *Fictions and Models: New Essays.* Munich: Philosophia Verlag (forthcoming).

Giere, Ron (1988). *Explaining Science: A Cognitive Approach* Chicago: Chicago University Press

Giere, Ron (2009) Why Scientific Models Should not be Regarded as Works of Fiction, In Mauricio Suárez ed. *Fictions in Science: Philosophical Essays on Modeling and Idealization* Routledge: 248-258

Godfrey-Smith, Peter (2006) The Strategy of Model-Based Science, *Philosophy of Biology* **21**:725–740

Godfrey-Smith, Peter (2009) Models and Fiction in Science, *Philosophical Studies* **134**:101-116

Goodman, Nelson (1976) *Languages of Art: An Approach to a Theory of Symbols* Indianapolis: Hackett Publishing Company, Inc.

Hesse, Marry (1966) *Models and Analogies in Science*, Notre Dame: University of Notre Dame

Keller, Evelyn Fox (2002) *Making Sense of Life: Explaining Biological Development with Models, Metaphors, and Machines* Cambridge, MA: Harvard University Press.

Levy, Arnon (2012) Models, Fictions, and Realism: Two Packages, *Philosophy of Science* **79**: 738-748

Levy, Arnon (2010) Fictional Models *De Novo* and *De Re* (preprint)

Levy, Arnon (2012) Models, Fictions, and Realism: Two Packages *Philosophy of Science* **79**:738-748

Morgan, Marry S. & Morrison, Magaret eds. (1999). *Models as Mediators*, Cambridge: Cambridge University Press

Rosen, Gideon (1994) What is Constructive Empiricism, *Philosophical Studies* **74**: 143-178

Rowbottom, Darrell P. (2009) Models in Biology and Physics: What's the Difference? *Foundations of Science,* **14**: 281-294

Suárez, Maurice (2003). "Scientific Representation: Against Similarity and Isomorphism," *International Studies in the Philosophy of Science.* **17**, 226-244.

Suárez, Maurice (2004). "An Inferential Conception of Scientific Representation," *Philosophy of Science.* **71**: 767-779.

Suárez, Maurice (2009a). Fictions in Scientific Practice, in *Fictions in Science: Philosophical Essays on Modeling and Idealisation*, Routledge, pp. 1-15.

Suárez, Maurice (2009b). Scientific Fictions as Rules of Inference, in M. Suárez (ed.), *Fictions in Science: Philosophical Essays on Modeling and Idealization,* Routledge, pp.158-178

Suárez, Maurice (2010). Scientific representation, Philosophy Compass 5 (1), 91-101

Teller, Paul (2001). "Twilight of the Perfect Model Model," *Erkenntnis.* **55**: 393-415

Toon, Adam (2010a) The Ontology of Theoretical Modelling: Models as Make-Believe *Synthese* **172**: 301-315

Toon, Adam (2010b) Models as Make-Believe, in Roman Frigg & M. Hunter (Eds.), *Beyond Mimesis and Convention, Boston Studies in the Philosophy of Science*, Dordrecht: Springer, 71-96.

van Fraassen, Bas (1980) *The Scientific Image* Oxford: Clarendon Press

van Fraassen, Bas (1994) Gideon Rosen on Constructive Empiricism, *Philosophical Studies* **74**: 179-92

Yablo, Stephen (1998) Does Ontology Rest on a Mistake? *Supplement to Proceedings of the Aristotelian Society* **72**/1: 119-37

Yablo, Stephen (2010) Go Figure: A Path through Fictionalism, in *Things: Philosophical Papers 2* Oxford: Oxford University Press, 177-199

Re-inflating the Conception of Scientific Representation

This article argues for an anti-deflationist view of scientific representation. Our discussion begins with an analysis of the recent Callender–Cohen deflationary view on scientific representation. We then argue that there are at least two radically different ways in which a thing can be used to represent: one is purely symbolic and therefore conventional, and the other is epistemic. The failure to recognize that scientific models are epistemic vehicles rather than symbolic ones has led to the mistaken (deflationary) view that whatever distinguishes scientific models from other representational vehicles must merely be a matter of pragmatics. It is then argued that even though epistemic vehicles also contain conventional elements, they do their job of demonstration (or showing) in spite of such elements.

1. Introduction

In this article I provide an argument for re-inflating the conception of scientific representation. The argument begins with a critique of the recent deflationary approaches to scientific representation, which has its most minimalist version in Callender and Cohen (2006). We find another version of it in Suárez (2003, 2004, 2010); and there is an earlier version that is closer to the Callender–Cohen deflationism (Teller 2001). Van Fraassen's view of scientific representation (van Fraassen 2008) may also be regarded as a deflationist view; only, he insists that his is not an account, *per se*, of scientific representation. Before I begin to argue for the re-inflation, let me give a summary of my argument in its entirety.

One of the cornerstones of deflationism on scientific representation is that there is really nothing more to a scientific model than an object — which may be a physical thing or an abstract entity — that can be used to symbolize another object as its target for certain practical purposes. To use *A* to *symbolize B* is understood in the deflationary view as *A*'s being used as a sign, a label or a model for *B* by an (in-principle) convention among the users of *A*. Some symbols may depend more on the specific practical purposes beyond the conventional agreement, as scientific models do, but ultimately they are no more than symbols whose constitutive relationship with their targets is a conventional relationship.

My argument against this view comprises the following components. (1) The additional conditions, if you will, that make a model a scientific model are not practical purposes, although practical purposes do play a role. They are instead epistemic virtues. (2) The epistemic virtues are not reducible to pragmatic virtues. (3) Any scientific models must fulfil at least two functions: they must allow their users easy identification of what their targets are (denotation/reference) and they must show to their users what the modelled are like in the relevant aspects (showing/demonstration); missing either, the objects cannot serve as models. (4) The deflationists would agree with (3) but their views fail to provide the right way to capture it.

My view in this article provides a hybrid view of scientific modelling, namely, in *most* scientific models two types of elements are combined to represent target systems. The symbolic elements, which include names, labels, and other sorts of symbols, denote (or refer to) the components in the target systems. And the modellistic elements are either mathematical and other abstract entities or physical objects, and they are put together in the models to show the likeness in the selected aspects of the target systems. Whether one takes modellistic elements to be abstract entities or physical objects depends on one's ontological commitments, but either version works with my account of scientific representation. My view may be called a hybrid view, which re-inflates the conception of scientific representation.

In this article, I first discuss the Callender–Cohen version of deflationism. I then defend my account of scientific representation. For that I first defend the distinction between the symbolic elements and the modellistic elements in a model and then the hybrid view. I then defend the epistemic role that modellistic elements play. Finally, I compare my view with the existing views on scientific representation that have similar features.

2. Deflationism

Controversies over the nature of scientific representation have generated a sizable literature in recent years (Morgan and Morrison 1999; van Fraassen 2008; and references therein). A recent work, Callender and Cohen (2006), gives a judicious summary of the controversies and offers what I shall henceforth call a deflationary resolution. Callender and Cohen begin their argument by separating and distinguishing the different questions one may ask about a scientific representation. The most fundamental among the questions is the 'constitution question': what constitutes a representational relation between a vehicle and its target? And then there is the 'demarcation question', a question concerning distinguishing legitimate scientific representations from non-legitimate ones (something presumably analogous to Popper's question about how to demarcate scientific from pseudo-scientific theories). This question should be distinguished from the 'explanatory/normative question', which asks about the requirements for a correct or adequate representation in science.

Callender and Cohen point out that much of the confusion in the debate derives from people's mistaking the latter two questions for the constitution question; for instance, people have been trying to figure out in what sense a model could be said to resemble — in terms of similarity or isomorphism — its target as if it were a constitutional relation, as if 'resemblance' could constitute the basic relation between a vehicle and its target (Callender and Cohen 2006, pp. 8–9). They argue that this is an example of confusing the demarcation question with the constitution question, namely, confusing the question of what must hold for something to be a legitimate scientific model with what must hold for something to serve as a representation at all.[1] To illustrate the point they use the 'lowly stop sign' as an example of a representation vehicle to show that just because the stop sign does not resemble the action of stopping, it cannot be said not to constitute a representation; and yet if one's purpose is to question whether it can serve as a good scientific model, the answer may well be negative (Callender and Cohen 2006, p. 10).

---

[1] After a fairly thorough discussion of the different positions in the literature, Callender and Cohen conclude, '[I]n our view, running these issues together is conducive to *confusion*' (Callener and Cohen 2006, 9; emphasis added).

To go on and argue for their deflationary view, Callender and Cohen point out first that scientific representation should naturally be regarded as a species of representation in general, and second that what philosophers of language, such as Grice, have worked out for representation in general should apply also to the case of scientific representation (Callender and Cohen 2006, pp. 10–11). Grice's theory of speaker meaning/representation — or what they refer to as 'Specific Griceanism' — is a *reductive account* of how a speaker's utterances get their meaning from a communal guarantee of the conveyance of the speaker's meaning intention. In other words, speaker *A* means (and hence qualifies as representing) that *p* by uttering *s* if and only if (i) *A* intends that the listeners of *s* form the belief that *p*, and (ii) that belief is reliably formed by the listeners of *s*. The mental/belief states so invoked are the 'fundamental representations', while the words (or other vehicles) that are used to invoke them in communications are the 'derivative representations'. The latter accomplish their task of representation in virtue of the former, and the reliability of such successful invocation is a matter of social convention (Callender and Cohen 2006, pp. 12–13; also see Grice 1989).

Building on that, Callender and Cohen argue that representational vehicles in science and technology, of which models are a subset, do their job in accordance with General Griceanism, which is a natural extension of Specific Griceanism. The basic argument is the same, and it gives a unified account of how any derivative vehicles do their jobs in representing the world to us, scientifically or otherwise. To illustrate their point, Callender and Cohen mention such acts of representation as a certain number of lanterns being raised in a church steeple to represent the movements of enemy troops, or more dramatically, having a salt shaker on one's dinner table represent one's favourite geographical region, e.g., Madagascar (Callender and Cohen 2006, pp. 13–14). The only condition of adequacy that legitimizes the vehicles is that the right belief states are reliably *intended* and *evoked* among the users, and for that, the right social convention (understood broadly) is in place. No other constraints are necessary since the success of a representational attempt, as illustrated above, requires no other. Given the wide-open possibilities of establishing representational conventions, anything can in principle be used to represent anything else (perhaps also including self-representation).[2]

In this respect, Teller may also be regarded as a deflationist, especially when he says,

> I take the stand that, in principle, anything can be a model, and that what makes a thing a model is the fact that it is regarded or used as a representation of something by the model users. Thus in saying what a model is, the weight is shifted to the problem of understanding the nature of representation. (Teller 2001, p.397)

Here, the talk of being 'regarded' or 'used' as a representation clearly implies that what makes something a model depends exclusively on a stipulation/convention in the community of model users; and 'the problem of understanding the nature of representation' clearly concerns the fundamental or natural representations.[3] And when Teller says 'the weight is shifted to...', he is obviously alluding to a reductive account. Given the year in which this statement and its accompanying arguments were made, Teller should be regarded as an early, if not the earliest, deflationist on scientific representation.

And so, perhaps to a lesser degree, is van Fraassen (2008). He observes that if one is to have a theory of representation (which he doesn't want for his account), one must accept what he takes to be the *Hauptsatz*: 'There is no representation except in the sense that some things are used, made, or taken, to represent some things as thus or so.' (van Fraassen 2008, p. 23). If not an explicit statement of deflationism, van Fraassen's pragmatic view of scientific representation is surely consistent with the deflationism of Callender–Cohen

[2] I was alerted to such a possibility by a referee, who gave an example of someone appearing in a play to play herself as an act of self-representation. I am however not convinced by this example. If Bill Clinton, let's say, appeared in a play in the role of Bill Clinton, is he as himself representing himself or as somebody else, a actor, representing Bill Clinton? If he is himself, he cannot 'act' in the role, and there is no representation; he must be 'somebody else' to play Clinton, or Clinton is playing a role of Clinton, which is not Clinton, the man. Anyway, there seems to be some deeper issue here that I cannot further explore, but the bottom line seems to be this: for anyone, if she is being herself, she cannot be representing herself, and if she is representing 'herself', the self she is representing cannot be her.

[3] Fundamental representational vehicles, such as belief states, are generally assumed to be non-conventional or natural. This may be questioned in philosophy of mind, but I shall here hold to this assumption.

and Teller (see also Suárez 2004).

Callender and Cohen are by no means dismissive of the earlier efforts in the literature on the nature of scientific models and modelling (see the detailed discussion in Callender and Cohen 2006, pp. 8–11), but they believe that the answers to all the other questions, such as the demarcation and the explanatory/normative questions, are really a matter of the pragmatics of vehicle uses. Embracing deflationism effects a re-orientation of the approach to those hotly debated issues concerning models and modelling in science. If the constitutional relation is a purely conventional one, all the other questions can only be a matter of what is useful or effective or high on some such pragmatic virtues. For instance, to see whether similarity of some sort between a model and the modelled is essential to scientific representation, one would search in vain in metaphysical or epistemic considerations over the intrinsic or relational properties of the two. The answer could come only from a determination of whether being similar to its target is more useful (in the broadest sense) to the vehicle users in a community than other relations (e.g., being convenient or beautiful; Callender and Cohen 2006, pp. 15–16). Here are a couple of quotations from Callender and Cohen in this connection:

> But note that, just as in the case of similar questions about non-scientific representations, the questions about the utility of these representational vehicles are questions about *the pragmatics* of things that are representational vehicles, not questions about their representational status *per se*. Thus, if the drawing or the upturned right hand should happen not to rank highly along the dimensions of value considered so far, this would, on our view, make them non-useful vehicles that do represent, rather than debar them from serving as representational vehicles altogether. (Callender and Cohen 2006, p. 15; emphasis added)

> Presumably scientific contexts come with their own set of pragmatic constraints, and these may drive the choice among possible scientific representations in ways that are idiosyncratic to science. (Callender and Cohen 2006, p. 16)

To summarize: (i) the constitution question should not be confused with other questions; (ii) the constitution question admits a deflationary answer; and (iii) all the other questions are really questions of a pragmatic sort. According to General Griceanism, how we represent reality is reduced to (1) how conventionally selected external vehicles are related to the belief states (internal vehicles),[4] and (2) how those belief states represent primarily. Only the former question concerns the discussion over scientific representation; the latter question is one partly of empirical science and partly of philosophy of mind.

The simple fact that most scientific models do not represent their targets as conventional symbols such as names and labels do and something more must be present for them to qualify was recognized in the literature even before it was raised as a point of criticism for the Callender–Cohen work (Suárez 2004; Contessa 2007). And others also raised the point as such (Frigg 2010; Toon 2010; see also Contessa 2010; Suárez 2010). Suárez's case is a bit more complex, for although he notices the same point (i.e. a mere symbol is not enough to be a model), he nevertheless regards his own view as a deflationary view. And as we will see later, his deflationism may indeed be ultimately consistent with the Callender–Cohen version. The authors all use similar examples, such as looking all you could into the physical structure of the word 'hydrogen', you would not be able to tell what hydrogen atoms are like in any relevant aspects. But the question is: so what? Why couldn't Callender and Cohen say that scientific models are a special category of symbols that carry the likeness-in-relevant-aspect feature as their utility? Therefore, the challenge in giving a fair criticism of deflationism is to go beyond the point that models are unlike words and labels and answer the question: how does a model, which is not like symbols that denote by convention, actually accomplish the job of secure denotation or reference? Different authors offer different theories of representation and different answers to the question are supposed to be implicit in the theories. Suárez's inferential account requires that models serve as inferential surrogates in the practice of science; while in Frigg's tripartite conception of modelling, symbols and models live in different parts of that enterprise. Contessa's interpretational account (Contessa 2007) continues Suárez's line but makes much stronger ontological commitments to models, treating them as highly idealized abstract entities, as does Toon's account. And despite the differences, they all subscribe to the fictionalist conception of models.

I shall explain in detail later the differences between my conception

[4] I shall henceforth use 'external vehicles' and 'internal vehicles' in place of fundamental vehicles and derivative vehicles, respectively.

and theirs. My analysis of what is wrong with deflationism covers similar grounds as their criticism does, and yet I draw from it different conclusions.

3. The Epistemic Vehicles

The first step in arguing against deflationism about scientific representation, at least in its Callender–Cohen version, is to point out that the argument for it — the General Griceanism argument — is misplaced or misapplied. Then, we argue for the essential *epistemic* role that scientific vehicles play in representing the world around us, and how that affects the answer to the constitution question.

General Griceanism, as an extension of Specific Griceanism on linguistic symbols, is applicable only to more general symbolic systems, which of course include all non-linguistic ones; and it is true that any object can in principle be used as a symbol. We argue that scientific representations, or at least scientific models as understood by their primary users, the scientists, are *not* non-linguistic symbols and *cannot* be regarded as something similar to them. They may serve on certain occasions as symbols, but if they do, they do it not as scientific models *per se.*

To argue for this point, we begin by noticing some general differences between *symbolic vehicles* and, for lack of a better term, *epistemic vehicles.*[5] Symbols, whether linguistic or non-linguistic, are essentially arbitrary physical objects that people use by agreement as tags or marks for the things they want to refer to. The main purpose of using symbols is for the convenience of memory, record, communication, commemoration, and so on, and the reliability of their usage is determined by the enforcement of the agreement. Symbolic vehicles typically include names, tags, signals,

[5] Other pairs of terms that are used to refer to similar distinctions exist. One of these is that of 'linguistic' versus 'pictorial' means. These mark the difference between descriptive versus pictorial representations. It is not the distinction we want to make here. Another is the distinction between purely 'conventional' and 'natural' means. Broadly construed, the conventional means represent in virtue of a convention that establish the representation relation between a vehicle and its target, while a natural vehicle is supposed to represent in virtue of some natural connection between the vehicle and its target. This pair is also different from what we have here.

emblems, etc. and they can be made of artificial as well as natural objects. Besides the pragmatic role, symbols have no other use in the community. For example, symbols in general cannot take on an epistemic role because they are typically not used to show their users what the represented are like (in the broadest sense of 'like').

The foregoing explains why Callender and Cohen are right to argue that the usual examples of non-linguistic symbols, such as the lanterns and the salt shaker, cannot be conceived of as carrying out any role other than the pragmatic one (which is fixed by a social convention that ties the symbols to their targets), and may include anything as a legitimate vehicle.

Epistemic vehicles, those vehicles that we usually see in acts of modelling things or events in the world, scientifically or otherwise, are not used as symbols. Rutherford's model of hydrogen atoms or the Crick–Watson model of DNA molecules does not serve as a symbol of its target system, very much unlike the lanterns or the salt shaker in the Callender–Cohen examples. A cursory look at the usage of such scientific models should reveal two rather obvious points: (i) they cannot be had, unlike the symbolic vehicles, through a stipulation among the users, and more importantly, (ii) they have features that are regarded as relating to their targets in such a way that they can be used to show or *επίδειξη* the targets or to learn about them.

The usage of symbols is very different. For instance, when I use the word 'hydrogen' to represent hydrogen, the thing, i.e., word, is a mere symbol. There is no information in the individual letters or their concatenation that necessarily tells us what the stuff the word represents is like. This is also true of other sorts of symbols we use to represent things and events in reality. A raised lantern may be used to represent by convention the way in which enemy troops approach or a salt shaker Madagascar (Callender and Cohen 2006, pp. 12–13), but nothing about the lantern gives any information about the composition or other aspects of the troops nor does the salt shaker tell us anything about Madagascar. Symbolic vehicles are what people sometimes call 'conventional means' of representation, and because it's conventional, anything can be used as symbol for anything else, and the General Gricean argument applies.[6]

[6] Daitz (1956) in arguing against Wittgenstein's 'picture theory' of meaning in *Tractatus* makes a distinction between 'conventional representation' and 'iconic representation'. Linguistic/symbolic vehicles of representation, according to Daitz, belong to the former while

It might be difficult to give a complete account of what this notion of 'showing what (the target) is like' amounts to, and yet it is a notion we are familiar with in ordinary as well as scientific/expert contexts. In fact, without such a notion, a great deal of what we call 'learning', on the one hand, and 'showing' or 'demonstration', on the other, becomes impossible. I have never been to Africa and yet I can acquire a good sense of what an African savannah looks like through vehicles of model representation. That is not possible if I can have access to symbolic vehicles only if even I know and consent to all the symbols that relate to African savannahs: for example, if I have access only to names of objects in and of savannahs.

Lest I be misunderstood, an important fact about how most scientific models actually represent their targets needs to be stated. Although we learn from the structure of the model what its target is like in the chosen aspects we study, the model has to be securely 'fastened' to its target in order for us to do that. The model structure is not usually what modellers use to do the job of denotation or reference. Very much like how maps work to represent cities and towns, this job is usually done by the names and labels — the symbolic elements — that are attached to the structural components in the model. Therefore, unless special circumstances make it otherwise, most models contain both the symbolic and the modellistic elements, and there is a sort of division of labour going on in the use of models. The symbolic elements work to secure the *what* in the representation, while the modellistic elements work to show the *like-what* or *how* (more on this point later).

With this caveat, we can roughly conclude what is essential to *epistemic vehicles* that show us what their targets are like.

1. They cannot be vehicles whose connection to their targets is *essentially* established by a convention.
2. They do not function as tags or names that are used simply to pick out what they represent.
3. They are used primarily to allow their users to have access, in however simplified or specialized manners, to aspects of their targets, which

pictorial representations, such as paintings and photographs, belong to the latter. The conventional/epistemic distinction I use here has its origin in Daitz's distinction, and yet I have modified it and especially extended the scope of the pictorial vehicles to the notion of epistemic vehicles.

fulfil in a broad sense an *epistemic role.*

4. When used in communication, the vehicles' function is not to invoke via explicit conventions the right belief states among interlocutors but to create such belief states without explicit conventions.

In other words, epistemic vehicles (which include most if not all scientific models) do contain symbols but cannot just be symbols. By this I mean they cannot be conventional signs for given pragmatic purposes.

4. Defending the Distinction

Some may argue that by separating epistemic representations from symbolic ones in this way, we may have smuggled in through the back door the notion of similarity or isomorphism as a necessary condition for scientific modelhood, which has been soundly banished by such luminaries as Goodman, van Fraassen, and Teller. This, however, is not the case. Our account of the essential epistemic role for modelling, especially scientific modelling, which contains the notion of showing likeness, does not imply any requirement for the logical notion of resemblance in terms of qualitative similar or isomorphism. Depending on the type of epistemic access that is required, this notion of likeness, broadly construed, could be cashed out in a great variety of forms and degrees. If a model is required to show the quantitative structure of some aspect of an economy, there is no need to demand that the model society resembles its target in terms of its people and non-economic aspects of their lives; and yet the model must tell us what that economy 'is like' in the relevant aspects. Or take the case of Locke's theory of secondary qualities in his theory of perception (Locke [1690] 1979). The ideas in our mind of such qualities as colour and sound bear no resemblance, according to Locke, to any intrinsic qualities of the objects that produce them (for there is no colour or sound in the objects themselves), and yet we can and do acquire knowledge of what the perceived objects *are like* in colour, sound, etc. Locke's case, to the extent that it approximates how we perceive external objects, provides us with a clear instance where representation without similarity or even isomorphism can supply us with good epistemic (internal) models.

Some may also object by saying that scientific models sometimes also serve as symbols, and so the dogmatic distinction between symbolic and epistemic representations as mutually exclusive modes of representation is unjustified. No, the modes are clearly conceptually distinct, and yet one particular thing may be used to serve as a model on one occasion and as a symbol on another. In fact, when that happens, it usually happens in different contexts when the same thing is used for different purposes of representation. The double-helix model of DNA as a model showing the relevant structure of the molecules is rarely used as a convenient symbol in the context of scientific research and pedagogy. When it is used as a symbol, it may be used as a symbol for a great scientific achievement; and then, and usually only then, is a fully realized replica of that model (perhaps in some precious metal) necessary to be placed somewhere to do the job of a symbolic representation.

Some may again argue that it is not possible to fully and dogmatically separate the symbolic and the epistemic representations because, when highly idealized and atrophied, a model may appear entirely symbolic. Or a symbol may appear so complicated and full of structural details that it is no different from an epistemic vehicle. Two observations can be made on this objection. First, whether an object when used as a vehicle of representation is symbolic or epistemic does not depend on how simple or complex its features are, although it is true that, in ordinary situations, pragmatic concerns often make people who use objects as mere symbols opt for as simple an object as possible. An epistemic vehicle could also be very simple because what it is used to demonstrate is an aspect of the target that appears simple after desired idealization. Second, and more importantly, the distinction between symbolic and epistemic vehicles has nothing to do with what sort of objects can or cannot be used for one or the other type; no, it depends only on how objects are constructed or used and what relationship they bear with their targets. Symbolic vehicles are, again roughly, tags for invoking the proper belief states in memory and communication among users who share the convention, while epistemic vehicles are showpieces that display, in a broad sense of 'displaying', their targets.

Some may argue that scientific models are sometimes linguistic items, such as a block of text that describes a system. Don't some scientists, for instance, regard with good reason the set of equations, known as the Maxwell equations, as a 'model' of the electromagnetic field? If this is so, how can we maintain the distinction between symbolic, which includes linguistic, representations and model representations? There are two alternative ways to respond to this objection. Which response one chooses depends on one's ontological view on models. (1) If a model is regarded as an abstract entity, or as that which descriptions and statues describe or realize, then there is no problem here. What is fundamentally symbolic, the dots, line segments, and words, are not components of the model, but components of its descriptions or (physical) realizations.[7] This point may be connected to the long-standing dispute between the 'syntactic' vs. the 'semantic' approach to scientific theories (van Fraassen 1980; Suppe 1989). According to the semantic approach, a scientific theory should be identified not with the actual linguistic formulations (including mathematical ones) we see in journal articles and textbooks but rather with the abstract (relational) structure that satisfies the theory. In the same way we would say that the Maxwell equations are not, *qua* linguistic items, the model; Maxwell's model for electromagnetism is instead that which satisfies the equations.[8] (2) But if models are physical things, such as scale models and statues, then the parts of a model (i.e. modellistic elements) are the objects that directly show what its target is like. They are not symbolic elements because they are not selected or put in place in the practice of model building by conventional stipulation. They are selected and put together by a process of idealization, which includes the selection of the required aspects for model representation and then the act of neglecting the negligibles. This process of putting together the modellistic elements is markedly different in nature from how symbolic elements are selected and put together. And the key difference, as we should have realized by now, is the difference between an epistemic virtue and a

---

[7] To adopt this ontological view, one does not have to say that the word 'model' refers *only* to the abstract entity — it may refer also to the realizations in our daily usage (Keller 2002; Rowbottom 2009); but the daily usage is ambiguous and when the ambiguity is revealed, most people, according to this view, will agree that the model refers to the abstract entity rather than its realizations (as people would agree that it doesn't matter of what material a model is made of or with what language it is described).

[8] There is a small but important detail regarding models in the form of differential equations, including the Maxwell equations as a typical example (Liu 1997). The equations express the dispositions of systems (i.e. spatiotemporal objects) that are distinguished by initial and boundary conditions; namely, the models are actually what make the solutions of those equations true.

pragmatic one.

Now, as briefly argued earlier, most models contains names and labels for their parts, and these items serve as the symbolic elements in a model. In fact, it is because of the presence of these symbolic elements (and the division of labour between denoting and showing) that the modellistic elements can be free from any restrictive oversimilarity relations to fulfil their modelling role.

Now, let me address point (2) mentioned in the introduction section: why can't the epistemic aspect of representation be reducible to the pragmatic aspect, and therefore the deflationists are still correct in saying (i) anything can be used to represent anything else, and (ii) what qualifies something as a scientific model is a matter of pragmatics (which includes the matter of episteme)? In fact, one of the main points of Callender and Cohen may be taken as amounting to something very similar. In section 4 of their article, (Callender and Cohen 2006, pp. 17–19) an argument is made to show that most of the controversies in the existing literature, some of which obviously concern knowledge, is caused by mistaking pragmatic issues with the constitutional one. I agree with Callender and Cohen on the point that no pragmatic aspects should be included in the constitutional question of modelhood, but I disagree with them that all other aspects are pragmatic.

In response, one might recall that epistemic virtues have always been pitted against pragmatic ones when it comes to how scientific theories are to be viewed and evaluated. The most famous case may be seen in Kuhn's objections to logical positivism regarding theory choice during crises and revolutions in the history of science (Kuhn 1977). On the one — the logical positivist — side, new scientific hypotheses or theories are to be chosen according to their epistemic merits, whether it was truth, approximate truth, or agreement with empirical evidence, while on the other — Kuhn's — side, it was the cluster of a theory's pragmatic virtues, such as its scope, fruitfulness, and accuracy, that determines the choice. To say in this context that the epistemic aspect of scientific theory is a subspecies of the pragmatic aspect can amount only to a fundamental confusion. Scientific models are important ingredients in the process of theory construction and choice in science, and therefore, the distinction must apply to them as well.

Let me summarize what has been argued for so far. We argue against the Callender–Cohen deflationism on scientific representation primarily because we see that they misapplied the argument from General Griceanism.

That argument can get one only from Grice's pragmatic theory of meaning for languages to a general theory for *symbolic* vehicles of representation. It is a mistake to identify vehicles used in scientific modelling as generally symbolic vehicles. They are *epistemic* vehicles instead. The distinction articulated above between symbolic and epistemic vehicles, if it holds, blocks the General Griceanism argument for deflationism; and further it argues for identifying scientific vehicles/models with the epistemic ones. Most, if not all, epistemic vehicles contain symbolic elements to help secure the denotation of parts of the vehicles with the parts of the target systems, but the main parts of such vehicles are there for epistemic purposes.

5. Defending the Re-inflation

The Callender–Cohen argument for deflationism is purportedly to point out a basic fact about all vehicles of representation, scientific or otherwise; and the argument — the General Griceanism argument — is supposed to show that for whatever object, it can be used to represent just in case it succeeds in completing the practical purpose of communication (as briefly explained in section 1). How can this be otherwise, one may argue on behalf of Callender and Cohen, for what we called 'epistemic vehicles'? Isn't it true that (by whatever means) a Rutherford model or a DNA model is also invented and used to fulfil the representational role we have mentioned just now? It might indeed be true that how such a model brings about in people's mind those belief states whose content shows some sort of likeness of a hydrogen atom or a DNA molecule *is very different from* how the names of those models bring about the same or similar states. The epistemic models can teach or show people what the modelled are like while names have no such capacity, and yet the end result is the same, namely, the invoking (by one way or another) of the right sort of belief/mental states for which the vehicles are used; and therefore General Griceanism holds universally and deflationism about representation still follows.

Several observations can be brought against this 'last-ditch effort' to rescue deflationism. To begin with, the mere fact that belief states are brought about, although in very different ways, in both cases is not reason enough to conclude that epistemic representation is fundamentally no different from symbolic representation. Furthermore, whether in the specific version or the general version, Griceanism is a theory of meaning, a semantic theory of symbol use by cognitive agents. There is no good reason to regard the question of how scientific models or modelling works as a question of meaning or semantics. The philosophical analysis of the nature of modelling should not be viewed as part of a generalized enterprise of philosophy of language, and hence the Callender and Cohen effort to 'rein in' the discussion of scientific representation under the tent of philosophy of language is to my mind wrong-headed in the first place.

At the very least, an explicit argument needs to be given to establish the truth about identifying models with non-linguistic symbols before one can even think of applying the General Griceanism argument to scientific representation; and yet such an argument is not forthcoming. In fact, if what is argued in the previous section is correct, such an argument cannot be found; and moreover, we have shown in the previous section that modelling is possible only when the semantics is already in place for the constituents of models. Scientific models may well make ample uses of symbols and yet they themselves are not such.

But wait — aren't there cases of vehicles that are clearly used as symbols and yet what is symbolized can be gleaned directly from the symbols' features? Imagine being in the international terminal of JFK Airport in New York City and looking out of the window. Jumbo jets from different nations are lined up at the gates with colourful symbols on their fuselages. Some of those symbols are words or word-like symbols, but some are stylized pictures. In particular, if you are from an Asian country and see a red crane with outstretched wings on a plane, you could unmistakably tell which nation's airline that plane belongs to. Such examples, we admit, abound in the world of symbol usage. Are they symbolic vehicles or epistemic vehicles, the deflationists may ask rhetorically? Other similar examples that might be used in the same challenge include such things as hieroglyphics, pictograms, totems, etc.

This objection seems to be suggesting two slightly different things. One is that there exists for forms of representation a spectrum that spans different degrees and combinations of the symbolic and the epistemic. On the one end we have the purely symbolic, such as words in English, and on the other end, we have scientific models that detail the features of the represented systems. The above examples, understood as partly symbolic and partly epistemic, should be located somewhere between the two ends in the spectrum.

We think that this is the wrong idea about the relationship between symbolic and epistemic vehicles. Cognitive agents as symbol users may employ all sorts of techniques to construct and maintain types of symbols for their convenience. Which objects are to be selected to symbolize something may well depend on how their features help their users to recall or communicate what is symbolized. Resemblance or other sorts of likeness may well be what the symbol users are after when they make their selections. It may well be that there is a spectrum of symbols going from pictograms, such as the ancient Chinese characters, and merely conventional words, such as the modern-day English, with decreasing degrees of resemblance; but the existence of such a spectrum is obviously irrelevant to the difference between symbolic representation and epistemic representation.

Representations that are essentially symbolic are different from those that are essentially epistemic for another important reason, which the deflationists have so far not touched upon. There is a legitimate ontological question about scientific models that makes no good sense when directed to symbolic vehicles. The question is: are scientific models *per se* physical things or abstract entities? One may think that the answer is obvious and that models cannot be anything other than physical particulars; but at least no-one can deny that it is a legitimate metaphysical question — no less legitimate than similar questions about, e.g., numbers. It does not seem sensible to pose the same question for symbols. There is no good reason to think that alphabets, hieroglyphics or pictograms might be abstract entities, such as numbers and propositions. If this point stands, if it makes sense to raise the ontological question for models but not for symbols, the epistemic vehicles should be regarded as radically different from the symbolic ones.

I am not here arguing that models are abstract entities and therefore they are different from symbolic vehicles; I am content to say that, since the question of whether or not they are abstract makes sense, which cannot be said for symbolic ones, there is enough of a difference to distinguish the two.

Lest I be misunderstood, let me make the following point clear. It would be a mistake to think that since scientific models/representations are epistemic rather than symbolic, there are no pragmatic/conventional degrees of freedom in them. That is by no means implied by our account of scientific models as epistemic vehicles, and it is certainly not true. Here is a simple illustrative example. Suppose a group of scientists are investigating a rare form of pancreatic cancer. They name the cancer cell 'PC207R' and nickname it 'the Devil'. They have a description of it in plain English and also in the precise language of biochemistry. They have a computer model that shows what the cell looks like and also made a plastic model of the cell.

Within each group of these representations — names, descriptions, and iconic representations — there exist pragmatic degrees of freedom. It is a matter of pragmatic preferences whether to use this or that name, or a description in plain English or in the language of biochemistry, or a computer model or material model; it is not such a matter when one has to choose, for instance, between using 'PC207R' or a plastic model of the cell. The existence of the internal pragmatic degrees of freedom in matters such as whether to construct a model for the cancer cell by using stainless steel or plastic or papier-mâché should not distract us from realizing that scientific representation is essentially evaluated along epistemic values.

Moreover, what matters to the selection of epistemic vehicles is not convenience or expediency or even explanatory power (if not also coupled to some extent with truth and objectivity), but objectivity or truth-conduciveness or truth-approximation (Nozick 2001); and the same point can be seen in the earlier example about Kuhn's pragmatic criteria for theory choice during a crisis or revolution. Regardless of what your belief states are about things around you, for the sake of communications, conventional vehicles of one kind or another have to be used to represent them externally. And presumably that is accomplished only when the model user's mind recognizes that the belief states that the models evoke are the same type of belief states that the targets evoke; and this should be true whether or not one believes that scientific models are ultimately abstract entities.

As I cautioned earlier, one should not think as a consequence of the foregoing that one's belief states of PC207R must be some sort of mental pictures of the cells, whatever that may mean, nor should the external vehicles used to represent the cells be necessarily similar to them. Whatever their nature, these two relationships are likely to be less than straightforward, and discoveries in cognitive science and neuropsychology are likely to revise continuously whatever philosophical accounts we have of them. There will always be a notion of faithful representation, but it may turn out to have little to do with similarity or resemblance between the vehicle and its target.[9]

[9] The belief states don't have to be 'iconic' or 'pictorial', but they must have representational

And finally, let me bring back the simple observation mentioned earlier. Most models contain symbolic elements to help secure the reference relation between model components and real-system components. This would not be necessary if the model components themselves were also symbolic in nature. It is safe to observe that no symbols of whatever complexity need to use other symbols in them in order to fulfil their functions. For instance, one may see a pictorial 'model' of the layout of a castle above the entrance of the castle, and usually it is put there to serve as a symbol of the castle. If it serves as a symbol, then it will not contain any other symbols, such as names or labels inside. If it does contain such symbols, then should it not be regarded as a map or a model, rather than a symbol, of the castle?

Therefore, my theory of scientific models, as briefly stated in the introduction, is that they are epistemic vehicles that in most cases contain symbolic elements. There is a division of labour inside such models such that the model can fulfil its denoting and showing functions simultaneously. Without securing the denotation, the showing elements in the model cannot enjoy the liberty via idealization and abstraction from the target components. But since that function is usually fulfilled by symbolic elements, such liberty is an essential feature of scientific modelling.

Again, to summarize, there is a special category of external vehicles, *epistemic vehicles*, which are distinctly non-conventional vehicles. *These are the vehicles that cognitive agents intend to use as showing (and/or learning) what the targets are like.* These vehicles bear a special relationship to the internal vehicles and the targets. No simple or single relationship between the vehicles and their targets suffices to define the epistemic vehicles, nor is it definable solely by the relations between the vehicles and the mental states alone. In most cases, it is with the addition of symbolic vehicles that the denotation/reference relations are secured. We may get a glimpse of the complexity of how they are characterized by reading chapter 1 of Goodman's *Languages of Art* (Goodman 1976, pp. 3–43; see also van Fraassen 2008, pp. 1–57). But however complex the relationships are, the claim above seems right because it marks the crucial difference between a vehicle used

---

content. Despite the suggestive examples used in this paper, there is no suggestion that all belief states that serve as the fundamental representations must be iconic. For different theories of how we represent the world through perception, see Freeman (1991), Siegel (2011); Harman (1973) argues for a widely discussed representationist theory of our thoughts.

primarily for labelling and one primarily for showing; and only the latter are fit for scientific purposes, or at least for the kind of scientific representations that has been discussed extensively in the literature.

6. Comparison with Other Accounts

Let me in this concluding section spell out to some extent how the re-inflation of our notion of scientific representation is related to some of the major insights in the literature on scientific models that deflationism aims at rejecting.

There was a time when the semantic view of scientific theory was all the rage (Suppe 1989). The central contention of the view is the idea that a scientific theory is independent of how it is formulated in any particular language. The syntactic formulations of scientific theories do not matter in terms of representing their objects to us. What matters is the models that a theory satisfies; and a model is a relational structure of the general form, $<D, P_1, P_2, \ldots, R_1, R_2, \ldots>$ (where *D* is the universe of discourse, the *P*s are sets of objects in *D*, and *R*s ordered pairs of objects in *D*, and so on), which makes the sentences in the theory true. This is a general notion of model that strictly speaking includes all scientific models in the sense we have been using in this article (and in the literature), *if one regards them as abstract entities.* And in the same ontology, our view of epistemic vehicles would certainly be consistent with the semantic view; and since the view is about scientific theories as models, there is little doubt that models are epistemic vehicles rather than symbolic ones, unless one regards scientific theories as merely conventional vehicles symbolically representing natural phenomena.

When understood broadly enough, analogy can cover all types of modelling, scientific as well as non-scientific; and finding things in common through analogy is a higher-order search for invariant structures, which are fundamental to scientific representations. Hesse's pioneering work on models and analogy (Hesse 1966) gives us some fundamentals of the logic of analogical reasoning. As a result, modelling as an act of scientific representation should obviously be viewed as relying on structural similarities of some sort. A model is regarded as a legitimate representation of a target (or *explicandum* in Hesse's term) because, all things being equal, it resembles the target in some salient and known aspects. And because of this, inferences can be made to unknown aspects of the target by observing causal connections among analogous aspects in the model.[10] It is true that not all models are analogical models, but there is little doubt that analogical models are epistemic vehicles rather than symbolic ones. The purpose of constructing models mainly through analogy for systems that we for some reason have no direct access to is precisely to figure out what those systems are like in the relevant aspects. Moreover, the use of analogy and analogical models has been condemned as an illegitimate method for science (*à la* Duhem 1954), again, precisely because such models are allegedly used to show us in perceptual terms what is not perceivable, such as in the cases of the optical ether and of subatomic particles and fields. The criticism, more than the endorsement, of analogical models speaks more effectively about their epistemic nature.

What about Hughes's DDI construal of scientific representation (Hughes 1997)? The DDI account is spelled out as comprising the three components of 'Denotation', 'Demonstration', and 'Interpretation'. If 'denotation' (1st D) can be interpreted as the 'encoding' by some conventional vehicle, and 'demonstration' (2nd D) as inferences within or between such vehicles, and 'interpretation' (I) as 'decoding', in the sense of getting that which is invariant among conventional vehicles 'back', the DDI account fits well with our anti-deflationary account. Hughes's analysis of Galileo's example should be quite illustrative: a kinematic problem of uniformly accelerated motion is first denoted (or encoded) by geometric figures, demonstrations within geometry are then carried out, and finally the result in geometry is interpreted back (or decoded) to yield a kinematic situation, which provides a solution to the problem. It is obviously a question as to whether the DDI account covers all cases of scientific modelling, but for those that it fits, there is little doubt that the model as a result of denotation must reveal the relevant structures of the target so much so that the subsequent steps in D and I would make sense. And what the denotation reveals is in a very strong sense what the target under study is *like*.

[10] Or more precisely, suppose a model represents a target by analogy, and it has ABC but the target has ABD, where A and B are the known aspects of both. If C is causally connected to A and B in the model, we have reason to expect that D is causally connected with A and B in the target if the model is a good representation.

Frigg's tripartite account of scientific representation

Model Descriptions =(*p*-represent)=> Model Systems =(t-represent)=> Real Systems

greatly clarifies the natural and structure of modelling, and I have no problem with the basic elements of his account. However, Frigg appears to hold that the three parts of modelling are nearly independent of each other, especially when it comes to how elements are referred to in the systems in different parts. The descriptions and mathematical equations together with the verbal and mathematical symbols used in them all point to components in model systems, as part of the *p*-representation, while the model components in turn refer and show parts of real systems through the *t*-representation. But how is that done? How does the *t*-representation work? Frigg has used the map metaphor for his 'first stab' at articulating this relationship. My account differs emphatically from his on this very point. There are no separate denotational or referential devices in the first two parts of modelling. As argued above about the division of labour of the symbolic and the modellistic elements in epistemic models, symbols such as names and labels exist in models as well as in the descriptive part, and they function in the same way: denoting or referring to the components in the real systems.

This realization of the hybrid nature (a mixture of the symbolic and the modellistic) also explains the difference between Contessa's account and mine. For like Frigg, Contessa regards idealized models with their components as autonomous entities including the names or labels that come with them. For instance, in a Rutherford model for hydrogen atoms, there are the idealized 'electron' and 'nucleus', etc. that compose the abstract hydrogen atom. My account disagrees with this conception about how referring terms are used in modelling. I have argued exhaustively elsewhere (suppressed for the review of this paper) that models are not particular objects such as 'ideal electrons' or 'point-mass planets' that populate a fictional realm of imaginary entities. They are rather generic abstract or physical objects that are used by being given labels or names (referring terms) to model real systems and show what they are in the chosen aspects of representation. The modellistic elements are often basic things that can be used in a great variety of modelling jobs; and only when symbols are attached to them in a model that they are tied to the referred parts as showing their likeness.

Suárez's inferential account of scientific representation requires as necessary conditions that an object has the force of pointing to its target and the potential of serving as inferential surrogate in the practice of science in order to qualify as a legitimate representational vehicle in science. The force is a sort of disposition of an object that has the potential of being used by some community as a representational vehicle. And the inferential capacity is certainly a pragmatic virtue of an object in the practice of science. Despite its difference of appearance, it is quite plausible to argue that Suárez's deflationism boils down to Callender–Cohen deflationism. By convention, if we follow Callender and Cohen, anything has the force to be used to point to anything (including itself). So, the force requirement is nearly trivial and universal for representational vehicles, as it ought to be. Now, Callender and Cohen agree that symbols used as scientific models need extra requirements as long as they are pragmatic ones; and surely being an inferential surrogate is a pragmatic role. Hence, Suárez's deflationism is not essentially different from Callender and Cohen's, and therefore Suárez is right to regard his own account as a deflationary one. This is emphatically different from my account, as the above should have amply demonstrated.

Among all the criticisms, Toon (2010) gives, as far as I can tell, the first in-depth discussion of the Callender–Cohen deflationism (it was called a 'derivative account' there). Toon, like the others, is correct in pointing out that, for example, while a salt shaker (as in one of Callender and Cohen's examples) may be used to symbolize Madagascar, it cannot be said to depict or provide a scientific model for it, in any reasonable sense of representation; but instead of going on to respond to the question of why extra pragmatic requirements are not sufficient, he proceeds to develop his own Waltonian conception of modelling as make-believe. I take such an approach to be a variation of the fictionalist approach to scientific modelling, and I have no space in this paper to launch a critique of the approach (my anti-fictionalism arguments are given in another paper). Suffice it to say that there may be independent reasons why modelling in science should be taken as analogous to fiction-making in general, but fictionalism does not seem to naturally emerge from an in-depth criticism of deflationism. In other words, to re-inflate the conception of representation, all we need is at most what I have put in my hybrid account: clearly recognizing the conceptual distinction between symbols and models, and realizing that most scientific models utilize both elements to complete the job of representation. With the 2-D freedom of modelling — choice of symbols and choice of modellistic elements for demonstrating a target — my conception of modelling gives, I believe, a better portrait of the practice of science.

In conclusion, I have argued in this paper that scientific representation, whose core activity is modelling, is fundamentally different from finding and using symbols. Its purpose is not primarily 'presenting' but 're-presenting', namely, showing or demonstrating, the target systems. Its chief aim is therefore to gain epistemic values rather than pragmatic ones. The debate over whether models, at least those with perceptual qualifies, ought to be outlawed from science makes sense only if seen as a debate over the epistemic values of modelling. The reason Duhem (1957) condemns the mechanical models for the luminiferous aether is that, *despite their enormous pragmatic attractions*, they couldn't possibly add anything to our knowledge of the electromagnetic phenomena that is not already contained in the Maxwell equations.

Acknowledgements

The author would like to thank Craig Callender for reading and commenting on an earlier draft of this paper. He would also like to thank the audience at the session of EPSA11 in Athens where a short version of this paper was read. Special thanks are given to the two referees of this journal for their constructive criticisms.

References

Callender, C., and J. Cohen. 2006. "There Is No Special Problem About Scientific Representation." *Theoria* 55: 7–25.

Contessa, G. 2007. "Scientific Representation, Interpretation, and Surrogative Reasoning." *Philosophy of Science* 74: 46–68.

Contessa, G. 2010. "Scientific Models and Fictional Objects." *Synthese* 172: 215–229.

Daitz, E. 1956. "The Picture Theory of Meaning." In *Essays in Conceptual Analysis*, edited by A. Flew, 53–74. London: Macmillan.

Duhem, P. 1954. *The Aim and Structure of Physical Theory*. Translated by P. P. Wiener. Princeton, NJ: Princeton University Press.

Freeman, W. J. 1991. "The Physiology of Perception." *Scientific American* 264 (2): 78–85.

Goodman, N. 1976. *Languages of Art*. Indianapolis, IN: Hackett.

Grice, P. 1989. *Studies in the Way of Words*. Cambridge, MA: Harvard University Press.

Harman, G. 1973. *Thought*. Princeton, NJ: Princeton University Press.

Hesse, M. B. 1966. *Models and Analogies in Science*. Notre Dame, IN: University of Notre Dame Press.

Hughes, R. I. G. 1997. "Models and Representation." *Philosophy of Science* 64 (Proceedings): S325–S336.

Keller, E. F. 2002. *Making Sense of Life: Explaining Biological Development with Models, Metaphors, and Machines*. Cambridge, MA: Harvard University Press.

Kuhn, T. S. 1977. "Objectivity, Value Judgment, and Theory Choice." In T. S. Kuhn, *Essential Tension: Selected Studies in Scientific Tradition and Change*, 320–339. Chicago, IL: University of Chicago Press.

Liu, C. 1997. "Models and Theories I: The Semantic View Revisited." *International Studies in the Philosophy of Science* 11: 147–164.

Locke, J. [1690] 1979. *An Essay Concerning Human Understanding*. Oxford: Clarendon Press.

Morgan, M. S., and M. Morrison, eds. 1999. *Models as Mediators: Perspectives on Natural and Social Science*. Cambridge: Cambridge University Press.

Morrison, M. 2008. "Models as Representational Structures." In *Nancy Cartwright's Philosophy of Science*, edited by S. Hartmann, C. Hoefer, and L. Bovens, 67–88. New York: Routledge.

Nozick, R. 2001. *Invariance: The Structure of the Objective World*. Cambridge, MA: Harvard University Press.

Rowbottom, R. P. 2009. "Models in Biology and Physics: What's the Difference?" *Foundations of Science* 14: 281–294.

Siegel, S. 2011. *The Contents of Visual Experience*. Oxford: Oxford University Press.

Suárez, M. 2004. "An Inferential Conception of Scientific Representation." *Philosophy of Science* 71: 767–779.

Suárez, M. 2010. "Scientific Representation." *Philosophy Compass* 5: 91–101.

Suppe, F. 1989. *The Semantic Conception of Theories and Scientific Realism*.

Chicago: University of Illinois Press.
Teller, P. 2001. "Twilight of the Perfect Model Model." *Erkenntnis* 55: 393–415.
Toon, A. 2010. "Models as Make-believe." In *Beyond Mimesis and Convention: Representation in Art and Science*, edited by R. Frigg and M. C. Hunter, 71–96. Dordrecht: Springer.
van Fraassen, B. C. 1980. *The Scientific Image*. Oxford: Clarendon Press.
van Fraassen, B. C. 2008. *Scientific Representation: Paradoxes of Perspective*. Oxford: Clarendon Press.

Part 2
Idealization and Approximation

The papers included in this part, some of which are previously published, are all dealing with the issues regarding idealization and approximation in the practice of science. They belong to the investigations of the methodology of science.

The use of idealization and approximation is ubiquitous in scientific and technological works; and a commonly held view is that idealization is no more than a method of neglecting the negligible and approximation is the result of such acts of idealization. The papers in this collection argue against such a view. If we could brand this commonsense view as a "Euclidean" view of idealization/approximation, the argued view in the papers may be called a "non-Euclidean" view. The "space" of approximation as a result of idealization is not always "flat" or uniform. There are regions, especially around the areas laws of nature sanction, where the metric of approximation is "curved."

Paper 7 also makes an argument for the necessity of making a connection between how approximate an experimental result is to its supposed true value and how likely that result is to appear.

For more recent works in the area of idealization and approximation, see John D. Norton (2012) "Approximation and Idealization: Why the Difference Matters." *Philosophy of Science* 79 (2): 207-232, in which the notions of idealization and approximation are defined in a novel way so as to provide a better conceptualization of the infinite systems frequently encountered in quantum statistical physics and quantum field theories. Also see Ilkka Niiniluoto (2011) "Revising Beliefs Towards the Truth," *Erkenntnis* 75 (2): 165-181 for a discussion of the models for approximation as "truthlikeness" and their connection to belief revision and its models. Another important recent work is Darrell P. Rowbottom (2011) "Approximations, Idealizations and 'Experiments' at the Physics-Biology Interface." *Studies in History and Philosophy of Biological and Biomedical Sciences* 42 (2): 145-154, in which difficulties of model-building when condensed matter physics meets molecular biology are discussed. Biologists appear to regard the computer simulated models from physicists not as "experimentally worthy," and yet those biologists who promote the program of system biology seem to be breaking new ground in biological research. The author critically evaluates the viability of such a program as an alternative to the physicists'.

Idealization, Confirmation, And Scientific Realism

The paper first raises the problem concerning the confirmation of idealized theories in science and its relationship to scientific realism. Then a solution by Laymon is discussed. It is then argued that two different types of idealization need to be distinguished and that only one of them produces false theories. But then, such "theories" are really theory-maps, which point to theories at the end of improvement paths. Finally, Laymon's account is modified in accordance with the above insight.

I. The problematic

In the traditional debate between realism and antirealism, or more broadly speaking, in any traditional systematic account of aspects of science, such as scientific explanation and confirmation, scientific theories or their constituent statements are assumed to be simply either true or false. This is understandably derived from the benchmark in philosophy of language in the analytic tradition, which may be traced back to Russell and the logical positivists, among others, in the early part of 20th century. Scientific theories are or aim at factual statements about natural phenomena, and therefore the statements are either true of false. The notion of idealization or approximation was rarely invoked. More recent participants of this debate seem to have noticed this neglect, and phrases such as 'approximately true' began to appear in place of "true" in the literature; but the change is merely cosmetic: the phrases are added merely to extend the applicability of conclusions obtained by arguments that take no notice of approximation or idealization. What I can gather in reading philosophers such as Richard Boyd (1983), who uses 'true or approximately true' rather than simply 'true' consistently in his writings, is either of the following two senses. A theory or a law statement in question either (1) is not true but very close to being true or (2) is not yet confirmed to be true but may be so later. Similarly, philosophers who acknowledge the fact that most, if not all, scientific theories or statements of law are idealized usually take such theories or law statements to differ from the true ones in only negligible degrees (cf. McMullin 1985; Nowak 1972, 1980; Niiniluoto 1986, 1987). It has been convincingly argued (Liu 2004) that such a notion of idealization is fatally flawed.

Similar situations occur in approaches of confirmation (or theory testing): most results in the major approaches of confirmation are regarded as easily extendable to approximately true theories or law statements. For example, if an account of confirmation works for the testing of a 'true' theory like Einstein's theory of gravity in general relativity, then it also works for testing the approximately true Newtonian theory of gravity in the appropriate context (cf. Schwartz 1978; Ramsey 1990, 1992).

As I have argued elsewhere, such nonchalant attitudes towards idealization and approximation are not justified, nor are they entirely harmless. Therefore, I explore in this paper what we can say about such questions as: what does it mean to confirm or disconfirm an idealized theory? Could an answer to such a question provide an account that is amenable to scientific realism? These are big and complex questions, whose comprehensive answers are beyond the scope of this paper. What I shall do to begin the exploration is to critically discuss a particular account (Laymon's account) that provides, so to speak, a triangulation of approximation/idealization, confirmation, and scientific realism, and figure out where we are and what we still need to do in answering the question about the relationship between confirming idealized theories and scientific realism. The result is an improvement of the Laymon account by an updated understanding of the conception of idealization and approximation. With this improvement, it is argued, the account is better able to be used in defending scientific realism.

To begin with, we have the following problem for idealized theories and realism: if idealized law statements are false, how can there be an independent reality whose goings-on are regarded as obeying such laws? Moreover, if idealized law statements are false, what sense does it have to try to test and confirm them? Furthermore, if scientists spend their lifetime looking for laws, how can such laws, being necessarily untrue because of being idealized, deserve their efforts? To contemplate these problems, one must remember that, on the one hand, idealization is indispensable in theory construction and, and on the other hand, the practice of science contains many cases in which everyone involved knows that the idealized theory/model is not really true of the phenomena s/he is trying to study.

At least two types of idealization coexist in the practice of science. Knowing the danger of formalization, I shall not attempt at giving definitions in terms of necessary and sufficient conditions; rather I provide here simple examples to illustrate their differences. Hypothetically, when idealization is used to draw up the model of ideal gas, from which Boyle's law is obtained, scientists know that the model is false of real gases and Boyle's law is no real law of physics. There might be alternative attitudes towards the model, and one of the more plausible attitudes may take it only as a guide in the direction towards the real law, but no one would take the model as a true reflection of real diluted gases. However, when idealization is used to obtain the notion of inertial motion, from which the first law of Newtonian mechanics is obtained, scientists know that the model of motion on a frictionless plane is true and real (cf. Jevons 1874; Schwartz 1978; Barr 1971, 1974). (This latter example is not the most ideal because of complications with the Mach principle and general relativity. A better example is perhaps the law of motion of a charged body, which is under the electromagnetic influence alone, or some such cases[1].) The point here is of course about the truth, not just the idealized or approximate truth, of the linear compositionality of causal influences. Because of the enormous complexity of actual phenomena, idealization is needed in the case of inertial motion to 'decompose' causal factors, each of which may be governed by a unique set of laws alone (cf. Liu 2004). This case is at least on the surface different from the idealization that "invents" ideal gas from real diluted gases.

First let us observe that neither is the ideal gas model the result of a trivial act that merely neglects the negligible (or adds the imperceptible) nor are completely isolated systems and the laws they instantiate regarded as false just because they are results of idealization. Both are typical examples of large classes of idealized models widely used in science and technology. Therefore, when idealizations are introduced, the results are neither necessarily trivial nor necessarily false.

Moreover, three philosophical attitudes seem to be possible regarding the models that are exemplified by these two examples and the theories that they spawn. One may think and argue that all idealizations are essentially of the

[1] But of course, this is also not completely problem-free because gravitational interaction will always be present.

first type, and if so, one may be more inclined towards antirealism regarding scientific theories. Or one may think and argue that all idealizations are essentially of the second type, and if so, one may be more inclined to realism regarding scientific theories. Or one may think and argue that there are two types of idealizations and they are fundamentally different, and if so, one's understanding of the nature of idealized theories does nothing to incline one towards either 'ism.

We should be clear at the outset that the notion of scientific realism in this discussion is a rather circumscribed one. A complaint may be lodged against this discussion by pointing out that a debate over metaphysical positions can be made independent of whether theories are idealized or not. Hence, to avoid confusion, let me state as clearly as I can what I take to be the topic at issue regarding scientific realism. It is not purely a metaphysical dispute about whether any systems studied in science exist independent of the study; rather I take it that both realists and antirealists agree that objects such as the sun and the moon exist independently of how they are known; and also they agree that the ideal gas as given in Boyle's model does not exist in the same sense. The dispute is rather about the existence of theoretical or unobservable entities and statements about them. Do they exist in the same sense as the sun or the moon exists? Further, are the truth-makers of statements about such entities facts that obtain independently of our conception of them?

II. Laymon's account for confirming idealized theories

More than anyone in the philosophy of science literature, Ron Laymon (cf. Laymon 1980, 1984/1982, 1985, 1989) has contributed amply to the above problematic, and he is therefore an exception. From the 1980s on, Laymon has single-handedly identified the various issues related to the general problem and provided ingenious ways to resolve them. His account is an attempt to triangulate the conception of idealized theories with the problem of their confirmation and of scientific realism. In Laymon 1980, the project is announced as follows.

> My theses in stark unqualified form are these. First, explanations... consist of two parts: (1) an idealized or simplified deductive-nomological

> sketch...and (2) an auxiliary argument or set of arguments made showing that if the idealizations of the sketch are improved, i.e., made more exact or realistic, then the prediction of the idealized sketch will be correspondingly more exact and realistic. I shall refer to this second component as the *modal auxiliary*, since the argument purports to show that an improved idealized sketch is *possible*. My second thesis is that confirmation and disconfirmation occur in the realm of the modal auxiliaries. That is, a theory is confirmed if it can be shown that it is possible to improve its idealized sketches (Laymon 1980, p. 338).

On the realism connection, Laymon says,

> I shall now consider the consequences of the above sort of *converging counterfactual* theory of confirmation for the issue of scientific realism. Realist and antirealist perhaps can agree on this methodological point: proceeds *as if* one were developing ever more accurate descriptions of an existing reality. Given this agreement, an argument *for* realism is that cases of successful convergence to better experimental fit are *miraculous coincidences* for the antirealist (Laymon, 1984, p. 118)

Laymon has had several later occasions to revise and enhance his account on the confirmation of idealized theories and its contribution to the support of scientific realism, and my summary below comes from this body of his work (cf. Laymon 1984/1982, 1985, 1989).

The general form of an idealized theory is given as follows. If a system were in *I*, it would be *T*, where *I* is a set of ideal conditions and *T* the theory that tells how the system would behave under *I*. For example, if a block moving on a hard smooth surface with an initial speed were to be subject to no friction, it would continue to move along the same direction with the same speed forever. Or, if a tank of gas were ideal gas under normal circumstances, its thermal states, states described by values of the gas's pressure, volume, and temperature, would be determined by Boyle's law (Laymon 1984, 1989; see also Weston 1987, 1992). Ideal conditions — the ones that circumscribe the exact extent to which the system in question behaves exactly according to the theory/laws — are not initial/boundary conditions ("i/b conditions," henceforth), although i/b conditions are usually also idealized for the convenience of inferences. They are the imaginary conditions in which the idealized theory "works perfectly." With friction and other complicating factors, such as the different sizes and interactions among real gas molecules, in place, whatever law statements or theory one has, they would not apply exactly to concrete cases with definite i/b conditions. Once such factors are removed in the act of idealization, they would apply exactly.

Although Laymon sometime speaks about idealizing i/b conditions, I think it is better to keep them apart. By a stipulation in the current discussion, if a theory were not idealized it would give realistic predictions under physically possible i/b conditions. It seems right to say that i/b conditions can also be idealized, but they are certainly not ideal conditions, *I*, that produce idealized theories. It might be correct to say that the distinction between ideal conditions and (non-idealized) i/b conditions is that the latter are always physical possible while the former may not. If there are deeper problems with this distinction, they are for other occasions.

As to the question of whether the theory, *T*, is simply false because of the ideal conditions, Laymon's answer is given most clearly in his critical response to Nancy Cartwright's view. Cartwright (1983) argued under the slogan "how laws of physics lie" that laws of physics can only be understood as true in the models that are made for them and the models, being idealized, never actually exist. Laymon's response (1989) is that the laws are true, but the models are false to natural systems, and the two together do not make practical predictions (even when given true i/b conditions) that produce the 'experimental fit' that is meant to confirm the theory (more on this last point later).

To the question of how an idealized theory is to be tested — confirmed or disconfirmed — Laymon's account proceeds as follows. Suppose that our theory, *T*, consists of law-statements that, when supplied with initial and boundary conditions, yield empirically testable predictions; and suppose again that *T* is idealized, which means that there are some set of ideal conditions (statements) Id such that *T* is true with Id and not true without it. In other words, given the appropriate i/b conditions, *T* and Id implies *P*, a prediction that we can test in an experiment. Suppose again there is a set of theories, $\mathbf{T} = \{T^j \mid j = 1, 2, \ldots, N\}$ where $T \in \mathbf{T}$ and N is a natural number; and correspondingly, there is a set of ideal conditions $\text{Id} = \{I^j \mid j = 1, 2, \ldots, N\}$ that *forms a strict descending order of idealness* such that, given the same set of i/b conditions, T^i and I^i yields P^i, where $P^i \in \mathbf{P} = \{P^j \mid j = 1, 2, \ldots, N\}$ is a testable result in the same experiment. This prediction-making scheme does not have to be understood as a simple Hypothetical-Deductive (H-D) scheme, although it could be; it represents any scheme of theory testing that goes from to-be-tested theory together with all the auxiliaries to the testing results. This, roughly, is what Laymon refers to as the "deductive-nomological sketch" in the first of the two quotes above. The descending order of idealness is also defined very broadly such that it ought to be able to accommodate any practical acts of gradually removing or weakening an idealization. Next, let P^E be a set of data from the experiment that tests *T*. P^E is almost certainly not the result of a single trial but of some standard statistical treatment of a set of trials; and it may or may not be a member of **P**. The confirmation relation is one between the **T** and P^E such that to confirm idealized theories is to take confirmation as the improvability of approximation under idealization. In other words, an idealized theory is found to be confirmed if it is shown that relaxing its idealness (making it more realistic) improves upon the approximation of its predictions (towards P^E). One way of making this more precise (as in what he calls 'having monotonicity' (Laymon 1985, 213-4)) is to have the following.

[L] T^k, for any 1 £ k £ N, is confirmed if for any T^i and T^j in **T**, where $k < i < j \leq N$, we have $|P^j - P^E| < |P^i - P^E|$ (i.e. P^j is closer to P^E than P^i whenever T^i is less idealized than T^j); otherwise, T^k is disconfirmed.

This version essentially says that **P** must form a strictly ordered set that matches the same strict (descending) order in **T** (which is induced by the original descending order of **Id**). This is precisely what is meant for an idealized theory to *approbate* approximation *monotonically*.

Some initial comments on this account are in order. It appears to be too strong; the requirement of monotonicity does not seem necessary. What one needs is something weaker; but how to weaken it with rigor is not an easy matter. One simple modification of [L] is to say that for T^k to be confirmed is for *some* T^i and T^j in **T**, where $k< i < j \leq N$, we have $|P^j - P^E| < |P^i - P^E|$. But is this too weak? It depends on how **Id** is constructed. Suppose we look at the ideal gas case. Two kinds of idealization are essential to the model: vanish the size of gas molecules and banish the interactions among molecules. Suppose it turns out that Boyle's law approbates approximation when the stipulation on molecule sizes is relaxed but it does not do that when we only reintroduce collisions and/or interaction forces. Would we still regard Boyle's law as confirmed since it obviously satisfied the weaker version of [L]?

Here is a plausible proposal: we can simply use the notion of a limit to characterize it, namely, for any positive number, ε, there is a natural number

N after which the difference between the N+1 term of the prediction and P^E is equal or smaller than ε. This does not require monotonicity, and the improvement process may suffer frequent deviations; but as long as the process forms a dense sequence, approximation approbation is counted as successful.

There are other problems with this account which I shall discuss shortly, but if we dismiss it as unrealistic because there may be actual cases of theory confirmation that do not fit it perfectly, we are doing philosophy of science a disservice. It has often happened in the history of philosophy of science, that certain formalized accounts of a practice in science, such as Hempel's D-N model of explanation, are rejected wholesale, namely together with the intuitions that ground the accounts in question, just because there are actual cases that do not fit into their formal details. In other words, they are rejected simply because they are idealized account of some practice of science. I regard this as an unfortunate situation. In this aspect, scientists are much wiser than philosophers, although it sounds a very odd thing to say. If one compares many philosophers' treatment of Hempel's D-N model with physicists' treatment of such laws as Boyle's ideal gas law, one will see that one should not simply abandon the earlier and more idealized accounts of explanation or confirmation just because they are too idealized, but use them in situations that they approximate.

It is obvious from the second quote above that Laymon means to use his account of confirmation to mount an inference to the best explanation (IBE) argument for a realist stance for idealized theories (see also Brown 1985). [L] supports scientific realism because otherwise, it would be difficult to understand why confirmation of an idealized theory should approbate approximation. If it is not a fact in a world whose goings-on are independent of us that a body moves in a straight line with constant speed if isolated from all external influence, how can we explain the fact, the observable fact, that by gradually adding friction and by checking its effects on the moving body, the predictions match the experimental results? Hence, approximation approbation supports realism via IBE.

In connection with the realism issue, Laymon offers two further important points. First is the point that the realism/antirealism debate is really about theoretical entities. It is difficult to take an antirealist view seriously if it says that ever more accurate mechanical account of the earth movement in the solar system is not really about an existing earth but a theoretical construct. But if we replace 'the earth' with 'an electron' in the statement, the view cannot be so easily dismissed. And the second is an easily anticipated objection raised by Laymon himself. What if '[t]here were,..., increasingly more realistic theories of phlogiston' (Laymon 1984, p. 119)? Shouldn't the realists have to say that phlogistons are real and the phlogistic theory of combustion is approximately true?

III. Assessing Laymon's account

First, some friendly amendments to Laymon's account are in order to make it more robust. One obvious feature of this account of confirmation is that it is a theory that depends on a more basic conception of confirmation. Laymon offers us through his modal auxiliary scheme an account for confirming idealized theories, but it will not work unless we are clear about what it means for each consecutive theory, $T^k \in \mathbf{T}$ to be confirmed or disconfirmed. Otherwise, we will not understand what it means for confirming a theory by a set of modal auxiliaries that specifies counterfactually how the theory would behave when the idealization is relaxed. Therefore, Laymon's account of confirmation, even if correct, must be supplemented by a confirmation theory of the normal kind, namely, a H-D, a Bayesian, a Boot-strapping, or some such model that tells us what it takes to confirm or disconfirm each $T^k \in \mathbf{T}$. Let us look at the simplest example of the basic confirmation scheme, namely, the H-D scheme. Since this is only an exercise in showing how Laymon's account must ultimately depend on a basic account, it should not matter very much which basic account we select to offer such a demonstration. Take any one theory from $\mathbf{T}$, T^k. It together with i/b conditions and auxiliaries produces a prediction, P^k, which is not found among the observations but is not too different from them. Depending on where T^k is in the ordered set $\mathbf{T}$, the corresponding closeness of P^k to the actual experimental results can be estimated. If the closeness matches the expectation, we can regard T^k as confirmed. If this is so for all k, $1 \leq k \leq N$, we say that there is confirmation. If too many of the k's in which T^k's prediction P^k deviate from a gradual progression towards P^E, then we do not have confirmation. So, at least the H-D scheme can be introduced as the basic scheme to complete Laymon's account. I leave the questions of whether using a Bayesian account or a Boot-strapping account may also be worked out to support Laymon's account as an exercise for the reader. The intuition is that it should work out because there does not seem to be anything in either of those accounts that goes in principle against incremental closeness of matching experimental findings to predictions.

Further, this part does not have to be conceived as entirely comprising modal auxiliaries, although Laymon is right to insist that it is *sufficient* to have established the possibility of approximation approbation. An idealized theory or law statement is 'confirmed' in this part of confirmation if the relaxation of the ideal conditions is conceivable and the approximation approbation is demonstrable under that conception. *However*, the opposite direction also and often serves as the objective for theory testing. Although they may in principle be unrealizable in a lab, ideal conditions in many cases can certainly be asymptotically neared. Perhaps it is already implied in Laymon's account, for it hardly needs any originality to see it, but we can certainly imagine situations in which experimental scientists pursue a strategy of inching ever closer to the ideal conditions and see if results approach the ideal prediction monotonically. Rather than reintroducing friction in a gradual progression and see how a certain system behave as a result, experimenters often do the opposite, by gradually removing friction and see if the results approaches the ideal asymptotically.

Here, the term, 'direction of fit,' in epistemology seems to be just right for illustrating this point. Laymon's account only looks at one direction of fit in the business of confirmation, namely, the desire to fit one's belief to the already experienced external world. One may desire to pursue the opposite direction, namely, the desire to fit the world to one's belief by way of arranging one's environment so that one experiences results that asymptotically approach what one firmly believes. This latter point lends support, as we will see in a moment, to the idea, to which Laymon has partially signed on, that an idealized theory may be TRUE as the asymptote, rather than only approximately true. This is one main reason for rejecting Cartwright's idea that all idealized theories or models are false by default.

And because of the added other direction of fit, let us first note the following simple and important point. It makes sense to gradually reduce the strength of friction, for example, but it does not make sense to gradually shrink the sizes of gas molecules (or to find actual gases with asymptotically smaller molecules). So, the distinction between the two types of idealizations mentioned in the beginning of the paper becomes prominent.

With this point broached, let us now return to Laymon's first point I mentioned earlier, namely, the realism/antirealism debate is really about theoretical entities; and therefore an idealized molecule or electron may, but an idealized earth does not, threaten realism. A number of questions arise from this point. First, as mentioned in section I, we can separate two types of idealized objects: one type results in theories or laws, such as Boyle's law, that "no one" would regard as true of nature, but the other type results in theories or laws, such as the first law of Newtonian mechanics, that scientists usually regard as true of nature. Does this mean that Laymon's rejection of Cartwright's view in defense of realism applies only to certain types of idealized theories but not to all? It seems reasonable to think those theories or law-statements that scientists themselves regard as false are results of *physically impossible* ideal conditions. Molecules cannot be without size nor can (actual) electrons be all identical, where 'can not' means physically impossible. But completely isolated objects or systems may not be *actual* but are nonetheless physically possible. Secondly, an idealized theory of an observable system does not seem to threaten realism because what is claimed as real is the system that we can all observe; it is not the idealized model, which is an artifact we use for convenience to discuss the real thing. And this appears to be how we should understand a theory of our solar system (containing idealized models and laws applicable to them) and why it is not really a thread to realism. The need for idealizing the system is that otherwise we can only either say something general and vague about it so that we say something true or say something exact about it that is not exactly true.

This does not seem to be the case at all for electrons or other theoretical/unobservable objects. We may wonder with good reasons whether electrons under our current theory are results of idealization. Do we know what the ideal conditions are for electrons? Because of this, the realist may be hard put to answer the question: what exactly is the thing that is supposed to be real or exist independently of our conception of it? Given how empirically underdetermined our models for the unobservables often are, it is difficult indeed to argue that there independently exist electrons or other unobservables that our idealized theory approximately depict. Therefore, the antirealists may have a case against the realists on theoretical entities because in order to be a realist about them, one has to either believe what exists is exactly like the model in our theory or know what the thing is like from which the model in our theory is an idealization. Since we cannot in principle know what factors from which and what degrees to which the idealization is drawn in these cases, neither of the above is a tenable position.

The distinction between observables and unobservables has been a controversial issue in philosophy of science. Some argue that the distinction, though useful on many occasions, has no philosophical grounds, but others argue for the opposite. Even among those who believe in the distinction, some do not know how, or think it is possible, to argue for it. Van Fraassen (1980) has an argument for the distinction in terms of possible direct observations; but it seems here is another argument for the distinction. Observables have or may have models that are explicitly idealized objects. We have a good idea of why and how the idealized observables in our theory got there. This is not true at all with the unobservables. We can't even be certain which features of the unobservables are the results of idealization and which are not. For instance, we have identical elementary particles in our theory. Is being identical an idealization or a real feature of the particles? It could be sensibly regarded as an idealization, essentially the same as the idealization in Boyle's ideal gas in which not only molecules are made sizeless; they are also made identical. However, it may also seem impossible to regard it as such. Here is one of the reasons why it seems so. If we compare classical gases (of which Boyle's ideal gas is one) with quantum gases, we know that the former obey the Maxwell-Boltzmann statistics and the latter the Bose-Einstein statistics, and they are fundamentally different conceptions of statistics. Classical gases have idealized identical particles (which are in fact not identical), and therefore they keep their identity while not having any relevant differences registered in the theory. If a quantum gas's particles' lack of identity is also an idealization, then the statistics should be the same as in the classical case; but no, the Bose-Einstein statistics is incompatible with the Maxwell-Boltzmann statistics, which seems to strongly suggest that the identity of elementary particles is not an idealization but a "real" feature of them.

Let me now explain an unexpected benefit of Laymon's account. If we think of idealization as an act that neglects the negligible in theory construction, Laymon's modal account matches well with an approach in the philosophical literature on approximate truth. Risto Hilpinen (1976) characterizes an approximately true proposition as one that is true in some other possible world that is *close* to the actual world, where the sense of closeness is essentially a Lewisian notion (á la David Lewis). One can certainly imagine a metric and a cutoff point in some formal sense that supply a criterion that separates approximate propositions from not even approximate ones. Complications aside, one can in general regard good idealized law-statements as expressing propositions that are (at least) approximately true and bad ones as propositions that are not even approximately true. I assume that the reverse is not true for the obvious parallel reason that all law statements are true but not all true statements are law-statements.

There is a problem with Hilpinen's conception of approximate truth, namely, when a proposition is about some theoretical objects, how is it possible to even in principle come up with a measure of closeness between the world in which the approximate proposition is true and the actual world? Since, for example, it is not possible to "witness" what molecules or electrons are *actually* like, how do we know whether the idealized ones are close to the actual ones or not? I think Laymon's account at least goes in some ways towards solving this problem. Of course, we can never directly experience what electrons are like; but if we have a model for them and the model approbates approximation in the observable predictions the model of electrons yields via a theory, then the possible worlds in which electrons are like the ones in our model is closer to the actual world than the ones in which electrons are different. Still, the radical underdetermination problem threatens such a realist solution, since it assumes that the idealization involved in our model of electrons is essentially of the same type as the idealization involved in our model of the moon (cf. Popper 1976; Oddie 1981, 1986).

IV. Two Types of Idealization and the Unobservable

Let us go back to one of our original questions. If idealized theories or law statements are false, in what sense can we regard it as confirmed or disconfirmed? Doesn't the notion of confirming something simply mean confirming that it is true? If like most people who, when confronted with this question, say that confirming something really means confirming that it is true or approximate true, where the latter means 'true enough,' why do we even need Laymon's account then? In fact, if the theories or law statements are obviously true enough in the sense that the deviation is negligible, then the idealization is necessarily so trivial that it is often not even mentioned. For example, in ordinary circumstances, the difference between non-relativistic mechanics and relativistic mechanics is so small that whatever confirms the one confirms the other. In other words, the confirming evidence in such circumstances is not likely to be sensitive enough to distinguish the two. In this case, scientists do not even consider the non-relativistic mechanics as an idealized theory. But frictionless planes or regions without gravity are not just neglecting the negligible, and such assumptions will certainly create considerable discrepancies in the testing results. In such cases, what does it mean to confirm idealized theories that are false? Now the question is: are statements about frictionless objects or fields in regions without gravity necessarily false statements? As I have discussed earlier, Cartwright (1983) thinks they are indeed false, but Laymon thinks that some of them are not, and I have supplied earlier the reasons for why I agree with him.

What about law statements such as Boyle's law, which we should all agree are not true in any sense of being true? The best way to regard statements of this sort is to think of them not as law-statement but as "law-maps." It is certainly not false, as a good map of a city is not false, but it is not true either, as the line segments on the map with street names attached are not true of the streets in the city that bear those names. A good map contains true information about the streets and buildings in the city it represents; it is their spatial relations and orientations, etc. that must be displayed correctly for a map to give such true information. And the notion of confirming a map, or one map being better confirmed than another, is not hard to define either. Some philosophers want to argue that all scientific theories are like maps. I think the idea is not so much wrong as is too sweeping to be literally correct. One often ends up, in trying to argue for such a sweeping claim, distorting the sense of maps so much so that the analogy loses its grip on our imagination. But law-statements such as Boyle's law are like maps in the most straightforward sense.

Therefore, harking back to the insight of the two opposite directions of fit I mentioned earlier, one can be a realist and maintain that there is a clear distinction between the two types of idealization. And then theories involving one kind can be true in accordance with the correspondence theory of truth and theories involving the other kind can never be true; and the latter theories are not real theories but rather theoretical maps that are intended to lead us later to real theories (which may only involve idealization of the previous kind). Let us call the theories or lawlike statements of the first kind 'theories' or 'lawlike statements', and theories or lawlike statements of the second kind 'theomaps' or 'lawmaps'. Opposite attitudes should be adopted when regarding these two types of theories. For theories and lawlike statements, the experimental effort should be primarily given to approximate the conditions under which a theory or lawlike statement in question is literally true. In our case for the law of inertia, approximation may simply mean getting rid of the friction as much as possible. For theomaps or lawmaps, we do the opposite: we relax as much as possible the ideal stipulations in them so that new and less idealized theories can be found. Such theories may have a chance to be true and may produce predications that approximate results in possible experimental conditions. Both confirmation procedures involve asymptoticity, and yet they are of opposite directions.

Keeping this distinction in mind, let us now reexamine Laymon's IBE argument for realism. The argument works if idealization is no more than what Laymon wanted, namely, just to approbate approximation. It appears to work well for such cases as an idealized earth or other planets in Newtonian mechanics. In fact, it works well for any cases where the idealized item is an object of which direct or semi-direct observations give us good impressions. In such cases, what Laymon's argument concludes is actually this: the object of which we make an idealization for the sake of theoretical expediency must be a real object; it must be something that exists independent of us and our theory. We cannot say that the earth does not exist because its idealized copy in Newtonian mechanics happens to have unreal aspects. On the contrary, if the real earth does not exist, it would not make much sense to call the earth in Newtonian mechanics an idealization, would it? This bit of Laymon's reasoning seems harmless enough. However, scientific realism claims much more than that. If the earth exists, then there is no reason to think that molecules or electrons do not, for the same reason; and there is also no reason to think that frictionless objects are not real and there is no reason to think that the weak and the strong interactional forces do not actually operate among elementary particles. Can Laymon's argument be carried over to these cases?

To recapitulate: the earth exists independently, but what exists is the earth, not the Newtonian earth, which is nothing but an idealized copy, a fantasy, an idea, no more. But when we say that a frictionless object exists, what do we mean? Suppose we replicate Galileo's incline experiment, in which a block is sliding down an incline and then continues its motion on a level plane. We reduce the friction among all parts and every time we reduce it a little more, the block moves more smoothly and stops further away from the incline. When we conclude with an idealization that if the friction were completely removed, the block would continue to move with a constant speed forever, what do we claim to be real? It looks like we have to be saying that the frictionless block is real. In other words, the idealized block rather than the actual one is real. This is the sense in which we say that the law of inertia that operates when there is no friction is actually true, not approximately true.

However, that is not what the above extension of the earth/Newtonian earth case is telling us. If a frictionless object could be real, why could not a dimensionless molecule be real? But if the dimensionless molecules could be regarded as real, we arrive at a rather bizarre argument in accordance with Laymon's modal auxiliaries, which must go as follows. Because the van de Waals modification of the ideal gas model and of Boyd's law is empirically successful, the ideal gas is real and Boyle's law is a true law-statement! Obvious this is not right since van de Waals' "real" gas model is better than the ideal gas model and the van de Waals law of real gases is closer to the true law for gases. The way out is to put such idealized objects as dimensionless molecules with the Newtonian earth rather than with frictionless blocks. Hence, what are real are the molecules in real gases of which the dimensionless molecules in an ideal gas are but mental constructs. What should we say about electrons and other elementary particles? It is not clear at all whether Laymon's monotonicity or asymptoticity argument can work with them. It will depend on whether it is feasible to have future quantum theories of elementary particles that portray the particles "more realistically" in the sense that it is possible to view them as concrete particular or individual things existing in nature.

Here are a few things regarding the realism-antirealism debate that we can clarify already. The question in the realism/antirealism debate is about whether the postulated entities in scientific theories do exist in nature. Since the objects are inevitably represented with idealization, we need to be sure which objects we mean when we try to figure out whether to believe in realism or antirealism. Are we talking about the idealized objects or are we talking about the objects from which we have made an idealization? Two types of idealization need to be distinguished: (1) one type ignores what objects *could not exist* or be real *without* - such as size and gravity — and (2) the other ignores what objects *could exist* or be real *without* but do in fact possess — such as friction and electricity. This is another, or more general, way of characterizing the same two types of idealization I discussed from the beginning. For type (1), the realists and the antirealists should agree that the idealized objects could be the objects that exist because what exist are the objects from which idealization is made. As I argued from the beginning, I shall take antirealism to be a position that grants existence or reality to the obviously observable, such as the ordinary objects. For type (2), the dispute between realists and antirealists is on the question of whether postulated theoretical/unobservable entities or factors exist or are real in the way our theories describe them. With idealized entities, the realists cannot say that such entities exist exactly as they are described in the theory, and yet they still can say that they exist in the form when the idealization is removed. The antirealists of course want to deny (or at least be agnostic about) this, saying that those entities, *as they are described or as they are when the idealization is removed*, have no existence in reality. What exists for them are only the ordinary objects or the observables; the rest are mental constructs that theoreticians enlist to organize or "explain" the former.

The antirealists now have two arguments against realism in this context. First, given the reasons explained above, it seems that what the realists meant to argue, when they argue for the reality of any postulated objects, is that such objects really exist; they are not just a figment of our imagination. But these objects are highly idealized. Do the realist mean to argue that such idealized objects really exist? Do they believe that electrons and quarks as identical objects, for example, exist in reality? Or rather they actually meant to say that they don't believe that electrons — the stuff that are talked about in quantum electrodynamics (QED) — are real because they are idealized objects, rather they believe as real whatever out there of which electrons are an idealization? If so, we really do not know what is it that exists, given my earlier argument for the impossibility of knowing what conditions of idealization are taken to come up with the electrons described in QED. Without that possibility, our idea of that which the model of electrons represent is not clear and substantial enough to help a realist, the same way that Kant's granting the existence of the noumena doesn't make him a realist. The realists need to come up with a plausible way of fixing what it is of which electrons in QED are idealized copies.

A problem that subsumes the above problem is anticipated by Laymon himself as we mentioned in the beginning, and it is this: what could the realists say about those theoretical objects that were abandoned in the history of science, objects such as phlogistons (Laymon 1984, p. 119)? There could well be a theory of phlogiston that approbates approximation, and yet there are no such objects. Laymon's response to this problem on behalf of realism (one of the several he thinks are available to the realists) is simple and in my opinion effective. The approximation approbation scheme only works within a paradigm or regime; whenever there is paradigm or regime change and, for instance, the phlogiston theory of combustion is replaced by the oxygen theory, the latter is able to explain why the former is approximately true and capable of approbate approximation in a limited and out-of-date scope. The antirealists would have a hard time explaining why such an explanation is always available for the later theory that replaced an earlier one. Of course, the realists do not have the last word on this, since it is surely not true that the later theory of a certain phenomenon always explains why the earlier theory works. Can a later physical theory of lightening and thunder explain why an earlier anthropomorphic theory of them works? However, at least within the history of mature sciences, what Laymon identifies seems to be prevalent (see also Krajewski 1977; Newton-Smith 1981; Niiniluoto 1984, 1990).

Second, I think a real problem with Laymon's version of the IBE argument is this: the convergence of approximation approbation underdetermines which of many possible idealizations regarding a given target is the right, if approximation is understood in terms of factual closeness. To all practical purposes, there is no difference between the law of inertia given by Newton and an alternative law that says if all influences are removed, the object will move in a speed with a damping factor of e^{-30} (or a damping factor as small as one likes).

IV. Conclusion

What I have argued in this paper, in stark unqualified form, is the following: (1) Laymon's account is good in general in that it obtains a stable triangulation of approximation/idealization, confirmation, and scientific realism that is at least in the ball park. (2) To improve upon the account, especially to deal with the problems for its alleged support for realism, two different types of idealization must be clearly distinguished, namely, the idealization to get real but perhaps non-actual physical systems, and the idealization to get unreal systems. The distinction is as much a matter of attitude as is a matter of logic. The unreal systems that idealization creates are regarded as such by scientists and/or are physically impossible. (3) With the first type, confirmation is achieved when ideal conditions are asymptotically approached by experiments, and with the second type, confirmation is achieved when a relaxation of the ideal conditions produces an asymptotic approximation of predictions. (4) Laymon's IBE argument for realism is most problematic when it concerns unobservable entities, first because it is not clear how a relaxation of ideal conditions can be achieved and secondly because it is not clear which — the ideal objects or whatever the ideal objects are models for — are claimed to be real.

References

Barr, W. F. (1971). "A Syntactic and Semantic Analysis of Idealizations in Science." Philosophy of Science **38**: 258-272.

Barr, W. F. (1974). "A Pragmatic Analysis of Idealizations in Physics." Philosophy of Science **41**: 48-64.

Boyd, Richard (1983) On the Current Status of Scientific Realism. *Erkenntnis* **9**: 45-90

Brown, James Robert (1985) Explaining the Success of Science. Ratio **27**: 49-66

Cartwright, Nancy (1983) *How the Laws of Physics Lie* Oxford: Clarendon Press.

Hilpinen, R. (1976). Approximate Truth and Truthlikeness. Formal Methods in the Methodology of Empirical Sciences Eds. M. Przlecki, K.

Szanianwski and R. Wojicki. Dordrecht, Reidel, 19-42.

Jevons, W. S. (1874). The Principles of Science: A Treatise on Logic and Scientific Method. London, Macmillan.

Krajewski, W. (1977). Correspondence Principle and the Growth of Knowledge. Dordrecht, Reidel.

Laymon, R. (1980). Idealization, Explanation, and Confirmation. PSA 1980 Eds. P. D. Asquith and R. N. Giere. East Lansing, Philosophy of Science Association. **1**: 336-350.

Laymon, R. (1984/1982). The Path from Data to Theory. Scientific Realism Ed. J. Leplin. Berkeley, University of California Press (1984) 108-123. (The same article as Laymon 1982 in PSA 1982.)

Laymon, R. (1985). Idealizations and the Testing of Theories by Experimentation. Observation, Experiment and Hypothesis in Modern Physical Science Eds. P. Achinstein and O. Hannaway. Cambridge, MA, MIT Press. 147-173.

Laymon, R. (1989). "Cartwright and the Lying Law of Physics." Journal of Philosophy **86**: 353-372.

Liu, C. (2004). "Laws and Models in a Theory of Idealization," *Synthese* **138**: 363-385.

McMullin, E. (1985). "Galilean Idealization." Studies in the History and Philosophy of Science **16**: 247-273

Newton-Smith, W. H. (1981). The Rationality of Science. London, Routledge & Kegan Paul.

Niiniluoto, I. (1984). Is Science Progressive. Dordrecht, Reidel.

Niiniluoto, I. (1986). Theories, Approximations and Idealizations, in Logic, Methodology and Philosophy of Science Eds. R. B. Marcus and e. al. Amsterdam, North Holland. 255-289.

Niiniluoto, I.(1987). Truthlikeness. Dordrecht, Reidel.

Niiniluoto, I. (1990). Measuring the Success of Science. PSA 1990 Ed. M. F. A. Fine and L. Wessels. East Lansing, Philosophy of Science Association. **1**: 435-445.

Nowak, L. (1972). "Laws of Science, Theories, Measurement." Philosophy of Science **39**: 533-548.

Nowak, L. (1980). The Structure of Idealization: Towards a Systematic Interpretation of the Marxian Idea of Science. Dordrecht, Reidel.

Oddie, G. (1981). "Verisimilitude Reviewed." British Journal for the Philosophy of Science **32**: 237-265.

Oddie, G. (1986). Likeness to Truth. Dordrecht, Reidel.
Popper, K. (1976). "A Note on Verisimilitude." The British Journal for the Philosophy of Science **27**: 145-159.
Ramsey, J. (1990). Beyond Numerical and Causal Accuracy: Expanding the Set of Justificational Criteria. PSA 1990 vol. 1, Eds. M. Forbes and A. Fine. East Lansing, MI, Philosophy of Science Association. 485-499.
Ramsey, J. L. (1992). Towards an Expanded Epistemology for Approximations. PSA 1992 vol. 1, Eds. D. Hull, M. Forbes and K Okruhlik. East Lansing, Michigan, Philosophy of Science Association. 154-164.
Schwartz, R. J. (1978). "Idealization and Approximations in Physics." Philosophy of Science **45**: 595-603.
van Fraassen, B. (1980). *The Scientific Image.* Oxford: Clarendon Press.
Weston, T. (1987). "Approximate Truth." Journal of Philosophical Logic **16**: 203-227.
Weston, T. (1992). "Approximate Truth and Scientific Realism." Philosophy of Science **59**: 53-74.

CHUANG LIU

LAWS AND MODELS IN A THEORY OF IDEALIZATION

ABSTRACT. I first give a brief summary of a critique of the traditional theories of approximation and idealization; and after identifying one of the major roles of idealization as detaching component processes or systems from their joints, a detailed analysis is given of idealized laws – which are discoverable and/or applicable – in such processes and systems (i.e., idealized model systems). Then, I argue that dispositional properties should be regarded as admissible properties for laws and that such an inclusion supplies the much needed connection between idealized models and the laws they 'produce' or 'accommodate'. And I then argue that idealized law-statements so produced or accommodated in the models may be either true simpliciter or true approximately, but the latter is not because of the idealizations involved. I argue that the kind of limiting-case idealizations that produce approximate truth is best regarded as approximation; and finally I compare my theory with some existing theories of laws of nature.

> We seem to trace [in *King Lear*] . . . the tendency of imagination to analyse and abstract, to decompose human nature into its constituent factors, and then to construct beings in whom one or more of these factors is absent or atrophied or only incipient.
>
> A. C. Bradley, *Shakespearean Tragedy*, p. 264.

1. INTRODUCTION

The central challenge of modern science is to discover invariant laws of nature in a sea of ephemeral phenomena. I say 'modern science' because ancient astronomers, for instance, were not dealing with the ephemera among the starry spheres, and by 'invariant', instead of 'universal', I mean laws that apply in diverse circumstances. This is especially true ever since the kinetic theory of gas dawned the search for invariants in the murky realm of thermo-statistical phenomena. Just as experimental design or control is essential to experimental sciences, idealization – the counterpart of that – is indispensable for theory construction.

This paper may be seen as a continuation of the work done in (Liu 1999), where a critique of the traditional theories of approximation and idealization is given and a possible alternative suggested; a brief summary

Synthese **138:** 363–385, 2004.

of that critique is given in Section 2. And after identifying one of the major roles of idealization as detaching component processes or systems from their joints, a detailed analysis is given of idealized laws – which are discoverable and/or applicable – in such processes and systems (i.e., idealized model systems). Then, I argue that dispositional properties should be regarded as admissible properties for laws and that such an inclusion supplies the much needed connection between idealized models and the laws they 'produce' or 'accommodate'. And I then argue that idealized law-statements so produced or accommodated in the models may be either true simpliciter or true approximately, but the latter is not because of the idealizations involved. I argue that the kind of limiting-case idealizations that produce approximate truth is best regarded as approximation; and finally I compare my theory with some existing theories of laws of nature.

2. IDEALIZATION: A BRIEF SUMMARY (cf. Liu 1999)

Idealization usually refers to acts of theorizing by which *ideal-condition* claims are stipulated and *idealized claims* obtained as a result, where ideal-condition claims are of physically possible but non-actual conditions and idealized claims those that are true under the conditions. The existing theories (or conceptions) of idealization further take such acts to be ones which either *neglect the negligibles* – the ideal-condition claims are approximately true – or *approbate approximation* – when the ideal-condition claims are relaxed towards the truth, the law-statements become more and more approximately true. Here, p is *approximately true* if the possible world in which it is true is *factually close* to our world. The factual distance may be most simply defined as a metric in a predicate space that gives exhaustive description of all the possible relevant states of affairs (cf. Hilpinen 1976; Niiniluoto 1984, 1998).

To make the above idea a bit more precise (cf. Laymon 1980, 1985, 1989; Nowak 1972, 1980), let us suppose that we have a generic theory T, and a set of n ideal-condition claims: $I = \{I_i \mid i = 1, \ldots n\}$, such that I_j is less idealized (or more realistic) than I_i, if $j > i$. For all I_i's, we have I_i: $T \Rightarrow T_i$, and $T_I \mid= P_i$, where $P = \{P_i \mid i = 1, \ldots n\}$ is a set of predictions. Suppose again that the non-idealized theory, T^U, predicts the true P^U. And let $\Delta^- I = |I_j - I_i|$ and $\Delta^+ P = |P_j - P_i|$ (where $j > i$) denote the factual distance between two idealized conditions and two predicted states of affairs, we then have,

[A] T is a good idealized theory if it is always true that T: $\Delta^- I \Rightarrow \Delta^+ P$; otherwise, it is not a good idealized theory.

In other words, when $I \to 0$, we have $T \to T^U$ and $P \to P^U$, if T is a good idealized theory. This is how the measure of idealization is given by approximation approbation. Here the notion of factual distance, such as $\Delta^+ P = |P_j - P_i|$, is not something easily definable in general; but when dealing with quantitative statements about singular states of affairs, the distance is supposed to be (Niniiluoto 1984) the distance with an Euclidean metric.

The theory and its variations are not so much wrong as are too simplistic in their depiction of idealization. For some limited domain of applications, Laymon's theory is just right, but if one looks at the whole range of things for which idealizations are needed – a major portion of which are for discovering scientific laws via the construction of models – one should realize that approximation approbation does not even come close in covering them. To show this I here give one example (see Liu 1999 for extensive discussion).

Suppose that we are in Newtonian regime so that Newton's law of gravity is the law, i.e., $F = G(mM)/r^2$, for two objects with masses, m and M, and separated by a distance of r. Let us see which of the following is more approximate.

[B] Given I, $F = G(mM)/r^2 + a$;

and

[C] Given I', $F' = G(mM)/r^2 + b\sin(\omega r)$, where $|b| < |a|$ and ω is a frequency term.

If the factual closeness – approximation approbation – is the measure, [C] is a better fit than [B], especially if the frequency term, ω, is large. While [B] consistently misses the law by $|c|$, [C] not only misses less – $|b| < |a|$ – but also has frequent 'hits' on the correct values because of $\sin(\omega r)$. Hence, F' given I' is a better idealized law-statement than F given I. But we all know that [B] is a better 'approximation' *qua law* of Newton's law even though it is not so *qua facts*. This and many other examples show that the notion of a law-statement being close to but not the true law-statement has not one but two dimensions, namely, the closeness to the form of the law in question and the closeness to the singular facts which follow from the law.[1] They obviously do not always go together, and the evaluation of whether a law-statement is approximate may involve checking out the balance between the two dimensions. However, it is not clear at all how or whether a measure of such a balance can be defined in general for all approximate law hypotheses.

3. AN ANALYSIS OF IDEALIZED LAWS

Idealization, as mentioned earlier, is too complex a notion of theory construction to be captured in a single logical formulation – as in Nowak's – or by a simple connection to approximation production – as in Laymon's. To see a few of its diverse applications we note, first, that many acts of idealization are indeed for neglecting the negligibles; for instance, making $\sin\theta = \theta$, when the swing angle θ of a pendulum is very small, or taking some edge of a solid to be a continuous line.[2] Second, idealizations play an important role in experimental data collection and analysis (cf. Kyburg 1990, ch. 10). Here, the idealized curves drawn among the data points, for instance, are the result of combined considerations of approximation and optimal statistical regularities. The third kind is one of the most important categories – and the richest – of idealizations whose sole purpose is, to use a catchy phrase, *to carve nature at its joints*. What scientists would like to do with nature in their theories is what that fabled butcher in *Chuang Tzu* can do with an ox: he can cut up (all the joints of), and so kill, an ox without anybody around him realizing it. (If someone but touches the ox, it will then collapse with all of its parts in perfectly clean separation.) There is a fourth kind – the class of limiting-case idealizations, which I will discuss later (cf. Liu 2001).

Let me leave the metaphor and begin to sketch a theory for idealization. But first, it seems better to call those abstractions which do nothing but to neglect the negligibles acts of approximation rather than of idealization. They are in fact so called by scientists. Now, our conception of idealizations as acts of coming up with idealized models in which idealized laws can be discovered and/or applied requires that we analyze first the notion of idealized laws and their relationship to the models. Here we run into a dangerous philosophical territory, for there are many competing theories of laws of nature and even if they can all be fit into three classes – the (sophisticated) regularitarian, the necessitarian, and the universalist – a cursory but meaningful discussion of them is beyond this paper. Fortunately,[3] to understand the *nature* of idealized laws, it is not necessary to assume a particular conception of the nature of laws, which is in dispute among the above rival theories; instead, I only need to assume what is not controversial, namely, law-statements are such that if it is a law that all Fs are Gs then not only all Fs are Gs in the actual world but also all Fs in the naturally possible worlds[4] are Gs (which is needed for evaluating counterfactuals). The tough question asks: what is the nature of that relation between the Fs and Gs, to which a law-statement refers, whose obtaining entails the above? Is it a special kind of regularity relation between the actual Fs and Gs or a contingent relation between the two universals, F-ness and G-ness? We do not need to commit to one of these conceptions in order to investigate how laws in ideal circumstances should be understood. Similarly, I will use the possible-world language whenever notions of necessities need to be unpacked, but withhold any commitment to the reality of such objects.

First, we need to know what kind of properties are admissible relata of laws of nature. This, as we will see later, is crucial for the proper understanding of idealized laws. It seems plausible to think that which property is admissible is ultimately for science to determine – which however does not mean that it is only a matter of empirical findings. But given the kind of laws that are discovered and used in science today (cf. Armstrong 1983; Lewis 1983), we may obtain some plausible generalizations. Some (e.g., Armstrong and Lewis) want to rule out dispositions and only allow categorical properties for laws. The reason for this exclusion (including the exclusion of propensities) is multi-faceted. There is the Humean aversion to the notion of power, which – understood in a broad sense – is what all the dispositional properties have in common; the idea that laws are first and foremost true regularities, which implies that the admissible properties ought to be *occurrent* to the systems which possess them; and the idea that dispositions are dispensable because they supervene on categorical properties, for example, the fragility of a vase is completely determined when all of its structural properties are, which are categorical.

I find this restriction utterly indefensible and mainly have two arguments against it. First, the categorists seem to take the fundamental magnitudes – positions and momenta – to be categorical, which uniquely determine the states of a system of particles in classical mechanics. But are they? Could the instantaneous velocity of a non-uniformly moving particle be considered a categorical property? I do not think it can. What the particle has at any instant is precisely the *tendency* of moving with the velocity it possesses at that moment but not actually *moving* with that velocity. The latter is a categorical property which it does not have, and the former – which it does – is a *bona fide* disposition. Here, the categorists seem to have been saddled in a dilemma: either they admit that momenta are admissible properties for laws of physics, and hence unwittingly admit that categorism is false; or they regard momenta as either illegitimate for physics or dispensable because they supervene on positions. They cannot really grab the first horn, but they cannot grab the second either; for either to deny the fundamental role of momentum in the description of physical states or to think that momentum supervenes on position is to reject physics, at least its classical portion, altogether. One of the most basic tenets of physics is that a system's *degrees of freedom* are determined by the independent components of the positions and momenta of its constituents – e.g., the 6N phase space for a system of N particles. If momenta supervene on positions, the number of degrees should be cut in half. Theories, such as classical mechanics, may not be true, but they cannot be regarded as false in such an elementary way.

Second, there is a large class of lawlike hypotheses in exact sciences which relate nothing but dispositional predicates. That suggests that they are statements for dispositional laws – laws relating dispositional properties (or classes of individuals sharing such properties). For instance, all laws expressed in the form of differential equations are laws of this kind. One may argue that the *law* that a differential equation represents is not the relation among the differentials (or derivatives), but rather that which is represented by the solution – if the equation is solvable – under certain initial and/or boundary conditions, which is in most cases a curve or trajectory of some sort. And the solution is indeed an equation relating categorical predicates. But it is a mistake to take in general the solutions rather than the differential equation as representing the law. The law of motion in e.g., Newtonian mechanics is represented by the differential equation relating the second-order time-derivative of a position function and the first-order (partial) space-derivative of a potential, not the infinite number and varieties of trajectories as the solutions of the equation under all the possible initial/boundary conditions. The categorists cannot even claim in this case that the law of motion comprise the set of all possible solutions of the equation of motion, because then the law is neither definable nor knowable. Therefore, I shall admit dispositions (and propensities) in my ontology, which we shall see is important in helping us to resolve a difficulty in understanding the nature of some idealized laws.[5]

I now turn to the discussion of idealized laws. By the end of this analysis we shall see that the admission of dispositions and how dispositional laws (or law-statements whose relata are dispositional predicates) are discovered or applied in idealized models help us to resolve the difficulty of how idealized laws may be true not only in their models but true simpliciter. I take an idealized law to be a law that obtains under certain stipulated conditions, which may be expressed by a hypothetical claim, such as

(1) if D, then it is a law that all Fs are Gs,

where D refers to the ideal conditions and F and G are arbitrary predicates. One sometimes see idealized laws formulated as

(2) it is a law that given D, all Fs are Gs,

which is certainly not adequate. First, it is equivalent to 'it is a law that all Ds and Fs are Gs', which is in turn equivalent to 'it is a law that all Hs are

Gs', where H is a conjunctive property of D and F; this would trivialize idealized laws. Second, and more importantly, it seems obvious that when we want to treat a statement or an equation, L, which is only true under ideal conditions as a law, we mean that L is a law only under those ideal conditions; we do not mean that L together with the ideal conditions is law. This last would mean that in all nomologically possible worlds, L is only true in those ones in which the ideal conditions obtain, whereas (1) says that in the set of all possible worlds in which the ideal conditions hold, W_D, there is a subset in which L obtains. In other words, 'it is a law that all Fs are Gs' defines a nomic necessity relation only in W_D. If ideal conditions are understood as never obtaining in the actual world (as they are usually so understood), the actual world is a member of W_D; but if we understand the ideal conditions as obtainable in the actual world, the actual world should be included.

I hold the latter, for I think one should take (1) to be a 'hypothetical' rather than a mere material conditional or a counterfactual. It differs from a material conditional in that it is not automatically true whenever D is false, for D is false in most cases – that's why it is a set of ideal conditions, but it is not a simple counterfactual (e.g., if D were the case, then it would be a law that all Fs are Gs). One may find this last claim puzzling, for it is usually held that ideal conditions can never hold in the actual world. But I disagree; and I shall argue in detail in the next section that since idealization is a means for model-building and experiment-designing, it is better to understand D as hypothetical rather than simply false.

Moreover, one should also note that idealized laws are not the same as, or even a subset of, *ceteris paribus* laws (cf. Earman and Roberts 1999). Many (Cartwright 1989; Hausman 1992; Silverberg 1996) have included idealized laws in the category of *ceteris paribus* laws, thinking that the ideal conditions are the *ceteris* that must be *paribus* in order for the corresponding law to hold. Earman and Roberts disagreed mainly for two reasons: (i) while the *ceteris paribus* conditions in a hedged law must actually obtain sometimes, the ideal conditions are by conception never actually obtain; and (ii) while the *ceteris paribus* conditions are usually not precisely and completely statable, the ideal conditions usually are. Because of what I have just said – treating ideal conditions as hypothetical rather than non-actual better agrees with what scientists regard such conditions when they use them to build idealized models – I cannot accept (i). I however think that (ii) is correct, because holding a law *ceteris paribus*, such as one in social sciences, is precisely holding a law without an explicitly specified model, which is contrary to what happens with holding an idealized law. Some *ceteris paribus* laws, if there are any such laws, may involve idealizations; for instance, the ideal condition of having perfectly rational agents is perhaps assumed by many such laws in social sciences. But the *ceteris paribus* conditions are usually not such conditions. In other words, a law may at the same time be idealized and a *ceteris paribus* law, but its *ceteris paribus* conditions should not be regarded as its ideal conditions.

Dispositional laws are relations relating *exclusively* dispositions. One should note that the relations between a disposition and its categorical displays are not themselves laws. However, there are laws involved in such relations; but the laws involved must be laws relating categorical properties. For instance, the relations between being fragile and the various displays of such a disposition involve laws of the form: given a set of initial velocities of parts of the fragile object, a certain configuration of the parts results (which may be regarded as the object being shattered). These laws have no obvious connections to laws relating dispositions, such as it is a law that all industrious people are jealous (false, no doubt, but lawlike nonetheless).

Let us now see what idealized law hypotheses express. Without D, we have lawlike statements that tells us that in every naturally possible world all Fs are Gs. There are plenty of such statements which if true are laws. 'All human are mortal', 'all material objects have mass', and 'whenever an object is affected by a force with a nonzero net, its behavior is altered in some way', and . . . ; all these if true are said to express *exceptionless* or *covering laws*.[6] What then is the case for the following example:

(3) It is a law that all objects move uniformly with rectilinear trajectories?

The conventional wisdom says that it is not true in the actual world and therefore is not a law unless it is fitted with a clause which says 'given they are not affected by any non-zero external force'. Then it expresses an idealized law, a law that is only true with the provided condition (cf. Cartwright 1983, 1989; also Kyburg 1990, ch. 10)). Figure 1a gives a possible situation in which (3) is not true.[7] There, the inertial motion of an object, *o*, with initial velocity $\mathbf{v}_0$, is affected by a force, **f**, so that it ends up having Y (= a curved trajectory of such-and-such a shape). The following seems another example of the same kind:

(4) It is a law that whenever an object of a mass, m, is acted upon by a force, $\mathbf{f} = m\mathbf{g}$, it moves with an acceleration, $\mathbf{g}$.

Similarly, this is said to be true only of an object under the condition that no other forces affect it. Figure 1b gives a possible situation in which (4)

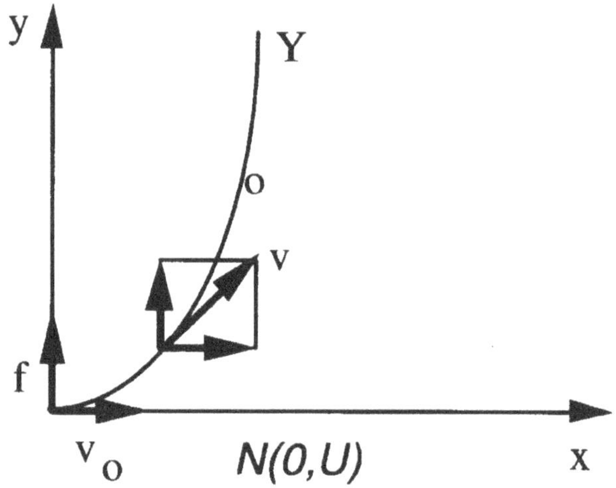

Figure 1a. A situation for (3), O = being an object with an initial velocity, $\mathbf{v}_0$, at $t = 0$; U = moving with that velocity along the x-axis.

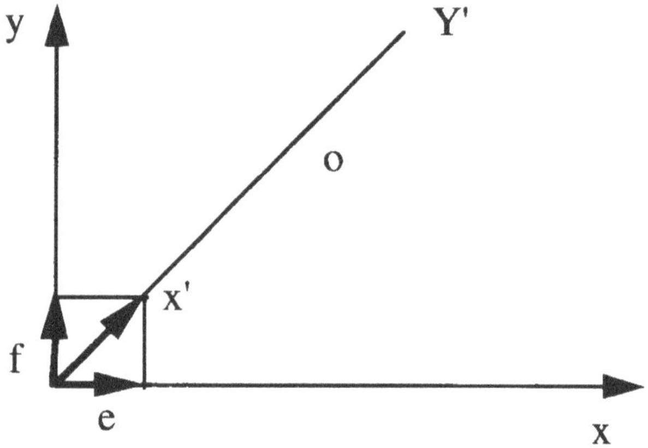

Figure 1b. A situation for (4).

is not true. When another force, $\mathbf{e} = (m/e)\mathbf{a}$, is also acting on o, it fails to move in the direction of **g**. Instead, it moves along the direction of **x** (or ends up having **Y**′).

Some, such as Cartwright (1983),[8] have argued that these, as idealized law-statements, cannot be true even though they may have enormous explanatory power, while many true law-statements – such as the covering laws given above – do not explain at all. And yet one cannot deny that our intuition strongly suggests that (4) can be true without the idealization while (3) perhaps cannot. Is such an intuition simply false? Now, why is (3) apparently false without the ideal condition? Because if so, an instance of

(3) as given in Figure 1a, i.e., $N(O, U)$, would be true, where O expresses the property of being an object with an initial velocity, $\mathbf{v}_0$, at $t = 0$, and U the property of moving with that velocity along the x-axis. Obviously, $N(O, U)$ does not appear to be true since any object possessing O in this case ends up having Y, namely, tracing out a curved trajectory as shown. One may say that for essentially the same reason (4) is also false in the situation of Figure 1b. This is how I understand Cartwright's view on such idealized laws.

However, there is a difference between

(5) the property of having described a trajectory of a certain shape,

and

(6) the property of tending to move at t with a $\mathbf{v}$ and/or $\mathbf{a}$ towards a certain direction.

(5) and its likes are full-bodied categorical properties while (6) and its likes can be seen as dispositions. If (4) is understood as a law which relates the actions of forces to dispositions of motion, it is possible to regard (4) as true in Figure 1b: at every moment of its motion, the object, being acted upon by **f** (and also by **e**), has the disposition of moving with **g**. It is only when one demands that the admissible properties, at least in such cases as regarding an object's motion, be categorical do we run into the problem, and such laws as (4) become false without the ideal conditions. Hence, whether one should agree with Cartwright on the truth of such laws depends on whether one admits dispositions as possible relata of laws (see also Hüttemann 1998).

To further clarify the situation, let us see another case of co-instantiated properties where it is clear that no relevant idealization clauses are needed. In Figure 2, we have a hot body moving inertially along the x-axis and radiating heat energy at the rate of **r**. Suppose that two laws govern the behavior of such objects: one is (3), and the other a law: $R = \mathbf{r}t$. Supposing again that uniform motion and heat radiation do not affect each other, we should have no qualm in asserting that the two laws are true in this situation and true with no exception. By the end of a time interval, the object describes a certain distance along x-axis and radiates a certain amount, R, of heat energy. If we compare this case with the case of (4) in Figure 1b, we could not help making the following observation. If there is a law in Figure 1b which is obviously true without exception among the relevant factors, it should be $N(X', Y')$ where X' = being acted upon be a force, $\mathbf{X}'$. For this law, whether one use Y' – which is categorical – or

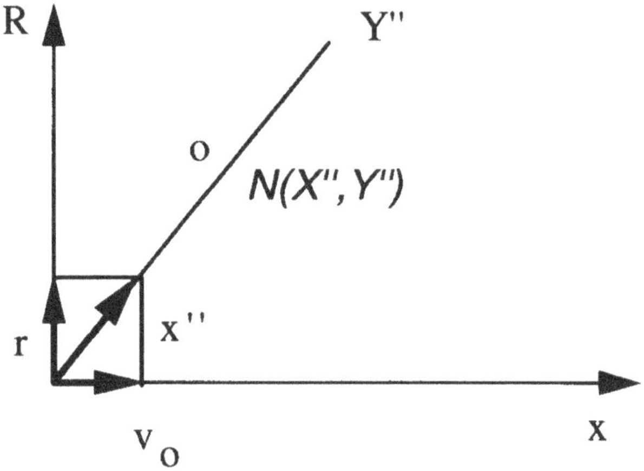

Figure 2. A hot body moving and radiating energy at the same time. X'' = moving with $\mathbf{v}_0$, at $t = 0$ *and* hot with $\mathbf{r}$; Y'' = having moved a distance of $\mathbf{v}_0 t$ *and* having radiated an amount of energy $R = \mathbf{r}t$, for some $t > 0$.

a dispositional equivalent, the law is true without exceptions. However, it seems odd to think that there also is a non-idealized law, $N(X'', Y'')$, in the case of Figure 2, where X'' = being moving with $\mathbf{v}_0$ and hot with $\mathbf{r}$ at $t = 0$, and Y'' = having moved a distance of $\mathbf{v}_0 t$ and radiated an amount of energy $R = \mathbf{r}t$, for some $t > 0$. One must realize that since $\mathbf{v}_0 t + \mathbf{r}t = (\mathbf{v}_0 + \mathbf{r})t = \mathbf{X}''t$, $N(X'', Y'')$ is a true generalization if each of the two statements of inertia and of radiation is separately. What worries us is the consequence of an unnecessary proliferation of laws: presumably any conjunctions of any number of co-instantiated (or co-instantiable) causal laws will give rise to new and distinct laws.

Let us now further analyze the three examples in terms of how different types of properties are co-instantiated and what results such co-instantiations produce.[9] In Figure 2, every moment since $t = 0$ sees the instantiation in the object, *o*, of the properties of being an object and of moving in its initial velocity and of being hot and of radiating energy. And any complex properties consisting some or all of these four properties are also instantiated in *o* at that moment. The two laws, one relating being an object and moving in its initial velocity and the other relating being hot and radiating heat energy, hold separately on *o*, because the co-instantiations of the antecedent properties – being an object and being hot – on *o* causes the co-instantiation of the consequent properties – moving at some constant velocity and radiating heat energy. We can generalize this result to all laws that relate compatible categorical properties and to any number of their co-instantiations; the laws, if they hold, hold without exception in all such circumstances. And for the discovery and/or application of such laws, idealization is rarely needed, even if approximation is unavoidable. Such laws include causal and non-causal laws; for the latter we have those synchronic laws that determine the states of a system.

In Figure 1b, every moment since $t = 0$ sees the instantiation in o of the properties of being affected by a force $\mathbf{X}'$ and of moving with an acceleration, $\mathbf{c} = \mathbf{g} + (e/m)\mathbf{e}$. Since these two properties may be regarded as complex, the two pairs of properties – being acted on by an $\mathbf{f}$ *and* by an $\mathbf{e}$, and moving with an acceleration of $\mathbf{g}$ *and* of $(e/m)\mathbf{e}$ – are also instantiated. Even though one cannot say that after time $t > 0$, o has in fact moved a distance of $\frac{1}{2}\mathbf{g}t^2$ *and* a distance of $(e/2m)\mathbf{e}t^2$, one can say that at each moment it moves with $\mathbf{g}$ *and* it moves with $(e/2m)\mathbf{e}$, because it is indeed true that at that moment it has the disposition of moving with $\mathbf{g}$ (or with $(e/m)\mathbf{e}$) regardless of where and how the other force is directing it to move. Although the final 'categorical' results over time cannot be a combined display that is *decomposible* into individual displays, the dispositions are uniquely and exactly decomposible at each moment. Therefore, if the two force laws are taken as laws relating dispositional properties, they are equally invariant laws, just as the ones in Figure 2, and are both true in the circumstance.

Here the relationship between accelerations (dispositional) and trajectories (categorical) is the typical relationship between a disposition and its *categorical display*. For instance, a fragile body – one that possesses the disposition of fragility – displays its fragility when it is shattered – possessing a set of categorical properties – after being subjected to a set of prescribed conditions. Similarly, an accelerated body displays its acceleration by describing over time a certain trajectory after being subjected to a set of initial and/or boundary conditions. In many cases, a co-instantiation of dispositional properties produces a 'combined' display that is not the simple sum of, or not decomposible into, the separate displays, each of which belongs to one of the co-instantiated disposition, such as the case in Figure 1b. However, the combined display are predictable, and even derivable, when rules or schemes are provided by the relevant theories. In mechanics, the combined effect which displays the co-action of forces is precisely derivable via the mathematical techniques of linear algebra.

This result can be generalized to all laws that causally relate dispositions, at least when the causes and their joint effects are unambiguously decomposible. In this respect, there is no good reason to believe that such laws are fundamentally different from the ones we discussed above. Idealization is essential in the discovery of such laws through decomposition. It is only when other laws and the properties they relate are 'idealized away' can we find out the desired law in its full display. More on this later.

In Figure 1a, every moment since $t = 0$ sees the instantiation on the object the properties of being a object that moves initially at $\mathbf{v}_0$ and of being acted upon by a force **f** and of moving in a velocity **v**. After some time $t > 0$ it also has the property of having moved a distance of Y. The motion of the object at each moment is completely determined in this case by two dispositions: the inertial tendency to moving with **v** and the tendency to accelerate with **g** (which changes **v** through time). With them are instantiated two laws: the law of inertia – as in (3)[10] – and the law of gravity – as in (4), and each is true regardless of the other: if **f** is removed at some moment, the object will move with **v** from that moment on in accordance with (3), and the same is true if $\mathbf{v}_0 = 0$, *mutatis mutandis*. This is an example of a kind in which co-instantiated properties change each other over time, so that even though there are unique decompositions into sets of the same *kind* of properties at each moment in the process, the magnitudes – i.e., the property variables – change over time.

Here we have the co-instantiation of two laws in which the co-instantiations of the respective dispositions produces not only a non-decomposible effect but also a change in some of the dispositions over time. However, the combined displays may still be predictable or derivable, even though the calculation may be more complex, such as the case of Figure 1a.

4. LIMITING-CASE IDEALIZATION AS APPROXIMATION

With the above analysis of idealized laws – or, better, laws under idealizations – we can now return to discuss the notion that idealization is to carve nature at its joints. At this point, one may have the impression that the laws I discussed in Section 3 and their likes exhaust the category of idealized laws. Given what I have said at the beginning of that section, this is far from the case. Moreover, there appear to be two other important classes of idealized laws that do not fit into what I have discussed so far.

There are those laws which are the results of idealized systems or their constituents. Examples include Boyle's law of ideal gas, mechanical laws of rigid bodies, hydrodynamic laws of continuous fluids, etc. (cf. Kyburg 1990, ch. 10). And there are those laws which are the results of idealizations on the domains (or regimes) of their application. Examples include all the non-quantum and/or non-relativistic laws which can be seen as the *limiting cases* of the quantum and/or relativistic laws; for instance, laws in Newtonian mechanics can be seen as idealized laws in the regime of

$v/c \approx 0$ and $\mathsf{h}/\epsilon \approx 0$, where v and ϵ are, respectively, the speed and energy of the system in question.

Despite their apparent differences and the fact that scientists regard them as of different classes, I believe they really belong to the same category. Is Boyle's law not the 'limiting case' of van der Waals's law, where the limit may be understood as $a \to 0$ and $b \to 0$ in the van der Waals equation: $(p + a/v^2)(v - b) = kT$, which results in the Boyle equation: $pv = kT$?[11] In practice, when the gas is diluted and at a high temperature, Boyle's law – just like Newton's law of motion in contrast to the law of motion in special relativity – is a good approximation. In both cases, the van der Waals equation and the special relativistic law of motion are 'truer', rather than true, law-statements. In general, it is plausible to understand the idealizations taken for such laws as taking 'limiting' values of the relevant (or salient) structural properties of the model systems in question. And so, we can also understand the idealized structures in a model as the 'limiting' regime in which a simplified but still approximate version of a true or truer law-statement holds.

In contrast to the kind of causal/dispositional laws discussed in the previous section, these idealized laws and their models do not seem to be the results of separation or disentanglement of processes or systems. Hence, the image of carving nature at its joints does not seem to apply to them. Also the idealized laws in these two classes are false without their idealized models. Further, there are no rules or schemes via which we can predict any *true* effects of any co-instantiations of such laws; whatever we can predict or derive from them must be at most approximately true.

As mentioned in the earlier part of Section 3, it seems better to call those idealizations which neglect the negligibles 'approximation'. And we now see that the limiting-case 'idealizations' are of the same kind as those and can also be seen as acts of neglecting the negligibles. It would bring considerable conceptual clarity and simplicity if we call all these cases 'approximations' rather than 'idealizations of a special kind', which would allow us to clearly separate them from the true type of idealizations that enact decompositions and help to discover or accommodate, *inter alia*, dispositional laws. In principle (or from a purely conceptual point of view), idealizations produce idealized laws and approximations limiting-case laws; but there is no reason to think that a particular law-statement cannot be both an idealized and a limiting-case law. In the practice of model-building in science, the two types of abstraction may very well be employed at the same time for the study of a phenomenon or in the discovery of a law. In order for Newton's law of gravity (not in the form it is used in Figure 1b, which is an even more simplified version of Newton's) to be true and fully displayed, an isolation of gravity from the action of other forces is not sufficient; one must also assume that the gravitational force in question is sufficiently weak, so that we are in a regime of 'weak-field' limit. Therefore, what we have in (4) is both idealized and approximate, while its counterpart in relativistic mechanics may be regarded only as idealized (at least, it is less approximate, if one worries about the quantum effect). Such a situation is in fact quite common in all branches of science, where most of the laws we now have are likely to be only approximately true; but one should note that what are responsible for the falsity of such laws are not the idealizations that disentangle or detach but approximations that set up the limiting-case regimes.

5. THE ROLE OF IDEALIZED MODELS

I now return to our discussion of the dispositional laws and their models. A large number of laws in science belong to the types of the two examples (i.e., (3) and (4)) discussed earlier. Unlike the approximation in the limiting cases, these laws are idealized not in the sense that they themselves are unreal or not true or only approximately true but in the sense that to display them in their pure form one usually needs to construct idealized models. These are certainly not *ceteris paribus* laws, if by such laws we mean laws that are not true except under the ideal conditions. In this respect, I disagree with Cartwright (1999, 25–26, 28–29) who consider such laws as typical *ceteris paribus* laws, namely, they are not true unless the *ceteris* are *paribus*. (In the next section, I shall explain where the disagreement really lies.) This is because, as argued above, the properties – dispositions in most cases – the laws relate are instantiated fully in those non-ideal circumstances and thus the laws obtain. What is not always true for such laws when they are co-instantiated with other such laws is that they can show themselves in their full categorical display. Therefore, *idealized models (or ideal conditions) are necessary for the laws to show their full categorical glory*. In a staged drama where imaginary situations 'occur', certain typical characters which daily drive all of us to commit certain acts show themselves in their 'true' – which means fully displayed – color. It is no coincidence that we usually only regard those essential dispositions of us, not our behavior, as our characters.

Modeling in science, whose essential means is idealization, does a similar job. The need for idealization and its difficulties increase when we go from cases as in Figure 2 to those as in Figure 1a. Little idealization if any is needed to discover the kind of laws operating in cases such as in Figure 2 because the antecedent and consequent universals, especially the latter, are separately instantiated categorical – and *separately observable* – properties. In other words, both the dispositions and their displays are decomposible in such cases. The situation is different in the case of Figure 1b and its likes. It would not be so absurd for someone to argue that the law operating in this situation is really not the one for **f** or for **e** but the one for **X**′; and one does not need any idealization – given these are the only two relevant forces – to discover this law. But one needs idealization to discover that in this case two laws are in operation and the result is a co-instantiation of the two (recall Galileo's effort in separating friction from natural motion). In other words, $N(X', Y')$ expresses a mere regularity between X' and Y' as a result of the combined operation of the two laws. One may be skeptical about this conclusion, but one can be so only at the risk of doubting what most scientists have been doing for centuries (at least since Galileo's time when he inaugurated the method of idealization (cf. McMullin 1985; Suárez 1999)). It is precisely this kind of 'disentanglement' of simple (or simpler) processes, or 'detachment' of simple (or simpler) parts, from actual phenomena that produces models in which invariant laws can be found, whose co-instantiations in different combinations produce the great variety of empirical or phenomenological regularities we see in the actual world. And if one wants to know whether certain law is indeed operating in a mixture, one imagines ideal conditions in which the law operates relatively alone – not that nothing else co-instantiate but rather that the situation be like in Figure 2.

Models are imagined in this fashion and experiments are often constructed to *approximate* the modeled circumstances. Here is the real connection between idealization and approximation. One must realize that not all good idealizations in science are directly approximatable; many have to be reduced via theoretical reasoning to some less abstract or more phenomenological situations where approximation in experiments is possible. We all know that whether a play reveals to us the true color of some characters does not depend on the verisimilitude of the circumstances designed to show them in the play ('actualistic' plays are usually not good plays). The merit of an idealization is rather measured in how well it offers the circumstances in which properties as admissible relata of laws are easily separated and identified.

In sum, both idealization and approximation involve acts of abstraction, where the former is necessary for the discovery and/or application of causal/dispositional laws and the latter for that of limiting-case laws; but the former is mainly for the discovery and display of a law, while the latter only provides the truth-conditions (or the truth-regimes). The two types of abstraction, which are indiscriminately called 'idealization' before, should therefore be regarded as two distinct kinds. Failure to make the distinction, I think, is responsible for the erroneous inference from the observation that all such idealized laws need idealized models for their discovery and/or application to that such laws cannot be true except in their models. If my arguments so far are sound, the inference goes through with approximation but not with idealization.

Is it still possible, if one insists, to conceive both types of idealized models as results of carving nature's joints? For the former, the carving consists of locating the places in nature that different component processes or systems join and detaching them into idealized models. As I explained earlier, some of such components may actually exist in isolation, but typically they do not, and hence the need for idealization. For the latter, the 'carving' might be understood as an act of isolating proper levels of stable phenomena. In thermodynamics and hydrodynamics, for instance, the stable macroscopic phenomena are shown to us by nature because of the rarity of large fluctuations. The idealization may be seen as an act that ignores the discrete nature of bulk matter in these disciplines and therefore theoretically insulates us from the microscopic world. And so does the main idealization for the ideal gas model – i.e., the sizeless and forceless molecules – for the diluted gases at high temperature. However, the notion of joints is no longer appropriate here and, because of it, the notion of recombination of different levels or domains would not make any sense.

6. OF LAWS, CAPACITIES AND DISPOSITIONS

In this section, I examine my proposal for understanding models and laws under idealizations in contrast to theories of laws and related notions by Armstrong and Lewis and by Cartwright. The fundamental, and perhaps the only relevant, difference between my conception of laws and that of Armstrong's and Lewis's is that I admit non-categorical properties whereas they do not. Although Armstrong and Lewis disagree on what laws of nature are – Armstrong takes them to be relations among universals and Lewis takes them to be the axioms of a system of regularities which has the best balance of strength and simplicity – they agree on the categoricity of admissible properties. Hence, I am not a Humean; but nor is Armstrong; however, he and I are non-Humean of different stripes.

Now, one may worry that by admitting dispositional properties, I am admitting properties that are not actually possessed by objects in question. One may even think that Armstrong's exclusion of dispositions from his universals is necessitated by his commitment to actualism (Armstrong 1983, 7–9), for universals must always be instantiated in actual individuals. Is my theory incompatible with the idea (a form of actualism) that all admissible properties for laws of nature must be possessed by the actual objects involved? The simple answer is 'no'. Dispositions (and propensities) – at least those ones which find frequent uses in physical sciences – can be just as actually possessed by an object as its categorical properties are. The fragility seems to be just as good an actual property for a vase to have (whenever it is indeed fragile) as the property of its actual shape or mass. One should not confuse the disposition itself with the categorical properties which constitute a display of it and which are connected to it by some laws under a set of circumstances. Such properties, which an object only has the tendency to possess, are indeed not actually possessed by the object; but the disposition – the tendency – the object undoubtedly does possess. Even if dispositions supervene on categorical properties, which I have argued cannot be true in general, the above remains true. (Since what a disposition may supervene on must be a set of categorical properties that an object actually possess, the object actually possesses the disposition precisely because it actually possesses those categorical properties.)[12]

Cartwright (1983, 1989, 1999) also rejects the Humeanism mentioned above (i.e., there are laws but not among dispositions) and argues that we should also, and more importantly, admit *capacities*. Capacities are not dispositions (or propensities), according to Cartwright (1999, 59, 64), because while dispositions 'are tied one-to-one to law-like regularities' (p. 59) or are 'identified with [some] particular manifestations' (p. 64), capacities are not. They are entirely general and 'can have endless manifestations of endless different varieties' (p. 64). And most important of all, they are not bound by laws of nature, which do bind dispositions. This distinction is crucially to Cartwright's conception of laws. For some reasons, which is not clear to me, Cartwright chooses to remain a categorist (á la Lewis) when laws and the admissible properties for their relata are concerned (cf. Cartwright 1999, 49–50), for she maintains that laws are regularities among occurrent (i.e., categorical) properties. This explains why she believes that most laws of nature are *ceteris paribus* laws in the strong sense that they are false when the *ceteris* are not *paribus*.[13] Moreover, when the *ceteris* are not *paribus* and the laws are thus inoperative, capacities take over, or fill out, what is previously thought the role for laws. So, what happens when a situation as in Figure 1b arises? The law of gravity is no longer true there because of the interference of the electric force, but gravity as a capacity is still in operation since the object is indeed affected by it, and the tendency to move along **g** – if it can be regarded as a capacity – is also in operation since the object does tend to move that way at every instant when it is affected by the gravity. But all these operations no longer obey the law of gravity!

What is it that connects the capacities so that they can be causally efficacious? Can such connections be anything but nomic (= lawlike) ones? They cannot be logical, nor can they be merely factual. In fact, I think Cartwright does consider them to be lawlike in some sense of the word, but not in the sense she is using: 'We may ask, though, "Are not the laws for describing the capacities themselves just further regularity statements?" But we know from chapters 3 and 4 that the answer to this is "no" ' (1999, 142). This may be right, for it is perhaps not defensible to regard laws relating capacities and dispositions as regularities; but if what I have argued so far is correct, there is no reason to so narrowly construe laws as regularities among occurrent properties.

I do not pretend to have completely understood Cartwright's notion of capacity; but it seems obvious that they are dispositional rather than categorical properties. It also seems that capacities are essential to whatever possesses them – i.e., they are the natures of things – and are highly abstract or general. But if they are, as Cartwright demanded, not to be related by laws of nature, they are not the type of dispositional properties that are admissible to my ontology. But since Cartwright's notion of laws, as mentioned above, is such a restricted one, not relatable by laws in Cartwright's terms does not mean the same as not relatable by laws in my terms.

Therefore, when I open the door to let in dispositions and propensities for laws, there is no sound general philosophical reasons for barring laws among capacities. There may or may not be good empirical reasons – in terms of scientific practice – to include laws of capacities, but I do not see how one could reject them on purely metaphysical grounds while accepting dispositional laws.

Therefore, I disagree with Cartwright's theory of idealized laws – to which I obviously owes a great deal – on two major points: a whole class of idealized laws are true, not true *ceteris paribus*, and laws connect (admissible) categorical as well as dispositional properties, perhaps even capacities which I see as a special type of dispositional properties.

7. CONCLUSION

Let me now give a summary of the main features of this theory of idealization(/approximation) which articulate a relationship between laws and the idealized models in which they are true and/or displayed. There are two main types of idealized models as results of two different procedures: one consists of models, as results of approximation, which insulate levels or specify regimes in which simplified laws (or theories) are true as limiting cases of the true or truer laws (or theories), and the other consists of models, as results of idealization, which disentangle or detach component processes or systems in which (true) laws – mostly dispositional ones – operate in their full display. A law-statement may be the result of a combined use of these two types of models, and if it is false (but approximately true) outside the models, it is because of the use of a model of the former type, not one of the latter.

Why do we need idealization in the sense of carving nature at its joints? Few, if any, genuine properties are instantiated (or exhibited) alone or in separation in our complex phenomenal world – where the meaning of 'instantiation in separation' is exemplified by the case in Figure 2 – and some are never so instantiated. To discover them and the laws relating them for the construction of scientific theories, one has to separate them, which means that one has to stipulate conditions under which they are displayed purely and simply. This is the work of idealization. Moreover, nature's joints are not always sharp: some properties may overlap in the sense, for instance, the extensions of 'cat' and 'dog' may not be completely clear of each other. Therefore, '*carving*', as one of the chief tasks of science, is to disentangle and set boundaries, which yields properties that are candidates for the relata of laws of nature.

Does not this talking of carving up nature (rather than finding out what nature itself is) commit me to the course of social constructionism? No, not really. I do believe that there are constructional elements in our understanding of nature (i.e., in our *true* theories). The reason is simply this: idealization is indispensable in theoretical science and the result of idealization is the construction of models that may not obtain in nature. Theoretical arguments and experimental devices are often used to *approximate* the idealized models, but the models may never be literally realized. Hence, science as giving a description of what nature itself is, taken literally, is a false image. On the other hand, I am by no means forced by this view of idealization to regard scientific theories and their truth as social constructions. In fact, the very fact that I take approximation and idealization seriously excludes me from such a view: the very idea of social constructionism, however one finesses it, excludes the notion of approximation and idealization. If it is fundamentally up to the scientific community to construct a coherent and practical system of theories which does not have to answer to any specific features of reality, why should, and how could, there be anything to approximate or to be idealized away?

ACKNOWLEDGEMENTS

I would like to thank Jossi Berkowitz, Roman Frigg, Carl Hoefer, Michael Redhead, Julian Reiss, and Sang Wook Yi, for questions and comments during and after my Popper Seminar at the London School of Economics. I want to give special thanks to Carl Hoefer, Roman Frigg, Peter Lewis, and Michael Redhead for reading the penultimate draft and giving me detailed comments. I also thank the two anonymous referees for their constructive criticisms.

NOTES

1 In this paper, I shall sometimes use the term 'law' to refer to both the law and the statement that represents the law when there is no danger of confusing about which is meant in the context.

2 Note that in the former we subtract something, i.e., $\delta = \sin\theta - \theta$, while in the latter we actually *add* something to the edge to make it continuous; but for convenience I shall call both types of idealization neglecting the negligibles since adding things to fill up the vacancies can be seen as acts of removing the vacancies.

3 I am in favor of the universalist theory of laws of nature, but I am aware that it has serious problems; however, the problems do not affect my discussion here because I do not have to assume a universalist conception of laws.

4 If one is a physicalist, then natural necessity is the same as physical necessity; but if one is not, the natural may be broader than the physical. Psychological laws, if there are any, are of natural, but not of physical, necessity.

5 See Mellor (1971, 1974) for arguments on the legitimacy of dispositional properties in laws of nature. Mellor maintains that all properties are dispositional, a view I do not share. I cannot see how one can make sense of the display of a dispositional property if one does not also have categorical properties. But, as Mellor rightly pointed out, dispositions can be possessed actually and nomic connections among them are no different from ordinary nomic connections, and dispositions (and propensities) may change through time in accordance with the laws which bind them (without ever being displayed), just like categorical properties do.

6 Cartwright (1983) has argued for the scarcity of covering laws, which I think is not quite true. It is perhaps true that covering laws are usually too general, as our examples seem to show, to be useful in scientific explanation; but it certainly not true that they are scarce.

7 Here, the law of inertia in (3) should be understood in its more precise formulation: it is a law that for any object, if it is initially in a state of motion with $\mathbf{v}_0$, it will remain in that state.

8 For criticisms of this view of Cartwright's, see Creary (1981), Kline and Matheson (1986), and also Laymon (1989); and for Cartwright's response, see Cartwright (1983, 54–73).

9 My view of idealized laws of the kind discussed so far is in many respects inspired by Mill's view of laws and causation (cf. Mill 1874, 229ff, 237ff, 266ff, 317ff). A discussion of the relation between Mill's view and mine calls for another occasion. It suffices here to

point out that my view is by no means the same as Mill's. As a starter, I do not subscribe to Mill's idea of what a law of nature is. There are serious problems with reconciling Mill's professed (sophisticated) regularitist conception of laws of nature and his use of dispositions and tendencies in his theory of causal laws. This and other complications in Mill's theory of laws will be examined in another place.

[10] I may be charged here with a gross equivocation between the law of inertia which governs objects that move under no influence and an intrinsic disposition – inertia – that all massed objects possess in resisting any changes of their states of motion. This is an important point, which is what makes me uncertain of how exactly to categorize laws like (3). On the one hand, it is a law concerning the universal intrinsic disposition of change resistance that all massed objects have; but on the other hand, the traditional formulation of the law is about objects' moving uniformly and rectilinearly unless being prevented from doing so. The latter cannot be true without the 'unless . . . ' clause, while the former is true at every moment of every massed object with no conditions required. Since there are not many law-statements of this kind in scientific theories, I shall leave the analysis of this case as is.

[11] Here, we have the following: p is pressure, v volume per molecule, T temperature, k the Boltzmann constant, and a, b are the two correctional constants, whose values are determined by experiments.

[12] One may still wonder whether there is a problem of admitting all dispositions and insisting that an object actually possesses all of its dispositions. A vase may be actually fragile, but can we say that it is also actually cruel, given that it is not physically impossible to become a person of that character? I do not know how to judge whether a modal property, such as cruelty of a fragile vase, can be properly regarded as the vase's disposition; but I do not see any problem of regarding all such properties as admissible relata of laws. After all, it is not up to us to say which related pairs of dispositions are in laws and which are not.

[13] I have explained in Section 3 my view on the difference between *ceteris paribus* laws and idealized laws. Here I follow Cartwright's use of the former as closely as possible.

REFERENCES

Armstrong, D. M.: 1978, *Universals and Scientific Realism, vol. 2: A Theory of Universals*, Cambridge University Press, Cambridge.

Armstrong, D. M.: 1983, *What Is a Law of Nature?* Cambridge University Press, Cambridge.

Cartwright, N.: 1983, *How the Laws of Physics Lie*, Clarendon Press, Oxford.

Cartwright, N.: 1989, *Nature's Capacities and Their Measurement*, Clarendon Press, Oxford.

Cartwright, N.: 1999, *The Dappled World: A Study of the Boundaries of Science*, Cambridge University Press, Cambridge.

Creary, L. G.: 1981, 'Causal Explanation and the Reality of Natural Component Forces', *Pacific Philosophical Quarterly* **62**, 148–157.

Earman, J. and J. Roberts: 1999, '*Ceteris Paribus*, There is No Problem of Provisos', *Synthese* **118**, 439–478.

Hilpinen, R.: 1976, 'Approximate Truth and Truthlikeness', in M. Przlecki, K. Szaniawski and R. Wojicki (eds.), *Formal Methods in the Methodology of Empirical Sciences*, Reidel, Dordrecht, pp. 19–42.
Hüttemann, A.: 1998, 'Laws and Dispositions', *Philosophy of Science* **65**, 121–135.
Kline, D. A. and C. A. Matheson: 1986, 'How the Laws of Physics Don't Even Fib', in P. D. Asquith et al. (eds.), *PSA 1986*, Philosophy of Science Association, East Lansing, **1**, 33–41.
Kyburg, H. E. Jr.: 1990, *Science and Reason*, Oxford University, Oxford.
Laymon, R.: 1980, 'Idealization, Explanation, and Confirmation', in P. D. Asquith and R. N. Giere (eds.), *PSA 1980*, Philosophy of Science Association, East Lansing, **1**, 336–350.
Laymon, R.: 1985, 'Idealizations and the Testing of Theories by Experimentation', in P. Achinstein and O. Hannaway (eds.), *Observation, Experiment and Hypothesis in Modern Physical Science*, MIT Press, Cambridge, MA, pp. 147–173.
Laymon, R.: 1989, 'Cartwright and the Lying Law of Physics', *Journal of Philosophy* **86**, 353–372.
Lewis, D.: 1983, 'New Work for a Theory of Universals', *Australasian Journal of Philosophy* **61**, 343–377.
Liu, C.: 1999, 'Approximation, Idealization, and Laws of Nature', *Synthese* **118**, 229–256.
Liu, C.: 2001, 'Infinite Systems in SM Explanations: Thermodynamic Limit, Renormalization (semi-) Groups, and Irreversibility', *Philosophy of Science* **68**(S), S325–344.
McMullin, E.: 1985, 'Galilean Idealization', *Studies in the History and Philosophy of Science* **16**, 247–273.
Mellor, H.: 1971, *The Matter of Chance*, Cambridge University Press, Cambridge.
Mellor, H.: 1974, 'In Defense of Dispositions', *The Philosophical Review* **83**, 157–181.
Mellor, H. and A. Oliver: 1997, *Properties.* (Oxford Readings in Philosophy), Oxford University Press, Oxford.
Mill, J. S.: 1874, *A System of Logic*, Harper & Brothers, New York.
Niiniluoto, I.: 1984, *Is Science Progressive*, Reidel, Dordrecht.
Niiniluoto, I.: 1998, 'Verisimilitude: The Third Period', *The British Journal for the Philosophy of Science* **49**, 1–29.
Nowak, L.: 1972, 'Laws of Science, Theories, Measurement', *Philosophy of Science* **39**, 533–548.
Nowak, L.: 1980, *The Structure of Idealization: Towards a Systematic Interpretation of the Marxian Idea of Science*, Reidel, Dordrecht.
Suárez, M.: 1999, 'The Role of Models in the Application of Scientific Theories: Epistemological Implications', in M. S. Morgan and M. Morrison (eds.). *Models as Mediators: Perspectives on Natural and Social Science*, Cambridge University Press, Cambridge, pp. 168–196.

Approximation and Its Measures

I

I shall at the outset distinguish a metaphysical notion of approximation from an epistemological one. The metaphysical question is; given a true (or theoretical) value or statement, is the found (e.g. observed, measured, or estimated) value or statement approximate or approximately true? And the epistemic question is: if the true value or statement is unknown, what does it mean to say we are justified to call a found value or statement approximate or approximately true? And if we are justified to call a found value approximate, what does it tell us about where the true value lies?

II

What do people usually mean in practice (especially in science) when they use the notion of approximation? In daily usage, when we have good reason to say, 'that bridge is approximately 30m long,' we mean that we do not know the bridge's true length, but we are confident that it is close to 30m. Or we mean if we know its true length, r*, r (= 30m) is close to r*. Statistical claims often go as follows: when a pollster says that a fair sample yields a probability, $p = k$. with a margin of error, $\varepsilon > 0$ — i.e. p is approximately k — he means that he does not know the true probability in the population, $k*$, but he is 95% confident that it lies in $[|k|\text{-}\varepsilon, |k|+ \varepsilon]$. The same, *mutatis mutandis*, is true with the mean value of repeated measurements. We often use polynomials to approximate functions (by interpolation). Smooth functions can always be approximated by a polynomial of n degree. The larger the n, the better the approximation. Note, the metaphysical question is almost never asked. What is required instead is to figure out how, when the true value is unknown, we know whether a found value is approximate and what that implies. In most cases, the true value is in the neighborhood of a found value we are justified to call approximate.

III

Here is a very brief summary of the existing conceptions of approximate truth.

1. Possible-world conception (Helpinen, Weston, Smith)

> [P] A claim φ is approximately true in world *w* if, and only if, there is a world *w′* which is *close* to *w* and in which φ is true.

The closeness of worlds: the notion is never well defined. One can use Lewis's theory (Helpinen, Smith) of comparative overall similarity among possible worlds. Or one can construct a descriptive measure (Weston) for the closeness of worlds.

2. Descriptional conception (Niiniluoto, Marquis, Oddie, Kyburg)

Suppose we have an affine metric space *X* that represents all possible propositions about a certain subject, among which only one proposition $p*$ is true. And suppose we can form the quotient space $T = X/\approx$, which is an ordered topological space, such that all the points in *X* that are of equal distance from $p*$ are mapped to one point in *T*, and the point in *T* mapped from $p*$ is *1* and the point mapped from all the propositions that are at the limit of approximation is *0*. The subspace *[0, 1]* in *T* is the (partial) truth space for approximation.

IV

First the *de dicto* versus the *de re* conception of approximate true.

1. The *De dicto* case:

> [D] 'a is P' is approximately true.

2. The *De re* case:

   [R] a is approximately P.

In [D] the sentence, 'a is P,' is false but close to being true, while in [R], a is not really P but close to being P, and if so, 'a is approximately P' is true, otherwise, it is false. To think of approximation as *de dicto* is to think of modifying the concept of truth or introducing the notion of partial truth, while to think of it as *de re* is to think of modifying quantities — to make them have different than true values — that are otherwise attributed to entities simplicitus. I propose that we switch from the *de dicto* to the *de re* approximation.

But first, let us see how a similar proposal fails. Peter Smith (1998a, b) proposes that we translate

[D'] '*s*' is approximately true,

to

[R'] Approximately *s*;

and argued that

[*] '*s*' is approximately true only if approximately *s*.

This proposal borrows from the Tarski theory of truth, where we have

[**] '*s*' is true if, and only if, *s*.

Here, '*s*' is a sentence, such as 'a is P,' while *s* is the fact, e.g. a is P (P is, e.g. some property or value of a quantity). But if this interpretation is maintained, what could 'approximately *s*' mean in [*]? An approximate fact? But how can a fact be approximate? Moreover, the switch from [D'] to [R'] has little effect on the theory and measure of approximation. One is still dealing with a modification of either the truth or the fact. Contrarily, the switch from [D] to [R] has great effect: it changes from a modification of a single and universal notion, truth, to a modification of the attribution of properties of all kinds. If one adopts [D], as the people in **III** do, one must first come up with a space for all propositions in question and then find a translation of it to a (partial) truth-space. Even if the truth-space is simple, the proposition space and the translation map are not. But if one adopts [R], as I do, one can develop different approximation manifolds or spaces for different types of properties and their individual measures.

V

Here are some examples of approximation manifolds (henceforth, AM).

1. A single classical particle on one dimension.

Quantities: position, q, and momentum, p. The AM, E, for the states of a particle a is the phase space with the true value $s^* = (q^*, p^*)$ and the limit of approximation, δ, such that 'a is approximately $s' = (q', p')$' is true if

$$d = \sqrt{(q' - q^*)^2 + (p' - p^*)^2} \leq \delta;$$

and false, otherwise (see Fig. 1). With *de re* approximation, the metric for p is independent of, and may be different from, the metric for q (see Fig. 4).

2. A multi-particle system.

N particle system has a $6N$ dimensional phase space with $3N$ positions and $3N$ momenta. If the true state is given by s^*, a $6N$-tuple of p^*s and q^*s, then 'the system is approximately in state s''' is true if

$$d_{6N} \leq \delta_{6N};$$

and false, otherwise, where δ_{6N} is the limit of approximation which bounds the $6N$ dimensional open ball that defines the region of approximation (see Fig. 1).

3. Quantities of physical dimensions, quantities of matter and energy, rates of motion, etc. can be similarly dealt with by constructing the corresponding AMs, which are all Euclidean spaces with standard metrics and limits of approximation.

4. Quantities of intensities, such as shades of color, can be translated into spectra or frequencies, which have similar AMs.

VI

Let's now look at some problems with the Euclidean AMs.

(a) The AM is not Euclidean in the neighborhood of 'limit' laws, such as the speed of light and the absolute zero temperature (see Fig. 2).

(b) The AM is not Euclidean in the neighborhood of any (or most) laws with a given set of i/b conditions (more on this in a moment) (see Fig. 3). What should the AMs be in these and other cases? Figure 4 shows us what the manifolds for (a) should look like (roughly); and Figure 5 for (b).

However, there is a simple objection to such an approach, namely, AMs should always be Euclidean. What limit laws or laws in general do is to tell us whether it is possible, or how likely it is, to obtain a value in a region of an AM. But possibilities or probabilities are irrelevant to the measures of approximation. Here are four rejoinders. (i) Knowing the limit, how can a value over the limit be said to be approximate? Given that Einstein's light postulate is true, does it make sense to say that a particle is moving at a speed of approximately *v*, which is only slightly greater than the speed of light? (ii) Even within the limit, the asymmetry seems true. Suppose it's a law: $\forall h[H(h) < 3m]$ and for most *h*, $H(h) < 2m$ (where $H(h)$ is the height of any human). Let $H(p) = 2.4m$. It would be unreasonable to say that '$H(p) =_{app} 2.9m$' is as good as '$H(p) =_{app} 1.9m$'. The former is a bad approximation even if the latter a good one. (iii) If we take the epistemic notion into account (see **I**), the following seems correct: when we have high confidence to say that *Q* is approximately *r*, we mean that we are pretty sure that the true value of *Q* lies somewhere in a neighborhood of *r* defined by, e.g. δ. But the neighborhood cannot be a Euclidean ball of δ, if part of the ball lies beyond a limit law (see Fig. 4). Because it seems absurd to say that we have equal confidence that the true value of *Q* lies at a equal distance on both sides. The same seems true, *mutatis mutandis*, for laws in general (see Fig. 5). (iv) Taking the possible-world conception (**III**.1), we should say the following. Suppose that *Q* is approximately *r* in *w** (in which *Q* is *r**). There must be a possible world *w* in which *Q* is *r* and which is close to *w**. But if the value, *r*, lies beyond a limit law, there is no sense in which *w* can be said to be close to *w**. The same seems true, *mutatis mutandis*, for laws in general (more on this later).

VII

Now, more possible objections and my responses. To (i) above, one can simply say: were the value beyond the limit law found, it would be equally approximate. But consider the possible-world version of this (as in (iv) above). If w is the world in which the value beyond the limit law is found, it must be a possible world that is radically different from $w*$. To (ii), it may be true that knowing $\forall h[H(h) < 3m]$ and for most *h's*, $H(h) < 2m$, and seeing p standing in front of us, we have good reason to say that $H(p)$ is approximately $1.9m$ rather than $2.9m$. But if the found value of $H(p)$ is $2.9m$, there is no reason to believe that it is not approximate while $1.9m$ is. This may be right if only the metaphysical notion is considered. But we should also include the epistemic notion. If we ask the epistemic question, then $1.9m$ is, while $2.9m$ is not, an approximate value for $H(p)$. Moreover, this may be right if approximation is regarded *de dicto*: both propositions: $H(p)$ is $2.9m$ and $H(p)$ is $1.9m$ should mapped to the same point in T (see **III**.2). But when regarded as *de re*, there is no reason to think that the predicate 'being $2.9m$' is equally attributable with approximation to p as 'being $1.9m$'. To (iii) one may object that there is an equivocation (or a mistake). If we have that $Q = r*$ and a found value r, such that $|r - r*| < \delta$, is approximately Q, we can't say that we are justified to say that any value r', such that $|r' - r| = |r - r*|$, can be regarded as the true value ($r*$) (see Fig. 6). This is correct, again, on the metaphysical notion: if $r' \neq r*$, $|r' - r| = |r - r*|$ is meaningless. But from a epistemic perspective we can say the following: if $r*$ is unknown and the metric is truly Euclidean and we only know that r is such that $|r - r*| < \delta$, then we are entitled to infer that r may be any point on the ball with a radius of $d = |r - r*|$. If so, we are justified to think that if Q is approximately r, $r*$ may lie anywhere in the neighborhood of r with radius d. And hence, what is said in (iii) is not incorrect. Finally, even if it holds for limit laws, what is said in (iii) and (iv) does not apply to laws in general. If the law is $pV = RT$ (R - constant), the true value of $p*$ given $V*$ and $T*$ must lie on the curve; and how approximate another value is depends solely on its distance from $p*$. It will not depend on how likely it is found in $p*$'s neighborhood. Right, if the law is deterministic, because no true values can lie off the law-curve, given the i/b conditions. And hence, the likelihood of the true value lying off the curve when a found value is justifiably said to be approximate is zero. Therefore, we may conclude that when a law is deterministic, there is no probability distribution in its neighborhood that makes the metric of the AM non-Euclidean. However, if the law is probabilistic or statistical, there must be such a distribution and the AM in its neighborhood is non-Euclidean. Suppose in the example in **VI** (ii), there is no limit law; but it is a statistical law that for most *h*, *H(h) = 1.75m*. It seems obvious that *being 1.72m* is more approximate than *being 1.68m* for *H(p)*, given *H(p) = 1.70m* (see Fig. 5). The same appears also true for empirical generalizations.

To summarize, the simple metaphysical question is not interesting nor does its answer clarify the conception of approximation that is commonly understood when the term is used (see **II**). When the epistemic notion of approximation is included, we need to consider the effect of laws and the non-Euclidean nature of the AMs.

VIII

What does the measure of approximation depend on?

> [M] The measure depends on the distance and orientation of a found value from the true value in the AM and the likelihood of that value being the true value.

In other words, the measure of approximation is a weighted distance. And the likelihood of a found value being the true value in a given AM is determined by a variety of factors; among them laws (including limit laws), empirical generalizations, statistical means, etc.

Now some remarks on [M].

1. Given [M], if Q has values in $[r_0, r]$ and $Q = r*$ $(r_0 \leq r^* \leq r)$, then it's possible that '$Q =_{app} r_1$' is true and '$Q =_{app} r_2$' is false, while $r_1 \neq r_2$ and $|r_2 - r^*| = |r_1 - r^*|$. Or r_1 is a more approximate value than r_2 of Q.

2. Given [M] and a deterministic law, *L*, the likelihood of a found value off the L-curve being a true value is *0*; hence, the AM is Euclidean.

3. Given [M] and a probabilistic (or statistical) law, *P*, the likelihood of the true value lying on the P-curve is the probability given by *P*. One can

calculate the approximation of a found value near *P* as the distance from it to the true value weighted with the probability given by *P*.

4. Given [M] and an empirical generalization, *G*, the likelihood of the true value lying on the G-curve is the probability given by the confidence level within the error term. One can calculate the approximation of a found value near *G* as the distance from it to the true value, which is limited by the error term and weighted by the confidence level.

5. [M] divides all measures of approximation into two kinds: if a found value is either theoretically determined to be true or false or entirely 'free' — not predicated by any means — the AM in the relevant neighborhood is Euclidean; but if there is a likelihood attached to it, the AM is non-Euclidean and the approximation metric is weighted by that likelihood. At this point, I leave the question of whether the likelihood is epistemic or ontic open.

6. More needs to be said to the question: what happens when a deterministic law (with certain i/b conditions) produces a 'trajectory' that goes through the neighborhood of a limit law?

Figures

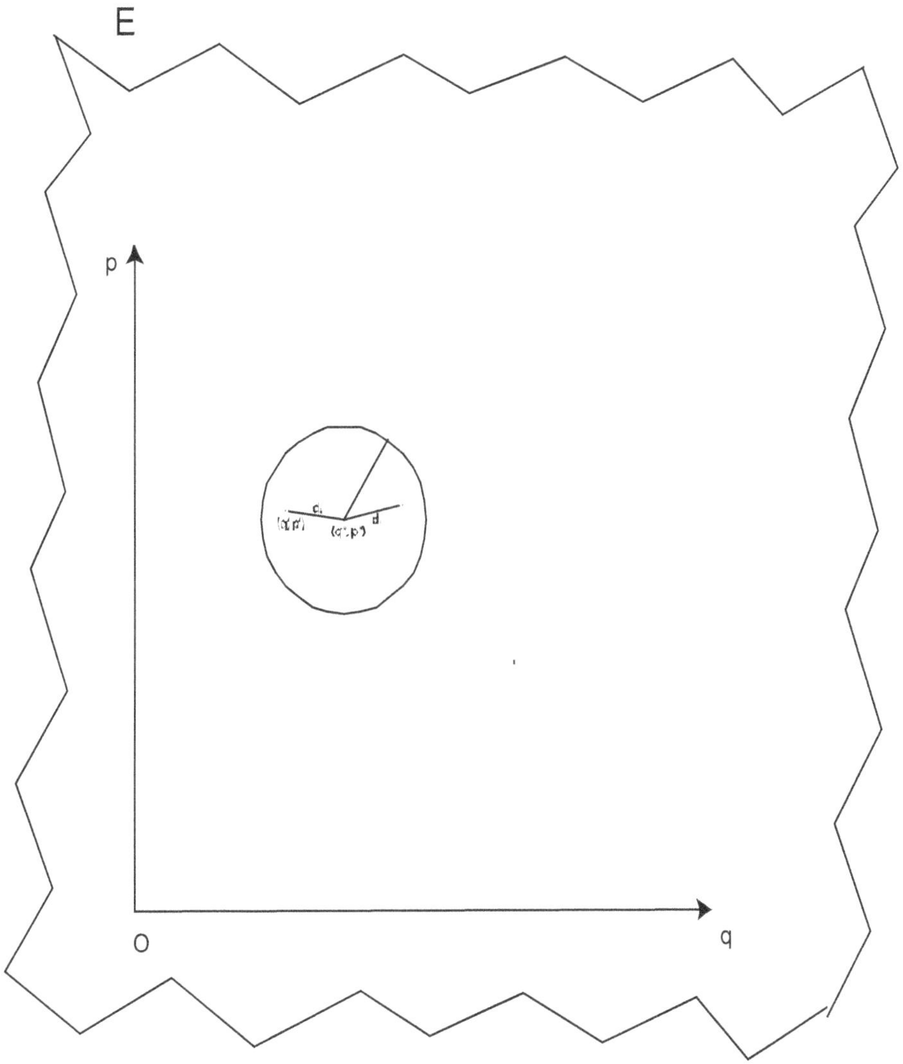

Figure 1

(The circle inside (in all figures) circumscribes the limit of approximation with a radius δ)

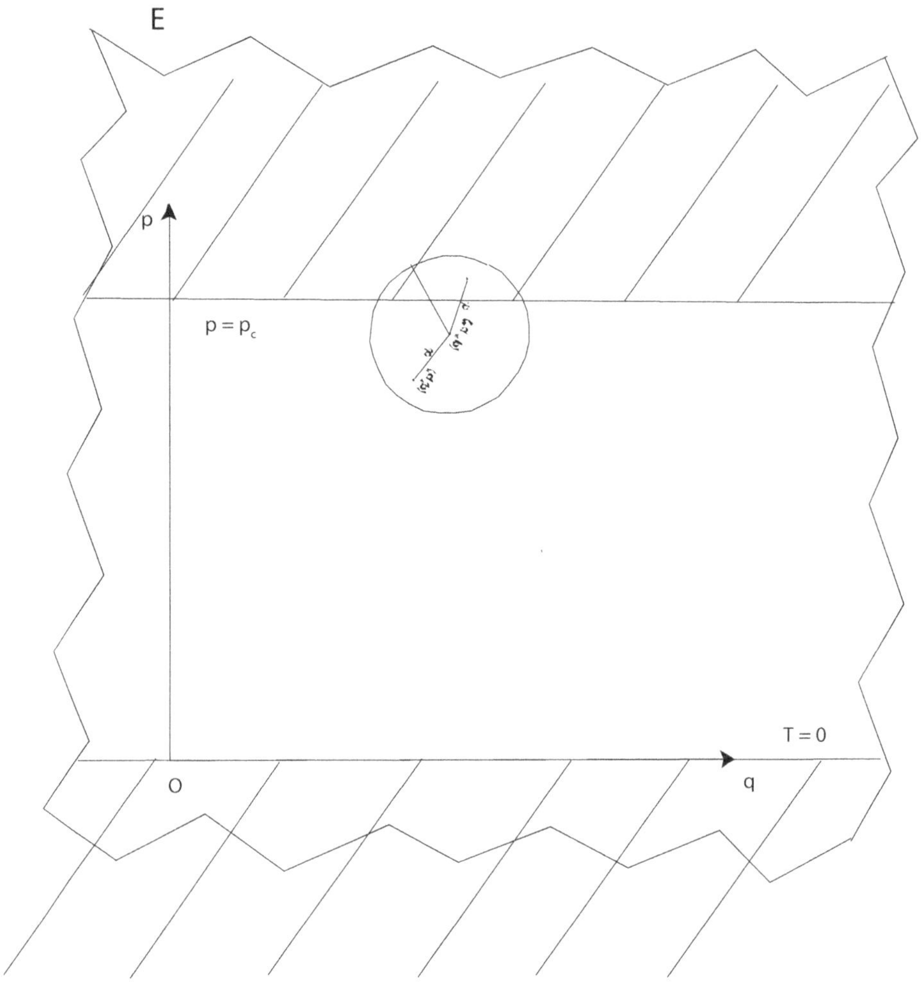

Figure 2

(p_c - the momentum of light, T - temperature)

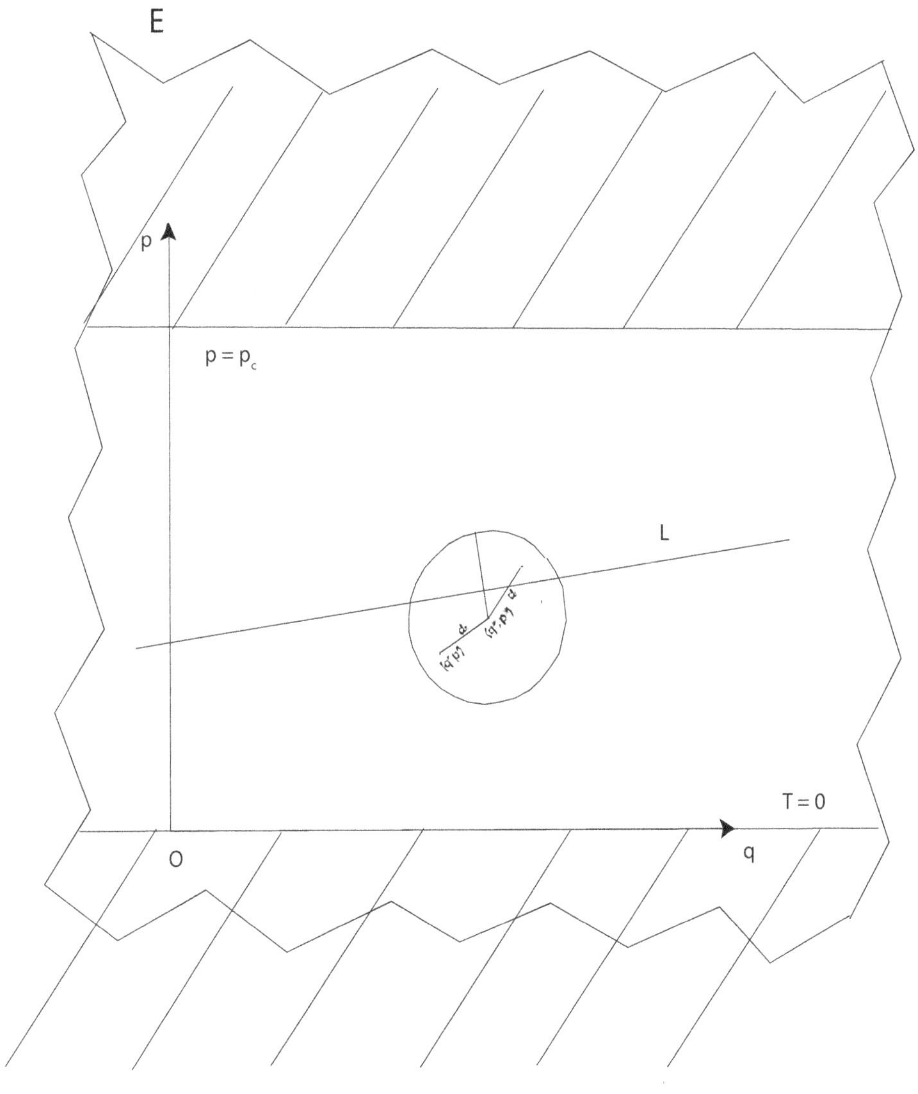

Figure 3

(L is a law)

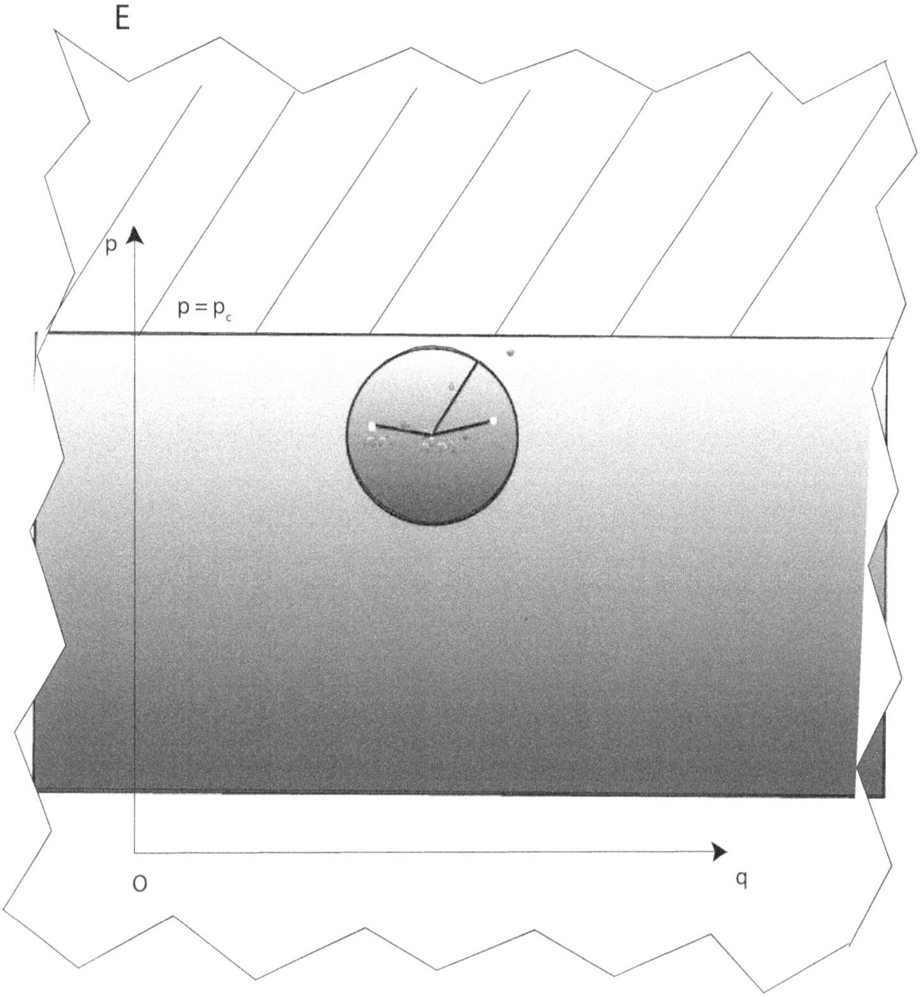

Figure 4

(The darker the shade the higher the likelihood; the same in Fig. 5)

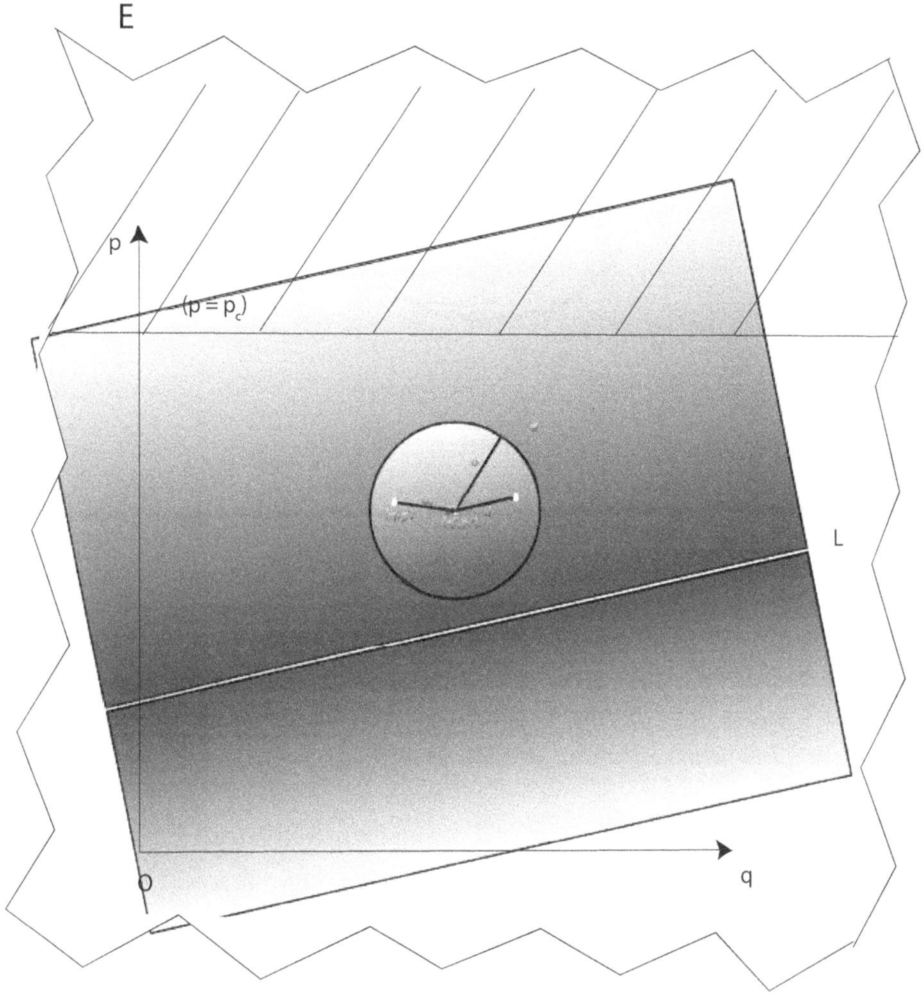

Figure 5

(L is a probabilistic law, a statistical law, or an empirical generalization)

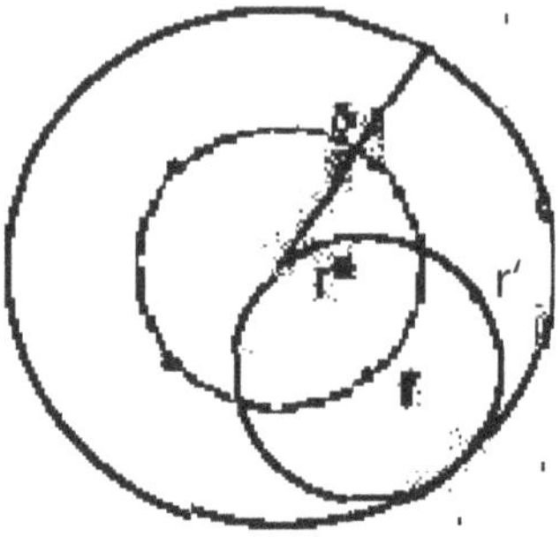

Figure 6

References

Hilpinen, R. (1976). Approximate Truth and Truthlikeness. *Formal Methods in the Methodology of Empirical Sciences* Eds. M. Przlecki, K. Szanianwski and R. Wojicki. Dordrecht, Reidel, 19-42.

Kyburg, H.E. Jr. (1990). *Science and Reason*. Oxford: Oxford University.

Lewis, D. (1973). *Counterfactuals*. Cambridge, Mass, Harvard University Press.

Lewis, D. (1986a). *Philosophical Papers*. vol. II, Oxford, Oxford University Press.

Marquis, J-P. (1991). "Approximations and Truth Spaces." *Journal of Philosophical Logic* **20**: 375-401.

Niiniluoto, I. (1986). Theories, Approximations and Idealizations. *Logic, Methodology and Philosophy of Science* Eds. R. B. Marcus and e. al. Amsterdam, North Holland. 255-289.

Niiniluoto (1987). *Truthlikeness*. Dordrecht, Reidel.

Niiniluoto, I. (1998). "Verisimilitude: The Third Period." *British Journal for the Philosophy of Science* **49**: 1-29.

Oddie, G. (1981). "Verisimilitude Reviewed." *British Journal for the Philosophy of Science* **32**: 237-265.

Oddie, G. (1986). *Likeness to Truth*. Dordrecht, Reidel.

Smith, P. (1998a). *Explaining Chaos*. Cambridge: Cambridge University Press.

Smith, P. (1998b). "Approximate Truth and Dynamical Theories." *British Journal for Philosophy of Science* **49**: 253-277.

Weston, T. (1987). "Approximate Truth." *Journal of Philosophical Logic* **16**: 203-227.

Weston, T. (1992). "Approximate Truth and Scientific Realism." *Philosophy of Science* **59**: 53-74.

CHANG LIU

APPROXIMATION, IDEALIZATION, AND LAWS OF NATURE

ABSTRACT. Traditional theories construe approximate truth or truthlikeness as a measure of closeness to facts, singular facts, and idealization as an act of either assuming zero of otherwise very small differences from facts or imagining ideal conditions under which scientific laws are either approximately true or will be so when the conditions are relaxed. I first explain the serious but not insurmountable difficulties for the theories of approximation, and then argue that more serious and perhaps insurmountable difficulties for the theory of idealization force us to sever its close tie to approximation. This leads to an appreciation of lawlikeness as a measure of closeness to laws, which I argue is the real measure of idealization whose main purpose is to carve nature at its joints.

1. INTRODUCTION

We accept claims in science which are only approximately true because accurate and/or precise and true ones are difficult if not impossible to come by; and we embrace claims of causal laws which only obtain under idealization because none may be had otherwise in as complex and entangled a phenomenal world as ours. It is often claimed, sometimes with arguments and sometimes without, that idealization is a method of theory construction which, if used properly, produces approximately true theories; thereby the adequacy of an idealization depends on how approximate it makes the corresponding theories or laws. And a statement is approximate if it is truthlike, namely, it is in fact false but close to being true, while the measure of closeness is given by a theory of approximation[1] (cf. Niiniluoto 1981, 1982, 1984, 1986, 1998; Hilpinen 1976; Weston 1987; Marquis 1991). Such a relation between approximation and idealization is also used to articulate a theory of confirmation which is thought to have lent natural support to scientific realism (cf. Newton-Smith 1981; Laymon 1982, 1989).

In this paper, I shall examine the notions of approximation and idealization against the following intuitions about our scientific theories. First, many, if not all, of our claims about the external world, while accurate and/or precise, are only approximately true. Second, many, if not all, of our scientific theories – whose essential ingredients are laws of nature[2] –

Synthese **118:** 229–256, 1999.

are idealized in the sense that they are only true under conditions which do not obtain in the actual world; and such idealizations can usually be removed – in most cases only asymptotically but never completely – in the applications of theories so that the theories become almost true. Third, there are sequences of theories in the history of science in which earlier theories are taken as approximations of the later ones, such as in the case of the succession of mechanics from Aristotelian to Newtonian to Einsteinian and quantum.[3] And finally, some of our scientific hypotheses are often 'not even false' – a phrase said to be frequently uttered by W. Pauli, which means that a hypothesis is so ill-formed that it should not be considered a candidate for approximate truth. I take it a truism that whatever theory of approximation and idealization one adopts, it has to square with such intuitions, especially the last one whose implications scientists know so well and yet are barely noticed by philosophers.

I shall first survey the major existing theories of approximation and idealization, explaining what kinds of difficulty they face. Then, I shall examine the root (or source) of such difficulties, which happens to be an intrinsic feature of laws of nature (hence the last item in the title of this paper). Then, I propose a resolution which calls for a separation of idealization from approximation. I shall argue that we keep approximation as the measure of *de facto* closeness of either two statements or two possible worlds. As such, some available account of it may be adequate with some modification, but it cannot straightforwardly be used to construct a measure of idealization. Finally, I analyze the situation of idealization: by contrasting the notion of *lawlikeness* with that of truthlikeness, the former being a measure of closeness to laws while the latter to truth, I argue that the purpose and measure of idealization is how well it *carves nature at its joints* – a concept that gets the essence of idealization right but is impossible to quantify and universalize. In the conclusion, I discuss briefly some related questions whose complete answers go beyond the scope of this paper.

2. THEORIES OF APPROXIMATION

Approximation in general is a relation of relative closeness defined on a set which forms a strict or partial ordering. It may be defined as a purely comparative notion, so that for any element in a set with such an ordering, one can say that one element is more (or less) approximate to it than another, but one cannot say whether or not any element is approximate to it *simpliciter*. For the latter, one must, or it must be possible to, add to the set a measure and a criterion on top of the ordering. The measure maps the distance of approximation to real numbers and the criterion specifies how far the distance is beyond which an element is no longer approximate.

When the relation is defined as *closeness to truth* among a set of statements,[4] we have the notion of relative approximation to truth, or its brethren: *truthlikeness* and *verisimilitude*.[5] The core of a theory for approximate truth is a measure of distance from, or closeness to, the true statement in question, and there are two ways of doing it: one uses a measure on a space of descriptions (hence, a *descriptional* theory) and the other one a space of possible worlds (hence, a *possible-world* theory).[6] If one suspects that the thorny problems in the logical theory of truth may mar the success of a theory of approximate truth, one can rest assured that no such threat exists. In fact, the notion of approximate truth can be articulated independently of whichever theory of truth one happens to subscribe to. In other words, what makes a statement close enough to a true statement to be considered approximately true has nothing do with what makes the latter true, because given an adequate theory and a statement, φ, if φ is true – in whatever sense of being true – we should know whether or not another statement which is intended to approximate φ is indeed approximately true. It is argued, partly because of this fact, that a quest for a theory of approximate *truth* is a misguided effort.[7] I shall return to this point in Section 7.

The most comprehensive descriptional theory, to my best knowledge, is given by Niiniluoto (1984, ch. 7, 1986; also Marquis 1991), according to which, the approximation to truth is measured by the metric distance between two cells or points (corresponding to the S-predicates) on an Euclidean space, depending on whether the space is discrete or continuous. The following simple example shows how the S-space is constructed and why it is naturally Euclidean. Suppose that a universe contains only one particle with two properties: position and momentum, and our language for such a simple universe has only two predicates or variables: P (for position) and Q (for momentum), and a name, α. An S-space is spanned by the Q-axis and the P-axis, so that each S-predicate is a pair of values (q, p), where $q \in Q$ and $p \in P$. The metric of S can then be defined in general as $D = \sqrt{(q - q')^2 + (p - p')^2}$, for all values of P and Q. If the particle has (q', p') (i.e., 'α is at q' with p'' is true), then the sentence 'α is at q with p' is approximately true if $D \leq \delta$, given that δ is the limit of approximation for this space.

More generally, we let **S** be a set of alternative states each of which is a possible state of the world and L a monadic first-order language which is sufficiently rich to describe the states in **S**. We construct a predicate space, S, which fully describes **S** by devising n families of one-place predicates:

$\{C_{11}, \ldots, C_{1k^1}\}, \ldots, \{C_{n1}, \ldots, C_{nk^n}\}$, such that all the predicates in each family are *mutually exclusive* and *jointly exhaustive*, and the n families can be seen as the n axes which span the S-space (cf. Niiniluoto 1986, 262ff). Such an S-space is then partitioned into $q = k^1 \cdot k^2 \cdot \ldots \cdot k^n$ cells, each of which is described by an S-predicate of the form: $s = C_{1j^1}$ & C_{2j^2} & $\ldots$& C_{nj^n}, where $0 \leq j^1 \leq k^1$; $0 \leq j^2 \leq k^2$; $0 \leq j^n \leq k^n$. If the C's admit continuous values, then instead of families of predicates, we have n C-variables whose values are real numbers. And each S-predicate corresponds to a point (rather than a cell) in the **S**-space. Now, each object in the world occupies one and only one cell or point in S, and the actual world consists of occupied cells or points. In such a space, the distance between two states can be defined by the Euclidean measure,

$$(2) \qquad D(s, s') = \sqrt{\sum_{i=1}^{n} d_i(C_{il^i}, C_{im^i})^2},$$

where $d_i(C_{il^i}, C_{im^i}) = |C_{im^i} - C_{il^i}|$ represents the distance between the two states on the ith dimension of **S**. If the space is simply an n-dimensional Euclidean space R^n, then each C-predicate maps onto R and d_i is the numerical difference between C_{il^i} and C_{im^i}.

With this general S-space (which is first given by Carnap), one should be able to say everything about the world. A simple statement about the obtaining of a single state would correspond to a point or a cell, and a complex statement about the obtaining of more than one state would in turn correspond to a region or a curve in **S**. And the Euclidean distance between two regions (a generalization of D) can be defined in several different ways, the most convenient of which would be the distance between the geometric centers of the two regions; and the distance between two curves may also be defined by several methods: one would be to use the area between two segments of the curves and another the maximum distance between the two curves (cf. Niiniluoto 1986, 267–275). Once such a metric space is set up, all one needs is to specify the limit of approximation (a function of δ) beyond which no statement can be considered approximately true.

The possible-world theory (cf. Hilpinen 1976; Weston 1987, 1992) is also quite elegant and general. As a semantic notion, truth is relational: a sentence φ is true or false only relative to an interpretation. For example, if φ is a sentence which says 'Tom is 6 feet tall', then it is true if and only if the person 'Tom' denotes possesses the property to which the predicate 'being 6 feet tall' refers. And the world, w, in which the person named 'Tom' is 6 feet is a model for φ. Suppose another sentence, φ', which says

'Tom is 6.01 feet tall'. It is false in world w but true in another, w', which is exactly the same as w except the person named 'Tom' is 6.01 feet. It is easy to see that w and w' are indeed very close or similar – they may hardly be empirically distinguishable. Hence the definition:

> A sentence φ is approximately true in world w if and only if there is a world w' which is close to w and in which φ is true.

Analogous to the descriptional theory, the central element in this theory consists of a measure of the closeness or similarity of worlds. David Lewis (1973) has a proposal for such a measure which Hilpinen adopts for his purpose. This is done by defining a set of *nested* spheres of worlds, N^i_w, with respect to any particular world w, and a particular property, C^i (in a set of mutually exclusive but jointly exhaustive properties), such that (1) a world is less similar to w the further away the sphere to which it belongs is from w; (2) w belongs to every sphere in N^i_w; and (3) any world which belongs to a sphere, $K \in N^i_w$, belongs to all the spheres of which K is a subset. Let $N_w = I_i N^i_w$ be the intersection of all such nested spheres so that all properties are taken into account, and using N_w we measure the distance from w with the set:

$$E_w(\varphi) = \{K | K \in N_w \quad \text{and} \quad K \cap |\varphi| = \emptyset\}$$

where $|\varphi|$ is the set of φ's models (or worlds in which φ is true). The smaller the E, the closer to w the worlds that makes φ true. Hilpinen then defines:

> A sentence φ is approximately true (in w with respect to C^i) if and only if $E_w(\varphi) = U \subset N_w$, or a sentence ψ is at least as close to the truth as φ iff $E_w(\psi) \subseteq E_w(\varphi)$.

What insures that φ be approximately true is that $U = \emptyset$, which means that the worlds in which φ is true belong to w's own sphere. This, of course, leaves N^i_w a primitive notion. Hilpinen (1976) mentions in passing that one can 'introduce a similarity metric on w, and represent the similarity between worlds by a numerical distance function d' (27).

Weston (1992), who does believe in the importance of formulating such a measure among the possible worlds, defines approximate truth as follows (using our notations).

> A sentence φ is approximately true (under Σ) in w if and only if $\exists w'$ such that it is α-close to w and φ is true in it.

Here, Σ, a sense of approximation, is a measure of distance between two worlds and α, a limit of approximation given Σ, is a number beyond which two worlds can no longer be considered as close. Weston does not, and thinks that one should not, produce a set of alternative senses of approximation from purely formal considerations. Instead he argues that both (i.e., Σ and α) depend on some background theories (which may include the one that φ is in), and concludes that the notion of approximate truth is theory-laden (although not viciously so) (cf. Weston 1992, 59, 63ff).

It should be obvious that the descriptional theory and the possible-world theory can be made equivalent if a 1–1 correspondence exists between the D-functions in the former and the distance-measures between worlds in the latter. Such a connection is indeed tried out by Niiniluoto (1986); however, it is done as a natural extension of his descriptional theory, namely, by transferring the relationships among sentences to those among the sentences' models (cf. Niiniluoto 1986, fig. 8 on p. 272). It is not obvious that the model-theoretic extension is faithful to the Hilpinen–Weston theory. On the other hand, it is equally easy to come up with a descriptional extension of the possible-world theory of the Hilpinen–Weston type which may not be the same as Niiniluoto's theory. Therefore, the disagreement, if any, between the two approaches we have so far examined is not one of formulation – i.e., mainly syntactic or mainly semantic – but one of substance – i.e., different conceptions of the closeness relations either between statements or between worlds (or models).

One should also note that there are descriptional theories that do not require an affine space – a space with a metric – theories such as Marquis's (1991) and Laymon's (1987). In the former, a topological space without an affine structure is devised to represent approximation, and in the latter, the method of Scott's domains is used to do the same. Both have the benefit of being able to offer partial orderings and other features less restrictive than found in an affine space. It is quite possible that a 1–1 correspondence between such theories and the Hilpinen–Weston type theories can be found so that an equivalence between the two can be had without having to impose distortions on either side.

3. DIFFICULTIES WITH THE THEORIES

One of the unstated but fundamental assumptions of the Niiniluoto type theories is that the state space, **S**, is everywhere *broadly uniform* or flat so that a global metric, such as the Euclidean metric in Equation (2), can be attributed to it. This assumption however cannot hold for some very simple reasons. Suppose that the theory of special relativity is true and the speed of light is precisely $c = 299{,}900$ km s^{-1}. And an object x accelerated – regardless of how difficult – to a speed of $v = 299{,}860$ km s^{-1}. Consider the following two statements:

(1) x is accelerated to a speed of $v = 299{,}800$ km s^{-1};

(2) x is accelerated to a speed of $v = 299{,}920$ km s^{-1}.

Which statement, (1) or (2), is more approximately true? Both are equally approximate since (1) and (2) are of equal distance from the truth if the metric of approximation is Euclidean. But one can easily see that (2) cannot possibly be approximately true; it is not even physically possible given one of the main postulates of special relativity (i.e., no object can be accelerated beyond the speed of light).

This is just one of the many examples which show how laws of physics may break, so to speak, our state space, **S**, such that it is not only non-flat but also not simply connected. In fact, every law in physics produces in **S** such a region or line where the space breaks or is disconnected. While many laws of physics set limits for what is physically possible, some, such as the one above, only breaks the space. (I am of course assuming that superluminal objects, such as tachyons, are physically possible, even though none *can* – i.e., physically possible – be created by accelerating ordinary objects beyond the speed of light.)

There are also less catastrophic breaks given by other kinds of laws. Among such laws we have laws of radioactive decay which specify for all radioactive elements how large a given aggregate of their atoms can be before its stability is gone, and laws of phase transition which specify for various substances at what temperature a slight change of pressure or volume may cause catastrophic changes, such as evaporation or condensation. These laws specify highly probable discontinuous changes in the behavior of some systems so that two states in **S** lying very close to each other because of the minute factual differences between them may in fact not be close at all in regard to their truth.[8]

Now, even if one can dodge this difficulty by restricting the **S**-space to those regions where the above mentioned 'breaks' do not occur, one must still face the next subtler and yet more basic difficulty. The difficulty is well illustrated in Figure 1. Here we have a two dimensional space with a simple metric – a simple token of Equation (2) – which can be taken as a state-space with pairs of $\langle x, y \rangle$ determining the states. And λ is a particular instantiation of a simple law of strict correlation between x and y, or in other words, it is such a law with a specific boundary condition, $\langle 0, 0 \rangle$. The three marked points in Figure 1 are so chosen that p_1 is closer than

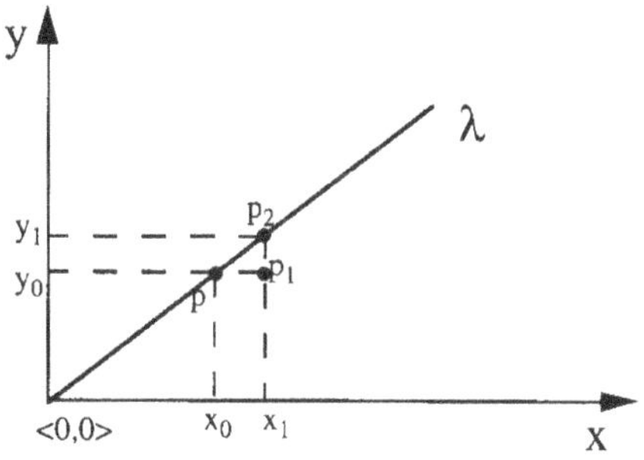

Figure 1. Approximation space with a law λ.

p_2 to p according to the metric of the space. Hence, given that p is true, p_1 is more approximate than p_2. But given the condition which instantiates λ_1, it should be the case that, *ceteris paribus*, p_2 is closer, and hence more approximate, than p_1 to p. This is so because given the actual correlation between x and y, if an object is in x_1, it is more likely to be in y_1 than in y_0. Therefore, if an object with a name α is actually at p, the sentence, 'α is at $\langle x_1, y_1 \rangle$' is more approximately true than the sentence, 'α is at $\langle x_1, y_0 \rangle$'. To make it less abstract, if human's weight and height is correlated by a strict law, then a statement saying that I am of such weight and such height which obeys the actual instantiation of that law will be more approximate to the truth than a statement in which both of my said weight and height are closer to my actual weight and height but together they disagree with the law. Since the metric measures the *de facto* distance of states, the latter statement should be more approximate than the former according to the descriptional theory.

This example shows that for a measure of approximation one must consider, in addition to the metric, at least two other indispensable factors: (i) what the laws are and (ii) how the laws are instantiated in the actual world (with initial and/or boundary conditions). And there does not seem to be any simple way in which such factors can be incorporated into a single metric measure.

Now, whether the same problems will trouble the possible-world theories depends on how one views the relationship between the descriptional and the possible-world theories. If the possible-world theory is simply a model-theoretic version of the descriptional theory (as Niiniluoto (1986, 271–5) has argued), both problems equally apply. However, Hilpinen's nested spheres, N_w, seem to have taken the effects of laws into account such that for any approximate sentence, φ, $|\varphi|$ will always be within the innermost sphere of N_w (because $E_w(\varphi) = \emptyset$). In other words, the differences between $|\varphi|$ and w will be factual and well within the limits of all laws of nature. Hence, the first problem is avoided but not the second one. And even for avoiding the first problem, the cost seems a bit too high: it excludes certain statements which we would regard as perfect candidates for approximate truths. Are Newton's law of motion not approximately true in the actual world even if it diverges widely from the truth when applies to objects which move with a speed close to that of light? And yet the worlds in which this law is true are certainly not in the vicinity of w. Other similar difficulties are pointed out by Lewis (1986, 24–27) which centered on how best to settle the 'messy business of comparative similarity' among possible worlds (24). Weston's amendment to Hilpinen's theory does not really help on this point either. The determination of the sense of approximation (Σ) is left with so much liberty (or so dependent of the contexts of a theory whose statements are under evaluation) that one cannot be sure whether the problems apply. For some senses, e.g., the absolute difference and the relative error (cf. Weston 1988, 60), our problems certainly apply when considered in a space of possible worlds.

4. IDEALIZATION AND REALISM

To prevent confusion, let me first define some terms I shall be using in the following discussion. An *ideal condition* is one which does not obtain in the actual world; and therefore, a statement describing the condition (which we may also call 'ideal condition' unless the context makes it ambiguous) cannot be true in the actual world but is true in the possible world in which the condition obtains. An *idealized claim* is a counterfactual statement of the form: if p were true, then q would be true, where p is an ideal condition. And if the whole claim is true in the actual world, q must be true in those p-worlds which are accessible to the actual world. All idealized laws or theories have the logical form of idealized claims. And an *idealization* is an act of theorizing by which ideal conditions are devised and idealized claims obtained as a result.

The challenge is to construe an approach of idealization which has adequate answers to questions in confirmation, explanation, the progress of science, and – for realists – scientific realism. At the center of all the approaches we find in the literature is the relation between idealization and approximation: the evaluation of an idealization is said to rest on how approximate it makes the resulting claims, or some variations of such (cf. Jevons 1874; Popper 1976; Barr 1971, 1974; Nowak 1972, 1980; Krajewski 1977; Schwartz 1978; Laymon 1980, 1985, 1989; Newton-Smith 1981; Oddie 1981, 1986; Niiniluoto 1984, 1986, 1987, 1990; Brown 1985; McMullin 1985; Weston 1987, 1992; Ramsey 1990, 1992).

Since a thorough review of the above-cited literature goes far beyond the scope of this paper, I shall be content with a sketch of its main line. Let me begin with Laymon's approach (1984, 1989; cf. also Weston 1992), which takes confirmation as the improvability of approximation under idealization and shows how such an approbatory function of idealization lends strong support to realism. Suppose that our theory, T, consists of laws which when supplied with initial and boundary conditions yield empirically testable predictions. However, most of the obtainable initial and boundary conditions are idealized and the predictions yielded by the theory are approximate. To formulate the idea for the confirmation of T, we have the following. Suppose again that we have a set of ideal (initial and boundary) conditions: $\mathbf{I} = \{I^i \mid i = 1, \ldots, n\}$ which is a partially ordered set against an R-scale (i.e., some scale of being Realistic); and for all I^i's, our theory T produces a set of predictions, i.e., $T\colon I^i \mapsto P^i$, where P^i – as in $\{T, I^i\} \models P^i$ – is the ith prediction in $\mathbf{P} = \{P^i \mid i = 1, \ldots, n\}$; and the elements of $\mathbf{P}$ approach P^E, the observed datum, also as a partially ordered set. Then T is confirmed if for any I^i and I^j where $i < j$ we have $|P^j - P^E| < |P' - P^E|$ (i.e., P^j is closer to P^E than P^i whenever I^j is more realistic than I^i); otherwise, T is not confirmed. We must note that for this concept of confirmation, Laymon needs neither the notion of approximate truth (e.g., to determine an α-limit) nor that of a true theory without idealization. All one needs is to see 'if the use of more realistic specification of [ideal] initial and boundary conditions in fact leads to more accurate predications' (Laymon 1989, 359). (For Laymon as a matter of conceptual analysis not of definition, a good idealization is no more than coming up with a set of initial and boundary conditions which approximate the real ones.) In other words, a good idealized theory (i.e., a set of idealized claims) must be 'approximation improving' (or what he calls 'having monotonicity' (Laymon 1985, 213–4)). To use again the above notations and let $\Delta^+\mathbf{I}$ be a positive increment on the R-scale in $\mathbf{I}$ and $\Delta^+\mathbf{P} = |P^j - P^i|$ (where $i < j$), we have that

> T is confirmed if it is always true that $T\colon \Delta^+\mathbf{I} \mapsto \Delta^+\mathbf{P}$; otherwise, it is not confirmed.

If one finds the above formal characterization of scientific theories too narrow (e.g., some theories are not just sets of laws which depend on initial and boundary conditions to make predictions), another more general formulation can be given as follows. Let T^U be our *unidealized* theory which can yield predictions, e.g., $T^U \models P^E$. However, the derivation of P^E from T^U is so complex that an idealization is called for. Suppose T^n is a result of idealizing T^U and with it a precise derivation of prediction P^n can be made. In other words, $T^n = (T^U \,\&\, \mathrm{Id}^n) \models P^n$, where P^n is close but not identical to P^E. Now, let $\mathbf{Id} = \{\mathrm{Id}^i \mid i = 1, \ldots, n\}$ be a set of ideal conditions such that for each Id^i, we have a corresponding T^i which in turn entails a P^i, while $\mathbf{P} = \{P^i\}$ may be the same partially ordered set as defined before. In other words, $\mathbf{T} = \{T^i\} \Rightarrow \mathbf{P}$, meaning each element in $\mathbf{T}$ yields a corresponding element in $\mathbf{P}$, where $\mathbf{T}$ forms a partially ordered set of theories which approach to T^U in descending degrees of idealization. Then, we have,

> T^f, for any $1 < f < n$, is confirmed if for any T^i and T^j, where $f < i < j \leq n$, we have $|P^j - P^E| < |P^i - P^E|$ (i.e., P^j is closer to P^E than P^i whenever T^i is less idealized from T^U than T^j); otherwise, T^f is not confirmed.

In other words, idealized theories are confirmed if they improve upon the degree of approximation of the predictions they produce when the ideal conditions are relaxed towards the actual (or the real) on some scale of realness. This is taken by Laymon to support realism in that the truth of our laws and theories (and so the realness of the objects posited in them) is the *best explanation* for their having the approximation improving property.[9] One should note that although approximate truth is not needed for Laymon's theory of confirmation, it is needed for his defense of realism. A well confirmed idealized theory must be approximately true, for otherwise, his argument for realism fails. However, this is problematic, as will be shown towards the end of the next section. The added benefit of this second construal is that it brings out the similarity between Laymon's approach of idealization and the one given by Nowak (1972, 1980) (see also Krajewski 1977; Niiniluoto 1986, 1987 and Cartwright 1989, ch. 5), except that the latter deals exclusively with idealized claims of a rather standard form:

$$(3)\qquad \forall x(Cx \supset h_1(x) = g_k(h_2(x), \ldots, h_n(x); w_1(x), \ldots, w_k(x))),$$

where C is the background, $h_i(x)$'s are variables, and $w_i(x)$'s conditions which become ideal conditions when their values are set to zero. Without too much loss of accuracy, Laymon's approach can be applied to this case as follows. The ideal conditions will be, $\mathbf{Id} = \{\mathrm{Id}^i \mid \mathrm{Id}^i = [w_i(x) = 0 \,\&\, w_{i+1}(x) = 0 \,\&\, \ldots \&\, w_k(x) = 0];\ i = 1, 2, \ldots, k\}$, and our theories will be, $\mathbf{T} = \{T^i \mid T^i = [T^U \,\&\, \mathrm{Id}^i]\}$, where T^U = Equation (3).[10] The predictions, $\mathbf{P} = \{P^j \mid P^j = T^j[x/\alpha]\}$, are instantiations of T^i's while $P^E = T^U[x/\alpha]$. Moving from T^1 to T^U via the T^i's is called *concretization*, and such a process is *approximation improving* if and only if that to which the process is approaching, T^U, is true. Therefore, '[t]he ability of scientific theories to give approximately true descriptions of ideal cases gives us an argument for critical scientific realism' (Niiniluoto 1986, 285). Therefore, while the widespread use of idealized theories seems to pose a threat to scientific realism, the threat is never real as long as those theories approbate approximation; and no idealizations are justified if they fail to produce approximation improving theories when they are relaxed.

5. A PROBLEM WITH LAWS

The theories of approximation and idealization have been subjected to some serious objections (cf. Ramsey 1990, 1992; and also Laymon 1985, 1989, 370–1): (i) they are too simplistic and static, and (ii) the criteria are external rather than internal to the praxis of science.[11] These may be true enough, but there is a simpler and more exact problem about the theories in question, namely, the problem with laws. Laws, just as factual statements, may also be approximately true and the theories, if adequate, should provide measures for approximate laws. For the moment, I shall call 'lawlike' those statements which resemble or approximate true laws. Now, what does it mean for a hypothesis of law to be approximately true? What is the measure of its distance from the law?

We can represent without much loss of generality lawlike statements as follows

(E) Given **I**, $\mathbf{L}(= \forall x[Fx \Rightarrow Gx])$,

where, **I** is a set of conditions, and $p \Rightarrow q$ means that p physically necessitates q.[12]

Now, let us say that such a statement is approximately true if and only if all of its logical *de facto* consequences are approximately true in the sense given in any, or the best, of the theories we discussed in Section 2. Or more precisely, such a statement is approximately true if and only if when the conditions in **I** are close to the actual (or approximately hold), all the *de facto* consequences of L are also approximately true. This clearly agrees with the approximation improving scheme in the Laymon type approaches. However, the following example shows why this is wrong.

Suppose that **I** holds, is the above answer adequate for the approximate truth of L? The problem is most easily seen in a descriptional theory of approximation. In Figure 2, we have three statements, λ, λ^*, λ^{**}.[13] Suppose that λ represents the law. The question now is which of the other two, λ^*, λ^{**}, are more approximate (i.e., closer) to λ as lawlike statements. λ^{**} is clearly a more lawlike statement given λ is the law than λ^*, but λ^* is *de*

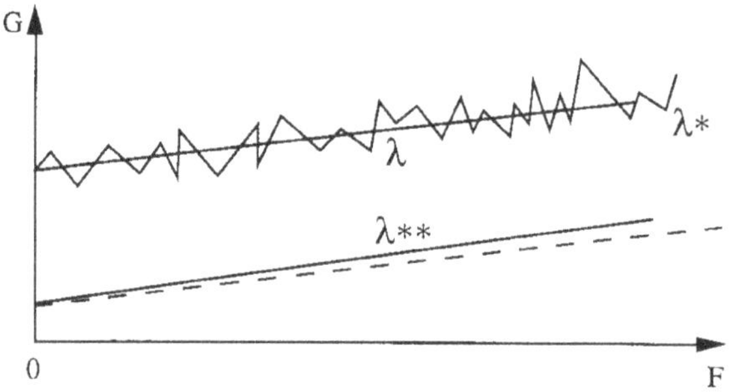

Figure 2. Approximation space for lawlike statements.

facto closer to λ than λ^{**} in that it has more consequences that are in fact closer to the corresponding consequences of λ. In other words, factually, $\lambda^{**} < \lambda^{*} < \lambda$, but as a law, $\lambda^{*} < \lambda^{**} < \lambda$ (where '<' = 'farther from the truth'). And by Niiniluoto's metric measures (e.g., by either the smallest mean distance or the smallest area between two trajectories), λ^{*} is certainly a better approximation, even though as a law it does not even remotely resemble λ. The irony is that in most cases (as in this example) we can easily tell which trajectory is a better approximation by other means. The same is also true in a possible-world theory if the similarity or closeness relation among possible worlds are given in terms of the number of shared states of affairs. To see this, let us suppose that $A(\lambda)$ is the set of all possible worlds in which λ denotes a law. Then, we obviously have that $A(\lambda^{*})$ share more actual states than $A(\lambda^{**})$ does with $A(\lambda)$, and therefore, λ^{*} is more approximate than λ^{**} does to λ, which again is the wrong answer.

For a less dramatic but rather ambiguous example, let us see the following. Suppose that we have a world, w, which is finite and extremely regular. We have three statements:

(A) For all x, if x is F then it is necessarily G;

(B) For all but a few x, if x is F then it is necessarily G;

(C) For all x, if x is F then it is necessarily G^{*};

where (A) denotes the law in w, and G^{*} differs from G slightly. If the exceptions in (B) are in w, then (B) is closer factually to (A) than (C), but (C) seems to be more lawlike. However, it is not as clear as in the previous example that (C) is more approximate to (A) since well defined exceptions are common in scientific laws; for instance, all systems except those at the absolute zero temperature behave in such-and-such a way. In this case, it is a real question as to which statement is more lawlike. One may note that this example is similar to the one given by Lewis (1986, 24–25) on Hilpinen's theory, by which he reached a similar conclusion.

Further, because of the close connection between laws and counterfactuals and because some of the latter define dispositional predicates, the same problem plagues counterfactuals and dispositions (and propensities) as well.[14] Suppose we have,

(D) if x were F, it would be G.

For (D) to be true, all the Fx-worlds in which x is also G must be closer to the actual world, $w^{@}$, than those Fx-worlds in which x is not G. What does it mean for (D) to be approximately true? It is not clear. One may say that it means that there is a possible world, w, which is not $w^{@}$ but close to it, in which (D) is true as a counterfactual. But since such worlds are rarely, if ever, unique, how can one define the notion of one counterfactual statement being more approximately true than another? One cannot do this by taking any two possible worlds in each of which one of the two counterfactuals in question is true and compare their distances from $w^{@}$. There may very well be other such worlds which will reverse the order of closeness.

Suppose now that **I** does not hold but holds approximately. It should be no surprise that the same problem remains. For example, let us suppose that there are two possible worlds, w^* and $w^{**} \neq w^{@}$, which are defined by two sets of ideal conditions, I^* and I^{**} and are such that the λ^* and λ^{**} in Figure 2 are true in them respectively. Clearly, w^* is in fact closer than w^{**} to $w^{@}$. However, we can improve on this case in two different ways. One is to say that λ^{β}, given I^{β} (where $\beta = {}^*$ or **) is approximately true if and only if whenever I^{β} is *removed*, λ^{β} is *true*. This would not help because when I^{β} is removed, we return to $w^{@}$, and λ^{β}, being true in the actual world, needs no notion of closeness, *de facto* or otherwise. However, this is too strong. In fact, if this condition is satisfied, we would not need the ideal conditions in the first place: λ^{β}, all by itself, already represents a law in $w^{@}$. The other begins with a weaker claim, namely, λ^{α} given I^{α} (where $\alpha = 1$ or 2) approaches the truth if and only if as I^{α} is gradually removed – i.e., it is made more and more realistic (in Laymon's term) or concrete (in Krajewski's term) – λ^{α} approaches the intended law of which it is an approximation. This move of course brings us back to square one. Suppose that when I^1 and I^2 reach the same degree of realness, $\lambda^1 = \lambda^*$ and $\lambda^2 = \lambda^{**}$, where λ^* and λ^{**} are given in Figure 2. Following the previous analysis, we will have to say again that by its factual consequences, λ^1 is closer to the law than λ^2 does, while by its being lawlike, the opposite is true.

It should be obvious by now that the defense of scientific realism examined in Section 4 cannot stand. The notion of approximation improvability is one about *de facto* approximation if it is to have a measure at all. But we now know that improving on getting singular facts right is no guarantee that we will get the right lawlike statements, and therefore, no guarantee for approaching the true description of nature. Not that I think scientific realism is false, but that I regard such arguments as from these accounts of approximation and idealization unsound.

6. LAWLIKENESS AND IDEALIZATION

It should by now become clear that the main problem – of laws – besetting approaches or theories examined above arise from a misunderstanding of the nature of idealization and, hence, its relationship with approximation. In this section, I shall suggest, via a discussion of the main *purpose* and *measure* of idealization, that the assumed inextricable connection between idealization and approximation be severed. Such an assumption is the result of mistaking an accidental connection for an intrinsic one. It is true that there are acts of idealization which seem no more than taking already small and negligible differences to be no differences at all, as in the case of assuming $\sin\theta = \theta$ when a pendulum swings in a very small angle, θ. But this is by no means the same *type of reasoning* as assuming that the universe contains only the sun and the earth or is without gravitation. Nor are both the same type as assuming c (light speed) $= \infty$, h (Planck constant) $= 0$, and V (volume of a vessel) $= \infty$, etc. Nor are these similar to the use of vague predicates as if they are precise ones.

Our analysis of the problem with laws in the previous section shows that the trouble seems to stem from running *lawlikeness* and *truthlikeness* together. One may of course follow Popper (1959, 1963, 1976) and insist that the notion of truthlikeness (or verisimilitude) must be one of 'the whole truth' which combines the measure of approximation and of content (and/or form) of a (singular or universal) statement. But we know from the literature that such an avenue seems by all indications to have a dead end. What we should do is to separate the two notions – i.e., lawlikeness and truthlikeness. For statements of singular facts and de facto generalizations, approximation is in general no more than a measure of their closeness to that which is actual, but when a statement of law is concerned, 'approximation' is more of a measure of the closeness to the corresponding *law* – i.e., lawlikeness. And idealization is clearly a method by which scientists go about achieving such an 'approximation'.[15] In other words, the essence of idealizing a condition is not merely to assume zero of an already very small difference; but rather to conceptually imagine or design a setting (which happens not to obtain in the actual world) in which a truly lawlike hypothesis can be discovered and (in some cases) tested. Is there any wonder that there is such a close connection between *thought experiments* and idealizations? It is important to note at this point that idealization, as one of the most useful methods in the praxis of science, is not regarded merely as a method of pure conceptualization. Some ideal conditions are in fact realizable in a controlled experiment. Many would even say that all scientific experiments are attempts to realize ideal conditions postulated under theoretical considerations. This of course cannot be so since no ideal conditions can be literally realized in any actual experiments; most of them can only be realized approximately (in the sense of closeness to facts). However, this is a minor point that should not distract us from realizing the fact that idealization is indeed a guide for setting up experiments, and such experiments are usually for discovering or testing hypotheses of scientific laws. I shall come back to the question of the measure of idealization in a moment, but first let me examine the purpose of idealization a bit further.

What I have said so far on the purpose of idealization can also be found in McMullin's (1985) excellent exposition of what he calls the 'Galilean Idealization'. As an interesting aside, let us note how the emergence of idealization was accompanied by the effort to mechanize and/or mathematicize terrestrial phenomena given by Galileo et al. The method of idealization could not have emerged in the ancient science of astronomy, though precise and mathematical as it was, partly because a manipulation of heavenly bodies in the true sense of that word was out of the question, and more importantly, because the 'idealization' was done for the astronomers by nature – i.e., by separating those bodies from their observers with large distances, or more precisely, by having the sizes of those bodies and of their observers much smaller than the distances which separate them. (It is not hard to imagine how different our scientific knowledge would be were we as small as electrons.) And at the other extreme, it is also easy to understand why a simple application of the mechanical method of isolating scientific laws in a science of human society is met with such difficulty and scepticism.

The complexity and variability of natural phenomena and our desire to explain them with a set of simple and invariant laws demand the use of idealization, whose ultimate aim is to *carve nature at its joints*. It is only when the actual world is carved up at its joints by idealization, genuine lawlike statements can be formulated. Therefore, the more complex and/or variable the phenomena and the deeper the posited mechanisms by which the lawlike are to be formulated, the wider the gap between the ideal conditions and the facts. And the gap should not affect the truth of the laws or theories in question. If one can conceptually (and sometimes even experimentally) cut the world apart at its joints, one can better understand how it actually operates – i.e., giving explanations – and predict how not yet existing phenomena may occur (out of a composition of causes which obey the laws).

Two further points need to be emphasized. First, when I said that how distant an ideal condition is from the fact does not affect the truth of the law in question, I meant the following. Many laws as idealized claims are often true if not false, but rarely approximately true. So, there is no necessary connection between a law's being true and how much its idealizations approximate the actual conditions. The law of inertia, for instance, would simply be false without its antecedent's supposing that the object is not under the action of any external force, while with it, it is true, not approximately so. The fact that no object in the actual world is free of influence from other objects does not affect the truth of the law. Some may argue that there are nearly free objects to which the consequence of the law of inertia is approximately true. This may well be so but is beside the point. Would this law be less true were the actual universe filled with liquid? Would Boyle's law of ideal gas be less true were the actual universe a block of solid? They may never be discovered in such universes, but that is clearly beside the point. Now, this is very different from the case of approximation: 'Alpha Centauri is free of external influence' perhaps is approximately true in the actual universe, but not if it were filled with liquid; .... This implies that into what simple components the world should be carved up depends not on how it actually is but on how it can possibly be. This leads to my second point. If we are to follow further the metaphor of carving nature, we would see that the joints of nature, just as those in an organism, are not necessarily those places where most empty space fills between two parts. If approximation is always the guide, to 'cut up' nature, the knife should go through as much of those places as possible so that the result will be factually the closest to the original. But scientists know better: one should cut apart those places that give the possibilities of recomposition.

There is a kind of idealization which McMullin calls 'construct idealization' that includes what he again calls the 'formal' and the 'material' idealizations. Formal (material) idealization is a method by which known or suspected relevant (irrelevant) aspects of a system or a process is omitted to simplify the model in question. An ideal gas, for instance, is the result of a construct idealization with both the formal and the material aspects: the formal idealization takes away the size and the interaction (other than elastic collisions) of the molecules, while the material idealization allows us to ignore, *inter alia*, the internal structure of the molecules. McMullin seems to have set this type of idealization in contradistinction to another type: causal idealization which is essentially subjunctive or counterfactual – as in his category of subjunctive idealization. By a causal idealization, one cuts the causal network of natural phenomena apart and searches for laws that govern each individual type of causal processes. In other words, one cuts nature at its joints.

However, I see the classification of idealization differently than McMullin. It seems right to say that material idealization is different from the ones whose purpose is to cut nature at its joints – e.g., causal idealization – but formal idealization is not. Neglecting relevant factors such as the motion of the sun caused by the action of the planets is precisely an act of cutting apart entangled causal processes or structures. In this case, one first separates the causal influence of the sun on the motion of the planets from that of the planets on the motion of the sun; and then consider the former and neglect the latter, which constitutes the act of idealization. Therefore, formal idealization should belong to the category that includes causal idealization. Further, I do not think material idealization is really an idealization in its proper sense, even though McMullin might be perfectly correct to have attributed it to Galileo. As I said earlier, what is already provided by nature needs no, and should not be considered as, idealization. If the internal constitution of the moon is in fact irrelevant to its orbital motion around the earth, then it is no act of idealization if astronomers neglect moon's internal constitution. However, this would be considered as material idealization by McMullin's definition. Therefore, I conclude that the real purpose of idealization as a scientific method is to carve nature at its joints, to disentangle the complex causal nexus of natural processes so that theoretical laws – laws that support counterfactuals, not those *de facto* generalizations – can be postulated and discovered.

One may say that my view on idealization is also similar to Cartwright's (1989, 183ff). However, Cartwright spends most of her time arguing for capacities which according to her are obtainable via abstraction but not via idealization. Cartwright (1989, 183–191) agrees with McMullin (1985) that idealization is for obtaining causal laws which underlie and explain empirical or phenomenal generalizations, but she argues that it is not sufficient – or perhaps not the right kind of method – for obtaining capacities. The difference is that in an idealization the values of relevant parameters are *changed* – in most cases to zero – while in an abstraction such parameters are *eliminated* or *subtracted* all together. As a result, the products of the former – causal laws – are only true under the ideal conditions, while those of the latter – capacity claims – are true in general – i.e., without *ceteris paribus* clauses. Whether capacities are real or not is a metaphysical question that I shall not pursue here; and if changing values of parameters to zero differs only in metaphysics but not in methodology from subtracting the parameters all together, we can safely put the distinction aside and take idealization to be the method for obtaining both laws and tendencies (or capacities).[16] Viewed from such an angle, Cartwright's discussion lends strong support to my claim that the *aim of idealization* is not to obtain approximate truths (given in the first part of this paper) but to carve nature at its joints so that we can discover causal laws or tendencies. This of course does not imply that the *measure of idealization*, to which I will turn presently, also has nothing to do with approximation.

Let us now turn to the measure of idealization. How do we know which idealization is good and which bad, or that one idealization is better or worse than another? An answer to this question is given by the approaches I discussed in previous sections. To recapitulate, a good idealization must produce a claim or theory that either is approximately true already or will be so when the idealization is relaxed on some scale of realness towards the actual. And the sense of approximate truth is that of *de facto* closeness to what is the case. This of course cannot be adequate if what is said in Section 5 is correct. If the primary function of idealization is for discovering lawlike statements and the measure of lawlikeness is independent of that of truthlikeness, as I have just argued, then the measure of idealization must also be independent of truthlikeness. However, there is a sense of closeness or 'approximation' which is implied in the measure of idealization. If idealizing is taking leave of the actual, its result must be the approaching to something else. And this something else is *lawlikeness.* Here I borrow an idea from Cartwright, even though hers is for abstraction and capacities not for idealization and causal laws: the success or failure of an idealization is not measured by how well its result – i.e., that which is true under the given ideal conditions – does when the idealization is relaxed or removed, but how close it is to the real law when the idealization is in place.

Hence, this measure of idealization differs from the traditional one on two counts. (i) The quality of an idealization is measured not by how little the ideal conditions deviate from the actual ones, nor by how little that which obtains under the ideal conditions deviates from the actual situation, nor by how little that which obtains under the ideal conditions deviates from the actual situation when the ideal conditions are relaxed or removed. Rather it is measured by how close that which obtains under the ideal conditions is to the real law. From such a measure we can easily see why the idealization that produces λ^{**} in Figure 2 is a better idealization than the one which yields λ^{*} (given that λ is the law and λ^{*} and λ^{**} are results of idealizations). (ii) Law and its approximation – i.e., lawlikeness

– is neither identical nor reducible to truth and its approximation – i.e., truthlikeness. There are different views on what a law of nature is: for some it is a feature of reality (cf. Dretske 1977; Armstrong 1983), for others a feature of a perfect logical system (cf. Lewis 1973), and for still others a matter of pragmatics (Skyrms 1980). Some think that laws are necessarily true and yet others (e.g., Cartwright) regard them necessarily false. While the measure of truthlikeness does not depend on which theory of truth one adopts, it is likely that the measure of lawlikeness will depend on what laws are supposed to be.

At this juncture a caveat is in order. It would be a mistake to claim or argue that idealization can only be used to discover highly theoretical causal laws. As I said in the beginning of this section, idealization embodies radically different types of reasoning; getting causal laws being one, or a very prominent one, among them. There are experimental data or empirical generalizations the obtaining of which is impossible without some sort of idealization. The measure of such an idealization does seem to coincide with approximation in the most simple and straightforward manner, namely, the ideal conditions must be one which assume zero some otherwise very small and insignificant differences.

This must be so because if the differences are too big and important, the data or generalizations obtained as a result will be too far from the actual to be of any significance by themselves or of any use in hypothesis testing. Even with laws, there are cases in which the purpose and measure of idealization seem to be the closeness to facts in the sense of approximation approbation. To see the difference let us compare the following two cases. In the case of ideal gas which obeys Boyle's law, it is true that the idealization is a good one because the law can be improved upon, namely, when the ideal conditions, such as the molecules having no size and no interactions besides inter-molecular collisions, are relaxed, a better law – or a series of laws one better than the next – result (e.g., van der Waals' law). On the other hand, let us imagine a case in which the (unknown) gravitational interaction between two bodies is weak but the electromagnetic interaction is much stronger. Ideal conditions are introduced to obtain the law of electromagnetism, among which putting the strength of gravity to zero is one. Assuming all other interfering factors are taken care of by the ideal conditions, the hypothesis we obtain of electromagnetism should be the real law. But should we still say that the idealization is no good because the law is not improvable by recovering those factors which are taken away by the idealization? Adding the weak gravitational interaction back into the situation will not allow us to have a more 'approximate' law in any sense of approximation, for otherwise we would have separate laws for different combinations of gravitational, electromagnetic, and other interactions, as it is the case with natural phenomena.[17] However, despite this apparent difference between the two cases – and there may be other cases which differ still from these two – the aim of idealization, if it is for causal laws and structures, is still to carve nature at its joints. In the case of gas laws, van der Waal's law may be seen as the result of an idealization which carves nature better than the one which yields Boyle's law. In other words, if nature is carved at its joints, some portions of it may still very much resemble the actual or the phenomenal while others may not, whereas if resemblance of the actual or the phenomenal is made the criterion of good idealization, many causal laws would be given up because they would be results of bad or unimprovable idealizations.

7. CONCLUSION

Of all the relevant questions that I have not dealt with in this paper, here is a partial list.

1. Can there be a theory of truthlikeness which avoids difficulties given in Section 3?
2. Is approximation really about approximate truth (or truthlikeness) or is it about something else? Does it make sense to talk about statements being approximately true or nearly true at all?
3. What is lawlikeness? Is there a unique measure for lawlikeness? Can the account of lawlikeness be independent of the theories of laws of nature?
4. What are the implications of this separation of idealization and approximation in terms of lawlikeness and truthlikeness respectively?
   a. Is it necessarily for or against scientific realism? Will it require a new kind of realism?
   b. What does this say about the progress or the lack of it in the history of science?
5. What is the relationship between vagueness and idealization (cf. Fine 1975)?

A thorough discussion of any of these questions would require another paper. Here I offer some brief and tentative thoughts on some of them.

It seems that the answer to the first question should be positive. What one needs to do is to come up with a measure of semantic closeness of either states or worlds which takes into account the effect of laws. It may not be as simple as the ones we had so far, but I do not see anything against it in principle.

About the second set of questions, one must note that it seems true that instead of saying that 'Tom is 6 feet tall' is approximately true since Tom is in fact 6.01 feet tall, we can say that 'Tom is approximately 6 feet tall' is true if the limit of approximation assumed for the sentence includes Torn's being 6.01 feet and false if the limit does not include that. The latter seems to be expressing in a material mode the same thing as expressed in a formal mode in the former. If this is so for all statements concerning approximation, it should be possible and desirable to construe a theory of approximation not as one concerning the distances between states or worlds, i.e., a semantic notion, but as one defining the limits or ranges within which statements missing the facts can still be true. Without working out the details, I cannot say how well such an alternative theory of approximation will work. However, even if everything works out, the difficulties and problems raised in this paper will apply to this theory with equal force.

About the third set of questions, I do not believe that it is possible to construct a unique metrical measure for idealization, considering its diverse applications. Even if we limit ourselves to the chief purpose of idealization: to facilitate the discovery of causal laws by carving nature as closely to its joints as possible, the measure has to be tailored to suite one's view of laws of nature. If, for instance, we take Lewis' view (cf. Lewis 1973, 72–77, 1986a, xi–xv, 122–131, 1986b), the approach to the ultimate system of laws is done by balancing the strength and simplicity of each system on the way to that goal: one set of idealizations is better than another in view of the available facts just in case that it yields a system of laws that strikes a better balance of strength and simplicity than the system from the other set of idealizations does. It may be possible to construct a rigorous partial ordering for such a balance, and hence a partial ordering for the idealizations. However, is this an ordering for a particular idealization in connection with a particular lawlike statement or for a set of idealizations (or some idealizing scheme) in connection with a system of lawlike statements? It appears that only the latter is possible. Then, how are single idealizations to be evaluated? These are the kind of questions that one must answer if one is to follow the Lewisian approach for laws of nature. On the other hand, if one believes in the reality of universals and that laws are contingent relations among them (cf. Armstrong 1983), then the measure could be given as the degree to which an idealization has succeeded in isolating two connected universals.

Are statements of causal or theoretical laws true? They certainly have true consequences in that statements about singular facts and de facto generalizations can be derived from them. But are they themselves true? Cartwright takes the majority of them to be false, while others (e.g., the semanticists such as van Fraassen 1989; Giere 1988; Suppe 1989) think that they are true in models some of which are isomorphic to the intended phenomena. For my account since I separate lawlikeness from truthlikeness, whether law statements (not lawlike ones) are true or not does not matter to me. Lawlikeness is a measure of the closeness to (or the distance from) that which is a *genuine* law. Nevertheless, I think law statements are true, and lawlike ones do *in this sense* approximate truth. As I have argued in Section 5 that theoretical or causal laws are ones of the form: $\mathbf{I} \supset \mathbf{L}$. Such statements must be true in the actual world if they are genuine laws. So, statements of laws are true statements. The above controversy concerns only a part of what I consider a law, namely, **L**. It should be obvious that **L** is true with **I** given or in those worlds in which **I** is true – which does not include the actual world – and false without it, for otherwise we would not have needed **I**. The rest are metaphysical concerns that are interesting on their own but have no bearing on the subject of this paper.

This paper has revealed nothing that is necessarily against the spirit of realism; and in fact, when the nature of scientific laws is concerned, I have taken a moderate realist stand. Nevertheless, we did find out that a certain kind of argument for realism does not work, namely, the inference to the best explanation argument given by Laymon and others by appealing to the approximation improving or approbation capacity of idealization.

ACKNOWLEDGEMENTS

I wish to thank Gérard Emch, Risto Hilpinen, Nancy Cartwright, Ilkka Niiniluoto, Kirk Ludwig, Greg Ray, and James Derden for criticisms and suggestions. I am also grateful for the valuable comments of an anonymous referee.

NOTES

[1] With the exception of Barr (1971, 1974) and, in some sense, early Layman (1980, 1982, 1984), all the others treat approximation as prior. While Javons (1874) does not seem conscious that idealization is a method distinct from approximation, Schwartz (1978) faults Barr for dealing with idealization without first getting clear on approximation; Layman (1990) later argues that to accept an idealized theory is 'to believe that every idealization can be converted into a transformational approximation' which has the same structure 'as that of calculational approximations' (522) (also cf. Layman 1985, 1987, 1989); Niiniluoto (1986, 278–285) relates idealization to approximation in the concept of

'concretization' which represents the 'lowering' of an idealized theory to particular cases via approximation; and still others discuss exclusively approximate truth (or truthlikeness or verisimilitude) and only mention idealization in passing or when they generalized their conclusions (cf. Hartkaemper and Schmidt 1981; Hilpinen 1976; Marquis 1991; Oddie 1981; Oddie 1986; Popper 1959; Ramsey 1990, 1992; Rosenkrantz 1980; Weston 1987, 1992; Wallis 1994).

[2] In this paper, the distinction between laws of nature and the statements of them is assumed. In a scientific theory, we have law statements not the laws themselves, just as the theory may contain statements of events or states of affairs but not the events and states themselves. For this distinction, I do not have to commit myself to admitting more than that it is a matter of fact what is a law and what is not. However, I will use 'law' to refer to a law or a law statement unless it is not clear from the context which is intended.

[3] Here I would like to avoid the debate over whether such successive theories are even commensurable (à la Kuhn (1962)). This is easily done by not considering the three theories of mechanics in question as historical theories, namely, the mechanics given and understood by Aristotle and his followers, the mechanics of Newton and his followers, etc., but regarding them as reconstructed ones in the post-Einstein, post-quantum era. There should be no question that they can be and are viewed as a sequence of approximations (or of special cases).

[4] I shall use 'statements' instead of 'propositions' or 'sentences' throughout the paper. They may be taken to refer to declarative sentences in a language, and since a theory of approximate truth is given in a given language (i.e., the language of exact sciences), the usual problems besetting using sentences may be avoided.

[5] For some reason – perhaps following Popper's (1959) lead – philosophers, such as Hilpinen (1976), Newton-Smith (1981, ch. 8), and Niiniluoto (1987, 1997), take their task of producing a theory of approximate truth and the like to be one that should account for both features. They seem to think that a statement or a scientific theory is approximately true or truthlike only if it is both true enough and exact enough, while being exact enough is understood in terms of having enough empirical content. And therefore, an adequate account should provide criteria for measuring the degree of truth and the degree of contentfulness (or exactness). This move seems to me a bit misguided and it makes things unnecessarily complicated. Therefore, I shall in this paper take approximate truth or truthlikeness to be a one dimensional concept concerning a measure of the distance or closeness of a statement or a theory to the true statement or theory. The only reason I can see that one would require closeness both to truth and to exactness for verisimilitude would be if one is thinking of a faithful description of a system. One may say that a description of a system S is approximately faithful if and only if it is approximately true of S and its description of S is exact enough (with well-understood measures of truthlikeness and exactness). But faithfulness is no substitute for truthlikeness or verisimilitude.

[6] Niiniluoto remarked with respect to Hilpinen's (1976) possible-world approach to approximate truth, '[i]n his contribution to the conference of Formal Methodology in Warsaw, June, 1974, Risto Hilpinen presented an approach to truthlikeness which uses as a primitive notion a comparative relation of distance between possible worlds. I observed immediately ... that one could replace possible worlds by maximally strong descriptions of possible worlds, ..., and then specify a quantitative distance measure between constituents. Moreover, this measure could be easily defined at least in the case of constituents of a monadic first-order language' (1978, 282–3). And in his 1986 article, Niiniluoto tries to

show that it does not matter whether one uses descriptions or models that satisfy them, implying the equivalence of the two approaches.

[7] I thank Kirk Ludwig for explaining the alternative to me when he heard an earlier draft of this paper.

[8] One may object to the point in this paragraph as follows. **S**-space as the true space for all possible states in the actual world should have the laws built in, so to speak. In that case, two states which are separated by any of such law-governed catastrophic changes should not be close to each other in **S**. The idea sounds plausible but is not yet clear to me. Let us take the critical mass law of radioactive decay. Suppose our universe consists only of plutonium atoms. The actual state (s^*) is one in which all balls of plutonium are below the critical mass, so no catastrophic chain-reaction occurs. A possible state (s) of such a world is that everything is the same as the actual state except one of the balls is slightly over the critical mass (and no catastrophic chain-reaction occurs). There is another possible state (s') in which a chain-reaction does occur on the very ball that is slightly overweight. Counting the categorical properties of the balls, s is certainly closer to s^* than s', but s' is closer if you figure in the law for plutonium. But how is this simple state space to be constructed so that s' shows up closer to s^* than s? The usual response, which has the form: 'the rest of the world will be such because of the laws that the much changed local state will be a closer state to the actual state than the little changed local state', does not seem to work with this simple example, because the rest of this plutonium world is the same between s and s', while the little changed state s is still closer to s^* than the much changed state s'. Unless one insists, at the cost of being unreasonable, that s and all states of this kind are physically impossible, I do not see how the suggestion in question can be defended.

[9] Layman says, '(R2) If a set of fundamental laws is true, then we can make in principle *sufficient* corrections so as to yield better predictions. What makes R2 interesting from the realist point of view is that piecemeal improvability of idealizations and approximations with resulting improvements in predictions is a significant feature of actual science. Realists have a nice explanation of this practice of seeking increasingly more accurate and less idealized initial or boundary conditions' (Layman 1989, 359).

[10] One must note that all $w_i(x)$'s are assumed to be non-zero values to begin with. In other words, if any $w_i(x)$ has value zero before the idealization is taken, it should not be included in the original set in the first place.

[11] Some may think that (ii) does not, while point (i) does, apply to Weston's account. This is however not quite clear to me. The measures, i.e., the Σ's and α's, in Weston's account is about different interpretations or worlds, which may not be internal to the praxis of science.

[12] The customary formulation of lawlike statements only has **L** (or some variations of it). It is certainly possible to incorporate **I** into F since the conditions in **I** are also properties that x possesses in order for (E) to hold; for instance, to say that given that there is no external influence, an object moving with a certain velocity must continue to move with the same velocity indefinitely is no different from saying that an object moving with a certain velocity and is free of any external influences must continue to move with the same velocity indefinitely. Therefore, to separate **I** from predicates in F is for purely pragmatic reasons: it is for isolating the potential ideal conditions. Moreover, the conditions in **I** may obtain in the actual world; therefore, (E) also include lawlike statements that do not require (or involve) idealizations.

[13] One must note that here I do not and need not assume that the differences among these three trajectories are created by idealization, i.e., by imputing ideal conditions to the situ-

ations of λ* and λ**. The differences can simply be the results of initial and/or boundary conditions. Apart from how they are used or introduced, ideal conditions are in no way different from physically possible initial and boundary conditions.

[14] I will have no space to discuss in full length the relation between non-categorical predicates and laws and why the difficulties inherent in the latter are also there with the former. Suffice it to say that I take dispositional or propensional predicates to be defined by counterfactuals linking two sets of categorical predicates, the first representing possible conditions and the second outcomes that display the properties those predicates denote. The close tie between laws and counterfactuals should be well known, and that is why (pending the spelling out of details) the difficulties for laws also occur in those essentially containing non-categorical predicates.

[15] Here I use quotes around the term, approximation, to signify that I'm using the term in an unusual way. It is possible to think of the term 'approximation' as having two different connotations in our discussion. One is for the closeness to truth (i.e., singular facts) and the other is for the closeness to lawlike necessities, just as in the case of Figure 2. Since it would be confusing to use the term in this manner, I shall avoid it in general in this discussion.

[16] Even though I shall not belabor the point here, I find it odd to think that obtaining the lawlike behaviors of a system, which is in the domain of idealization, is not sufficient for us to find out the corresponding tendencies or capacities it possesses, which according to Cartwright is a job for abstraction. If the set of capacities is not 1–1, onto mapped to that of causal laws which display them, I do not know what capacities could possibly be. The metaphysical distinction between capacities and causal laws is as clear as the distinction between that which is displayed and that which is doing the displaying, and yet I do not see why methodologically, that through which we discover causal laws is not sufficient for us to discover capacities. Unless capacities can manifest themselves through means other than causal laws, I fail to see why discovenng the laws is not the best and only way to discover capacities.

[17] To be more precise, let us go back to Equation (3) and Nowak's approach of idealization and concretization. Suppose that T^1 is the most idealized theory – either Boyle's law or the law of electromagnetism – and T^U the least idealized one (or most concrete one) in an ordered set of theories, $\mathbf{T} = \{T^i | T^i = [T^U \ \& \ \mathrm{Id}^i]\}$, where $\mathrm{Id}^i \in \mathbf{Id} = \{\mathrm{Id}^i \mid \mathrm{Id}^i = [w_i(x) = 0 \ \& \ w_{i+1}(x) = 0 \ \& \ldots \& \ w_k(x) = 0]; i = 1, 2, \ldots, k\}$. Here for ideal gas, some of the $w_i(x)$'s are the sizes of, and the interactions among, molecules, and for electromagnetism, some of the $w_i(x)$'s are the amounts of other forces. In the case of ideal gas, the move from T^1 to T^U via the T^i's is certainly a process of approximation improvement or approbation, but in the case of electromagnetic laws, this clearly is not the case. T^U and some of the T^i's after the influences of other forces are recovered are heterogeneous statements which have no standing among laws of physics, even if T^1 is the correct law of electromagnetism. Hence, this approximation improving method advocated by Layman and Nowak et al. does not work with causal laws in general since it does not provide means by which causal entanglement can be sorted out.

REFERENCES

Armstrong, D. M.: 1983, *What Is a Law of Nature?* Cambridge University Press, Cambridge.

Barr, W. F.: 1971, 'A Syntactic and Semantic Analysis of Idealizations in Science', *Philosophy of Science* **38**, 258–272.
Barr, W. F.: 1974, 'A Pragmatic Analysis of Idealizations in Physics', *Philosophy of Science* **41**, 48–64.
Brown, J. R.: 1985, 'Explaining the Success of Science', *Ratio* **27**, 49–66.
Cartwright, N.: 1983, *How the Laws of Physics Lie*, Clarendon Press, Oxford, UK.
Cartwright, N.: 1989, *Nature's Capacities and Their Measurement*, Clarendon Press, Oxford, UK.
Dretske, F.: 1977, 'Laws of Nature', *Philosophy of Science* **44**, 248–68.
Fine, K.: 1975, 'Vagueness, Truth and Logic', *Synthese* **30**, 265–300.
Giere, R.: 1988, *Explaining Science: A Cognitive Approach*, University of Chicago Press, Chicago.
Hartkaemper, A. and H.-J. Schmidt, eds.: 1981, *Structure and Approximation in Physical Theories*, Plenum, New York.
Hilpinen, R.: 1976, 'Approximate Truth and Truthlikeness', in M. Przlecki, K. Szaniawski and R. Wojicki (eds.), *Formal Methods in the Methodology of Empirical Sciences*, Reidel, Dordrecht, pp. 19–42.
Jevons, W. S.: 1874, *The Principles of Science: A Treatise on Logic and Scientific Method*, Macmillan, London.
Krajewski, W.: 1977, *Correspondence Principle and the Growth of Knowledge*, Reidel, Dordrecht.
Laymon, R.: 1980, 'Idealization, Explanation, and Confirmation', in P. D. Asquith and R. N. Giere (eds.), *PSA 1982*, East Lansing, Philosophy of Science Association **1**, 336–350.
Laymon, R.: 1982, 'Scientific Realism and the Hierarchical Counterfactual Path from Data to Theory', in P. D. Asquith and T. Nickles (eds.), *PSA 1980*, East Lansing, Philosophy of Science Association, **1**, 107–121.
Laymon, R.: 1984, 'The Path from Data to Theory', in J. Leplin (ed.), *Scientific Realism*, Berkeley, University of California Press, pp. 108–123. (The same article as Laymon 1982.)
Laymon, R.: 1985, 'Idealizations and the Testing of Theories by Experimentation', in P. Achinstein and O. Hannaway (eds.), *Observation, Experiment and Hypothesis in Modern Physical Science*, MIT Press, Cambridge, MA, pp. 147–173.
Laymon, R.: 1987, 'Using Scott Domains to Explicate the Notions of Approximate and Idealized Data', *Philosophy of Science* **54**, 192–221.
Laymon, R.: 1989, 'Cartwright and the Lying Law of Physics', *Journal of Philosophy* **86**, 353–372.
Laymon, R.: 1990, 'Computer Simulation, Idealizations and Approximations', PSA 1990, East Lansing, Michigan, Philosophy of Science Association **2**, 519–34.
Lewis, D.: 1973, *Counterfactuals*, Harvard University Press, Cambridge, MA.
Lewis, D.: 1986a, *Philosophical Papers*, vol. II, Oxford University Press, Oxford.
Lewis, D. 1986b, *On the Plurality of Worlds*, Blackwell, Oxford.
McMullin, E.: 1985, 'Galilean Idealization', *Studies in the History and Philosophy of Science* **16**, 247–273.
Marquis, J.-P.: 1991, 'Approximations and Truth Space', *Journal of Philosophical Logic* **20**, 375–401.
Newton-Smith, W. H.: 1981, *The Rationality of Science*, Routledge & Kegan Paul, London.
Niiniluoto, I.: 1978, 'Truthlikeness: Comments on Recent Discussion', *Synthese* **38**, 281–329.

Niiniluoto, I.: 1982, 'On Explicating Verisimilitude: A Reply to Oddie', *British Journal for the Philosophy of Science* **33**, 290–296.
Niiniluoto, I.: 1984, *Is Science Progressive*, Reidel, Dordrecht.
Niiniluoto, I.: 1986, 'Theories, Approximations and Idealizations', in R. B. Marcus et al. (eds.), *Logic, Methodology and Philosophy of Science*, North Holland, Amsterdam, pp. 255–289.
Niiniluoto, I.: 1987, *Truthlikeness*, Reidel, Dordrecht.
Niiniluoto, I.: 1990, 'Measuring the Success of Science', in M. F. A. Fine and L. Wessels (eds.), *PSA 1990*, East Lansing, Philosophy of Science Association **1**, 435–445.
Niiniluoto, I.: 1998, 'Verisimilitude: The Third Period', *British Journal for the Philosophy of Science* **49**, 1–29.
Nowak, L.: 1972, 'Laws of Science, Theories, Measurement', *Philosophy of Science* **39**, 533–548.
Nowak, L.: 1980, *The Structure of Idealization: Towards a Systematic Interpretation of the Marxian Idea of Science*, Reidel, Dordrecht.
Oddie, G.: 1981, 'Verisimilitude Reviewed', *British Journal for the Philosophy of Science* **32**, 237–265.
Oddie, G.: 1986, *Likeness to Truth*, Reidel, Dordrecht.
Popper, K.: 1959, *The Logic of Scientific Discovery*, Basic Books, New York.
Popper, K.: 1963, *Conjectures and Refutations*, Routledge, London.
Popper, K.: 1976, 'A Note on Verisimilitude', *The British Journal for the Philosophy of Science* **27**, 145–159.
Ramsey, J.: 1990, 'Beyond Numerical and Causal Accuracy: Expanding the Set of Justificational Criteria', in M. Forbes and A. Fine (eds.), *PSA 1990*, vol 1, East Lansing, MI, Philosophy of Science Association, pp. 485–499.
Ramsey, J. L.: 1992, 'Towards an Expanded Epistemology for Approximations', in D. Hull, M. Forbes and K. Okruhlik (eds.), *PSA 1992*, vol. 1, East Lansing, Michigan, Philosophy of Science Association, pp. 154–164.
Rosenkrantz, R.: 1980, 'Measuring Truthlikeness', *Synthese* **45**, 463–488.
Schwartz, R. J.: 1978, 'Idealization and Approximations in Physics', *Philosophy of Science* **45**, 595–603.
Skyrms, B.: 1980, *Causal Necessity: A Pragmatic Investigation of the Necessity of Laws*, Yale University Press, New Haven.
Suppe, F.: 1989, *The Semantic Conception of Theories and Scientific Realism*, University of Illinois Press, IL.
van Fraassen, B.: 1989, *Laws and Symmetry*, Clarendon Press, Oxford.
Wallis, C.: 1994, 'Representation and the Imperfect Ideal', *Philosophy of Science* **6**, 407–428.
Weston, T.: 1987, 'Approximate Truth', *Journal of Philosophical Logic* **16**, 203–227.
Weston, T.: 1992, 'Approximate Truth and Scientific Realism', *Philosophy of Science* **59**, 53–74.

Part 3

Theory and Unification

The papers included in this part deal with two connected areas: the nature of scientific theories and the unity or disunity of scientific theories.

One may assume that a scientific theory should be identified with its standard formulation, if there is one, in technical treatises or textbooks. Such formulations are set of sentences, either verbal or mathematical, that express corresponding propositions. The semantic approach to scientific theories challenges this largely syntactic conception of scientific theories. The semanticists believe that theories should be identified with the class of models that satisfy their verbal or mathematical formulations. Taking the semanticist approach to theories, one may avoid those particular philosophical problems that beset the syntactic approach. It also allegedly offers a much clearer notion of what a theory is. The two papers on the subject give a detailed analysis of the controversy between syntacticism and semanticism on theories and propose a hybrid view of scientific theories.

There has been a literature on the application of the semantic approach (or view) of theories to various disciplines in science. For example, Elizabeth Lloyd is among the first to study biological theories from a semantic approach: "A Semantic Approach to the Structure of Population Genetics," *Philosophy of Science* 51 (1984):242-264 and "The Semantic Approach and Its Application to Evolutionary Theory," *PSA: Proceedings of the Biennial Meeting of the Philosophy of Science Association* (1988): 278-285. Also see Steven Cook (2005) "On the Semantic Approach to Econometric Methodology," *Journal of Economic Methodology* 12 (1): 117-123, for an application to economic theories. More recently, we have Peter Gildenhuys (2013) "Classical Population Genetics and the Semantic Approach to Scientific Theories,"

Synthese 190 (2): 273-291.

For more general philosophical discussions, see Steven French & James Ladyman (1999) "Reinflating the Semantic Approach." *International Studies in the Philosophy of Science* 13 (2): 103-121; and Jean Leroux (2001). "'Picture Theories" as Forerunners of the Semantic Approach to Scientific Theories.' *International Studies in the Philosophy of Science* 15 (2): 189-197, which explores the relationship between Wittgenstein's picture theory of meaning and the semantic approach; and also Otávio Bueno & Décio Krause (2010) "Scientific Theories, Models, and the Semantic Approach." *Principia* 11 (2): 187-201. Sebastian Lutz (2014) "What's Right with a Syntactic Approach to Theories and Models?" *Erkenntnis* 79: 1475-1493 argues for the rehabilitation of the syntactic approach, the archrival of the semantic approach.

Papers 9 and 10 deal with the conception of unity of science in two different settings. In "Coordination of Space and Unity of Science," the issue of whether science is intrinsically unifiable is given a "geometrical" representation. Various fallouts are evaluated within such a representation of the issue. The paper on the "grand" unification of gravity and other quantum fields in the quantum gauge field program introduces the attractive idea of unifying gravity with other gauge fields and discusses the various difficulties with such a program. The "gauge gravity" program has been in existence since the 1960s, and yet it has never become a hot topic in the physical literature. It was alleged that Richard Feynman made some disparaging remarks on the prospect of the program and that might have had some effect on its fortune. For recent works on unification in physics, see Michael Heller, Leszek Pysiak & Wiesław Sasin (2011) "Fundamental Problems in the Unification of Physics." *Foundations of Physics* 41 (5): 905-918; K. Karaca (2012) "Kitcher's Explanatory Unification, Kaluza-Klein Theories, and the Normative Aspect of Higher Dimensional Unification in Physics." *British Journal for the Philosophy of Science* 63 (2): 287-312; and Erik Weber & Merel Lefevere (2014) "The Role of Unification in Micro-Explanations of Physical Laws." *Theoria* 29 (1): 41-56.

Coordination of Space and Unity of Science

I explore in this essay the possibility of taking a well-understood problem in geometry — the coordination of a space (for the study of its geometric properties) — to set up a framework for organizing and clarifying the diverse positions in the debate over the unity and disunity of science. I show that it is a good framework for our purposes by going through the existing positions in the debate and showing how it brings clarity and precision to our understanding of them.

1. Introduction

This essay is not aimed at supplying another argument for or against the unity of science; it is rather a search for a simple framework in which the existing positions in the debate can be better analyzed and evaluated. It offers to show in a nutshell that such a framework can be found in a well-understood problem in geometry, namely, the coordination of a space. This is done by first articulating a sense in which the problem of whether science is unifiable is similar to that of whether a space admits a global coordination and then going through the existing positions in the debate, hanging them on the frame and taking a careful look at what they mean and how they are related to one another. My analysis, if correct, reveals more than the conclusion that whether science is unifiable is not up to us but is determined by the 'shape' of the phenomenal world. It also offers such insights as the fact that two or more areas of science cohere — being consistent — at their boundaries does not necessarily imply the possibility of an ultimate unification of such areas.

In section 2, to get things going, I give a brief sketch of the core idea of the unity of science. Section 3 introduces the problem in geometry, and section 4 explains how the problem can be borrowed to analyze the various problems about unification. The discussion is carried out in general, without referring to any specific positions in the actual debate. Section 5 gives a detailed analysis of the existing positions in the literature.

2. The unity of science

What does it mean to have a global unification in science? To obtain precision, it is best to begin by way of elimination. I shall not consider unity of science as a political achievement, e.g. the establishment of a global NASA; nor shall I consider it as a unification of a code of conduct for scientific research; nor as a unified ideology: e.g. strict testability by experiments of every hypothesis as the criterion for being scientific; nor as a purely methodological unification, e.g. the fact that Hamiltonian differential equations are used both in physics and economics does not imply that the two disciplines are unified; nor as a giant conjunction of all true scientific theories.

I take the core concept of unity of science to be derived mainly from two sources: the Carnapian (or logical positivist) reductionism: all scientific theories being reducible to a theory of the observables; and physicalism: one theory (i.e. physics in the most general sense) for all scientifically recognizable phenomena. Surveying the literature of the unity/disunity debate, which will be selectively examined in section 4, one cannot fail to notice that the target notion of unity of science — for a kind of explanatory unification of scientific theories — contains the following elements (cf. Oppenheim & Putnam 1957).

[U1] A metaphysical postulate: the world — at least the actual one — is composed of a limited number of kinds of constituents (including one kind).

[U2] A unified fundamental theory: a single theory that accounts for the behavior of the fundamental constituents.

[U3] Reductionism: a set of reductional statements (or bridge lawlike statements) that connect statements of all other phenomena to the theory of the fundamental constituents.

An example of [U1] would be that nature ultimately consists of particles and fields (of a few kinds) and spacetime. An example of [U2], consistent with the metaphysical picture, would be a set of terms describing all possible states of particles and fields in spacetime — i.e. their behavior — and a set of lawlike statements in those terms expressing laws that determine such behavior. And an example of [U3], consistent with the above two, would be a finite set of statements that relate the theories for non-fundamental phenomena to the theory of the fundamental constituents. All phenomena are to be explained by the fundamental laws via such reductional relations. There are other possible images of unity of science, such as the Carnapian notion to which I will return in section 4, that are variations of the above. Suffice it to say that mine, which is essentially derived from Oppenheim & Putnam's, rests on the belief that if unification is to succeed, it will do so along or near the current trajectory of science.

It is also obvious — judging from the current state of science — that such a unification can only be defended as a goal that science may reach in the distant future. The debate is essentially on the question of whether or not the above should be regarded as a goal of science. But whether we should regard something as such-and-such depends on whether it can possibly — as a matter of metaphysics — be a such-and-such. How do we approach the metaphysical question of whether science is unifiable? We need a framework for thinking about how the structure of the phenomenal world may determine the structure of science. I shall try to show for the rest of the paper how the problem of coordinating a space in geometry suggests such a framework.

3. The coordination of a space

Ever since Descartes discovered the algebraic method of studying geometry, we know that one needs to put a coordinate system on the geometrical object one is studying before one can subject it to any algebraic analysis. Obviously, not all objects are suitable for coordination; and for the purpose of our study, we begin with the topological spaces, which, among other things, are connected and have well-defined open subsets. A topological space is a pair, $\langle X, T \rangle$, where X is a set of points and T a subset of the power set of X such that (i) the intersection of a finite number of members of T is also in T; (ii) the union of an arbitrary number of elements in T is also in T; and (iii) both the null set and X are in T. A coordinate system, or a chart, of any open subset of X is a map (or function) that sends every point in the subset to a unique tuple of numbers in R^n, where $R^n = R \times R \times \ldots \times R$ is an n-fold Cartesian product of the real numbers. This is how properties of points — i.e. geometrical properties — can be studied algebraically by relations among tuples of numbers. For instance, the chart for the 3-dimensional Euclidean space is the Cartesian coordinate system, R^3.

One may think that every manifold — i.e. a topological space that can be covered by charts — can be covered by a single chart. After all, if a manifold is coverable by a set of charts, it may also be seen as coverable by a single chart, which is produced by extending one chart to replace the rest. This, however, is not true. An illustrative example of the failure of a single global chart is given by the case of coordinating an n-dimensional sphere, S^n. It can be proven that it is impossible to map S^n onto R^n with a single mapping function. Figure 1 is an illustration of why it is the case. The easiest way to see this is to imagine an (n+1)-dimensional space in which S^n is imbedded, where in Figure 1 the 2-dimensional surface of the page represents the (n+1)-dimensional space and the 1-dimensional circle S^n. We put R^n (R_1 in Figure 1) at one point on S^n — call it S — as its tangent space. Then we pick the point on S^n directly opposite to this point, call it N, as the vintage point from which we stereographically project every point on S^n to R^n. Every point of S^n can be so projected — by ϕ — onto a distinct point on R^n, except N, since the projection line of N, if we can call it that, is parallel to R^n.

However, we can put another R^n at N as its tangent space and then we have another chart — R_2 in Figure 1 — that coordinates every point of S^n except S; and if we could use these two charts together, we would have completely 'covered' S^n. That we can use the two charts together is guaranteed by their *compatibility*, namely, a transformation, say, from R_2 to R_1 with respect to the same open subset of S^n does not introduce any distortion in the coordinate description of the subset. Mathematically, this property is reflected in the requirement that the two charts are C^∞-related, namely, for any non-empty intersection of two arbitrary open subsets of S^n, $\phi \cdot \psi^{-1}$ and $\psi \cdot \phi^{-1}$ are infinitely differentiable functions. The two charts, R_1 and R_2, are obviously C^∞-related because ϕ and ψ are C^∞ and since they are 1-1 onto maps, ϕ^{-1} and ψ^{-1} are also C^∞.[i]

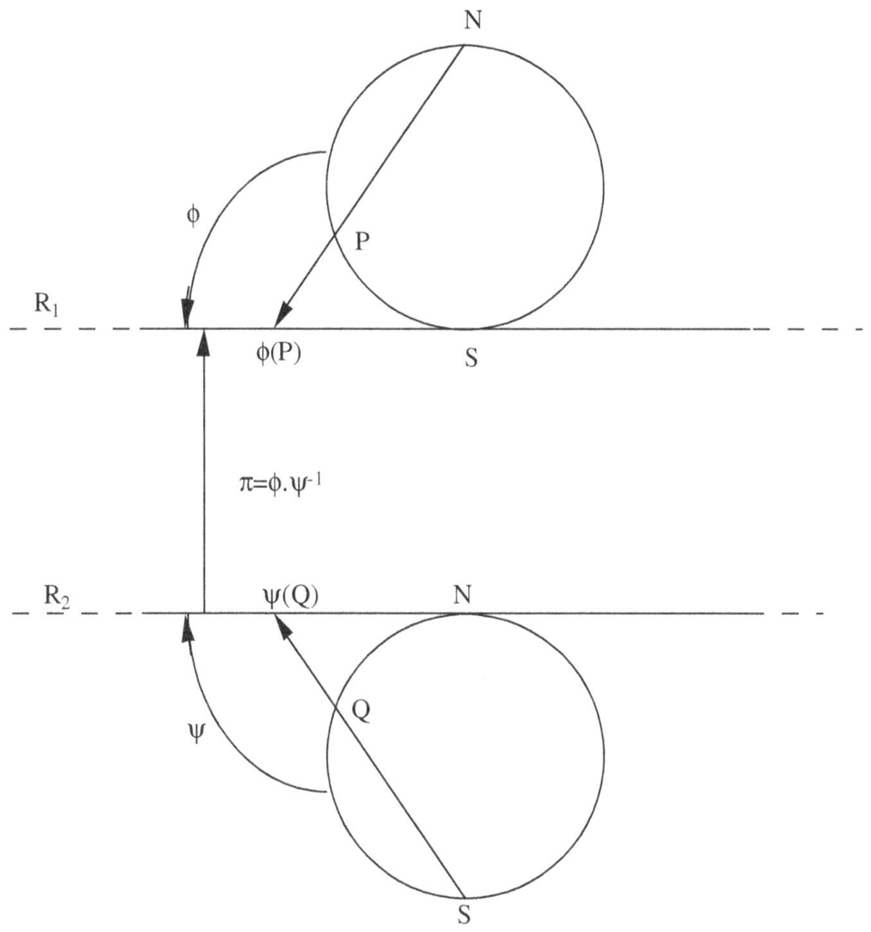

Figure 1: The coordination of an n-dimensional sphere in an (n+1)-dimensional space.

4. Setting up the framework by analogy

Let us now explore what the above problem in geometry may tell us about the unity of science. To properly set up the framework we must first examine the analogy between the coordination of a manifold and the epistemic representation of the world. How useful the framework may turn out to be of course depends on how much strength we may find in this analogy.

There is an obvious sense in which constructing scientific theories to describe the phenomenal world is similar to constructing charts to coordinate a manifold. However, one must be careful about how to pick out the counterparts of the manifold (and its geometrical properties), on the one hand, and the counterpart of the coordinations, on the other. Most importantly, the phenomenal world should not be thought of as the physical world that contains nothing but the fundamental stuff in spacetime. Because that would be presupposing the truth of reductionism, about which this study is supposed to shed light on rather than to presuppose. It, instead, should be viewed as the totality of all actual systems and events.[ii] Therefore, different open subsets of a manifold is analogous to different families of systems and events, whose various properties can be represented and studied by different theories in different branches of science. Different theories in science are in an obvious sense analogous to the coordinations of a manifold, and it may be understood as follows. A scientific theory — understood in the broadest sense — may be seen as the result of a map from a family of systems and events to a specific scientific language in which the behavior of the systems (under certain given circumstances, i.e. events) may be described. The map should be understood as an epistemic relation between the systems and the theories through which the theories may be tested for being true or false. It then follows that different charts for the same open subset of a manifold should correspond to different theories for the same family of systems and events (I shall hereafter use "systems" to mean "systems and events").

Here a slight complication arises. Different theories of the same system may be alternative theories of the same aspects of the system, e.g., Newtonian mechanics and special relativistic mechanics of the mechanical properties of a system; or they may be different theories of different aspects of the system, e.g., thermodynamics and mechanics of the same system. This difference, interesting though by itself, does not affect the validity of our discussion of unification.[iii]

It then follows that representing the same family of systems by different scientific theories is exactly analogous to coordinating the same open subset of a manifold with different charts. And the overlapping of differently coordinated open subsets of a manifold should then correspond to the overlapping of families of systems that are typically represented by different theories. We may conveniently take the term 'phenomena' to mean here, as a technical term, those systems typically represented by a theory or in a discipline. Therefore, mechanical phenomena are those systems typically studies by the theory of mechanics, and sociological phenomena those by sociology. And we can talk about the overlapping of phenomena in the same way as we talk about the overlapping of open subsets of a manifold. For example, let us suppose that biological phenomena obtain in systems belonging to B and economic phenomena in systems belonging to E, where B and E are respectively all and the only systems that exhibit biological and economic phenomena. B and E certainly overlap but they do not contain one another (assuming that robotic societies may have economies). On the other hand, it is probably safe to say that economic phenomena do not overlap with particle-mechanical phenomena.

Which such analogies in place, let us imagine that a group of people is given the task of coming up with charts to coordinate an unknown physical object that is larger than their sizes by many magnitudes. Suppose that these people are only interested in the purely geometrical properties of this object; so what they are doing differs from a similar task in geometry as instanced in the previous section not in essence but only in degrees of approximation. They divide themselves into several teams and disperse into several equally distant locations on the object and begin charting or mapping. Suppose again that from where each team starts, the surface looks flat. So they all begin with the straightforward Cartesian coordinate system for the respective tangent spaces. But as they move towards one another's neighborhood regions, they discover that the Cartesian coordinate system which fits so well at their own home-bases can no longer be used. Their first instinct is probably to try out other types of charts and see if more territory can be covered by a single chart. Let us suppose again that they obtain some success in doing this and by keeping on inventing new charts — i.e. using new mapping functions, they keep combining more and more previously incombinable regions into one under a single chart. (Let us remember that a chart is an arbitrary function from a geometrical object to R^n, where n is the number of dimensions of the object. So with enough ingenuity one may be able to do a lot more than what the simple example given in the previous section shows.)

To make the long story short, if the object happens to be one that is geometrically possible to be coordinated by a global chart, these geometers have every reason to believe that a global chart for the whole object will one day be discovered. The task may be difficult or requires a time longer than any of these people could possibly survive, if the object is geometrically complex and/or of an immense size, but we would not say that their push for a global chart is mistaken or wrong-headed. But on the other hand, if the object is something like S^n in the previous section, the task is impossible even though most of these people's effort will be rewarded. *Prima facie*, the same is true, *mutatis mutandis*, with scientists looking for theories to describe and explain the phenomenal world. The metaphysical question is: what is the 'geometry' of the phenomenal world?

What lessons, if any, can we draw from this analogy?

I. It is obvious that if the analogy holds, whether or not science can be unified — or whether or not it should be a goal of science — is not up to scientists and/or philosophers. It is determined by the properties of the world *and* the limits of our theory-construction capacities. The failure of a global chart for a manifold is caused both by the geometry of the manifold and the limitation of mapping functions. Something analogous should be true if unification is to be impossible. However, this does not mean that one has to be a realist in order to use this framework. The 'world' of which a unified theory is considered could be a world of 'experiences' or 'sense-data', if you will, and the question of unification may still be determined by the properties of this world and the limitation of our resources of theory construction. Therefore, being an empiricist cannot shield one from the above implications, nor can being an instrumentalist of scientific theories, because as long as theories are used as instruments to make predictions that have to answer to the world, whether or not they can be unified should still be a matter of the world and of the limitations of the instruments. One has to be a conventionalist or a social constructivist in order to escape this conclusion. If the real world or the experiential world is like a sphere and our scientists are like its geometers, their attempt to understand the world can never reach a global unification.

II. What happens if the geometrical object is proven to admit a global chart? It does not follow at all that the global chart should always be used in the algebraic study of the object, nor does it even follow that such a chart can always be found. The mapping function may be so complex that it is not practically discoverable or, even it is, not useful. The analogous case for science should be obvious. In fact, one may consider the following. Suppose that physicalism is established by some metaphysical arguments and, therefore, it is possible that the extended science of physics may have a single theory which accounts and explains, token by token, the entire set of states of affairs that comprise the phenomenal world. Should all scientists give up whatever they are doing and begin the pursuit and completion of such a unified theory? Would such a theory be useful, or necessarily superior to the current motley of special sciences, even if it can be formulated? Hence, the geometrical framework is more helpful with the notion of the impossibility rather than that of the possibility of a unified science.

III. The difference in the geometrical case between a chart and an atlas may dispel some of the confusion in the debate on unification. In geometry, there are spaces that are not coverable by charts. If a space is not a topological space, there is no guarantee that it can be covered by a set of compatible charts. Moreover, being covered by charts that are not all compatible with one another is another possibility, but a mere possibility only, for it borders on an abuse of the predicate, 'being covered.' Using charts to cover a space is to allow us to study every point and its neighborhood of the space by algebraic means; but if there are points that are coordinated by two incompatible charts, such a study of them would be impossible. And finally, the difference between being covered by a set of compatible charts — an atlas — and being covered by a single chart is precisely defined in geometry. The latter is the case when a single mapping function suffices to coordinate the entire space, while for the former more than one function are needed, where the transformations among the functions must be smooth, which means that they are compatible. The analogy with the case of science should lead us to three scenarios:

(a) a state of science in which one theory explains all (two senses of which will be discussed later);
(b) a state of science in which different theories explain different families of systems, while the theories are compatible (to be made precise later); and
(c) a state of science in which different theories explain different families of systems, while not all theories are compatible.

If our framework stands, the controversy between unity and disunity of science should be analogous to a choice between (a) and (b), while ruling out (c).

IV. In Figure 1, the two charts, ϕ and ψ, are almost global whether we use ϕ or ψ, only one point in each, N or S respectively, being missed. There presumably are many such cases in geometry where a manifold is practically covered by a single chart except for a few 'pathological' points or regions;

and the different charts we use are usually such that each of them makes one of such pathological points or regions appear normal — as S is normal in ϕ and N in ψ. The moral of this consideration is that even if a manifold is known to be non-coverable by a global chart, it does not follow that we must give up constructing and using a single chart for most of what we need to do. When confronted with the case in Figure 1, we may still have the freedom to choose either (i) to adopt one of the charts and extend it all the way until it reaches the neighborhood of this singular point, N or S; or (ii) to insist on using two distinct charts and do the transformations for the overlapping region, which by the way is the entire sphere except the points N and S.

However, one crucial difference remains between choice (i) — or rather its analogue in science — and choice **III** (a) above: a search for an 'almost' or 'approximately' unified theory while a genuine one is impossible can only be justified on purely pragmatic grounds. If such a theory is neither simple nor useful, we then have no reason to make it a goal of science. In other words, if analogous pathological regions can be found in the phenomenal world by transcendental arguments, the battle between the unificationists and the anti-unificationists can only be fought on purely pragmatic grounds. But if there is no reason to think that such regions exist, namely, a unified theory is metaphysically possible, the burden of proof rests then on the shoulder of the anti-unificationists. Unless it can be shown that it is practically impossible to reach that unified theory, the unificationists do not have to be moved by any other arguments.

V. If a manifold is coverable by a global chart, it by no means follow that only one such chart exist for that manifold. Therefore, we only talk about the existence of 'a' global chart, not 'the' global one. Similarly, if science is unifiable, it does not follow that there is only one theory which unifies science. So, we should only talk about the possibility of 'a' unified theory of science, not 'the' unified one.

On may think that the whole point articulated above is a trivial one since it is rather obvious that whether science is unifiable should depend on what the world turns out to be. I do not think this is so. The unity (or disunity) of science is not the same kind of concept as the truth (or falsity) of an interpreted sentence in a language or a theory. One may be justified in thinking that it is obvious that whether an interpreted sentence is true depends on whether what it says obtains in the world. The concept of a unified science is not like that. I have here two observations. First, there is no suitable analogue in the case of unity of science for the concept 'such-and-such obtains in the world.' It makes no good sense to use concepts such as 'a unified world or reality' so that one can say that whether science is unifiable ultimately depends on whether the world it represents is unified. For the second, we can perhaps use the geometrical case again to show that the case of unity is unlike that of truth. While it is obvious that whether an algebraic statement about some features of a space is true depends on whether those features of the space are indeed as what the statement says they are, it is not obvious at all whether a space can be coordinated by a single global chart has anything to do with any geometrical features of that space. It is rather more intuitive to think that if a space can be coordinated, it can be coordinated either by a global chart or by an atlas. It is rather surprising to realize that there are spaces which do not admit global charts. The same is intuitively true in the case of science as well. If we have theories which cover different kinds of phenomena and which are compatible in those areas where the kinds overlap, what prevents us from concluding that these theories can be molded into a single, unified theory?

To conclude, would it still seem as trivial — whether science is unifiable ultimately depends on what the world turns out to be — if it were the case that it is always possible to have a global chart for any manifold?

5. The unity vs. the disunity

Now let us make some contact with the existing philosophical literature on the unity of science.

For Carnap and the logical positivists (Carnap 1934, 1938; see also Neurath 1946; Morris 1946), the notion of unification should not carry any metaphysical commitment, namely, anything similar to [U1] would not make any sense. Carnap considers the notion of unity to consist of a unity of language and a unity of laws, but maintains that the former is more pertinent to science and more likely to be realized than the latter. Scientific theories are roughly divided into those of the physical and those of the biological, in the broadest senses of the two terms. The unification of language is to be achieved via reduction, though not of, for example, the biological to the physical (or vice versa), but rather of both to what he calls the (physical) 'thing-language.' The vision is an operationalist one in which all terms used in either the physical or the biological sciences are to be rendered meaningful by a specification of the conditions under which observations are obtainable that either confirm or refute those sentences in which the terms figure. Whatever is needed for describing the conditions and the outcomes comprises Carnap's thing-language.

There is a clear and simple sense in which the Carnapian conception of unity is very similar to our case in geometry. Is not the thing-language similar to a global chart? With a set of more or less homogeneous terms and a few rather simple rules of application, the thing-language, if capable of reducing all other terms in science, can in principle descriptively 'cover' the entire phenomenal world. Even the reduction from other languages to the thing-language finds its analogy in the geometrical case, where different local charts can be transformed to the global chart so that all geometrical properties are preserved under the transformations. One might say that for Carnap and the logical positivists, the 'space' — the object of science — is not the world or phenomena but the set of all possible experiences, which they claim is 'coverable' by a single thing-language. As for unity of laws (or of theories), it should be viewed as 'an aim for the future development of science. This aim cannot be shown to be unattainable. But we do not, of course, know whether it will ever be reached. (Carnap 1938, 61)'

The difference between the language and the laws with respect to unification (or the lack of it) seems to be one of whether or not a plausible general scheme is available by which reductions can in principle be carried out. In the case of the thing-language, a scheme was thought to be in hand while in the case of laws, no scheme of such specificity was nearly in sight. If the framework we set up in the previous section is correct, it should be clear that whether or not the thing-language can unify science, in terms of having a single language for science, depends almost entirely on the nature and structure of our experience, about which nothing less than a full-fledged metaphysical investigation can help us to find out. The positivist approach which shuns metaphysics and champions the logic of science — the syntax and the semantics of scientific languages — is clearly not an adequate approach. This version of the unity of science was justifiably given up, together with the whole doctrine of logical positivism/empiricism. I have no intention to rehearse the reasons, but it is gratifying to see that the geometrical analogy actually points to the right reasons for which the Carnapian notion of unification must fail. The possibility of a unified description of our experience is not merely a matter of the capacity of our language, just as it is not a matter of choosing the right mapping function that determines whether a manifold is globally coverable.

My earlier scheme of unification, [U1] - [U3], is obviously modeled after the Oppenheim and Putnam proposal (Oppenheim & Putnam 1958) (see also Margenau 1941), which hypothesizes a six-level hierarchy for science and emphasizes successive microreductions from higher to the lower levels. The sequence of microreductions ends with the physics of elementary particles. Unlike Carnap, the notion of reduction carries both metaphysical and epistemic obligations: not only theories at a higher level are reduced to those of a lower level in accordance with the standard account of Kemeny and Oppenheim (1956), but microreductions of one level of phenomena to another also imply compositional and structural (asymmetrical) relations between the reduced and the reducing. By such an image, the explanatory function of scientific theories — which goes in the opposite direction of reduction — becomes one of the key attributes of a unified science (which is exactly what scientists are after in a unification), and the reductional scheme roughly matches the one that the actual practice of science displays. Causey's version (Causey 1977) of unification is similar to Oppenheim & Putnam's, except it is a far more elaborate theory that goes beyond a simple and general hypothesis. It stipulates a logical structure for scientific theories, namely, it consists of fundamental laws, derivative laws, and identity relations (apart from logical terms), where the last item is mostly responsible for the microreductional relations; and because of this structure, it is able to offer a set of explicit conditions that a unified theory must satisfy (Causey 1977, 114-121). Such a vision of the unity of science was further enhanced by Friedman's theory of scientific explanation (or understanding) (Friedman 1974), in which convincing arguments are made at least for the case of *local unifications*. It also put in an extra dimension into the argument for unification, namely, scientific understanding of phenomena is mostly achieved only through reductive unification: we would not truly understand a set of unrelated phenomena until we see how they share the same underlying compositional elements and causal mechanisms.

Our framework provides a natural analysis for this notion of unification. Just as in the case of coordinating a manifold, whether or not science can in principle be unified must be ultimately determined by the properties of the world and the limitation of our resources of theory construction. Oppenheim and Putnam dismissed a trivial version of unification — science as a single conjunction of all theories — and avoided addressing the issue of a unified fundamental discipline (Oppenheim & Putnam 1958, 4). Our framework may help us see where exactly their position should be located, given these two moves. I shall discuss them in turn.

1. If the single conjunction of all theories accounts for everything and is true, it means that the theories are consistent with one another, especially when they deal with overlapping phenomena (whose meaning is given in the fourth paragraph of section 4). By dismissing this possibility, Oppenheim and Putnam is rejecting options **III** (b) and (c) in section 4. Hence, they are committed to **III** (a) which resembles the demand for a global chart. This is a significant point about the Oppenheim-Putnam position which may not be realized without the framework. A successful reduction of two theories which cover two overlapping sets of phenomena, either in the form of one reducing to the other or the two being reduced to a third, is quite similar to the merging of two maps of two overlapping open subsets of a manifold, which can also happen in either of two ways: one map may be completely replaced by the other — if one function can map the union of the two subsets in question — or a third map is introduced to replace the two. And the break-down of a reduction is therefore also similar to the break-down of a single chart in Figure 1: whenever neither the extension of the stereographic chart from N can cover point N (or S *mutatis mutandis*) nor a third chart can cover both S and N, we know that an atlas is required.

Fortunately, there is a simple case of an *apparent* break-down of reduction in science that I may mention briefly for an illustration (see Liu 1999 for more details). Thermodynamics of equilibrium had been considered successfully reduced to statistical mechanics in that all thermodynamical phenomena in equilibrium are accounted for by the theory of statistical mechanics via a general reductional scheme that also reflects the compositional structures of thermo-systems. However, it was then realized that certain phenomena, such as phase transitions (a transition, for instance, of a fluid from liquid to steam when boiling) and critical phenomena (e.g. opalescence), cannot be accounted for by the statistical mechanics of finite systems. Eventually, it was realized that one has to go the length of modeling the finite systems in which such phenomena occur as *infinite systems* before this 'hole' in the reductional scheme is plugged. When one expands to infinity — with certain restrictions — the size and number of particles of the system which is undergoing a phrase transition, what is said in thermodynamics about such a transition is rigorously reduced to what is said about it in statistical mechanics. Otherwise, a simple 'no-go' theorem proves the impossibility of such a reduction. This case is so similar to the case in Figure 1, it is almost uncanny. The stereographic chart from N, ϕ, is capable of covering the entire sphere except for the point N. One may be tempted to think that we may extend R^n to include the 'point of infinity' so that N can be mapped onto that point.

This case also presents an acute difficulty for the realists. Whenever, as in the above case, a reductional relation can only be rescued with certain maneuvres that appears to be purely formalistic, what should we believe about our world? Are the properties of phase transitions and critical phenomena as described by thermodynamics genuine properties or mere mathematical fluff? Either reduction fails — so fails unity of science — if they are mathematical fluff or it holds, and then we are forced to admit that some ordinary systems, such as a kettle of boiling water, are systems of infinite physical dimension.

2. Shelving the question of whether physics — or the fundamental science — is itself unified is in fact not as harmless as it seems. If one thinks that science is unified as long as everything else is proven reducible to physics, while physics itself may comprise different theories for different phenomena, one is quite mistaken. It might be thought that if everything else in the world is made of particles and every other theory is reducible to some or all of the theories of their behavior, science is adequately unified even if those theories of the particles are not. However, such a picture of physics is too simple-minded, and it does not do justice to what physics does and could do. Physics is not just a science for elementary particles and fields; it is also a science for ordinary objects, such as plants and people, and for large collections of such. If we have a unified theory of physics, we may then have a clear idea — e.g. a set of criteria — of which properties of such objects belong to physics and which to other special sciences, such as biology and psychology. But if we don't — which means all kinds phenomena which are not currently included in physics may eventually be included — then science may be unified quite trivially: work out all the reductions to physics that can be worked out among theories for different types of phenomena, and then take the remainder — which includes the non-reducible theories or theory parts — as also a part of physics (or the fundamental science). Science is therefore unified. For instance, if there is a part of the mental whose theory is apparently not reducible to the current science of physics, we can simply extend physics to include the theory. And if that is the only part in the whole enterprise of science that is not apparently reducible to the current physics, science is thus unified. It is only when we say that the physical must also be accounted for by a single theory, that such trivial cases of unification is ruled out. And this is why the drive within the community of physics for a unified theory is not a superfluous effort, (and also why I put [U2] in my scheme above). For what it means for physics to be unified and why it should be unified, see Erhard Scheibe's two-volume work on the 'Einheit der Physik' (Scheibe 1997/9).

In connection to the unification within physics, Morrison (1994) points out an interesting possibility that the theory seems unified while the things — particles and/or fields — are still 'disparate'. The actual unifications — such as the unification of the electric and the weak interactions — seem to give us nothing more than some formal or mathematical uniformities and/or similarities among the 'unified' entities; they do not tell us anything about their causal or substantival unity. If Morrison is right, it looks as if what I have dismissed in the beginning as one of the non-starters for unification, i.e. finding both physical and economical phenomena describable by Hamiltonian differential equations shows nothing about their unification, applies to physics as well. In sharp contrast, Einstein's original dream of reducing all natural forces to the 'geometry' of spacetime is a true unification in every ounce of its meaning.

Between the article of Oppenheim and Putnam and Causey's book, many discussions of the unity of science and, especially, of the possibility of reducing the mental or the social to the physical (which in principle affects the issue of unity of science) have appeared in the literature (some of which are cited and discussed in Causey's book), but none seems more significant than Fodor 1974, which threatens to shatter the dream of a reductional unification. It opened the floor gate for anti-reductionism arguments, especially regarding the reduction of biology and psychology (cf. Kitcher 1984; Rosenberg 1985; Kincaid 1990). After Fodor one must acknowledge that for any upper-level property in science, a reduction of it to some lower level properties is seriously threatened if it is multiply realizable at that lower level[iv]. With the realization that token-token identity does not imply type-type identity, one must see that metaphysical unity of the world no longer implies the unity of science via reduction. It might be true that all phenomena supervene on processes of the fundamental stuff of the world *and* that there exists a theory of that stuff that is capable of explaining the rest; but by these alone we cannot conclude that theories of upper-level phenomena will eventually be reduced to that theory.

It is important to note that Fodor's arguments are not necessarily against unification, but they are against *reductional* unification. Fodor alluded to his version of non-reductional physicalism as a revised or alternative version of unity of science (cf. Fodor 1974, 97ff), and Causey (1977, 142ff) called this version 'Tokenism,' which helps to accentuate the point. The difference may be illustrated by an actual example in science. In the case of 'reducing' thermodynamics to statistical mechanics I discussed above, it is not only possible but almost certain that all thermo-phenomena in equilibrium — including phase transitions and critical phenomena — supervene on configurations (or states) of molecules; the latter are accounted for by statistical mechanics of finite systems. And yet thermodynamics as a upper-level theory of thermo-phenomena is not reducible to statistical mechanics of finite systems for the aforementioned reasons. Phase transitions and critical phenomena as described in thermodynamic terms are only reducible to statistical mechanics of infinite systems, which does not seem to be reduction at all.

Now, the following seems to be a straightforward analysis of Fodor's token version of unity of science in our framework. Physicalism without reduction still does not imply the unity of science. (Therefore, it is perhaps too quick for Causey to grant Fodor's Tokenism as a genuine alternative conception of unity of science.) It is quite possible that everything else is token identical with the physical and yet the physical cannot be accounted for by a unified theory. Physics is not yet unified and may never be unified. Let me take some time to explain this point. The rationale for proposing reductive unification is that some things are obviously composed of other things as their parts. In the same way, the rationale for wanting to find a single theory that explain all fundamental physical processes is that they clearly show signs of being the instantiations of properties that belong to a single kind of stuff. But, again, if the geometrical analogy is apt, we must say that even if they are such instantiations, there is no guarantee that a single theory exists to account for all of them. The fact that all points belong to a single manifold — all topological spaces are connected — and that each point and its neighborhood can be mapped with a single chart onto R^n evidently does not guarantee that the whole object can be mapped onto R^n with a global chart. Similarly, the inference is not valid if it goes from everything being physical and each token physical event being nothing but a set of happenings at certain spatial and temporal locations — which appears to have already a definite description in physics — to there existing a globally unified science of physics. Whether physicalism, even if it is true, leads to the unity of science still depends on the 'contour' of the physical, not just on the describability of its parts.

Furthermore, even if the physical is accounted for by a unified theory — i.e. we have physicalism with the possibility of a single theory for everything, it may still be a long way from having the grand unification as a goal of science. Unlike a reductionist scheme, where the actual local higher-level explanations can still be done by the special sciences — knowing that such explanations can in principle be delivered from the bottom, the physicalist theory which has to reach higher-levels token by token for each actual explanation may have no practical plausibility. It may well take the whole machinery of science an unreasonable amount of time and resources to explain a 'simple' higher-level phenomenon, so that the theory is virtually of no use. One of the lessons from Fodor's article is that even if science is already proven unifiable via the 'token route', it would not help at all, because it would still be much better to go about one's business with the special sciences — i.e. to explain phenomena of different levels with theories at, or close to, those levels. In this sense, Quine (1969) is probably wrong in arguing that as science develops, more and more natural kind terms — we may substitute 'natural kind terms' with 'kind terms that are suitable and used by special sciences' — will be rendered useless and given up and replaced by explanations of a more homogeneous kind, such as by an account of the structures and functions of the ultimate constituents.

The framework may also be useful in analyzing other proposals for the unity or the disunity of science. At the same year when Causey's book appeared in print, two articles, one by N. Maull (1977) and the other by L. Darden and N. Maull (1977), came out, proposing a notion of unification 'without reduction', namely, unification via what they call 'interfield theories'. The idea is roughly this: between any two levels of phenomena, there are always phenomena which do not belong to either. The theory at each level however does reach down (or up) into them and can provide descriptions or explanations which, though compatible, are not the same. In such situations, an interfield theory — they regard levels of special sciences as consisting of 'fields' rather than 'theories' — is usually required and constructable. These interfield theories, rather than reductions, are the cement that join different fields of science and make it a unified whole.

Suppose, now, that we have two overlapping open subsets on a manifold, each of which has its own chart. Since the manifold is covered by an atlas, the two charts are compatible at the overlapping regions. This is analogous to the claim that in science all theories among traditional levels are at least consistent — i.e. if two claims from theories at two distinct levels are made about the same family of systems at an overlapping region between levels, they should not contradict each other. If none of the charts can well coordinate the overlapping region, then a third chart may exist that best coordinates it. If the manifold is not singlely coordinatable, it is quite possible that none of the three charts are replaceable by another. This, I believe, is an exact analogy of the situation that Darden and Maull considered. So, the kind of unification they propose is a unification of an 'atlas'. Note that their 'unity by interfield-theory connection without reduction' is exactly analogous to a complete covering by an atlas without the possibility of a global chart.

From our previous discussion, I believe that such a unification of science is a unification in name only. No one can prevent me from calling an atlas of a manifold a 'global chart', but .... Moreover, an unpleasant possibility appears for this notion of unity via interfield theories. Given reductionism or the physicalist unity via the 'token route', there is no need to be concerned about unification within a level of phenomena — except at the fundamental level. Even if biology or psychology is not unified within itself, science can still be unified when every piece in biology and psychology is reduced to a unified theory of physics (if reductionism is true) or when it can be discarded (if the token route works). But if the unity must be achieve by linking up the peripherals of each level of phenomena, it is no good if pieces inside some level are not 'connected' — i.e. unified in some sense. Whatever a unified science may turn out to be, a single piece of rag would not do.

The most recent advocates of the disunity of science, here I mention Dupré and Cartwright among others, are champions of metaphysical pluralism. Since their target is actually broader than unificationism — they also aim at fundamentalism (or foundationalism) (Cartwright 1999) and imperialism (Dupré 1994) — I can only lightly touch on the parts where the metaphysical thesis of unity of science are taken up. Dupré (1983, 1993) explicitly complained about the ineffectiveness of many previous anti-reductionist arguments because they only argue for epistemic or pragmatic impossibility of unification. For Dupré, if reality is metaphysically disordered or pluralistic, those arguments become superfluous. Here the geometrical analogy is useful again: if a manifold turns out to be one that is not coverable by a global chart, whatever other arguments concerning the power (or the lack of it) of mapping functions would be superfluous. Many people, including the ones whose works we have so far examined, believe that it is ultimately an empirical question as to whether reality can be captured by a unified science (which is fundamentally different from the case in geometry, where the possibility of a global chart is determined *a priori*). Dupré, if I understood him correctly, apparently thinks otherwise. Metaphysical pluralism is certainly not a thesis to be verified or falsified by empirical investigations. There are good reasons to think that if knowability of the world is a matter of metaphysics — how knowledge of any kind is possible — it is also a matter of metaphysics whether the world is knowable ultimately through a unified theory. Of course I do not suggest that there is any direct analogy between geometrically determining whether a manifold is coverable by a global chart and metaphysically determining whether the phenomenal world is describable and explainable by a single unified theory.

Also relevant is the difference between a non-coverable space — in the sense of no cover by atlas — and a coverable space which is not coverable by a global chart. Dupré does not always seem to be clear which counterparts of these alternatives he is arguing for. A metaphysically plural world could be either, but his arguments against Determinism — which move towards an image of the world as causally incomplete — seem to indicate that he is inclined towards the former — the analogue of an non-coverable space.

For Cartwright (1994, 1999), whatever resembles a global chart for a manifold is out of the question for science. She clearly supports anti-reductionism and equal rights for special sciences, and her arguments mostly derive from a deep appreciation of the central role of models in science. From the apparently impossibility of constructing a mechanical — more accurately, a compositional-mechanical model — for a thousand dollar bills blown apart by a gust of wind, she concludes that mechanical laws do not apply to the situation; but since a hydrodynamic model is still possible for it, the laws of hydrodynamics may apply. From that, she infers that hydrodynamics is not reducible to mechanics — more accurately, it is not micro-reducible to the mechanics of the constituents of the hydrodynamic systems in question. (Note, there is no metaphysical question about whether the system of the wind-blown bills is composed of a thousand dollar bills.) The same move also seems to block a unification via the token route, for according to Cartwright, no models, no laws, and hence no explanations by laws. A more urgent question is whether Cartwright, like Dupré, also does not care for the analogue of an atlas for science. Sometimes, her image of Nature — '[who] has a rich, and diverse, tolerant imagination (1994, 361)' — seems to indicate that Nature resembles more of a space not coverable by an atlas.

However, Cartwright has a move, especially in her discussion of the relationship between classical and quantum mechanics, that seems to deprive our framework of its relevance. Whenever there seems to be an inconsistency in assigning a quantum and a classical state to the same physical state of a system, given we are fully justified in doing so, we can deny that they are about the same state, and the contradiction disappears. 'There are both quantum and classical states and the same system can have both without contradiction. (Cartwright 1994, 362)' There is no analogue of this in our case of coordinating a space: if there are two open subsets on a space whose charts are not compatible at the overlapping regions, we cannot appeal to any relevant differences of the two subsets to make the charts compatible again. There must be tough cases in which making such a move may cause difficulties. For instance, what if according to a 'true' classical description the age of the universe is A, and according to a 'true' quantum description it is B, and $A \neq B$. Whether the answer is positive or negative determines whether or not Cartwright subscribes to the analogue of an atlas in science.

6. Conclusion

If our framework is correct, physicalism does not imply the possibility of unification; nor does it imply reductionism. For the latter one does not need the geometrical analogy (because of Fodor 1974) but for the former one does. If the world is unifiable, the framework further tells us the following.

Metaphysically speaking, there are two possible alternatives for unifying science: one is to have a horizontally unified fundamental theory — e.g. physics — and then to have every other phenomenon explained token-by-token; the other is to have special sciences for different levels of phenomena and achieve the unity by complete reductions to the (self-unified) theory at the fundamental level. The latter implies the former but not vice versa. However, epistemically speaking, the former is almost impossible while the latter is not. Unless our current conception of physics of elementary particles and fields are fundamentally mistaken, it is almost ridiculous to think of a direct explanation of, say, French Revolution, by laws and initial/boundary conditions of the elementary particles and fields involved.

But on the other hand, what constitutes a break-down of reduction? The toughest kind of questions that philosophers have to face regarding judgments on reduction is the kind instanced in the reduction of thermodynamics to statistical mechanics that we examined earlier. When in the end mathematical physicists proved that thermodynamics is only reducible to statistical mechanics of infinite systems, what should we say? *Ceteris paribus*, is science unifiable or not? In our geometrical case, if one can find a systematic way of extending the map of a manifold to its pathological points or regions, there seems to be no reason why the extended map cannot be taken as a global chart for the manifold. The same, *mutatis mutandis*, seems applicable to the scientific cases, at least to the case in question, where finite statistico-mechanical systems (and laws and types of conditions that go with them) are extended to infinite such systems so that the whole realm of thermo-phenomena, at least, is adequately 'covered' — via reduction — by statistical mechanics. But if no such extension is available, we probably should call it quits. For a fascinating discussion on how techniques of taking thermodynamic systems to infinity suggests a solution to the problem of multiple realizability, see Batterman 2000.

References

Batterman, R. (2000). "Multiple Realizability and Universality." British Journal for Philosophy of Science. **51**: 115-145.

Carnap, R. (1938/1955). "Logical Foundations of the Unity of Science." Eds.

O. Neurath, R. Carnap, and C. Morris, International Encyclopedia of Unified Science: Volume 1. Chicago, Chicago University Press, 1955, pp. 42-62.

Carnap, R. (1934/1995). The Unity of Science. Bristol, Thoemmes Press (reprinted 1995).

Cartwright, N. (1994). "The Metaphysics of the Disunified World." PSA 94, vol. 2, New Orleans, PSA, pp. 357-364.

Cartwright (1999). The Dappled World: A Study of the Boundaries of Science. Cambridge, Cambridge University Press.

Causey, R.L. (1976). "Unified Theories and Unified Science." PSA 74, Dordrecht, Reidel, pp. 3-13.

Causey, R. L. (1977). Unity of Science. Dordrecht, Reidel.

Darden, L. and N. L. Maull, (1977). "Interfield Theories." Philosophy of Science **44**: 43-64.

Dupré, J. (1983). "The Disunity of Science." Mind **92**: 321-346.

Dupré, J. (1993). The Disorder of Things: Metaphysical Foundations of the Disunity of Science. Cambridge, MA, Harvard University Press.

Dupré, J. (1994). "Against Scientific Imperialism." PSA 94, vol. 2, New Orleans, PSA, pp. 374-381.

Fodor, J. (1974). "Special Sciences (Or: the Disunity of Science as a Working Hypothesis)." Synthese **28**: 97-115.

Friedman, M. (1974). "Explanation and Scientific Understanding." The Journal of Philosophy **LXXI**: 5-19.

Isham, C. J. (1989). Modern Differential Geometry: For Physicists. Singapore, World Scientific.

Kemeny, J. G. and P. Oppenheim, (1956). "On Reduction." Philosophical Studies **7**: 6-19.

Kincaid, H. (1990). "Molecular Biology and the Unity of Science." Philosophy of Science **57**: 575-593.

Kitcher, P. (1984). "1953 and All That: A Tale of Two Sciences." Philosophical Review **93**: 335-376.

Liu, C. (1999). "Explaining the Emergence of Cooperative Phenomena." Philosophy of Science **66**(S): 92-106

Margenau, H. A. (1941). "Foundations of the Unity of Science." Philosophical Review **50**: 431-438.

Maull, N. L. (1977). "Unifying Science without Reduction." Studies in the History and Philosophy of Science **8**: 143-162.

Morris, C. (1946). "The Significance of the Unity of Science Movement." Philosophical and Phenomenological Research **6**: 508-514.

Morrison, M. (1994). "Unified Theories and Disparate Things." PSA 94, vol. 2, New Orleans, PSA, pp. 365-373.

Neurath, O. (1946). "The Orchestration of the Sciences by the Encyclopedism of Logical Empiricism." Philosophical and Phenomenological Research **6**: 496-507.

Oppenheim, P. and H. Putnam, (1958). "The Unity of Science as a Working Hypothesis." Minnesota Studies in the Philosophy of Science. vol. 2. Eds. H. Feigl et al. Minneapolis, University of Minnesota Press.

Quine, W. V. (1969). "Natural Kinds." in W. V. Quine, Ontological Relativity and Other Essays New York, Columbia University Press, 114-138.

Rosenberg, A. (1985). The Structure of Biological Science. Cambridge, Cambridge University Press.

Scheibe, E. (1997 & 1999). Die Reduktion physikalischer Theorien: Ein Beitrag zur Einheit der Physik. Teilen I & II. Berlin, Springer-Verlag.

[i] The actual proof that $\phi \cdot \psi^{-1}$ and $\psi \cdot \phi^{-1}$ are infinitely differentiable goes something as follows. First we imbed the stereographic maps into the higher dimensional space so that a normal Cartesian coordinate system can be given to them. Then we can prove that ϕ and ψ and their inverses as functions in such Cartesian coordinates are infinitely differentiable functions. For an example of such proofs, see Isham 1989, 3-4.

[ii] For the sake of simplicity I shall here use systems **and** events as comprising the world that the sciences study. For the purists on ontology, it may be better to think of the world as consisting of events alone, while systems can be regarded as abstractions from sets of events. Because we rarely see an actual scientific theory taking as its subject nothing but events, I shall use the phrase 'systems and events' to denote the items on the side of the represented, as distinct from those on the side of the representing.

[iii] It **may** become an interesting dimension for the problem if unification. For instance, the relationship between Newtonian and relativistic mechanics is not entirely unlike the relation between hydrodynamics and particle mechanics. It is possible to view hydrodynamics as an approximate mechanical theory of the molecules just as Newtonian mechanics is an approximate relativistic theory. And just as there are borderline systems to which it is a matter of taste or convenience whether Newtonian mechanics or relativistic mechanics should be used, there are also borderline cases (in the same sense) between hydrodynamics and particle mechanics. However, I shall not take up the development of such 'extra' dimensions of unification in this paper. Hence, I assume whatever theory is in question, it

is that which is currently recognized as the correct or the leading theory for the family of systems and events in question.

[iv] One may be tempted to say that reduction then fails, but this may be too strong, see Batterman 2000.

INTERNATIONAL STUDIES IN THE PHILOSOPHY OF SCIENCE,
VOL. 17, NO. 2, 2003

Routledge
Taylor & Francis Group

Gauge gravity and the unification of natural forces

CHUANG LIU
Department of Philosophy, University of Florida, Gainesville, Florida, USA

Abstract *Physics seems to tell us that there are four fundamental force-fields in nature: the gravitational, the electromagnetic, the weak, and the strong (or interactions). But it also seems to tell us that gravity cannot possibly be a force-field, in the same sense as the other three are. And yet the search for a grand unification of all four force-fields is today one of the hottest pursuits. Is this the result of a simple confusion? This article aims at clarifying this situation by (i) reviewing the gauge-field programme and its conception of unification of force-fields, (ii) examining the various attempts at a gauge theory of gravity, and (iii) articulating the nature of "gauging" and using it to explain the difference between gravity and the other force-fields.*

1. Introduction

If we ask what reality is like if the current theories of physics are true or approximately true or, at least, on the right track, an answer from the physics community may roughly run as follows. The physical world—in the broadest sense of "physical"—consists of matter in spacetime whose parts interact with one another via force-fields.[1] It seems likely that all objects in bulk are composed of a few species of elementary particles which affect each other's behaviour through one or more of the four fundamental force-fields (or interactions), namely, gravity (= the gravitational field), electromagnetism, the weak interaction, and the strong interaction. Gravity is identified with spacetime, if Einstein's general theory of relativity (GR) is true or approximately true, and the other three fields are gauge-fields, if the standard model of high-energy physics is true or approximately true.[2] Because of the wave–particle duality in quantum theory, elementary particles are also fields (i.e. matter-fields) and force-fields are also particles (e.g. photons and gluons), but they still belong to markedly distinct categories.

There is nothing inconsistent or intrinsically wrong about this world picture, but the following puzzle arises for the force-fields in the form of a dilemma. Either gravity is one of the fundamental forces in nature or it is not. If it is, it then makes sense to pursue the dream of a unified field theory—a theory which provides, or at least suggests, an ontology in which the four force-fields are different *manifestations* of a single field or substance. However, according to our best theories today, gravity alone—whether in its classical outfit or when it is eventually quantized—is identifiable with the 4-dimensional spacetime.[3] This implies many asymmetries among the relations between gravity and the other three force-fields. For instance, the other three interact with one another in the background of spacetime, while their interaction with gravity is nothing more than their

ISSN 0269-8595 print/ISSN 1469-9281 online/03/020143-17 © 2003 Inter-University Foundation
DOI: 10.1080/0269859031000160612

being in spacetime! Therefore, it is reasonable to believe that gravitational force is no force at all (more later on how "force" is traditionally construed). Hence, it may be better to exclude gravity from the category of fundamental forces of nature; but if so, the pursuit of including it in the grand unified field theory—which has ranked so high on the agenda of physics for several generations—would not make much sense.

The dilemma is resolved if one can reduce the other three forces to spacetime as well. There are then no forces in the traditional senses of the word, and our current theories of force-fields will be abandoned just as the theory of phlogiston was. This, as we all know, is the type of grand unification that Einstein envisaged and devoted most of his career pursuing (cf. Pais, 1982; Vizgin, 1994). The failure of Einstein's (and others') effort in finding such a unification has put a damper on, if not altogether stopped, the pursuit of such a dream.[4]

Partly because of such an asymmetry, several philosophers of science recently raised questions about the rationale of regarding gravity as a gauge-field. Maudlin argues:

> there seems to be a fundamental incompatibility ... between the basic approaches of gauge field theories and GTR.... Objects couple to the [gauge] field only if they interact through some charge that serves as a coupling constant.... But according to GTR, *gravity simply is not a force*.... Particles do not couple to the gravitational field, they simply exist in space-time. (Maudlin, 1996, p. 143, emphasis added)

Therefore, even if there is a unified theory with $SU(3) \otimes SU(2) \otimes U(1)$ for the three gauge-fields, it is questionable whether we should expect gravity as the next stop. Weinstein (1999) observes that even though there is a sense in which the diffeomorphism group—Diff(R^4)—of GR is a kind of gauge group (see Wald, 1984), the gauge groups for the gauge-fields are so radically different from it that to think of GR as a gauge-field theory would be misleading. Callender and Huggett (2000) pose the question, "Why quantize the gravitational field?", in the title of their article. Besides pointing out many seemingly insurmountable conceptual difficulties in the quantization of the gravitational field, they explode the myth that the quantization of gravity is required by the very consistency of physical theories. Even if one adopts the canonical formulation of quantum gravity from which the theory can be viewed as a gauge-field theory, one still runs into two types of problems. First, the canonical quantum gravity is not the same kind of gauge-field as the other force-fields in the programme. Second, there are grave problems in canonical quantum gravity involving (a) the nature of observables and (b) the very possibility of time and change, which are not present in quantum gauge-field theories of the other forces. (For details, see Belot & Earman, 2001 and the references therein.)

It would be naîve to think that it is entirely an empirical question as to whether and how a grand unification including gravity can be achieved. Little if any progress in settling fundamental questions in physics such as this one has ever been made without philosophical insights. While carrying out a full investigation of all the possibilities in the logical space for the question of the unification of force-fields needs the length of a monograph if not more, I shall focus my attention on the following possibility—which is diametrical to Einstein's original conception of unification. Could GR be modified or replaced by another theory so that gravity can be made to join the other force-fields? Weinstein is right in arguing that GR in its original formulation is no gauge-field of the Yang–Mills (YM) type, it does not, however, mean that a gauge theory of gravity—

which differs from GR—is not possible, which both meet the experimental findings for gravity and is a genuine YM gauge-field.[5]

2. The gauge-field programme

Let us begin with the programme in which a unification of the "other" fields under the YM gauge-field model is carried out. Just as in GR, where a term that can be used to represent the gravitational effect arises from the local Poincaré invariance of the laws, a term representing the effect of a YM gauge-field arises from the local gauge invariance of the laws of physics. I shall explain in detail the former in the next section. Here I briefly sketch the gauge-field programme that produces theories and unifications for gauge-fields of the YM type.

We begin with a picture of the physical world as having particles as quantum fields existing in a 4-dimensional Minkowski spacetime, M. The dynamics of these matter-fields are determined if the Lagrangian of each type is formulated and justified, a scheme virtually identical to the classical one, except here we must use *Lagrangian densities* of fields. Beside being invariant under the global Poincaré group of transformations—a requirement for any relativistic field theory—the Lagrangians also enjoy symmetries of other kinds. For instance, electrons enjoy the symmetry of the U(1) group of global transformations: $\psi(x)\rightarrow\gamma\psi(x)$, $\gamma=e^{-i\Lambda}$, where Λ is an arbitrary constant, and leptons enjoy the symmetry of the SU(2) group of global transformations of similar form, and so on; more on this later. These global symmetries produce conservations of the corresponding charges. Once the constant factors in those transformations are made functions of $x\in M$, an act commonly referred to as "*localizing*" or "*gauging*", new degrees of freedom open up. The aim of the gauge-field programme is to geometrize the "other" force-fields in these new dimensions (= degrees of freedom), the general scheme of which runs roughly as follows (for details, see Göckeker & Schücker, 1987; Frankel, 1997).

Given a spacetime M and a Lie group,[6] G, into which M is mapped by local gauge transformations $\gamma\in G^M$, a connection 1-form, A, is introduced (by the demand of gauge invariance) on G which transforms as

$$A\rightarrow A'=\gamma A\gamma^{-1}+\gamma d\gamma^{-1}. \tag{1}$$

Then, the exterior covariant derivative D is introduced for G as

$$D\Psi := d\Psi + A\wedge\Psi \tag{2}$$

where Ψ is any vector-valued p-form (i.e. it takes values in a vector space which is a representation of G) and d is the ordinary exterior derivative which maps a p-form to a $(p+1)$-form and "$\wedge$" is the wedge operation. With that, the curvature of G can be simply defined as: $F:=DA$, a 2-form. In other words

$$F=DA=dA+A\wedge A=dA+\frac{1}{2}[A,A] \tag{3}$$

where [...,...] is the Poisson bracket. The Lagrangian of a gauge-field then has the following general form:

$$(D\Psi)^{+}\wedge{}^{\star}(D\Psi)+\frac{1}{q^2}\mathrm{tr}(F\wedge{}^{\star}F)-m^2(\Psi^{+}\wedge{}^{\star}\Psi) \tag{4}$$

where q and m are respectively the charge and the mass of the matter-field which is represented here by Ψ; "$\star$" is the Hodge star operator on forms (analogous to the indices-raising operator for tensors on M), Ψ^+ the Hermitian conjugate of Ψ. The first is the dynamic term, the second the field term, and the third the "mass" term.

The symmetry group for a particular force-field is determined by the transformational property of the matter-field, i.e. of Ψ; for instance, for electromagnetism, it is U(1) because the charged particles are represented by a Ψ of complex singlet. For the weak interaction, the group is SU(2) because the Ψ of the leptons are complex doublets, and similarly we have SU(3) for gluons of the strong interaction because the Ψ of quarks are complex triplets. The force-fields corresponding to these groups also correspond uniquely to different charges, even though the particles may carry more than one charge. With G determined, gauge invariance under G makes it *possible* for the introduction of a connection, which in turn determines the exterior covariant derivative and the curvature. And with these the total Lagrangian, including the dynamics and the force-fields, is determined.

One should note that in the gauge-field programme a gauge invariance under G only provides the *possibility* of introducing a connection but by no means determines its form (see Cat, 1993; Liu, 1999; Teller, 2000). This point can best be seen in contrast with the analogous case in GR. The geodesic equation for a free particle in Minkowski spacetime reads simply as: $d^2u^\sigma/ds^2 = 0$ (where u^σ describes the particle's state in a Cartesian coordinate system, and ds is the proper time), and it is invariant under any rigid (i.e. global) Poincaré transformations. Let us now switch to a general spacetime, M, and let the coefficients of the transformations become functions of $x \in$ M, and in so doing we switch to arbitrary coordinate systems. Now, u^σ in a curvilinear system x^μ becomes a function of x^μ. By a simple application of the chain rule, we obtain the following geodesic equation:

$$0 = \frac{d}{ds}\left(\frac{\partial u^\sigma}{\partial x^\mu}\frac{\partial x^\mu}{ds}\right) = \frac{\partial u^\sigma}{\partial x^\mu}\frac{d^2x^\mu}{ds^2} + \frac{\partial^2 u^\sigma}{\partial x^\mu \partial x^\nu}\frac{\partial x^\mu}{ds}\frac{\partial x^\nu}{ds}$$

which, when multiplied by a factor $\partial x^\lambda/\partial u^\sigma$, turns into the geodesic equation in GR

$$0 = \frac{d^2x^\lambda}{ds^2} + \Gamma^\lambda_{\mu\nu}\frac{\partial x^\mu}{ds}\frac{\partial x^\nu}{ds}$$

where $\Gamma^\lambda_{\mu\nu}$ turns out to be the needed affine connection for the description of gravity. One may conclude that the "gauging" of the transformations induces and determines the connection that fits exactly the role of representing the effect of gravity. The connection for a gauge-field cannot be so introduced. Take the U(1) case in which the Lagrangian density of a charged scalar matter-field has the form: $(\partial_\mu\psi\partial^\mu\psi - m^2\psi^+\psi)$ in some suitable basis, where $\partial_\mu = \partial/\partial^\mu x$. It is invariant under the global U(1) transformations, but when they become local, as functions of x^μ, the chain rule of differentiation does not introduce the connection. To see this, suppose we have $\psi(x) \to \gamma\psi(x)$, $\gamma = e^{-i\wedge(x)}$, then

$$\partial_\mu\psi \to \partial_\mu\psi' = \partial_\mu(e^{-ie\wedge(x)}\psi) = (\partial_\mu\psi - ie\psi\partial_\mu\wedge)e^{-ie\wedge(x)}$$

and hence

$$\partial_\mu\psi\partial^\mu\psi \to (\partial_\mu - ie\partial_\mu\wedge)\psi^\star(\partial^\mu - ie\partial^\mu\wedge)\psi.$$

However much the right-hand side of the arrow resembles the result of the "minimal coupling", i.e. $\partial_\mu - ieA_\mu$, in the gauge-field programme, it is in fact not. If we define the connection as $A^\mu = \partial^\mu \wedge$ (i.e. $A = \mathrm{d}\wedge$ in the abstract formulation), what is introduced is an identically flat connection because the curvature in this case is $F = \mathrm{d}A = \mathrm{d}^2 \wedge \equiv 0$. Hence, the "gauging" of the U(1) manifold alone is not sufficient to introduce a term that may represent the effect of electromagnetism.

The quantum weak force-field operating among leptons and quarks is considered to be unified with quantum electrodynamics into a theory of electroweak force-field under the structure group, $SU(2) \otimes U(1)$.[7] What kind of unification is it, if it is one? Since its structure group is a Cartesian product of two groups, its exterior covariant derivative is the sum of the two separate minimal couplings, i.e. from (2)

$$\mathrm{D}\Psi = \mathrm{d}\Psi + (A_e \wedge \Psi) + (A_w \wedge \Psi)$$

and its Lagrangian is also a sum, i.e. from (4)

$$(\mathrm{D}\psi)^+ \wedge \star(\mathrm{D}\psi) + \frac{1}{e^2}\mathrm{tr}(F_e \wedge \star F_e) + \frac{1}{q^2}\mathrm{tr}(F_w \wedge \star F_w) - V(\psi^+ \wedge \star\psi),$$

where $F_e = DA_e$ and $F_w = DA_w$ are the corresponding curvatures (see (3)), $V(\ldots)$ is a quadratic function of $\psi^+\psi$, and

$$\psi = \binom{\psi_1}{\psi_2}$$

is a complex doublet. For the U(1) field of electromagnetism, from (1) above we have $A_e = \mathrm{i}eA$, where e is the charge and A the potential vector of the electromagnetic field; and $F_e = \mathrm{d}A$, since the U(1) field is Abelian, $A \wedge A = 0$. For the SU(2) field of the weak force, we have $A_w = \mathrm{i}qW^k \tau_k$, where q is the charge of the weak source, W^k and τ_k ($k = 1$, 2, 3), are respectively the potentials (a triplet) and the generators of the weak field; and the curvature is $F_w = \mathrm{d}A_w + (1/2)q[A_w, A_w]$.[8]

The SU(2) part of the electroweak field is the typical YM gauge theory for the weak isotopic spin space, where when electric charge is ignored there is a perfect local symmetry between the particles in a doublet, such as $\{e, v\}$, the electron and its neutrino. Besides the phenomenological charged weak bosons, $W^\pm$, where $W^\pm := (1/\sqrt{2})(W^1 \pm \mathrm{i}W^2)$, which were known before the theory was invented, the theory predicts a neutral weak boson, W^3, which is, moreover, *mixed* with the photon. One finds the mixing in the neutral interaction terms of the Lagrangian density, such as the term for a pair of neutrino and antineutrino. There, we have the phenomenological (or observable) neutral weak field as

$$Z^0 = (eA + qW^3)/\sqrt{q^2 + e^2}; \tag{5}$$

and then, the phenomenological photon—which must be "orthogonal" to Z^0—comes out naturally as

$$A^{em} = (qA - eW^3)/\sqrt{q^2 + e^2}. \tag{6}$$

Accordingly, the phenomenological electric charge and weak charge are different from q and e: the electric charge is $E = qe/\sqrt{e^2 + q^2}$, and the neutral weak charges are $G_v = \sqrt{e^2 + q^2}$ (coupling with neutrinos), $G_{eR} = e^2/\sqrt{e^2 + q^2}$ (with right-handed electrons), and $G_{eL} = (e^2 - q^2)/\sqrt{e^2 + q^2}$ (with left-handed electrons). (The difference of the last two is the result of the parity violation in weak interactions.)

To complete the picture, one needs to add the Higgs mechanism which spontaneously "breaks" the gauge symmetry of the original theory and therefore makes it possible for the W-bosons to have non-zero masses (cf. Higgs, 1974; Aitchison, 1982;

Quigg, 1983); and furthermore, the SU(3) theory for the strong interaction is also of YM type and the SU(3)⊗SU(2)⊗U(1) unification is very much like the SU(2)⊗U(1) unification.[9]

To summarize, the gauge-field programme offers a *two-step* procedure for the unification of force-fields: first we find out whether or not some fields are gauge-fields of the YM type (whose general scheme I sketched above), and then if they are, we find out whether or not we can formulate a single theory for them, whose gauge group is either a single group or a tensor product of their own gauge groups.

At this point, one may question in what sense I am regarding the gauge-field programme as one for unification. From a philosophical point of view, there may well be many different senses for the notion of unification. *Reductive unification* is a general category which contains different notions. Reduction may refer to an intertheoretical relation, or it may refer to an ontological one. With each there is a 2-place reduction and an n-place reduction. A 2-place intertheoretical reduction may be roughly understood as follows. A theory U is *reducibly unified* by another theory V if, and only if, V has broader applications than U and U can be accounted for by V—e.g. is derivable with bridge principles from V.[10] And an n-place intertheoretical reduction can be understood similarly. A set of $n-1$ theories, Σ, is *reducibly unified* by another theory V if, and only if, V has broader applications than any member of S and every member of Σ can be accounted for by V. There are also non-reductive notions of unification, one of which that is relevant for us is the notion of unification in which the theories being unified shared the same set of axioms or principles and are derivable from them with the help of different constitutive laws or conditions. Such a unification may be purely formal.

The gauge theory of the electroweak force-fields is not a 2-place reductive unification; but it may be regarded as a 3-place one, since one may take the SU(2)⊗U(1) gauge theory to be the theory that reduces and unifies the two previously separate force-fields. So can one regard the unification of the strong, the weak, and the electromagnetic in the SU(3)⊗SU(2)⊗U(1) gauge theory as a 4-place unification for the same reason. However, such unifications are not very ideal for two reasons. First, a SU(3) gauge theory for the electroweak and a SU(5) gauge theory for the three previously separate force-fields, in each of which only one coupling constant (or charge) is present, would be *better* unifications. Second, there are even better examples of unified field theories in physics which makes these unifications look pale in comparison. One of the best, and most well known, such cases is the unification of the electric and the magnetic field in Einstein's special theory of relativity (cf. Maudlin, 1996). There, the separate appearance of the electric and the magnetic field becomes purely a choice of inertial reference frames; and since the privilege of any inertial frame is completely removed in the theory, any residual independence of the two fields is also removed. The Minkowskian unification of space and time into a 4-dimensional spacetime may also be regarded as such a case. The YM gauge-field programme offers no such feats.

While it is important to clarify in what sense the YM gauge-field programme should be regarded as a scheme of unification, the following is, I think, more important. The real meaning of step one of the programme is not for unification (whose realization depends more on how step two turns out); it is rather to provide a "unified" answer to the question: what is it to be a force-field? I may even venture to say that it answers the question: whence does a force-field come? For the answer from the gauge-field programme seems to be that it comes from *gauging*: the production of extra dynamical degrees of freedom via localizing appropriate gauge symmetries. However, as I explained above, gauging does not determine uniquely the connection but only makes it possible for its introduction, and hence gauging of the appropriate dynamical degrees of freedom is not sufficient for a gauge theory of the corresponding field. It does, however, appear to be necessary, i.e. a gauge-field cannot be introduced without a localized gauge symmetry. Therefore, gauging (or local gauge symmetry) is indeed essential to the presence of a force-field. More on this in Section 4.

3. Gauge theories of gravity

Can gravity be unified with the gauge-fields in the programme? To answer it we need to first find out whether it can be made into a gauge-field as the others in the programme. Utiyama (1956) is widely regarded as the first person to have explicitly proposed a gauge theory of gravity. His theory was later improved by Kibble (1961) and Sciama (1962). Trautman later argued that the Utiyama–Sciama–Kibble theory can be recognized "as being essentially equivalent to Cartan's theory of 1923" (Trautman, 1980, p. 305). The Cartan theory—now known as the Einstein–Cartan theory (see also Trautman, 1972, 1973; Göckeler & Schücker, 1987; Prasanna, 1989)—is a theory of gravity on a Riemann–Cartan spacetime, U^4, which is a paracompact, Hausdorff, connected, C^∞, and oriented 4-dimensional manifold, on which are defined a local Lorentz metric g and a linear affine connection Γ, which transformed just like A in (1), with $\gamma \in GL^+(4, R)$—the set of all arbitrary linear transformations on the manifold with positive determinants and is *metric*, i.e. $Dg = 0$. From Γ a curvature and a torsion tensor can be obtained on U^4. If the torsion tensor vanishes, the Riemann–Cartan spacetime becomes the pseudo-Riemannian spacetime, V^4; if the curvature tensor vanishes, it becomes the Weitzenböck spacetime, A^4; and if both tensors vanish, it becomes the Minkowski spacetime, M^4 (see Hayashi & Shirafuji, 1979; Ivanenko & Sardanashvily, 1983). The pseudo-Riemannian spacetime, V^4 (i.e. with a signature of $(-, +, +, +)$), is the one used by GR to describe, *inter alia*, gravity.

For more details, let us begin with U^4 and its local tangent space, $T_x(U^4)$. A frame can be defined on it by a linear base, $\{e_i\}$ $(i = 0, 1, 2, 3)$. With the Lorentz metric tensor, g, we can also define a *coframe* for the cotangent space at $x \in U^4$ with a base, $\{\theta^i\}$, such that $\theta^i = g^{ij}e_j$. In local coordinate system x^μ, we have $\theta^i = h^i_\mu dx^\mu$, where the components, h^i_μ, are known as the tetrads or vierbeins. The two systems of indices—Latin and Greek—denote two different sets of transformational properties: the Latin indices for tensors in the local coframe which transform as the *G-tensors* (where "G" indicates the gauge group), and the Greek indices for tensors which transform as *covariant tensors* on U^4 under Diff(R^4) (the set of all diffeomorphisms). Given the group, which could be $GL^+(4, R)$ or one of its subgroups, a connection Γ may be determined; and it is independent of the metric in U^4. Then the curvature is similarly, see (3), defined as $R = d\Gamma + \frac{1}{2}[\Gamma,\Gamma]$ and $R' = \gamma R \gamma^{-1}$, and the torsion is defined as a 2-form with values in R^4, $T := D\theta = d\theta + \Gamma \wedge \theta$ and $T' = \gamma T$. In words, the torsion of a space is represented by the twisting of its local coframe. The field action for the Einstein–Cartan field can be written as

$$I_{EC} = -(1/32\pi G) \int_s \mathrm{tr}[R \wedge \star(\theta \wedge \theta)],$$

where G is the gravitational constant. The appearance of two different 2-forms is due to the need for an independent representation of the connection and of the metric; and R is a 2-form in terms of Γ and $\theta \wedge \theta$ a 2-form in terms of g. Now, $\delta I_{EC} =$ under the variation of θ alone gets us Einstein's field equation, while $\delta I_{EC} = 0$ under the variation of Γ alone gets us the torsion equation.

Different gauge groups, G, when applicable to the Riemann–Cartan manifold may produce slightly different versions of the Einstein–Cartan theory, but there is no gauge group (cf. Ivanenko & Sardanashvily, 1983) which directly produces the connection in GR (i.e. the Christoffel symbols) and, hence, GR itself. This is rather different from the YM type gauge-fields. Let us now see briefly what these different versions are.

The gauge group Utiyama (1956) used to generate the Riemann–Cartan spacetime is that of homogeneous Lorentz transformations (LTs), sometimes known as the Lorentz group.[11] Here, the tetrads comprise the local *orthonormal* coframe in which the transformations among bases are restricted to the 6-parameter LTs. The coordinates in each local frame—where $g = \eta = (-1,1,1,1)$—are related to the global curvilinear coordinates by the vierbein fields, $\theta^i = h^i_\mu dx^\mu$. The use of the Lorentz group is natural if one takes literally the idea of "gauging"—i.e. the localization of a previously global symmetry. If the gauge theory of electrodynamics is the exemplar to be followed strictly, GR should be the result of a simple gauging of the global LTs in SR. However, it was argued that the vierbeins are put in "by hand", because they are introduced by demanding that tensors in the local coframes be covariant under arbitrary diffeomorphisms across spacetime; and this is clearly a demand beyond the idea of gauging. There is no analogous move of this kind in the YM type gauge-fields.

Kibble (1962) and Sciama (1962) noticed the defect and discovered a remedy in the 10-parameter group of inhomogeneous LTs (the Poincaré group). The inhomogeneous LTs, written as infinitesimal transformations in Cartesian coordinates, have the form: $x^i \rightarrow x^i + \wedge^i_j x^j + \varepsilon^i$, where $\wedge^i_j$ is the rotational part and ε^i the translational part. Gauging the group means that the two parts become arbitrary functions of x. The result, to put it very roughly, is essentially a manifold which admits local coframes as copies of the Lorentz group, i.e. the rotational (or homogeneous) part, but which also admits general coordinate transformations, i.e. the translational part.[12] "On the other hand", Trautman pointed out, "curvature and torsion are related, respectively, to the groups of homogeneous transformations and of translations in the tangent spaces of a manifold endowed with a linear connection" (Trautman, 1973, p. 194). Once the local coframes are naturally related to the base via the vierbeins, $\theta^i = h^i_\mu dx^\mu$, the rest of the geometry of U^4 is derived as in Utiyama.

The Poincaré group approach was further developed by Hehl *et al.* (1976). In that work an active (rather than a passive) interpretation of the localized Poincaré transformations is used to derive the geometry of the Riemann–Cartan spacetime. The general idea of gauge-field programme—the invariance of physics under localized transformations—produces in this case two gauge potentials: the rotational potential for the connection and a translational potential for the tetrads. Hehl *et al.* further argued that the contortion (because of the torsion in U^4) is a physically meaningful magnitude; it couples to the spin of matter just as the curvature couples to the mass. We do not see the contortion part in GR because GR is a theory of the macroscopic, and at the macroscopic level the total spin of matter is usually unnoticeable (e.g. being averaged out); more on this in Section 5.

One may think that this is the end of the story for gauge gravity; what remains is the experimental verification (or falsification) of the coupling between spin and contortion. But that is far from the case, for as early as 1928, Einstein entertained the idea of characterizing gravity in absolute parallelism, i.e. in a spacetime with a vanishing curvature (see Pais, 1982, pp. 344–345).[13] The idea was later picked up and developed into a full-fledged gauge theory of gravity (cf. Hayashi & Shirafuji, 1979; Muench *et al.*, 1998). It turns out—details aside—that the Weitzenböck spacetime has enough geometry of the right kind to describe gravity. This result may appear surprising because gravity has always been associated with the Riemannian curvature; but here the curvature identically vanishes. However, if one begins with a Riemann–Cartan spacetime—whose connection has no intrinsic relation with the metric, there is no a priori reason that gravity be captured only by a curved spacetime. The so-called *teleparallel* tetrad fields can do the job equally well, according to Hayashi and Shirafuji (1979). Since the vanishing of the Riemannian curvature in a Weitzenböck spacetime does not mean a flat spacetime, it is entirely possible that it is the "*twisted*" rather than the "*curved*" manifold that accounts for gravity in the teleparallel version.

Along the same line, an argument is made recently by de Andrade and Pereira (1997, 1998) for the conclusion that gravity can be equivalently described by either the Riemann or the Weitzenböck spacetime. In other words, one can use either the Riemann connection (torsion free) or the Cartan connection (curvature free) to describe the effects of gravity.[14] For reasons I will discuss in the next section, this approach can be dismissed on physical grounds (though not on formal grounds).

Some other groups are also considered for the group of gauge gravity, e.g. the de Sitter group or the general linear group, $GL(R)$, or the general affine group, $GA(4, R)$ (see Trautman, 1980; Ivanov & Niederle, 1982; Ivanenko & Sardanashvily, 1983; Nounou, 1995). Unlike the case of the Poincaré group, which is intended to be a replacement of the Lorentz group, the $GL(4, R)$ and $GA(4, R)$ are generalizations. The local coframes with such groups as their gauge groups are more general than the Riemann–Cartan spacetime; and a contraction of $GL(4, R)$ or $GA(4, R)$ to either the Lorentz or the Poincaré group will have to be implemented before we get the Riemann–Cartan spacetime back.

There are also proposals (cf. Ivanov & Niederle, 1982; Mukunda, 1989) that treat the Poincaré group as an internal symmetry group when it is gauged—i.e. the elements of the group do not act on spacetime points at all. In other words, the full invariance group of the gauge gravity Lagrangian is the product of two groups: $P^{loc} \otimes \text{Diff}\ (R^4)$, where P^{loc} is the gauged Poincaré group and Diff (R^4) the set of all diffeomorphisms acting on the spacetime manifold. The point of doing this is to make the result of gauging in gravity have exactly the same structure as that in the YM gauge fields. In the later case, we have, $G^{loc} \otimes P$, where G^{loc} is the gauge group of internal symmetries and P is the global Poincaré group which acts on the Minkowski spacetime. I can see several problems with this approach, the most obvious of which would be this: what would be the group structure for an interaction between, say, electromagnetism and gravity? Should it not be $U(1) \otimes \text{Diff}\ (R^4)$, according to the approach? But according to the same approach, gravity disappears in the Lagrangian under this group.

All the above are radically different from the canonical formulation of quantum gravity (as briefly mentioned in Section 1). The formulation does not at all adopt the strategy of gauging the group (or some generalized group) of SR, for it appears to disregard the symmetry of the Poincaré group from the very beginning by making an explicit separation of the temporal dimension from the spatial ones,[15] and it returns to Einstein's GR in its classical limit. (For more on this see Wald, 1984; Earman, 2000; Belot & Earman, 2001; and the literature therein.)

4. The nature of gauging and the unity of force-fields

We now return to the question: can gravity be unified with the other fundamental force-fields in the gauge-field programme? An affirmative answer is supposed to be given by gauge gravity, at least with regard to the first step of the unification in the programme. The Einstein–Cartan gauge theory of gravity may not be true, but that is a possibility for all gauge-field theories in the programme. Hence, our question should be: if the Einstein–Cartan theory is true, is it a gauge theory that truly belongs to the YM gauge-field programme?

Two essential features of the YM gauge-fields are pertinent to our evaluation of gauge gravity:

[F1] Such fields are compatible with a Minkowskian (i.e. flat) spacetime background. (They are also compatible with curved background spacetime, such as the pseudo-Riemannian spacetime, but the important thing is that they do not require such a background.)

[F2] The existence of such fields is made possible by the gauging of the appropriate gauge spaces.

Gauge gravity cannot have the first of these features because if one "flattens" the background spacetime for the Einstein–Cartan theory, none of the geometric objects on the coframes can have the appropriate properties to describe the effects of gravity. Since the coframes with a Minkowski metric will be $\theta^i = h^i_\mu \, dx^\mu = \delta^i_\mu \, dx^\mu = dx^\mu$, i.e. they coincide with the global Cartesian frame (which is holonomic and orthonormal), both the curvature and the torsion will vanish.[16]

Perhaps this fact gives some explanation to the real difference between the *external* and the *internal* symmetries which were said to separate GR from the other gauge-field theories. While the other gauge fields are geometric properties of spaces over, though attached to, spacetime—hence, internal to the matter they affect—gravity cannot be so conceived: it is "soldered" to spacetime—hence, external to the matter it affects. And while for gauge-fields it is still possible to retain the sense of "force" as something that compels objects to deviate from their inertial motion in a spacetime, i.e. from their geodesics, that is not at all possible for gravity.

Does the gauge gravity have the second feature? From the works we discussed above, the answer seems to be a "yes". The question is: which group should it be, the gauging of which "automatically" yields the Einstein–Cartan theory of gravity? We have seen that it cannot be the Lorentz group, for its gauging in the coframe of a flat spacetime is compatible with the spacetime's remaining flat; but no gravity arises this way. As mentioned earlier, if one gauges the Poincaré, rather than the Lorentz, group, this defect can be removed. We now examine some details of this argument to see how it is done. To my best knowledge, Hehl *et al.* (1976) and Mukunda (1989) give the most detailed arguments. Hehl *et al.* begin their argument by claiming:

> We shall show ... that to require the invariance of a special relativistic theory of matter under local space-time rotations and translations (independent actions of the Poincaré group at every point) leads *inexorably* to the introduction of torsion and curvature. In view of the fundamental significance of the Poincaré group for physics, and of the local gauge approach in general, we take such a derivation as the strongest evidence that the geometry of the physical world is indeed a Riemann–Cartan geometry. (Hehl *et al.*, 1976, p. 400, emphasis added)

To show this, they begin with the global inhomogeneous LTs of the Poincaré group $x^i \to x'^i = \wedge^i_j x^j + a^i$, where $\wedge^i_j$ are the rotational, and a^i the translational, elements of the group. Making these infinitesimal transformations, we have, $\delta x^i = \omega^i_j x^j + \varepsilon^i$; and for a scalar matter-field, $\psi(x)$, we have

$$\psi(x) \to \psi'(x) = (1 + \omega^{ij} f_{ij})\psi(x), \tag{7}$$

where f_{ij} are the generators of the group. There are two ways of gauging the Poincaré group. If one treats the LTs as transformations among coordinate systems (i.e. the passive reading), then making the transformations local would blur the distinction between the rotational and the translational elements. It turns the group into a 4-parameter group of general coordinate transformations: $x^\mu \to x'^\mu = x^\mu + \varepsilon^\mu(x)$, where the Greek indices signify coordinates in arbitrary coordinate systems.

However, if one interprets the infinitesimal LTs as acts of re-orienting and displacing our experimental (or measuring) frames or apparati (i.e. the active reading) and *keeps the two actions separate*, one first writes (7) as

$$\psi(x) \to \Pi\psi(x) = (1 + \omega^{\mu\nu} f_{\mu\nu} - \varepsilon^\mu \partial_\mu)\psi(x), \tag{8}$$

where Π is a linear mapping of the field at the neighbourhood of x, which *re-orients* and *shifts* $\psi(x)$ with the following respective amounts:

$$\omega^{\mu\nu} = \omega^{ij} \text{ and } \varepsilon^\mu = \varepsilon^i + \delta^\mu_i \omega^i_j x^j, \tag{9}$$

where ε^i and ω^{ij} (and ω^i_j) are the same parameters as in (7), which are now restricted to an orthonormal basis, $\mathbf{e}_i$, at x. This operation is explained by Hehl *et al.* thus:

> Replace the fields ψ at a point x^i by fields which have first been rotated by an amount $-\omega$, that is, $\psi(x) \to [\wedge\psi](x) = (1 + \omega f)\psi(x)$, and then have been translated by an amount $+\varepsilon = \varepsilon + \omega\bullet x$, that is, $[\wedge\psi](x) \to [\Pi\psi](x) = [\wedge\psi]\ (x - \varepsilon)$. Then, experience shows, matter distribution $\Pi\psi$ and ψ are physically equivalent. (Hehl *et al.*, 1976, p. 401)

Because of the the equivalence of the active and the passing reading, nothing is changed in SR. However, when one gauges the symmetry in the active reading, one localizes the ε^i and ω^{ij} (and ω^i_j) in the orthonormal frames. As usual, when $\varepsilon^i \to \varepsilon^i(x)$ and $\omega^{ij} \to \omega^{ij}(x)$ (note that they are all G-tensors), we have the Lagrangian density:

$$L(\psi, \partial_i\psi) \to L(\psi, \partial_i\psi, \Gamma^{ij}_\mu, h^i_\mu)$$

where in the limit of a weak force-field, we have

$$\delta\Gamma^{ij}_\mu \cong -\partial_\mu \omega^{ij}(x) \text{ and } \delta h^i_\mu \cong \omega^i_j \delta^j_\mu - \partial_\mu \varepsilon^i(x)$$

as the results of non-vanishing differentials of ε^i and ω^{ij}. This is the first step through which the Poincaré group is said to have introduced the connection coefficients Γ^{ij}_μ and the vierbeins h^i_μ as coordinate vectors via the localization of ε^i and ω^{ij}. (Again, Latin indices denote tensors in the orthonormal coframes and Greek indices tensors in the general coordinate systems.)

Further advantage is taken of this active reading of the Poincaré group, for (8) and (9) may be seen as yielding a rigidity condition which says that Π changes "neither the distance between events nor the relative orientation of neighboring matter fields" (Hehl *et al.*, 1976, p. 402). More formally, the condition before gauging is

$$\Pi(\xi^i \partial_i \psi) = (\Pi\xi^i)\partial_i(\Pi\psi) = (\Pi\xi^i)\delta^\mu_i \partial_\mu(\Pi\psi)$$

for any infinitesimal vector field, ξ^i, that is bound to the field ψ. Here the insertion of δ^μ_i

in the last step is trivial with the global Poincaré group. However, when the group is gauged, the rigidity condition should hold as well. But then we have

$$\delta^{\mu}_{i} \to h^{\mu}_{i}(x) \text{ and } \partial_{\mu} \to D_{\mu} = \partial_{\mu} + \Gamma^{ij}_{\mu}(x) f_{ij} \tag{10}$$

such tht the rigidity condition holds in general as

$$\Pi(\xi^{\mu} D_{\mu}\psi) \cong (\Pi\xi^{i})(h^{\mu}_{i} + \delta h^{\mu}_{i})(D_{\mu} + \delta\Gamma^{ij}_{\mu}\mathrm{f}_{ij})(\pi\psi).$$

Here the first transition in (10) introduces the vierbeins which connect the orthonormal coframes to the general coordinate base, while the second transition in (10) is the result of a relative rotation, $\mathrm{d}x^{\mu}\ \Gamma^{i}_{\mu j}\ \mathbf{e}_i$, of the frame base, $\mathbf{e}_j$, at neighbouring points x^{μ} and $x^{\mu} + \mathrm{d}x^{\mu}$. Once the connections and the vierbein fields are introduced, the rest of the steps towards the full Riemann–Cartan manifold are rather standard, which most of the accounts of the Einstein–Cartan theory follow. For instance, the metric of the background spacetime is given by

$$g_{\mu v} = \eta_{ij} h^{i}_{\mu} h^{j}_{v}.$$

In other words, the pseudo-Riemannian metric is a natural result of the gauging of the translational element of the Poincaré group.[17] Although Weinstein (1999) is right in arguing that Diff (R^4) is not the gauge group of any fibre bundle that may describe gravity, gauge gravity is at least able to tell a story of how Diff (R^4) drops out from the background spacetime as a result of gauging the Poincaré group.

Therefore, with respective to [F2], gauge gravity—the Einstein–Cartan theory of gravity—appears to be just as good a gauge-field theory as any of the YM type; but it differs from them with respect to [F1]. This, I believe, clarifies the precise sense in which gauge gravity is or is not a YM gauge-field theory: it is with respect to [F2] but it is not with respect to [F1]. Note that the GR gravity has neither features.

Why should [F1] be an essential feature to the gauge-field programme (or why is it important that gauge fields are of YM type)? I have ventured a short answer to this earlier, namely, it allows us to retain the traditional sense of force for these gauge-fields. Geometrized or not, they can still be understood as agents which cause deviations of quantum particles from their inertial evolutions in spacetime. Such an interpretation is no longer available for gauge gravity because of the modified notion of geodesics in GR (or the Einstein–Cartan theory). This difference is also fundamental for another reason. Gauging is an act of making spacetime independent (global) symmetries into spacetime dependent (local) ones. If the symmetry is of a gauge space other than spacetime, such as the U(1) symmetry of phase space, it makes sense to say that the gauging of the symmetry produces a transition from a "flat" phase space to a "curved" one in the *same spacetime background*. We cannot use the same image for the gauging of the Poincaré symmetry, since it is a symmetry *of* spacetime *in* spacetime. As I discussed earlier, the gauging that changes a global coframe into local ones, $\theta^i{:}\delta^{i}_{\mu}\mathrm{d}x^{\mu} \to h^{i}_{\mu}\mathrm{d}x^{\mu}$, inevitably "unflatttens" the metric as a result of the *soldering* relation between the vierbeins, h^{i}_{μ}, and the metric.

We now consider the question: is the traditional notion of a force-field worth preserving, even though the YM gauge-fields do seem to some extent to preserve it? As mentioned above, the pre-GR unified notion of forces (or force-fields) takes a force to be an external interaction that alters the inertial state of a physical system. More accurately, if one uses the action principle of the pre-GR physics, one forms various actions—which are either time integrals of the Lagrangians or, in the case of fields, the space integral of the Lagrangian densities—for various forces involved, and add them on to the action of the physical system in question to form the total action. From this action one can derive, by the least action principle, the equation of motion of the system under the influence of the involved forces. Here, even though the concept of force is a unified one, the forces are still distinct. After GR, the quest is no longer for a unified concept which applies to different forces but for a unified theory for one substance (which is not a force), to which all the forces are reducible. Had Einstein's dream for a unified field theory been fulfilled, there would have been a unified theory for a single "geometric" field in which *the forces disappear*. If all fundamental forces were reduced to some geometric properties of spacetime—i.e. a particle freely moving in any or all force-fields is moving along its geodesic—it would be better to stop using the traditional sense of force all together.

Something very similar also holds in the gauge-field programme. The notions of reduction and geometrization are almost the same as in Einstein's original conception, except the spaces in which the force-fields are geometrized are different. Therefore, I urge that, at least for the YM force-fields, we see the first stage of the programme as an attempt to *replace* one unified concept of forces with another, perhaps more profound, one.

In view of the gauge-field programme, we now realize that the GR reduction involves two separable components, one is the geometrization of force-fields (i.e. gravity) and the other is the identification of the geometry with that of the pseudo-Riemannian spacetime. The sense in which the gauge-field programme provides a replacement of the traditional concept of unified forces only involves the first component—i.e. geometrization. Redhead (1999) gives a general sketch of the nature of such a geometrization in the mathematical modelling of physical systems. According to his account, surplus structures are often inevitable in the process of imbedding a physical system into a mathematical model. Such structures are always "controlled" by using groups of global transformations such that the contracted model (one without the surplus) becomes isomorphic to the physical model. But such surpluses are fertile grounds for gauge-fields, for once some of them are gauged—i.e. that the transformations are localized—their geometries may offer the grounding for the reduction of the appropriate force-fields. Because of [F1], the gauging of internal symmetries realizes the first component and therefore geometrizes the corresponding force-fields, but it does not quite eliminate them, since it does not realize the second component. There is still a proper notion of the geodesic of a system, whether in a flat or a curved spacetime, from which the deviation of the system's motion is accountable by the presence of some YM gauge-fields.

However, one may want to know, apart from the formal differences examined above, what the difference is between a force as traditionally conceived and one as reducible to some geometric properties of a fibre bundle that is of any *metaphysical significance*. A complete answer to this question goes beyond the scope of this paper. Here I can only point out the following aspect (cf. Liu, 1996). In the pre-GR regime, force-fields are represented either in the interactional Lagrangians, as mentioned earlier, or by potential terms, which represent *dispositions* of making systems under the fields' influence to deviate from their geodesics. The force-fields are unified in terms of the same *causal role* they share in nature: by how they *dispose* objects under their influence to behave under given initial/boundary conditions. They are not unified by a set of shared *categorical properties*, properties that can be defined independently of the causal roles their possessors play (or properties that objects possess independently of the causal roles they play). Even if one gives a substantival reading of a force-field as traditionally conceived, the substance is essentially understood as a "bundle" of dispositions, whose manifestations are seen in its effects on the objects it acts on. The geometrization of force-fields in the gauge programme seems to have fundamentally changed this (in the direction of Einstein's GR). Now the "force"-fields can be understood, under a realist interpretation, substantival (internal) spaces with certain "shapes", which are typical *categorical* properties. And the geometrical properties of such spaces are certainly definable independently of the causal roles the fields play in the dynamics of some systems.

5. Conclusion

Gauge gravity in the form of the Einstein–Cartan theory is found not to be exactly of the YM type, because it fails to satisfy [F1]. But one may argue that perhaps [F1] should not be regarded as essential to the gauge-field programme. More important is *the applicability of "gauging" by which a force-field is geometrized.* In this respect, the Einstein–Cartan theory is fully qualified for the gauge-field programme.

Now one of the remaining important questions is: can the Einstein–Cartan theory, or some variations of it, be experimentally differentiated from Einstein's GR? The main difference between the two is of course the appearance of torsion in the Einstein–Cartan theory. As mentioned in Section 3 (cf. Hehl *et al.*, 1976), the curvature in the Riemann–Cartan manifold is caused by the mass of matter and the torsion by the spin. The two main field equations can be schematically expressed as

Einstein tensor = $k \times$ energy-momentum
modified torsion = $k \times$ spin angular momentum (Hehl *et al.*, 1976, p. 400).

The spin angular momentum is from the sum of all the spins of the elementary particles which comprise the matter; it is different from the total rotational angular momentum the matter may possess. Unlike its mass, the total spin of a macroscopic object is usually negligible. Therefore, the effect of torsion is unlikely to be detectable in the GR regime. In other words, this can explain why we have not so far detected any significant deviation from Einstein's GR in its regime, even if the Einstein–Cartan gauge theory is the true theory. One does expect that a significant difference will be detectable in a microscopic system; but an estimation of the critical density under which the effect of spin on the torsion of spacetime is comparable to the effect of mass on its curvature is so high, in the order of 10^{47} g/cm^3 for electrons, that it does not seem realistic at present to ask for any reasonably direct confirmation.

Be that as it may, the Einstein–Cartan gauge theory of gravity is not an interpretational theory, which postulates a different substance or mechanism but does nothing more than recover the exact experimental results of Einstein's GR. It is an improvement upon GR as GR is an improvment upon SR. If it is true, spacetime will not be a pseudo-Riemannian but a Riemann–Cartan manifold, and gravity will not only be represented by the curvature in spacetime; it is also partly represented by the torsion in it. The latter may become significant only at the quantum level; therefore, the gauge theory of gravity if correct will be useful only in the regime of quantum gravity.

In fact, one of the main motivations for bringing GR in line with the YM gauge programme via the Einstein–Cartan theory is to make it easy for quantization. We know how to quantize a YM gauge field in general; and so if the theory of gravity can be made such a gauge-field theory, the quanitzation problem can be regarded as solved. It is not yet clear how much of a difficulty it is for quantization given that the gauge gravity whose theory we sketched in this article is not exactly a YM gauge-field. Partly because gauge gravity has not been occupying the centre stage in theoretical physics in recent decades, this and many other similar questions are for now left unanswered.

Acknowledgements

This article is dedicated to the memory of Wes Salmon. I would like to thank Gordon Belot, Richard Healey, Peter Kosso, Stephen Leeds and Paul Teller for listening to an earlier version of this article and for their questions and comments.

Notes

1. I shall consistently use the term "force-field" (or simply force) in contrast to the term "matter-field". In classical (i.e. non-quantum) physics, there is no matter-field, so the simple term "field" may suffice; but in quantum physics, where we have both matter and force-fields, the longer term is necessary.
2. If one finds the claim that gravity is identified with spacetime in GR too crude, here is a better way of stating it. In GR, gravitational effects can be identified with geometrical properties of spacetime regions in a way that the effects of the other three force-fields cannot (at least, not likely if the standard model is on the right track). This, I warn, should not be taken as being incompatible with the *metaphysical* claim that all matter and forces are properties of spacetime regions, e.g. a house may be thought of as some kind of property of the space it occupies, but it cannot be said to be a geometrical property of the space unless a theory about that is given.
3. This predicate is not superfluous here because the claim may not be true for a spacetime of more than 4-dimensions. See the next footnote.
4. There is a small complication of the otherwise clear dichotomy. If the "manifold" of nature is a Riemannian manifold of more than four dimensions, then it is possible to integrate the other forces into this manifold. One example of such a theory is the Kaluza–Klein theory of gravity and electromagnetism in the 5-dimensional Riemannian manifold. Does this theory, if made viable, count as a successful unification of gravity and electromagnetism? Perhaps it is; but it is certainly not a reduction of the two forces to manifestly 4-dimensional *spacetime*. Hence, even if the Kaluza–Klein theory is true, gravity still does not belong to the rest of the natural forces. Because the Kaluza–Klein approach to the unity of force-fields is not one in the gauge-field programme, I shall not discuss it in any detail in this article. See O'Raifeartaigh (1997) and Maheshwari (1989) and the literature therein for details.
5. Weinstein (1999, pp. S152–S154) acknowledges such attempts, and points out: "Though some of these constructions are interesting and illuminating in their own right, in none of them is the diffeomorphism group a gauge group." I shall address the last point in due course.
6. Keep in mind that a Lie group is a special kind of manifold; and in this case, having a copy of G sitting on each $x \varepsilon$ M (since M is mapped into G by local gauge transformations γ) provides the new degrees of freedom (or new space) to which gauge fields are reduced. For those who are familiar with the fibre bundle language, these are the fibres; and the A and F introduced later are the connection 1-form and the curvature 2-form of the fibre bundle.
7. In this article I will not introduce the quantization of these fields in any systematic way. The procedures of quantizing force and matter-fields have their own deep conceptual problems, but they are not at all pertinent to the question of whether or not the fields are unifiable. What I delineate in the following is of such generality that it is entirely preserved after the quantization of the fields in question. One may therefore think of the fields discussed here as quantum fields, e.g. interactional bosons. For problems in the quantization of gravity, see Belot and Earman (2001).
8. In a suitable coordinate system the components of this F reads: $(F_w)_{\mu v} = (\partial_\mu W^k_v - \partial_v W^k_\mu) - q[W^k_\mu, W^k_v]$, which is the familiar form one sees in most of the physics textbooks.
9. There have been attempts to reduce the coupling constants to a single one. For electroweak, one may imbed SU(2) U(1) into SO(3), but it is "ruled out by experiment" (Weinberg, 1996). For the strong and the electroweak unification, one may likewise imbed SU(3) SU(2) U(1) into groups such as SU(5) or SO(10). The jury is still out on such attempts (Weinberg, 1996).
10. I will not say anything about ontological reduction, since it will not be relevant to what is discussed in this article.
11. In Utiyama's (1956) article, the full Riemann–Cartan spacetime is never explicitly derived, nor is torsion

even considered. All results are derived under the stipulation that $Dg = 0$ and $\Gamma^{\lambda}_{\mu\upsilon} = \Gamma^{\lambda}_{\upsilon\mu}$. It was later realized that the natural result of gauging the Lorentz group is indeed the Riemann–Cartan spacetime.
12. Here I mostly follow Trautman (1973). But we will see the details of this point in Section 4.
13. Einstein appeared to be wanting to see whether absolute parallelism could carry electromagnetic field. But it is not clear how he could do it without curvature on the manifold for gravity. If the torsion tensor was used for electromagnetic field, which Einstein did so use, what could he use for gravity? The metric is still pseudo-Riemannian because $g_{\mu\upsilon} = h^i_{\mu} h^i_{\upsilon}$, but it is not clear whether this is sufficient for gravity.
14. De Andrade and Pereira give the following interesting argument. First, by entertaining what appears to be a principal bundle (think of it as a bundle of local coframes) whose typical fibre is a local translational group—i.e. the translational part of the localized Poincaré group—and the base space a Minkowski spacetime, they are able to derive a "gravitational analogue of the Lorentz force" (de Andrade & Pereira, 1997, p. 4691) that gives an adequate account of gravity in the YM fashion. Then, the vierbein fields so obtained (as the vector fields that set up the local translational coframes) are used to define, first, a teleparallel covariant derivative and, second, a Riemannian covariant derivative. Both eventually lead to a description of gravity that is then proven to be equivalent. However, as they show, the teleparallel description treats gravity as an external force-field—external to the Weitzenböck spacetime, while the Riemannian description identifies gravity with the spacetime (as is the case in GR). In other words, if we choose the Weitzenböck formulation, gravity is just like any other YM type gauge-field, whose presence is external to spacetime.
15. But see Crnkovic and Witten (1987) for an argument to the contrary.
16. There is a long-standing tradition consisting of attempts to reproduce results in Einstein's GR in a flat spacetime background. It is not at all clear to me whether they can be integrated into the gauge-field programme I discuss in this article. Interested readers may consult Gupta (1954) and Deser (1970).
17. See also Mukunda (1989) for a simpler and more intuitive but less rigorous discussion of the gauging of the Poincaré group.

References

AITCHISON, I.J.R. (1982) *An Informal Introduction to Gauge Field Theories* (Cambridge, Cambridge University Press).

BELOT, G. & EARMAN, J. (2001) Pre-Socratic quantum gravity, in: C. CALLENDER & H. HUGGETT (Eds) *Philosophy at the Planck Length* (Cambridge, Cambridge University Press), pp. 213–255.

CALLENDER, C. & HUGGETT, N. (2000) Why quantize the gravitational field (or any other field for that matter)?, *PSA 2000* (preprint).

CAT, J. (1993) A philosophical introduction to gauge symmetries, Lectures delivered at the London School of Economics, October 1993 (preprint).

CRNKOVIC, C. & WITTEN, E. (1987) Covariant description of canonical formalism in geometrical theories, in: S. HAWKING & I. ISRAEL (Eds) *300 Years of Gravitation* (Cambridge, Cambridge University Press), pp. 676–684.

DE ANDRADE, V.C. & PEREIRA, J.G. (1997) Gravitational Lorentz force and the description of the gravitational interaction, *Physical Review D*, 56, pp. 4689–4695.

DE ANDRADE, V.C. & PEREIRA, J.G. (1998) Riemannian and teleparallel descriptions of the scalar field gravitational interaction, *General Relativity and Gravitation*, 30, pp. 263–273.

DESER, S. (1970) Self-interaction and gauge invariance, *General Relativity and Gravitation*, 1, pp. 9–18.

EARMAN, J. (2000) Gauge matters, *PSA 2000* (preprint).

FRANKEL, T. (1997) *The Geometry of Physics: An Introduction* (Cambridge, Cambridge University Press).

GÖCKELER, M. & SCHÜCKER, T. (1987) *Differential Geometry, Gauge Theories, and Gravity* (Cambridge, Cambridge University Press).

GUPTA, S.N. (1954) Gravitation and electromagnetism, *Physical Reviews*, 96, pp. 1638–1685.

HAYASHI, K. & SHIRAFUJI, T. (1979) New general relativity, *Physical Review D*, 19, pp. 3524–3553.

HEHL, F.W., VON DER HEYDE, P., KERLICK, G.D. & NESTER, J.M. (1976) General relativity with spin and torsion: foundations and prospects, *Reviews of Modern Physics*, 48, pp. 292–416.

HIGGS, P.W. (1974) Spontaneous symmetry breaking, in: R.L. CRAWFORD & R. JENNINGS (Eds) *Phenomenology of Particles at High Energies* (New York, Academic), p. 529.

IVANENKO, D. & SARDANASHVILY, G. (1983) The gauge treatment of gravity, *Physics Reports*, 94, pp. 1–45.

IVANOV, E.A. & NIEDERLE, J. (1982) Gauge formulation of gravitation theories I: the Poincaré, de Sitter, and conformal cases, *Physical Review D*, 25, pp. 976–987.

KIBBLE, T.W.B. (1961) Lorentz invariance and the gravitational field, *Journal of Mathematical Physics*, 2, pp. 212–221.

LIU, C. (1996) Potential, propensity and categorical realism, *Erkenntnis*, 45, pp. 45–68.

LIU, C. (1999) Gravity and electromagnetism under the gauge-field program, Lecture notes, Conference on the Philosophical Significance of Gauge Theories in Physics, Department of Philosophy, University of Arizona, 18–21 March 1999.

MAHESHWARI, A. (1989) Kaluza–Klein theories, in: B.R. IYER, N. MUKUNDA & C.V. VISHVERSHWARA (Eds) *Gravitation, Gauge Theories and the Early Universe* (Dordrecht, Kluwer), pp. 423–447.

MAUDLIN, T. (1996) On the unification of physics, *Journal of Philosophy*, 93, pp. 129–144.

MUENCH, U., GRONWALD, F., *et al.* (1998) A brief guide to variations in teleparallel gauge theories of gravity and the Kaniel–Itin model, *General Relativity and Gravitation*, 30, pp. 933–961.

MUKUNDA, N. (1989) An elementary introduction to the gauge theory approach to gravity, in: B.R. IYER, N. MUKUNDA & C.V. VISHVERSHWARA (Eds) *Gravitation Gauge Theories and the Early Universe* (Dordrecht, Kluwer), pp. 467–479.

NOUNOU, A. (1995) *Gauge theories and gravity* M.A. thesis, Imperial College, London.

O'RAIFEARTAIGH, L. (1997) *The Dawning of Gauge Theory* (Princeton, NJ, Princeton University Press).

PAIS, A. (1982) *'Subtle is the Lord ...': The Science and the Life of Albert Einstein* (Oxford, Oxford University Press).

PRASANNA, A.R. (1989) Differential forms and Einstein–Cartan theory, in: B.R. IYER, N. MUKUNDA & C.V. VISHVERSHWARA (Eds) *Gravitation, Gauge Theories and the Early Universe* (Dordrecht, Kluwer), pp. 119–132.

QUIGG, C. (1983) *Gauge Theories of the Strong, Weak, and Electromagnetic Interactions* (Reading, MA, Benjamin).

REDHEAD, M. (1999) Lecture on symmetries and gauge principles, Lecture notes (preprint).

SCIAMA, D.W. (1962) On the analogy between charge and spin in general relativity, in: *Recent Development in General Relativity* (Oxford, Pergamon and PWN), p. 425.

TELLER, P. (2000) The gauge argument, *Philosophy of Science*, 67 (Suppl.), pp. S466–S481.

TRAUTMAN, A. (1972) On the Einstein–Cartan equations, parts 1, 2, 3, *Bulletin de l'Académie Polonaise des Sciences*, Série des sciences math., astr. et phys., 20, pp. 185–190, 503–506, 895–896.

TRAUTMAN, A. (1973) Theory of gravitation, in: J. MEHRA (Ed.) *The Physicist's Conception of Nature* (Dordrecht, Reidel), pp. 179–198.

TRAUTMAN, A. (1980) Fiber bundles, gauge fields, and gravitation, in: A. HELD (Ed.) *General Relativity and Gravitation*, Vol. 1 (London, Plenum), pp. 287–308.

UTIYAMA, R. (1956) Invariant theoretical interpretation of interaction, *Physical Review*, 101, pp. 1597–1607 (reprinted in O'Raifeartaigh, 1997, pp. 213–239).

VIZGIN, V.P. (1994) *Unified Field Theories (in the First Third of the 20th Century)* (Basel, Birkhäuser Verlag).

WALD, R.M. (1984) *General Relativity* (Chicago, IL, University of Chicago Press).

WEINBERG, S. (1996) *The Quantum Theory of Fields*, Vol. 2 (Cambridge, Cambridge University Press).

WEINSTEIN, S. (1999) Gravity and gauge theory, *Philosophy of Science*, 67(Suppl.), pp. S152–S161.

INTERNATIONAL STUDIES IN THE PHILOSOPHY OF SCIENCE VOL. 12, NO. 2, 1998 111

ARTICLE

Models and theories II: issues and applications

CHUANG LIU
Department of Philosophy, University of Florida, USA

Abstract *This paper is the second of a two-part series on models and theories, the first of which appeared in* International Studies in the Philosophy of Science, *Vol. 11, No. 2, 1997. It further explores some of themes of the first paper and examines applications, including: the relations between "similarity" and "isomorphism", and between "model" and "interpretation", and the notion of structural explanation.*

Introduction

In part I of this project (Liu, 1997), I argued for an enhanced (or hybrid) version of the semantic view for scientific theories. In part II I shall pursue other aspects of the semantic view, including some of those mentioned in part I. But before I address the issues and applications, let me list in the following scheme all the different types of theories and models that are used in science and the ways in which they are tested.

(1) *Theory* an axiomatized (or semi-axiomatized) collection of statements which is closed under deduction and which describes a set of phenomena. (Some axioms and theorems of such theories are known as statements of laws of nature which are often expressed in the form of differential equations.)
   (i) *Structural theory*: a theory that describes a system by characterizing its structural properties, such as a theory of atomic nucleus or of mammals.
   (ii) *Behavioral theory*: a theory that describes the behavior of a system by characterizing the possible states it may occupy, such as the paths of an subatomic particle in a cloud chamber or the graphs of a mammal's blood pressure.

(2) *Interpretation*
   (i) *The semantic*: to give meanings to quantities in a theory so that they can be directly or indirectly given values.
   (ii) *The philosophical*: to place the terms and principles of a theory in a certain philosophical framework or to evaluate the presupposed philosophical framework for such terms and principles.

0269-8595/98/020111-18

(3) *Models (from concrete to abstract)*
- (i) *Causal/structural models*: representations of systems whose causal and/or structural properties provide the explanation for observable phenomena.
- (ii) *Causal/structural analogue models*: analogical representations of a causal/structural kind.
- (iii) *Functional models*: representations of the functions of systems (while disregarding their physical compositions).
- (iv) *Mathematical models*: representations of the mathematical (geometrical or algebraic) structures of the phenomena (disregarding their true physical aspects).

(All of the above models have one crucial aspect in common, namely, they are independent of time. A structural model of an airplane, be it given by verbal description, by computer imagery, or by a physical replica, is a model not indexed by instances of time, although for some of these time limits may be necessary.)

- (v) *State-space models*: sets of states that a system occupies through time. These are behavioral models, trajectories of mechanical particles being the best examples, but if understood in their general sense, we may simply call models of this type trajectories, namely, a set of points in a state space.

(4) *Epistemic models*: theories or models that are distinguished merely by their epistemic status. A theory is a set of statements that truly and completely describes the intended phenomena, while a model is a *potential* theory that approximately describes them. Such models are entirely different from the above and do not even resemble semantic models in logic. Here exists no semantic distinction between a theory and its model.
- (i) *Idealized models*: theories that are true only under excessive idealization conditions.
- (ii) *Exploratory models*: theories that are true only of the known aspects of the phenomena while it is clear that unknown essential aspects exist.
- (iii) *Analogue models*: theories that are true of well-known phenomena that closely resemble the intended unknown ones.
- (iv) *Instantiation models*: theories that are true of specific systems which instantiate theories that consist of universal (or general) laws (or principles).

(5) *Theory testing*
- (i) *Test of consistency*: to see if a theory is satisfied by at least one model (of the structural, functional or the state-space kind).
- (ii) *Direct test of a (set of) statement(s)*: to see if a (set of) statement(s) is satisfied by a model whose universe contains systems that may resemble (be similar or isomorphic to) the intended real systems.
- (iii) *Indirect test of a (set of) statement(s)*: to directly test a (set of) statement(s) A, and see if the (set of) statement(s) which is tested is derivable from A.
- (iv) *Suitability of models*: to see if using a well-tested model one can derive those results that are obtained from a theory without the model (e.g. the molecular model for a thermodynamic theory of gases). (Epistemic models can also be so tested for their suitability.)
- (v) *Model testing*: to see if a model resembles within reasonable limits the intended phenomena.

According to the semantic view given in part I, the following seems a useful way of seeing the relations among the main items given above: a structural theory/model set is a triple: $< T, M, H >$ where M is a set of semantic interpretations with a universe S, which satisfies T that describes the *structure* of the members of S, and H (the theoretical hypothesis) relates the members of S to the intended phenomenal systems in the world. And a behavioral theory/model set is a similar triple, $< T, M, H >$ in which T describes in *behavior* of the members of S, and H relates the members of S to real state-space trajectories.

Semantic view and theory testing (or confirmation)

Opening either Sneed's seminal work (1971) of structuralism or the structuralist's most recent and ambitious milestone by Balzer *et al.* (1987) and browse the index, one will not even find any of the following entries: confirmation, justification, testing and experiment, nor is there any name associated with works in any area of confirmation theory or scientific methodology (see also Stegmüller (1973)).

This neglect seems surprising and unjustified. Confirmation of scientific theories is a major part of a philosophy of science and most approaches in the literature, such as Hempel's (1965) and the Bayesian (cf. Hawson & Urbach, 1993), are in a syntactic orientation. If the semantic view is to replace the syntactic view, should it not provide revisions to the old accepted approaches?

According to van Fraassen's (1989) succinct formulation of the semantic view, to test a theory is to test "a theoretical hypothesis, which asserts that certain (sorts of) real systems are among (or related in some way to) members of that class" (pp. 222–223). As stated, the theoretical hypothesis (TH) yields no general constraints on how it can be tested. For instance, suppose the class of models are of non-viscous fluids, and so a typical TH would be that "a certain sample of pure liquid helium is among or related to one of the fluids in the models". It tells us nothing about how we can find out if one of the model fluids does resemble liquid helium.

R. Giere's book (1988) (see also Giere (1994)) has chapters on justification and experiment, but they are not really connected to the semantic view. Lloyd (1988) and Suppe (1989) have formulated detailed accounts of theory testing, but as will be shown, they are only applicable to a particular type of theory testing.

Now, what are the possible ways of substantiating a TH? The simplest would be when the semantic models of a theory include simple mechanical systems, such as *model* airplanes. According to Apostel (1961), to test whether airplanes work precisely as our theory predicts is to build model airplanes and fly them in a controlled environment. If a model airplane performs well within desired limits and we know that merely altering its absolute scale without changing the proportions of its parts does not affect its performance, then the theory is confirmed (at least up to a scale change). This is perhaps also true for more complex but artificially constructable systems, e.g. chemical compounds (it may be technically complicated but not with respect to the logic of testing). Other kinds of models, such as those of natural biological organisms and abstract microscopic systems, are not so easily tested.

Suppe (1989) speaks of a theory being empirically true if "the class of theory-induced physical systems are identical with the class of causally possible physical systems" (p. 84). To test a theory is to find out empirically whether such an identity relation obtains. However, in the class of causally possible physical systems, Suppe clearly meant to include actual as well as non-actual (and yet physically possible) systems. Hence, the testing could not be a matter of ostensibly identifying through empirical means the members in that class in the actual world. Suppe's suggestion is that we develop an "experimental methodology [that] enables us to determine the states of a causally possible physical system by observing and measuring, and so on, the corresponding phenomenal system" (1989, p. 103). Despite its special terminology, Suppe's claim in fact captures the general idea of theory testing for semanticists, namely, to test a theory is to find out by empirical means if the behavior (i.e. sets of states through time) of an element in the model set *matches* that of some phenomena (i.e. the actual states of some real system).

Now, let us see if this idea works with the hybrid view. In view of Suppe's suggestion, it seems most suitable to first consider the case of behavioral theories and state-space models. The question then would be how one can establish the similarity or isomorphism relation between the state-space models and the phenomena. The laws that are satisfied (in the sense of Sat_1, cf. Liu (1997)) by the state-space models are laws of coexistence; or in practice, they are the equations by which a theoretical graph of consecutive states can be plotted. For instance, the Gay–Lussac law of ideal gas undergoing a quasi-static process is an example of such a law, which is defined as a process: $t \rightarrow (P(t), V(t))$ (where t is time, $P(t)$ the pressure and $V(t)$ the volume of the gas) such that $PV = RT$. To put it roughly, the testing of a set of state-space models defined by a behavioral theory is to construct experiments on the intended phenomenal system in which the system's states through time are obtained as measurement results and plotted. We then compare this graph with the one predicted by the law and see if they match. If they do, the theory is (partially) confirmed; otherwise, it most likely needs modification.

This kind or level of testing or confirmation corresponds to two items in our scheme above under *Theory testing*, namely, "direct test of a (set of) statement(s)" and "model testing" (cf. 5(ii) and 5(v) in the introduction). In the present context, the statement could either be an algebraic equation, e.g. again $PV = RT$ (where $P(t)$, $V(t)$ and $T(t)$ take positive values) or a conjunction of simple statements in the form "P at time t has value p_t". The models should all be state-space models (i.e. the testing of such models is direct testing).

Qualifications and refinements obviously need to be introduced in this picture before it can be expected to hold. But first, let us see how the original general statement of TH is given specific details by this rough sketch of a methodology. The TH in our specific case should probably be read as that the set of actual states of some systems when mapped to the real numbers agrees with some of our state-space models. If we identify the actual system with the collection of all of its actual states, which seems to be a sound idea of ontology, the sense in which the two kinds of systems resemble or are isomorphic to each other seems clear. However, the sense of a comparison between two kinds of systems is only metaphorical; what is really tested here are individual statements of the form, that for a system S, and time, t, the number attributable to the measured property of S at time t matches (with a certain degree of approximation) the number attributable to the corresponding property of M at t, where S is an actual system and M a model. The possibility of such a direct comparison is given by a theory of measurement whose foundations run deep in logic and mathematics.

With this point clarified, it is not difficult to discuss the qualifications and refinements. Since this aspect has been thoroughly discussed in the literature (cf. Suppe, 1989, pp. 118ff; Lloyd, 1988, pp. 145ff) and I do not have anything significant to add, I shall be brief. One should not expect perfect matches in comparing predicted results and experimentally recorded ones; approximations of some degree are always expected. In most cases, two kinds of factors play a role in such approximation. First, most theories, even if they are at the level of phenomenological laws, are only true under idealized conditions. Such conditions must be taken into consideration when one handles the goodness-of-fit problems between predictions and data. Second, interference and random noise always create some error in our experiments, which must also be estimated (e.g. through statistical means). But there are some cases in which the degrees of approximation are not precisely knowable because either the idealization is too radical or the system too complex. Hence, we distinguish *controllable* from *uncontrollable* approximations. The former consists of approximations in which the deviations in the initial/boundary conditions of a process are connected by some laws to those of the predicted outcomes. The laws do not have to be precisely known, but some assurance of their existence is necessary. When no such laws appear to exist between those deviations, the approximations are uncontrollable. Of course, specific experimental testings will require specific models of approximation, but we do not see any general philosophical problems arising in this respect.

At this juncture, two questions need to be asked. First, how different is this from a syntacticist theory of confirmation? Second, does this capture more than the syntactic formulation the notion of *theory testing* in science as listed in 5 of the introduction? I shall discuss them in turn.

One of the earliest and most popular theories of confirmation is the hypothetical-deductive (H-D) method, which although discredited in its simple form still forms the core of many more sophisticated theories. In its simple form it can be given by a micro-scheme: for any hypothesis H and a set of auxiliary assumptions A, if H and A yields a statement P which is capable of being tested in an experiment, and P agrees with the experimental results, then H and A survive the test. But if P disagrees with the results, then either H or A is false. Since A is usually already better confirmed or tested, H must be rejected. $\{H, A\} = P$ is purely a logical deduction (or a mathematical derivation). So stated, the H-D method seems a perfect theory for syntacticists: it is a theory about logical relations among sentences or statements. Models need not be invoked at all.

However, if one asks for the details of testing P (rather than H), the situation is different. Suppose, without loss of generality, that P is a conjunction of particular statements each of which states that a certain value (or set of values) obtains at time t of system S. Then one conducts experiments on the intended property (or properties) of S and sees if, with proper calibration and approximation, the data match the derived values of the property. Not much is different at this level of testing, whether one believes in the H-D method, the Bayesian method, the boot-strapping method, or any other similar method. That is because at this level, i.e. the lowest level of any confirmation process, most methods will come to more or less the same procedures by which claims are directly tested by experimental findings.

No syntacticist would deny that the above is (at least roughly) how P is found agreeing or disagreeing with the phenomenon, but she pays little attention to this aspect of confirmation. So Suppes (1967, 1988) is right in saying that the syntactic view may not be wrong, but is certainly too simplistic, omitting the rich details of, for instance, measurements and experiments, where the semantic view provides much illumination as shown above in how state-space models are related to experimental findings. And Suppe's (1989) theory of "experimental methodology" (cf. pp. 96–99, 118–151) could certainly fill the lacuna (see also, Suppes (1962)). But what is the semanticist alternative to the H-D scheme, especially when *H* is a set of highly theoretical laws about entities that are not accessible to direct observations? I do not think that she has a radically different alternative to the largely syntactic methods mentioned above.

As argued in part I (Liu, 1997), it is not legitimate to always take the models for law-like equations to be the ones for their solutions. A model for the Schwarzchild solution of Einstein's general relativistic equation of gravity, for instance, is not a model for the equation; they do not bear the same kind of semantic relation with each other as do the model and the Schwarzchild solution. Now, there are many other cases like this in which the distance between the laws and the testable generalizations are far greater than that between the Schwarzchild solution and Einstein's law of gravity. In those cases, from the syntactic viewpoint, many steps of derivation are involved with added auxiliary hypotheses and statements of singular facts. It is even more disagreeable there to call the state-space models for the testable generalizations the models of the laws. It is not that those laws do not have models of their own, but their models are as far from any testable models as the laws are from the testable generalizations. Any talk about isomorphism or similarity between such models and actual systems seem far-fetched if not wrong.

It is obvious from the characteristics of the H-D method given earlier that a confirmation theory of that kind can handle these cases very well. In fact, as much as it errs on the experimental aspects of theory testing, it makes up on the theoretical aspects. Moreover, whether one believes in some enhanced version of the H-D method, or a Bayesian approach to confirmation, or a boot-strapping strategy, one must be committed to the idea that to bridge the gap between a theoretical law and some set of experimentally testable generalizations, derivations of some kind (with the help of auxiliary hypotheses) between the sentence expressing the law and those expressing the generalizations must be given. Therefore, confirmation procedures are essentially arguments in which evidential support from the experimental data is distributed in accordance with the logical relations among their component sentences. Some may argue that propositions rather than sentences should be used in such considerations, but that would not change the nature of the process. The semanticists, it seems, owe us a theory, different from the above, about how a theoretical model can be tested and confirmed.

As an initial approach, let us consider an example. Suppose that we have a theory/model of superconductivity in which Cooper-pairs (of electrons) are postulated. Presumably, a set-theoretic predicate, *P*, can be defined which characterizes the structural properties of the Cooper-pairs, *Q*. According to the semantic view, the TH might be simply put as "some *S* are *P*", meaning some phenomenal systems are isomorphic or similar to the systems defined by *P*. Since systems of Cooper-pairs are only remotely accessible, if at all, to direct observations, the models must contain unobservable entities and theoretical properties (e.g. electrons and their momenta and spins). How is the TH tested or subject to confirmation?

One will be ill-advised to construct state-space models for *Q* and think that they can be used for testing the TH, because those are most likely to "exist" in some abstract mathematical spaces which have no direct connection to the kind of state-space that are observable. Superconductivity is a quantum phenomenon, so the state-space description of the Cooper-pairs will be in some Hilbert space (of wave functions). Even if Hilbert spaces do exist, it will not help us because they are not the spaces in which we can experimentally obtain recordings. (Wave functions, in general, are not the kind of functions that one can obtain values for in an experiment.)

It seems that the best way to unpack the meaning of the above TH is to borrow some components from the received view of confirmation which as I said before is essentially syntactic (or derivational). In fact, what Cooper (and later Bardeen, Cooper, and Schriffer) did was to derive under certain extremely idealized conditions approximate phenomenological (i.e. macroscopic) generalizations which agreed with the observed phenomenon of superconductivity. Here, the phenomenological generalizations are those of thermodynamics on the macroscopic level, while the theoretical laws characterize the microscopic interaction and structure among the electrons involved. While it is probably true that the semantic view gives a better characterization of the reductional relation between the thermodynamical models and the quantum statistical models, when it comes to which law statement in the latter is confirmed by which set of data statements in the former, model reduction is obviously not enough.

Besides the types of testing or confirmation discussed above, which are widely examined in the literature, there are other kinds of testing in science (cf. 5 (i)–(v) in the introduction). The testing of consistency (cf. 5 (i)) is an important enterprise in fields ranging from mathematical physics to econometrics. To see whether a theory is consistent is to find out whether there is (at least) one structure (specified in a language other than the one for the theory) that satisfies the theory. This activity is especially made intelligible by the semantic view because if a theory is a description of a set of models, what could be more important initially than finding out whether the set is empty?

Now, to directly test a set of statements in a theory (cf. 5(ii)), according to the semanticist, one only needs to find a model that satisfies them. In this case, the statements are true at least in the model. But to be found true in a model is not sufficient, because scientists need to know if they are also true of nature/reality. To insure this, one needs to carry out the second step of the testing, namely, to compare the model with an actual system or phenomenon. But this is really *model testing* (cf. 5 (v)). As analyzed above, whether this can be directly done depends on the kind of models one is testing. If they are experimental models, the semantic theory provides a close approximation of such a model testing, but if the models are highly theoretical, e.g. the Cooper-pairs model of superconductivity, a combination of syntactic maneuvers and model construction is needed for testing such a model. It is also obvious, therefore, that such maneuvers must be involved in the *indirect testing of statements* (5(iii)).

To find out the suitability of a model (cf. 5(iv)) is entirely a different matter. Although it frequently occurs in the practice of science, their use is controversial in philosophy. Duhem's (1954) condemnation of model building solely for explanatory purposes is one of the earliest examples of that controversy. Controversy aside, let us consider the case of using a molecular model for a thermodynamic theory of gases. The molecular model can be understood as a semantic model or an epistemic one. If the latter, it may be an instantiation model (4(iv) in the introduction) in that the kinetic theory of gases can be considered as an instance of the general theory of thermodynamics which is indifferent to the specific constraints of various thermo-systems. But this does seem right, for the kinetic theory is better regarded as an instantiation of statistical mechanics. For one law to be an instance of another (under specific constraints), the two laws must be of the same type (i.e. at least sharing a set of terms). This is true between kinetic theory and statistical mechanics but not between kinetic theory and thermodynamics. Between the molecular models and thermodynamics, there is a relation of *reduction* involved which lies beyond the scope of testing or confirmation. Similar considerations apply if the model is a semantic one. A molecular model cannot be said to satisfy (in the sense of Sat_1) a thermodynamic theory.

Thus, it looks as if when the confirmation of theoretical laws is concerned, a combination of syntactic and semantic methods (for lack of better terms) is used. What must, and in fact does, happen is that scientists take the linguistically formulated law statements and relate them via a certain derivation to those lower level (phenomenological) generalizations. What general features and conditions of adequacy such a derivation should have is the task of a confirmation theory, but whatever they turn out to be, they could not possibly be characterized solely in terms of relations among different models. It is only at the level of phenomenological laws that the semantic perspective makes a significant contribution. The semanticist contribution in this respect consists mainly in providing a detailed structure for the experimental testing of a set of predicted values, whose intricacy and significance have long been neglected by the syntacticists.

Semantic view and structural explanation

When Hempel (1965) laid out his deductive-nomological (D-N) scheme for deterministic explanations and the inductive-statistical (I-S) scheme for statistical explanations, logical positivism had waned, but the syntactical approach which rode to power with it was still going strong. So, it is natural that Hempel takes an explanation to be an argument relating explanans sentences as its premises with the explanadum sentence as its conclusion. One should note that Hempel is clear in the very beginning that a distinction should be made between an explanation of the occurrence of an event and that of an empirical generalization. This distinction has been by and large observed by the subsequent contributors to the subject. To explain an event's occurrence, one provides an account of the circumstances under which it must, or is very likely to, occur. And this, according to Hempel, is accomplished when all the relevant initial/boundary conditions are given in conjunction with the relevant covering laws. Whereas, to explain an empirical generalization, which is often called "theoretical explanation", one invokes laws of higher order or of theoretical characters from which the empirical generalizations are derivable.

Under Hempel's schemes, confirmation seems to be the exact opposite process of explanation and a kind of symmetry between the two holds. If $\{L, \alpha\}$ implies β, then β is explained by L and α, and L is (partially) confirmed by β. It is not difficult to imagine that when a new piece of datum is recorded and recognized, it provides new support for the theory that anticipates it and at the same time is explained by that theory as to why it should be as it is. This symmetry holds roughly on both levels of explanation, i.e. whether the explanadum is a single event or an empirical law. When it is the latter, we say that the theoretical laws that explain a newly discovered empirical generalization are further confirmed by it.

Hempel's theory of explanation has been criticized, modified and improved in the past four decades (cf. Salmon (1989) for a retrospective), and all but a few attempts are given along the same syntactic line. The approach of *structural explanation* is one of those few that takes a different path. So far two different versions of it are present in the literature. It was first given a voice by E. McMullin (1968, 1978) and then modified by R.I.G. Hughes (1989a, 1989b, 1990). In his argument, McMullin pitches his proposal for structural explanation against what he calls "nomothetic explanation" which purportedly refers to those models of explanation developed by Hempel, Salmon, and others. We shall first discuss McMullin's theory, keeping an eye at the same time on its real differences with the Hempelian paradigm.

However, the nomothetic explanation is defined as that in which "one appeals to an empirically determinable regularity, a 'law.' The *explanadum* is to appear as an instance of this regularity" (McMullin, 1978, p. 143). This is certainly more limited than Hempel's D-N or I-S scheme. No restriction is made in those schemes such that only empirical laws or observable regularities (i.e. empirical generalizations) are allowed in the premises. A distinction between explanations of events and those of empirical laws is made by Hempel and Salmon *et al.*, as mentioned above, but it should not be understood as justifying a separation of the two levels of explanation. It does not follow from the distinction that singular events can only be explained by lower-level empirical generalizations, while theoretical laws can only explain empirical generalizations. A single explanation can be an argument that cuts through the two levels; for instance, an increase of a gas's pressure when it is heated can be explained by the mechanical laws of molecules (together with some statistical considerations). This is possible partly because of the syntactic orientation of the schemes: theories consist of sets of sentences, and from a logical point of view, an argument for explanation may consist of sentences from different theories at different levels, as long as they form a deductively valid argument.

It is clear that when McMullin divides scientific explanation into three largely disjoint groups, i.e. nomothetic, structural and genetic explanation, the nomothetic is restricted in the domain of using the D-N or I-S schemes for observable phenomena alone, while the structural and the genetic penetrate the unobservable world of theoretical laws and entities (cf. McMullin, 1978, pp. 143ff). Hence, a successful criticism of the nomothetic scheme is not an adequate one for the Hempelian paradigm.

McMullin (1978) defines the structural explanation at the very beginning of his article: "[w]hen the properties or behavior of a complex entity are explained by alluding to the structure of that entity, the resultant explanation may be called a *structural* one" (p. 139). The theory or model is better called "hypothetico-structural" (HS), because the entities and their structures are in most cases postulated as something remotely connected to our observations. It is clear from the article, McMullin considers the postulation of "hidden" or "unobservable" or "theoretical" structures as crucial to HS explanation. I find this requirement unnecessarily stringent. Perhaps most of the historical cases of HS explanation happen to conform to this stipulation, but it is not necessarily so. One may presumably explain the working of a clock by the structure of the parts inside, all of which are directly visible (in another circumstance). Why should such an explanation be excluded from the group of HS explanations?

Otherwise, the concept seems clearly defined. The ways in which a structure of a system provides explanation to its behavior will presumably have to be specified for each specific case, but examples abound in science and daily life that show how this can be done. For the simplest cases, we have the swing of a pendulum being explained by the very structure of that contraption. More complex would be the case of explaining the behavior of a glass of water by invoking the molecular structure of the liquid that is relevant to such behavior. For these, the notion of models comes immediately to mind. *Prima facie*, the semantic view and the use of models should most befit structural explanation, but before we see if this is indeed so, we need to clarify a related issue.

What is needed, but McMullin does not supply, is an account of how exactly a structural explanation differs from a Hempelian one. McMullin takes "a theory to be a set of propositions, explanatory in its intent.... part of the theory will be a specification of the structure" (1978, p. 139). But how is this different from Hempel's largely syntactic schemes? If the explanadum is a type of behavior of a system, then according to the Hempelian schemes (either the I-N or the I-S), the generalization that describes the behavior is explained by theoretical laws that govern (deterministically or indeterministically) the microscopic components of the system together with the initial/boundary conditions, if it is derivable from them. Nothing is said to exclude a specification of the system's structure in the set of initial/boundary conditions. If this is right, McMullin's structural explanation is threatened with collapse into a species of Hempel's account of scientific explanation in general: it is a set of Hempelian explanations where statements about a system's structure at a certain time, *t*, appear in the initial conditions. However, we do think that there is a significant difference between McMullin and Hempel if McMullin's account is interpreted in the direction of the semantic view. But we need to discuss Hughes' version of structural explanation first, because it offers the needed element to argue for that difference.

In Hughes's version (1989a) of structural explanation, McMullin's theory is modified and extended. The modification is introduced by way of broadening the notion of theoretical structures. Besides physical structures of material objects, Hughes takes facts and their relations also as legitimate structures for explanations.

> A structural explanation displays the elements of the models the theory uses and shows how they fit together. More picturesquely, it dissembles the black box, shows the working parts, and puts it together again. "Brute facts" about the theory are explained by showing their connections with other facts, possibly less brutish. (Hughes, 1989a, p. 198)

To see what Hughes really means by these claims, let us look at his "structural explanation for the EPR correlations" (cf. Hughes, 1989a, pp. 202–207). The explanadum in that case is the correlation among pairs of particles that are prepared in a singlet state prior to their being measured at some distance apart from each other, and the failure of attributing classical probability distributions to the measured results. The explanans, according to Hughes, consists of four essential principles of quantum theory: representability, entanglement, non-orthodoxy and projectability. These principles may also be seen as facts about quantum phenomena that support a non-classical probability structure. It is this structure that Hughes argues to be capable of explaining the EPR correlations.

One may object, especially if one believes in McMullin's notion of structural explanation, that this is no structural explanation since it does not tell us anything about the physical structure of the system (and the measurement apparatuses). But I take the spirit of Hughes's theory to be that logical, mathematical, or some other abstract models (cf. e.g. 3 (iii) and 3 (iv) in the introduction) should equally be counted as capable of providing explanations. It is in this sense that I take Hughes's theory to be an extension of McMullin's and to have opened the door to the semantic view (which Hughes, but not McMullin, explicitly mentions in his writing).

Moreover, whether or not a structural explanation should involve physical or some more abstract structures seems to depend on the nature of the explanadum. Hughes's structural explanation of the EPR correlations, for instance, uses some pure mathematical structures of quantum theory because what is explained there is not the existence of those correlations but why they deviate from the classical ones. Some philosophers even argue against the belief that the non-classical nature of the EPR correlations is something that needs an explanation; they seem to think that it is an unjustified preference towards the classical world that lures people into thinking that quantum correlations are queer properties. However, it is clear, that Hughes's account is not for explaining the occurrence of some singular events or repeated phenomena. For another example, we have two harmonic oscillators: a pendulum swinging in small angles in nearly vacuous space and a certain kind of electric circuit whose current oscillates. If the explanation is why they behave similarly, we may cite the formal similarity between the laws that govern their movements. However, this is not an explanation of why a pendulum or the current in an electric circuit moves and does it in a certain way. For this, the physical structure of either is needed.

Now, let us see whether in light of the semantic view, structural explanation provides a substantial improvement over the Hempelian and later syntactic schemes. Let us suppose that it is harmless to extend the notion of structures in structural explanation to that of models in the semantic view (Hughes's modification, as we mentioned above, makes this plausible). If a scientific theory is essentially a description of a set of models (i.e. structures), it should be obvious that structures are the ones responsible for the success or failure of scientific explanation, not the law statements that describe the relations among the properties in the structures. Moreover, from our previous discussion of the kinds of models included in the semantic view, it is easy to see that the structures McMullin has in mind are mostly those advocated by Achinstein and Hesse (the AH models, cf. Liu (1997)). The arrow in Figure 3 of Liu (1997) between boxes II and IV, which is labeled "Explanation", seems to be adequately captured by this extended version of structural explanation. Let us examine below the benefits this account may have over the Hempelian.

First, the notion of structural explanation (when properly extended) provides a way of avoiding the defects of Hempelian schemes which take explanations to be arguments. Explanations have to be formulated as linguistic entities because they have to be deductively valid or inductively strong arguments. But to explain an empirical generalization about a certain system structurally, one does not necessarily need to derive that generalization from theoretical laws. It is sufficient if one knows the physical or biological structure of the system. Take, for instance, the case of explaining the thermo-behavior of a system by invoking the system's molecular structure. The understanding we have concerning the molecular structure may only allow us to know the characteristic (or greatly idealized) thermo-behavior of the system. The fact that an actual thermo-behavior pattern of the system does not deductively follow from any laws at the molecular level does not necessarily deprive us of an explanation, as the Hempelians would think. To generalize a bit further, an understanding of a system's structure, especially when it is complex, such as in the case of a biological organism, usually gives us an idea about the essential dispositions of the system, which in turn tells us how the system behaves under certain idealized conditions. It will more often than not fail to provide accurate predictions of what will actually happen in most actual circumstances. According to Hempel, no explanation is to be had in these circumstances, but that seems mistaken. Intuition tells us that we do have explanation in such cases, and the structural account provides the justification for such an intuition.

Second, one may provide an explanation for or understanding of all the behavior patterns of a system if one knows its physical or biological structure. So, explanation may exist on a level when explicit laws and their deductive relations to lower-level generalizations are not known or do not exist. For instance, we think with good reason that we can explain the running of a car or a clock if we know enough of its physical composition and structure. We may not be able to derive every possible behavior of the car or the clock based on our current knowledge of its composition and structure, but in each case, we know at least the direction towards which such a derivation can be had. So also is the case with providing a structural explanation of thermodynamic phenomena by using the molecular structures of the thermo-system in question. If one goes through the details of Boltzmann's H-theorem, one could not help but realize that a straightforward derivation of every thermodynamic behavior of a tank of gas by the known mass and position/momentum of every one of its molecules is not possible. But that does not prevent us from recognizing that Boltzmann's understanding of the structure of gases provides excellent structural explanations for those behavior patterns. In other words, to explain, one should not always require derivability.

Third, even for those theoretical explanations in which the empirical laws are derivable from the theoretical ones, producing the derivations may not be what explanations are about. Rather than saying that Kepler's or Galileo's law are explained by the Newtonian laws of mechanics because the former can be derived from the latter, as the syntacticists would have it, it is probably better to say that different mechanical phenomena, such as the movement of the planets, of a block on a smooth incline, and of a pendulum in a near vacuum, are explained by Newtonian mechanics because they are all Newtonian systems as accurately captured by the models that satisfy the Newtonian laws of mechanics.

Therefore, while the semantic view does not seem to have much of an advantage with respect to theory testing or confirmation, except in the domain of experimental models, it is conducive for a new notion of scientific explanation which may be closer to capturing the concept of explanation, on the one hand, and, on the other, avoid some of the problems that the syntactic formulation cannot. Moreover, as is well known, explanation is inextricably tied to the unification of scientific theories, which in turn is connected to the notion of theory reduction. The benefits for going semantic in the account of reduction are thoroughly explored in the literature (cf. Balzer *et al.*, 1987); I shall not pursue the topic in this paper.

Interpretations vs. interpretations

Under "Interpretation" in the scheme given in the introduction, two senses of that word are given: the semantic and the philosophical. In this final section, I shall begin to explore whether the notion of interpretation widely used in the practice of science is the former or the latter. The example I choose is the interpretation of quantum theory, but the analysis, I hope, will also be applicable to other areas in science.

Semantic interpretations are relational structures. When an interpretation satisfies a set of sentences, it is called a model of the set. Interpretations may contain universes of abstract objects, such as numbers and geometrical figures, or of real/actual objects, such as planets and elementary particles (real means physically possible, hence, whatever is actual is also real, but not vice versa). This seems straightforward and well defined.

The notion of philosophical interpretation is less well defined and much more involved. It is an offering of placing a concept or a theory in a certain philosophical framework so that certain metaphysical or epistemological questions can be answered for it. Now the question is how one can tell if a claimed interpretation is a semantic interpretation or a philosophical one.

The best way to approach this question is to select a case which everyone seems to regard as typical of philosophical interpretations and see if it indeed is. The interpretations of probability seem the best example of this kind. Since the case has been discussed in great length in the literature and my task is not to evaluate them or to propose a new interpretation, I shall be brief. It is commonly understood that the (purely syntactic) calculus of classical probability is established by Kolmogorov (1950), which provides rules of calculating probabilities in a universe of events that seem to have captured all or most of our intuitions about how such calculations should proceed. The philosophers' task, then, is to find the best interpretation for it. A variety of different interpretations have been proposed and debated in the literature; among them are that of limiting frequency, of degrees of personal belief, of logical relation between a hypothesis and a set of evidence, and of propensity (or chance).

One is gravely mistaken if one thinks that what Kolmogorov, in his seminal work, gives is no more than a purely syntactic calculus. If that were so, Kolmogorov would not have known whether the calculus is consistent. But, he knows that it is because he has shown that the calculus has a model, the model being supplied by a set theory over a universe of events. Of course, the model is an abstract one that is not directly connected to the world, but it is a model nonetheless, and so a semantic interpretation is given by Kolmogorov. All those attempts at coming up with *an interpretation* mentioned above should be understood as efforts to discover realistic models, ones that represent real features of the world.

Now, the question is whether they are semantic or philosophical interpretations. As an initial approach, let us compare probability with the notion of force. Force is an important function in the calculus of classical mechanics. In a model that takes the universe to comprise primarily a set of point-mass particles, the force is a vector function that maps particles and time to R^3 that satisfies certain relations with other functions. Now, take for instance the limiting frequency interpretation of probability measures. (The result of the following analysis should apply equally well to the other interpretations.) It can also be construed as a model of the calculus whose universe is the set of all actual and possible events. What is mapped on to by the symbol such as "*P*(*A*/*E*)" is the ratio of the number of *A* events appearing in all those *E* events when the number of *E*s approaches infinity, or "*P*(*A*/*E*)" means

$$\lim_{n \to \infty} (A_n/E_n),$$

where E_n are individual events sharing the same property, *E*, and A_n those E_ns which exhibits property *A*. Whether this interpretation is good is another question, but if the interpretation of force in the model of classical mechanics is a semantic interpretation, then so should this be.

One may ask that if the limiting frequency interpretation is just one of the models for the probability calculus, why is it so controversial? Moreover, why do we have the impression that none of the available interpretations are satisfactory? I think that the controversy is not over whether they are semantic interpretations but over which one, if any, of these models can best represent the notion of probability used in science and in everyday life. For instance, one worry about the limiting frequency interpretation is how it can capture the use of probability when finite sequences are involved. Defining probability as the value at the limit does not help us to understand how to obtain its expected values in finite sequences. More serious is the problem concerning the chance of singular events (i.e. events that seldom re-occur, such as historical events): it is at least not obvious that the limiting frequency applies here.

Now, this kind of problems does seem to be of a philosophical rather than of a logical nature. So, in a sense, the interpretation of probability as a whole contains a philosophical aspect. What should be realized is that the two aspects—the semantic and the philosophical—should be kept separate. Just because a certain interpretation is philosophically problematic, it does not mean that it could not be a good semantic model. The available interpretations of force in classical mechanics are also controversial, and there is a long history of debates over which if any of the interpretations of force is acceptable (cf. Jammer, 1957). But no one should argue that classical mechanics does not have good models to satisfy it.

It seems fair to conclude, then, that with respect to the interpretations of probability, the interpretations which satisfy the classical calculus of probability are certainly semantic models. But the suitability of such models is in question. Further, since the problem is whether the interpretations are good enough to represent our use of probability in science and daily life, one may locate the problem at the relation between models and reality. According to the semantic view, this is where the theoretical hypothesis comes in. To rephrase the problem in terms of the semantic view, one should say that all interpretations of probability, just as those for forces, are legitimate models for the calculus, but whether any system in the world is isomorphic to or bears a strong similarity relation with any of them is questionable. It seems that counterexamples abound, at least, in the case of limiting frequency interpretation. Hence, it may be that none of the models, although satisfying the calculus, can be taken as the one that correctly represents our use of probability.

However, certain problems are unique to the probability case. For instance, the different interpretations do not even share the same ontology with one another. Unlike the case of classical forces where the universe for different models is the same (a set of point-mass particles), the different models of probability do not share the same universe. For the propensity and the limiting frequency interpretation, the universe contains objects or events, whereas for the subjective or the logical interpretation, the universe comprises statements (or propositions) of generalizations and of particular events. It is difficult to imagine that a debate between the objectivist and the subjectivist can be settled through empirical means such that one interpretation is shown by empirical evidence to be superior to the other.

Now, let us turn to the interpretation of quantum theory, which consists mainly of two parts. One is about the meaning of the wave function, and the other about that of quantum measurement. The standard approach to the first task is (á la Max Born) to take the square-modulus of a wave function to represent the probabilistic distribution of a certain observable of a system. Therefore, the problem of interpretation, if there is one in this part, is a problem of interpreting quantum probabilities. The fact that quantum probabilities obey a calculus (unfortunately misnamed "quantum logic") that is different from the classical one should not alter the logical nature of the interpretations. If the interpretations of the classical calculus are semantic models, those of a non-classical calculus should not be any different in this respect.

However, there is a philosophical dimension to the interpretation of wave functions. One may understand Born's probabilistic interpretation in two different ways: the stronger construal takes it to be an interpretation of the wave function, and the weaker version takes it to be one only for the square-modulus of the wave function. If the former is taken, then there are incompatible alternative interpretations, such as Bohm's (1952) interpretation of the wave function, which takes it to be a physical field as substantive as the electromagnetic field. While if the latter is taken which means that one does not make any commitment as to what a wave function refers, then Born's interpretation of the square-modulus of the wave function is compatible with Bohm's and other interpretations. Whether a wave function represents some physical field or is just a mathematical device for calculating probability distributions is clearly a philosophical question of interpretation not a semantic one.

The meaning of quantum measurement is an entirely different problem. The challenge can be roughly sketched as follows. In quantum theory proper (i.e. without a theory of measurement), there is a postulate called the *projection postulate* (PP), which states what happens as a law of nature when a system is measured in any quantum state (given by a wave function). More accurately, although still roughly, the postulate says that whenever an observable, *O*, of a system, *S*, is measured in a state, Ψ, then at the end of the measurement process *S* will be "projected" from Ψ to one of the eigenstates of *O* and obtain a definite (eigen) value in that state. Quantum theory does not contain physical laws that explain this process of projection. Either such processes are governed by some laws that are beyond quantum theory or they obey the same law of unitary evolution that all other quantum processes obey. Both directions have been explored in the literature; the GRW theory (Ghirardi *et al.*, 1986), for instance, belongs to the former while the decoherence approach belongs to the latter.

Are the proposed models for quantum measurement also semantic models? To answer this question, one must be clear on the scope of the theory (or rather its calculus) under investigation. Whether models of quantum measurement are semantic models depends on what they are supposed to satisfy. If one of these efforts (of coming up with a theory of measurement) actually succeeds, it seems that we will have provided a better understanding of PP, since whatever is said by PP can be said better in the theory of measurement. Now, is not such a theory an addition rather than a model of the quantum theory proper? Not necessarily. If no new (or hidden) variables are introduced, it will not involve anything beyond the scope of quantum theory. It should be a description of processes that are governed exclusively by the Schrödinger equation (and some constitutive equations), and so whatever the model turns out to be, it will be a semantic model for quantum theory. In other words, if quantum measurements are a special kind of quantum process, not essentially different from, say, the process of quantum scattering, then the quantum theory proper will be satisfied by all quantum processes including all measurement processes.

Let us now see what such a model would involve if one were successfully built. Supposedly, the processes will be unitary evolutions, which means that the measured system and the measuring apparatuses are describable by wave functions and their behavior at any time is governed by the Schrödinger equation with the appropriate interaction Hamiltonian. Then by solving the equation we obtain the measurement results that the original PP predicts. If this is the right picture for establishing a theory of quantum measurement, the problem of interpreting the measurement process is a problem of getting the right semantic models for quantum theory.

However, if it turns out that a measurement theory goes beyond the scope of quantum theory (the other alternative mentioned above) so that it involves new (or hidden) variables and laws, the matter is then entirely different. The theory is not conceivable as a description of a set of semantic models, but a full-fledged addition to quantum theory. The GRW theory of spontaneous collapse is a theory of this sort, and so is the Bohmian interpretation where hidden variables are postulated to restore determinism to quantum phenomena. In the GRW case, a law is formulated by which a Gaussian is superimposed on the state function of a quantum system at regular intervals so as to collapse the wave packet, whereas in Bohmian mechanics, the position of quantum particles is given priority and definite values at each moment of a process and a new law of motion are postulated. Even though both save the same phenomena as the quantum theory does, they are different theories.

Furthermore, what would one say about interpretations such as the Copenhagen interpretation? Is there any semantic model given by that? Is it even an attempt to construct semantic models for quantum theory? Interpretations of this type are truly philosophical interpretations. The reason is quite simple: it should have been called the

Copenhagen *philosophy* of quantum theory, not an interpretation. Or if one insists on calling it an interpretation, it is one with an explicit ontological and epistemological commitment attached. It certainly embraces Born's probabilistic interpretation of the wave function, but it has no intention of resolving the problem of quantum measurement along the directions sketched above. Here is a list of the considerations that do go into the Copenhagen interpretation: the complementarity of mutually non-commuting observables, the wave-particle duality of quantum objects, the priority of macroscopic objects and the classical vocabulary, and the instrumentalist view of concepts and laws regarding microscopic objects.

Bohr's principle of complementarity is not given to attribute or clarify the meaning of any terms in quantum theory; it is rather an attempt to square the non-classical features of quantum phenomena with classical ones. One may think of the wave model and the particle model as two distinct semantic models for microscopic objects. However, they are not. The wave model represents those objects as classical wave packets, but it does not satisfy the classical theory of waves; nor does the particle model, which represents those objects as classical particles, satisfy classical mechanics.

They are models, if one can call them that, which represent different sets of features of microscopic objects: waves represent the probabilistic distributions of those objects, and particles their singular impacts. Moreover, the priority of the macroscopic and the instrumental view of theoretical entities are obviously philosophical views. At this level, the various philosophies, such as the Copenhagen "interpretation" and Bohmian mechanics, are not about the circumstances under which each statement in quantum theory is true, but rather about the ontological status of such items as the wave functions (whether or not they represent real substance) and observables (whether or not they represent real properties of microscopic objects).

Such philosophical interpretations are not unique with the quantum theory; other branches of physics (or of science in general, for that matter) also need this kind of interpretation. The meaning of force, for insistence, and whether spacetime is an absolute substance are interpretation problems of this kind. The difference between the classical and the quantum case is that while in the former, the semantic and philosophical interpretations are clearly separated and understood, they are not so well separated and understood in the latter. In fact, the philosophical interpretation is in part caused by a lack of success in the semantic part. By this, I do not mean that quantum theory is not successful or less so than classical theories in making predictions and meeting the challenges of experimental testing. Rather, my point is that the theoretical models of microscopic objects that are not directly testable by experiments have failed to find acceptable places in our philosophical view of reality.

This analysis seems to suggest that we redefine the theory of quantum mechanics. Instead of taking the mere calculus of quantum theory together with the measurement results (i.e. recordable data) to be quantum theory proper, we should insist on including all the possible semantic models, while leaving the truly philosophical part of the interpretation for the term "the interpretation of quantum theory". In so doing, non-relativistic quantum theory as we have it today is not a complete theory because we have not found an acceptable model of quantum measurement for the theory. No philosophical speculations could resolve the problem unless one finds such a model or proves a priori that such an model is impossible. But if the latter, then either quantum theory is again incomplete or it is inconsistent (in the straightforward logical sense of the term).

Conclusion

In this paper, I have discussed issues and problems concerning the applications of the hybrid semantic view of scientific theories. While remarking on its success in such a domain as structural explanation, I noted its inadequacy in theory testing (or confirmation). What emerges from this study is a justification of what I have said in part I of this project (Liu, 1997), namely, the concept of scientific theory is fully captured neither by being characterized as sets of sentences with inference rules nor by a description of sets of relational structures. Even though it appears that the original semantic view is more fruitful than the syntactic view, it cannot be the right view unless it is modified in ways which these two papers suggest.

References

APOSTEL, L. (1961) Towards the formal study of models in the non-formal sciences, in: H. FREUDENTHAL (Ed.) *The Concept and the Role of the Model in Mathematics and Natural and Social Sciences* (Dordrecht, Reidel), pp. 1–37.

BALZER, W., MOULINES, C.U. & SNEED, J.D. (1987) *An Architectonic for Science: The Structuralist Program* (Dordrecht, Reidel).

BOHM, D. (1952) A suggested interpretation of quantum theory in terms of 'hidden' variables. I & II, *Physical Review*, 85, pp. 166–178, 180–193.

DUHEM, P. (1954) *The Aim and Structure of Physical Theory* (Princeton, Princeton University Press).

GHIRARDI, G.C., RIMINI, A. & WEBER, T. (1986) Unified dynamics of microscopic and macroscopic systems, *Physical Review* D, 34, pp. 470–491.

GIERE, R. (1988) *Explaining Science: A Cognitive Approach* (Chicago, University of Chicago Press).

GIERE, R. (1994) The cognitive structure of scientific theories, *Philosophy of Science*, 61, pp. 276–296.

HEMPEL, C. (1965) *Aspects of Scientific Explanation* (New York, The Free Press).

HOWSON, C. & URBACH, P. (1993) *Scientific Reasoning: The Bayesian Approach*, 2nd edn (Chicago, Open Court).

HUGHES, R.I.G. (1989a) Bell's theorem, ideology, and structural explanation, in: J.T. CUSHING & E. MCMULLIN (Eds) *Philosophical Consequences of Quantum Theory* (Notre Dame, University of Notre Dame Press), pp. 195–207.

HUGHES, R.I.G. (1989b) *The Structure and Interpretation of Quantum Mechanics* (Cambridge, Harvard University Press).

HUGHES, R.I.G. (1990) The Bohr atom, models, and realism, *Philosophical Topics*, 18, pp. 71–84.

JAMMER, M. (1957) *Concepts of Force* (Cambridge, Harvard University Press).

KOLMOGOROV, A.N. (1950) *Foundations of the Theory of Probability* (New York, Chelsea).

LIU, C. (1997) Models and theories I: the semantic view revisited, *International Studies in the Philosophy of Science*, 11, pp. 147–164.

LLOYD, E.A. (1988) *The Structure and Confirmation of Evolutionary Theory* (Westport, CT, Greenwood Press).

MCMULLIN, E. (1968) What do physical models tell us? in: B. VAN ROOTSELAAR & J.F. STOAL (Eds) *Logic, Methodology and Philosophy of Science III* (Amsterdam, North Holland), p. 392.

MCMULLIN, E. (1978) Structural explanation, *American Philosophical Quarterly*, 15, pp. 139–147.

SALMON, W. (1989) Four decades of scientific explanation, in: P. KITCHER & W.C. SALMON (Eds) *Minnesota Studies in the Philosophy of Science* (Minneapolis, University of Minnesota Press), pp. 3–219.

SNEED, J.D. (1971) *The Logical Structure of Mathematical Physics* (Dordrecht, Reidel).

STEGMÜLLER, W. (1973) *Theorienstruktruen under Theoriendynamik* (Berlin, Springer). English Translation by WOHLHÜTER, W. (1976). *The Structure and Dynamics of Theories* (New York, Springer).

SUPPE, F. (1989) *The Semantic Conception of Theories and Scientific Realism* (Urbana, IL, University of Illinois Press).

SUPPES, P. (1962) Models of data, in: E. NAGEL, P. SUPPES & A. TARSKI (Eds) *Logic, Methodology and Philosophy of Science: Proceedings of the 1960 International Congress* (Stanford, Stanford University Press), pp. 252–261.

SUPPES, P. (1967) What is a scientific theory? in: S. MORGENBESSER (Ed.) *Philosophy of Science Today* (New York, Basic Books), pp. 55–67.

SUPPES, P. (1988) Representation theory and the analysis of structure, *Philosophia Naturalis*, 25, pp. 254–268.
FRAASSEN, B. VAN (1989) *Laws and Symmetry* (Oxford, Oxford University Press).

INTERNATIONAL STUDIES IN THE PHILOSOPHY OF SCIENCE, VOL. 11, NO. 2, 1997 147

ARTICLE

Models and theories I: the semantic view revisited

CHUANG LIU
Department of Philosophy, University of Florida, USA

Abstract *The paper, as Part I of a two-part series, argues for a hybrid formulation of the semantic view of scientific theories. For stage-setting, it first reviews the elements of the model theory in mathematical logic (on whose foundation the semantic view rests), the syntactic and the semantic view, and the different notions of models used in the practice of science. The paper then argues for an integration of the notions into the semantic view, and thereby offers a hybrid semantic view, which at once secures the view's logical foundations and enhances its applicability. The dilemma of either losing touch with the practice of science or yielding up the benefits of the model theory is thus avoided.*

Introduction

The study of the structure of scientific theories as a philosophical topic was, with the rise of logical positivism, strongly influenced by the achievement and early development of first-order mathematical logic. The syntactic view (or the received view) of scientific theories mainly exploited the axiomatic method in logic and went with logical positivism as it was later discredited. However, some philosophers took the failure of logical positivism to be a general indictment of the use of mathematical logic, the sterility of whose employment was assumed across the board. Such sentiment seems to have contributed to the undeserved neglect in the past decades of the semantic view, which rejects the syntactic view but makes use of even more mathematical logic. To see that, one needs only to note the disparity between the amount of attention given to van Fraassen's constructive empiricism or pragmatic theory of explanation and that given to his semantic view, despite the fact that they are articulated in the same book (van Fraassen, 1983) and that he insists on the latter's central role in the former positions. But the recent revival of the semantic view not only in the studies of physical theories, its traditional territory, but also in those of biological theories and theories in social sciences shows how far such an assessment is from the truth (cf. Giere, 1988, 1994; Hausman, 1992; Lloyd, 1988; Sloep & van der Steen, 1987; Suppe, 1977, 1989; Suppes, 1967; van Fraassen 1972, 1983, 1989).

However, because of the recent effort by its proponents, especially on this side of

0269-8595/97/020147-18

the Atlantic, to push the semantic view to the forefront of research and see its applications in non-traditional areas, a voice of criticism and an effort of refinement are conspicuously lacking. I am convinced that it is now time to step back and critically evaluate the semantic view. Some may argue that that is the job for the view's opponents, but here is the catch. Is there anyone who is an opponent of the semantic view? There are some, I am certain, who are skeptical about it, but no one seems to oppose it because there is no plausible opposite position to take. No serious philosophers today can defend the syntactic view. On the other hand, a lack of criticism or critical interest may also condemn a philosophical position. With such consideration in mind, I begin what I call a revisit of the semantic view. In this Part I of the project, I shall first contrast the semantic with the syntactic view in light of the fundamentals in model theory. Even though the syntactic view as a whole is false, it does not mean that it contains no true elements that the semanticists can borrow. Second, I investigate the concepts of models independently of the discussion of the semantic view, and then see if all or any of them fit in the view. Since one of the main virtues the semantic view is claimed to possess is that it makes sense of the ubiquitous model-building activity in the natural and social sciences, it would be a disappointment, to say the least, if the semanticist models in terms of which scientific theories are supposed to be given turn out to be something entirely different from what scientists in fact use to construct their theories. Third, I explain the shortcomings of the original construal of the semantic view, and then offer a hybrid view which combines true elements from the syntactic and the semantic view.

The syntactic vs. the semantic view

Model theory (of mathematical logic)

Both the syntactic and the semantic view of scientific theories are the offspring of (first-order) mathematical logic. It therefore requires that we give a brief review of the relevant materials in logic (for technical details, see Malitz, 1979; Monk, 1976), so that we can see in later discussions how close to their origin the views have remained, or how far from it they have strayed.

The language of first-order mathematical logic should be understood as a fragment of our natural language, and so are the scientific languages. In most cases, a scientific language is closer to a natural language than a logical one, but that is not necessarily so. In principle, a logical language can be made indefinitely close to a natural language, for all one needs to do for that is to sacrifice some of the unity and simplicity of one's syntax. The same thing can be said of scientific languages as well. In fact, most scientific disciplines begin with theories formulated in languages that barely differ from our natural languages, but subsequent development inevitably proceeds towards more and more abstract and general theories which demand languages with simpler syntax.

The sharp distinction and contrast between the syntax and the semantics of a language in logic provides a useful conceptual tool for analysing scientific theories. This does not imply that the two parts are easily separable in the practice of science. The syntax of a language L consists of a *vocabulary* and rules of formation by which any sequence of symbols allowable in L is formed, while the semantics of L provides interpretations which map symbols in L to entities and their properties in a relational structure, so that all elements in L are given meanings *via* some compositional rules. Such maps allow us to judge whether a sentence in L is true or false in the given structure.

Because of this division, it is possible to define two different but related notions of a "theory". A theory may be defined syntactically as a set of sentences T that is a derivational closure of a set of sentences Σ, i.e. $T = \{\phi \in L : \Sigma \vdash \phi\}$, where Σ comprises of T's axioms. And a theory may be defined semantically as a set of sentences that are true in a model A, i.e. $T = \{\phi \in L : A \models \phi\}$. A typical syntactic theory is the Euclidean geometry, which consists of five axioms and all the theorems derivable from the axioms via rules of geometrical construction (cf. Hilbert, 1902). An example for a semantic theory would be group theory, which consists of all the first-order sentences that a set of groups satisfies, the groups are definable by set-theoretic entities (cf. Sneed, 1971, pp. 9–10). Furthermore, if a theory is axiomatizable, to see if a structure is a model of it, one only needs to find out if it satisfies the axioms. This is so because of the completeness (and soundness) theorems of first-order theories. Hence, $A \models T$ iff $A \models \Sigma$, provided Σ is the set of T's axioms. It will be helpful to keep the notions of these two kinds in mind when we consider the concept of scientific theories from either the synactic or the semantic view.

Ideally, T's models are determined uniquely by the given interpretation. Once the relational structures are given, we can discuss items such as the *consistency* of a theory, one theory's being *equivalent* or *reducible* to another, and one structure's being *isomorphic* to or an *extension* of another structure, etc. Even though the consistency, equivalence and reduction of or among theories are primarily syntactic notions and can in principle be handled syntactically, it is much easier and more elegant (because of the completeness of first-order theories) to use the structures. For instance, to prove syntactic consistency of a theory one must make sure that no contradiction is derivable from it, which could be a daunting task if the theory is complex. But semantically, one only needs to make sure that at least one structure exists which satisfies the theory. Again, because of the completeness, the equivalence or reduction of theories can be formulated in terms of the isomorphism or extensions of their models.

One caveat should be observed at this point. The schemes reviewed above are only known to apply to first-order languages. Therefore, we know neither if they can be applied to theories which are not first-order but close to it, nor if any of our actual scientific languages, although not a ready-made first-order language, can be made into one.

The syntactic view

The view is at the core of the logical positivist philosophy and says much more than of what a scientific theory consists. The flow-chart in Figure 1 gives a rough sketch (cf. Suppe 1977, pp. 6–62); the upward direction marks the ascent of the theoretical ladder.

The difference between C_O and C'_O is that in the latter set there are universal or semi-universal generalisations into unobserved actual (i.e. universally quantifiable) or possible (i.e. modal) domains.

According to the syntacticists, a scientific theory is, or can be reconstructed as, a set of interpreted sentences some of which are axioms, and the others in the theory are derivable from the axioms through syntactic (including mathematical) rules. There are two types of interpretations which occur on two different levels. At the first (or lower) level, one has the simple interpretation, I, which relates terms in V_O to observational results. It supplies the semantics for C_O and C'_O just as what one does in mathematical logic. The second (or higher) level of interpretation is a partial interpretation via rules

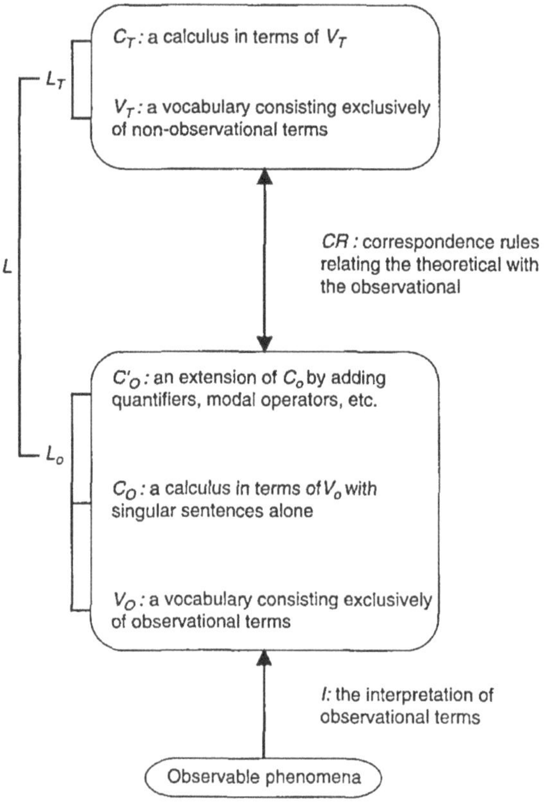

Figure 1. *The syntactic (or the logical positivist) view.*

of correspondence, *CR*, which relates semi-definitionally terms in V_T to terms in V_O. One must note that interpretations via *CR* are radically different from that at the lower level: they are partial definitions or reductional sentences in L so that the meaning of terms in V_T and of sentences/inferences in C_T can be given ultimately in terms of C_O and C'_O. The positivists consider "C_T (and V_T) $+ CR$" to be a scientific theory proper. A theory acquires its meaning and confirmation (positivists consider them one and the same) exclusively through matching the observational sentences (in terms of V_O) with data under the standard interpretation of V_O. The rest (especially terms in V_T and sentences in C_T) gain their meaning and/or confirmation when they are partially defined by, and/or connected with, (via *CR*) C_O and V_O.

In the positivist view, models of a theory, especially of its theoretical part, have little significance. While the intended model for sentences in V_O is deemed obvious, no terms and sentences in V_T directly admit models that are independent of experimental data. Instead of interpreting theoretical terms directly and discuss theoretical models, one first reduces them to V_O. Theories are semantically "sealed", so to speak, around their edges.

Take the theory of thermodynamics known to Gibbs or Planck, for example. The theory consists of two different kinds of functions: those that represent temperature, volume, pressure, etc., and those that are called "entropy", "enthalpy", and "Gibbs potential", etc. The former kind are observational and the latter non-observational, and the latter are explicitly defined in terms of the former. The axioms of the theory are the three laws and every other statements about any thermodynamic properties of a system can be derived from those and the system's state-equations. (The state-equations are necessary because the three laws are completely general and so applicable to all thermo-systems.)

One can now see that "the syntactic view", if not solely in contrast to "the semantic view", is a misnomer. It gives the impression that syntacticists only consider the uninterpreted and highly abstract calculus (such as C_T) as a theory, which is in fact not the case. Interpretations and correspondence rules are also essential components of a theory. In other words, a theory in the syntactic view is not a syntactic theory as defined above. The right way of conceiving this as a syntactic view is that it emphasizes the importance of a theory's linguistic structure at the expense of other elements. From now on, we shall use the term only in this more circumscribed sense.

The semantic view

Before I give a summary of the view, let us see what it appears to its chief proponents, among whom Beth (1949, 1961) probably conceived the idea first. Van Fraassen said in his early construal of the view:

> To present a theory is to specify a family of structures, its *models*; and secondly, to specify certain parts of those models (the *empirical substructures*) as candidates for the direct representation of observable phenomena.... the theory is empirically adequate if it has some model such that all appearances are isomorphic to empirical substructures of that model. (1983, p. 64)

He later adopted Giere's (1988) formulation and said:

> [A] theory consists of (a) the *theoretical definition*, which defines a certain class of systems; (b) a *theoretical hypothesis*, which asserts that certain (sorts of) real systems are among (or related in some way to) members of that class. [The systems's behavior is determined by the] *laws of co-existence* and *laws of succession* [and realized in] a space of possible states, ... [so] the presentation of the theory must proceed by describing a class of *state-space types*. (1989, pp. 222–223)

But according to Suppe:

> This suggests that theories be construed as propounded abstract *structures* serving as models for sets of interpreted sentences that constitute the linguistic formulations.... what the theory does is directly describe the behavior of abstract systems, known as *physical systems*, whose behaviors depend only on the selected parameters. However, physical systems are abstract replicas of actual phenomena, being what the phenomena *would have been* if no other parameters exerted an influence. Thus by describing the physical systems, the theory indirectly gives a counterfactual characterization of the actual phenomena. (1989, pp. 82–83)

Suppes (1967) is among the pioneers of the semantic view and perhaps the least dogmatic. First, his early attempt at an axiomatisation of mechanics (classical and

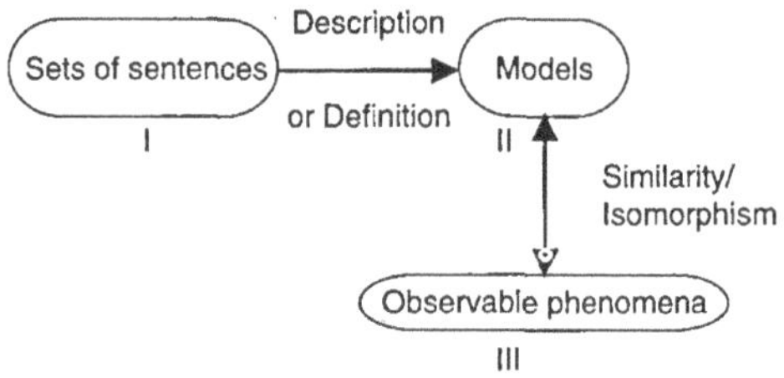

Figure 2. *The semantic view (cf. Giere, 1988, p. 83).*

quantum) champions an approach to scientific theories that includes both syntactic and semantic apparatuses in mathematical logic. Then, in defending the semantic view, he maintained:

> The view I want to support [a prototype of the semantic view] in this essay is not that this standard sketch [the syntactic view] is wrong, but rather that it is far too simple. (1967, p. 57)

A theory, according to Suppes, is given by a predicate (such as "being a mechanical system") which admits both an intrinsic and an extrinsic characterization: the former by a set of uninterpreted sentences and the latter by the set of models that satisfies those. The syntactic view, which emphasizes the former, is not "wrong" but "far too simple" (ibid.). By incorporating a theory's models, such issues as the theory's axiomatization, its consistency and its various symmetry properties can be discussed with much simpler means. This shows that Suppes' theory is modeled very closely to the relation between the syntax and the semantics in the first-order logic which I briefly reviewed above. The theory-model pair may also be applied to different levels of theorization. For instance, a theory of measurement of a particular kind may have axioms and theorems explicating the experimental procedures and sets of numbers (i.e. data) as models that they satisfy.

The flow-chart (Figure 2), which is a slightly modified version of Giere's, seems to have summed up what is in common in the above characterizations of the view. The first box of Figure 2 (containing sentences in a language) would be where a syntacticist says a theory lies. Since the syntactic features of sentences and their interdependent relations are no longer of primary importance to a semanticist, th. fine linguistic structure touted in Figure 1 is not appreciated. Instead, which set of sentences should suffice depends entirely on what kind of models are under consideration. The predicate is described or defined by the sentences, but the property that it picks out is specified by the structure of the models. So, the sentences should describe necessary and sufficient conditions for picking out the models. It then follows from this that two descriptions are equivalent and exchangeable so long as they are co-extensive (i.e. they pick out the same set of models).

The following might be one of the simplest examples of equivalent descriptions. Imagine that "there is a trajectory in a space spanned by x and y, which is a straight line going through (0,0) and having an angle θ to x such that $\tan\theta = a/b$, where $a, b > 0$". The quoted sentence is a theory that the trajectory satisfies, but the trajectory also makes "$(x)(y)[ax - by = 0]$" true, which is another theory that is equivalent to the previous one.

The relation between the second and the third boxes is understood as an empirical (or theoretical) hypothesis, which says that some phenomena (or phenomenal systems) are isomorphic to the empirical substructures in the set of models, or some real *systems* are similar to some of the models. It is notoriously difficult to make this relation clear except in the case where state-space is used for all model construction. In other words, if formulated in mathematical language, a model or the behavior of a system is given in a state-space, whose local dimension is determined by the number of degrees of freedom the main variables in the theory possess, and whose points hold objective correspondence with states of the system. So restricted, the hypothesis amounts to the following claim: configurations in the state-space, which are sets of phase points in that space, are given or predicted by the models. These points can be directly or indirectly compared with the observational results or data.

Suppose, for instance, that the state-space is the *PV* space of thermodynamics, where $P(t)$, $V(t)$ are variables representing, respectively, the pressure and the volume of a thermo-system. A model could simply be a set of processes: $t \mapsto (P(t), V(t))$, which represents the time-evolutions of the system. To see if an actual system is one to which this predicate is attributable, one only needs to compare the measured data of that system with the values in the models (provided all conditions of calibration and approximation are taken into account). Not all hypotheses are this simple, and so the comparison might be much more involved and indirect. But the semanticists believe that in the end, all such comparisons are essentially of the same kind. It will be argued later that this claim is too naïve.

Furthermore, a mathematical theory usually comprises two types of basic equations which correspond to laws of co-existence and of succession, the former of which determine the states, and the latter time-evolution, of a system in the state-space. As linguistic items, the equations or the lawlike sentences are among those in the first box that, together with other sentences, define models or systems. It is not clear what exactly are laws of nature in the semantic view. It seems that they should be sets of models, but how can such sets be determined independently of the sentences we call lawlike? I shall return to this question later.

A few things in this view calls for our immediate attention. First, Suppe uses the counterfactuals to capture the idealized nature of all models. If that is allowed, the possible-world talk lends itself conveniently to the semantic view, whether or not one is a realist about such worlds. A nonviscous fluid is simply a fluid in a possible world where, *ceteris paribus*, frictional forces do not exist. Similarly, it seems that all conditions of idealization and/or approximation can be thus understood or analyzed, and the models of a sentence are no more or less than the possible worlds in which it is true. Possible worlds may differ from the actual one by having different initial/boundary conditions, or by having or lacking some contingent but lawlike features. The latter worlds are more distant from the actual one than the former, but they are the ones that will most likely to account for conditions of idealization and approximation.

Second, the semanticists are not entirely clear on what the two arrows in the flow-chart in Figure 2 really are. These arrows seem to be semantic relations since they are between words and the world, but are they? Not really. If the above possible-world talk holds, we have the following:

$$T(\text{sentences}) = (\text{semantic}) = > M(\text{possible-worlds}) = (?) = > P(\text{actual world}).$$

In other words, M contains those possible worlds in which T is true and which resemble in some aspects the actual world P. Depending on whether one believes in the reality of possible worlds, the relationship between M and P varies: it is one of similarity for the believers, and one of isomorphism for the non-believers. (I take it that similarity is a relation holding between systems of the same kind, while isomorphism is one that may hold between those of different kinds.)

The advantages of the semantic view

They are said to mainly consist of the following aspects (cf. Giere, 1988; Suppe, 1989; Suppes, 1967; van Fraassen, 1989).

First, scientific theories should not be made dependent on their linguistic formulations. Otherwise, it will not only force an unnecessary proliferation of theories, but also obscure their structures. The aim of theorizing is to represent the properties and behavior of systems in question not the ways in which they are linguistically rendered. The syntactic view endorses, but the semantic view denies, such a dependence.

Second, the model-building activity is common in science, in both the natural and the social sciences. This is most conspicuous in theoretical model-buildings where instead of formulating abstract theoretical axioms which remain uninterpreted until observable consequences are derived, scientists build in their mind's eye systems of abstract objects whose properties or behavior satisfy certain constraints (including laws). The syntactic view does not, but the semantic view does, make sense of this practice.

Third, the syntactic view is riddled with difficulties such as the unattainable distinction between the theoretical and the observable (e.g. the distinction between V_T and V_O), while such difficulties do not affect the semantic view, for it does not recognize such a distinction.

Finally, the syntactic view greatly oversimplifies the structure of theory-testing by trying to connect directly the sentences in a theory with experimental data. One may see this point in, for instance, Popper's scheme for hypothesis (or theory) testing. Let "all Fs are Gs" be a theoretical claim. One can then derive a sentence such as "if a is F, it is also G". Suppose that F and G, although highly theoretical, are reducible to F^* and G^* which are sets of observable terms. One can then test whether some a which is F^* is also G^* by directly comparing it with experimental findings. For Suppes, this does not even remotely resemble what actually goes on in actual theory-testings. With the semantic view, however, theories of different levels of idealization (or approximation) may be characterized by the theory-model pairs and their relations by those among the models.

Conceptions of models

The term "model" is widely used in English language and bears several distinct meanings. According to the *Oxford English Dictionary*, as a noun, it has 15 distinct meanings grouped under three categories: as representation of structure, as type of design and as object of imitation, all of which have something in common with our scientific usage of the term, whose philosophical nature has been extensively discussed in the literature on models (cf. Achinstein, 1964, 1965, 1968; Braithwaite, 1962; Freudenthal, 1961; Hesse, 1970; Horgan, 1994; Hughes, 1990; Kroes, 1989; Leatherdale, 1974; McMullin, 1968; Redhead, 1980; Wartofsky, 1979). I will give a summary of various views on models in the literature (not exhaustive but representative) and then a precise delineation of them.

First, I shall call the models in mathematical logic *semantic models* which are simply interpretations that satisfy the theory in question. In this concept, a theory is defined as $ThA = \{\sigma : A \models \sigma\}$, which is not to be identified with the semanticist concept of a scientific theory.

Redhead (1980) (see also Emch, 1995) takes a model, especially in physics, to be an impoverished or an enriched version of a theory which is a set of interpreted sentences (which include, *inter alia*, laws and definitions). A simplified version of a theory is one of its models (an impoverished version) and so is a further articulated (e.g. with more details) version of a very general theory (an enriched version). Theories and models are entities of the same logical type, the difference being that they serve different purposes in the practice of theory-building.

For Achinstein (1964, 1965) and Hesse (1970) (A-H) (see also McMullin, 1968, 1978; Hughes, 1990), a theoretical model is an abstract and non-linguistic entity that bears analogy to some other familiar theoretical systems; it is equipped with an exactly describable *inner* structure or *mechanism* which *explains* the system's observable behavior, and is treated as an *approximation* useful for certain purposes. It is not clear whether each stated component in this view is required for an object to be a model. The most problematic is the analogy requirement, which Redhead (1980) strongly disputed. The idea is that a theoretical model is the product of borrowing a structure from a familiar branch and modifying it to best capture the systems in question so that its inner mechanism gives an adequate explanation of the systems' behavior. What is not clear is whether the borrowing part is necessary and whether the structures (and hence the models) are necessarily abstract. The molecular structures of crystals may be abstract if molecules are considered point masses but not necessarily so, unless there are no molecules (which does not seem at all plausible). Moreover, as emphasized by McMullin and Hughes, most models (especially in physical sciences) are unobservable microscopic structures of matter that explain the observable behavior. This concept is indifferent to how the models are specified: whether by a set of sentences in some language or by a computer generated picture or by some other means. This seems to imply that models can have multiple equivalent descriptions.

One can also formulate what I shall call Earman* concept (1989) (also Beth 1949, 1961, and early van Fraassen 1970, 1972), according to which a model is simply that which satisfies a solution or set of solutions of the (differential) equations of a scientific theory (under specific initial/boundary conditions). Furthermore, a theory should be identified with its model closure (i.e. all of the things that its solutions represent). One should note of course the distinction between the solutions which are either algebraic formulae or geometric figures (essentially linguistic items) and the models that they are used to represent. Given the state-space version (van Fraassen), these models are trajectories (understood in the broadest and most abstract sense of the term) in a state (or phase) space (limited by the parameters and given by the values of the variables). For Earman, models of space-time theories are solutions of Einstein's equations that are n-tuples of geometrical objects on a space-time manifold (in its most general form, $m = \langle M, g, O_1, O_2, \ldots, O_{n-z} \rangle$, where M is the manifold, g the metric, and the O_i's are the physical magnitudes defined on M. Any mutually consistently attributable set of values for g and the O_i's will be a physically possible model for space-time M.) (This is why I take the concept to be a natural extension of Earman's original idea on space-time theories, and why it is given the name "Earman*".)

A model may also refer to a real object, natural or constructed, which is either a scaled replica or an analogue of another object; for example, there is the model airplane (for a replica) and the mechanical model of a human heart (for an analogue).

All of these concepts are undoubtedly instantiated in science and daily life; the question is whether they are indeed distinct concepts. For scientists, the notion of models of whatever kind bears at least some connotation of them being instruments by which an access to the phenomena in question may be gained. In practice, there may be many ways to gain such access, the easiest and most common of which would be to use an actual and well-understood system whose similarity with the system under study is either known or easily projected. Then, one formulates a theory of such a system and take it to be approximately describing the behavior of the target system. This is especially helpful if the difference between the system and its model can be theoretically mapped onto a set of parameters which admits continuous values. Thus, the degree of difference and its possible reduction can be precisely represented and predicted. Many variations of such a strategy exist: one may, for instance, drop the "actualness" requirement and use a purely mathematical structure, if one only aims at capturing some structural properties of a system. "Well-known" does not have to mean "already familiar"; to be a well-known system might mean being a "well-defined" system the discovery of whose structures is only a matter of time and diligence.

Once this is clarified, exactly which item(s) in such praxis of science one calls model(s) does not really matter, as long as one does not confuse them and mistake one for another. Let us take M^A to be the class of systems or structures that approximate the target system or structure A and T^A a theory of M^A, i.e. $T^A = \{\phi \varepsilon L : M^A| = \phi\}$, such that T^A is an approximation of a theory of A. The logical conception obviously carries no sense of approximation, but given T^A, M^A are certainly its model set. Hence, the logical conception is compatible with scientists' strategies.

It seems that Redhead's models are by and large meant to be T^A, while those of A-H could be identified as M^A,[1] except of course the A-H models do not have the part of interpretation that goes with M^A. For instance, it makes no sense to write $M^A| = \phi$ if M^A only contains the bare structures. One must add the intended interpretation (which is usually presupposed in scientists' discussion of models and theories) to the A-H models to make them comparable to those in the logical conception. Henceforth, the intended interpretation will be assumed to be attached to the A-H models unless explicitly stated otherwise.

One may go either way on the Earman* conception, except items in M^A or T^A must be phase-space trajectories or solutions respectively and they are not supposed to be approximations of objects. This seems to be the main difference between the Earman* and the A-H notion. (Solutions can serve as models only if they are formulated gauge-free; but I shall not pursue this point further in this paper.) Now, models that are themselves real objects, although actual, play essentially the same functions as those of A-H. The fact that they are sometimes constructed with real rather than cyberspace materials is of little significance.

Suppes believes that the different conceptions of models used by scientists and mathematicians alike (when properly understood as in the above) share the same meaning as semantic models but differ in use[2] (see also Apostel, 1961; Wartofsky, 1979). Suppes is, and, in a sense, must be, right. Otherwise, either the semantic view fails to capture the use of models in sciences (if its models are different in kind from the ones scientists use) or it cannot use the results in model theory, namely, that the notions of isomorphism, embedding etc., among relational structures will not even apply (if its models are different in kind from the ones logicians use). This is a dilemma that semanticists cannot afford to be straddled on.

Models and the semantic view

To avoid the dilemma, one must make plausible that the A-H and Earman* models are indeed semantic models (i.e. interpretations that satisfy a theory) and admissible in the semantic view. The current formulation of the semantic view as described in II.3 seems to militate against such a move. Depending on which aspects of it one emphasize, the semantic view can be formulated in two different versions: the state-space vs. the system version.

> [V1] The state-space version: the domain of any set of models only contains trajectories (understood in the broadest and most abstract sense of the term) in a state space defined by the values of the variables.

Such models are certainly semantic models because the trajectories are exactly those that satisfy the quantified sentences (e.g. algebraic equations). The Earman* notion also works perfectly here: the trajectories are simply those that satisfy the solutions (under specific conditions) of a set of (differential) equations (representing laws of succession) which are usually the core of a theory. The state-space version naturally works only with quantitative theories that describe a system's evolution in some space, and the Earman* notion of models comes closest to such an idea.

The A-H notion does not seem to apply in this case, for the A-H models are idealized systems, not the state-space trajectories that they produce or that display their behavior. What is in the domain of such models are not points or numbers but molecules or people (even if they are not actual in most cases), and the relations that obtain among them are not mathematical but physical or causal. Since I consider those scaled replicas and analogues as models of the A-H kind (even with the idealization condition satisfied), what is said above should thereby equally apply.

> [V2] The system version: the domain of any set of models contains physically possible systems whose structural properties are given by the models. And such a system provides the extension for the predicate, "being a system of ...".

Such models can also be taken as semantic models if these systems or structures come with an interpretation. Giere certainly believes that they are semantic models when he says that "this terminology even overlaps nicely with the usage of logicians for whom a model of a set of axioms is an object, or a set of objects, that satisfies the axioms" (1988, p. 79). Surely, the object must be an interpretation in the sense as given in the earlier section on the model theory of mathematical logic if it is to satisfy any axioms.

Strictly speaking, the Earman* notion does not apply here because no solutions or trajectories are allowed by this version of the view. However, one may say that the systems, whose dispositions are represented by the differential equations, are the ones that *produce* the corresponding set of trajectories (i.e. models) which are legitimate items in the Earman* notion, so this notion still fits (even though somewhat indirectly). But the indirectness matters and gives rise to complications so as to render serious doubt on this rather hand-waving claim. More on this will follow.

The A-H notion fits, of course, nicely here. The A-H models are precisely those idealized structures or systems whose inner mechanism (represented by the equations) are causally responsible for the systems' behavior.

We now see that both versions of the semantic view are compatible with the logical conception, but each picks out a different type of relational structures: [V1] picks out the Earman* type and [V2] the A-H type. If the semantic view were conceived of as including both types, then it would be nearly correct. A scientific theory is then a set of sentences that define a class of models of the [V1] or the [V2] type. But the received view takes [V1] and [V2] to be alternative formulations of the semantic view (cf. Hausman, 1992, pp. 72–78; Lloyd, 1988, pp. 14–20). For those who choose [V1], all theories are (ultimately) sets of state-space structures or trajectories which satisfy a set of sentences thus "define" them. They seem to think that law statements are part, and indeed the core, of such definitions since the state-space trajectories are what satisfy such laws. For them, the A-H models are unnecessary since everything is eventually reduced to a state-space structure, or they can only be heuristic devices. Since [V1] now dominates the discussion and application of the semantic view, it is no wonder that the A-H type models are rarely mentioned. For those who choose [V2], the situation is less clear. Giere seems to think that once you get the right set of systems, their behavior as represented by the state-space trajectories is automatically given; so one should not worry; and Suppes does not even mention the state-space.

To illustrate this point, let me use a simple example in thermodynamics: the ideal gas. According to thermodynamics, the theory of ideal gas should be the result of applying the general theory of thermodynamics of equilibrium and homogenous fluids to a special kind of system (and so, as an aside, it should *ipso facto* be a model according to Redhead). Besides the statements that define the system, there are the laws that govern its dynamic process: (1) $\delta Q = dU + \delta W$ and (2) $\delta Q = TdS$; and the conditions, (3) $\{pV = RT;\ C_v = c\}$, under which (1) and (2) are satisfied. Solving the equations under the conditions, one obtains the general solutions:

$$U(T) = C_V(T - T_0) + U(T_0)$$
$$S(V,T) = C_v ln(T/T_0) + Rln(V/V_0) + S(V_0,T_0)$$

Obviously, both equations define trajectories in the V-T state-space. Strictly speaking, the models that satisfy (1), (2) and (3) (and the definitions) belong to the causal/structural category, and thus fit [V2], while those satisfying the above two equations belong to the state-space category, and thus fit [V1]. But those who choose [V1] take the state-space trajectories to be models of equations (1) and (2) (i.e. the general laws) under condition (3), and for those who choose [V2], the models are simply ideal gases (abstract objects) which satisfy equation (1) and (2), and condition (3). Although perhaps harmless in this special case, none of the two choices can hold in general. Let me now explain why it is so.

First, there is a dangerous equivocation in the above use of the term "satisfaction". A relational structure *satisfies* a sentence in L, and an equation *satisfies* another as the latter's solution. Both are legitimate usage, but with very different meanings. The former which I shall call Sat_1 is a relation between a sentence and a structure that make it true, while the latter, calling it Sat_2, is a relation between two sentences such that the one is derivable from the other. That a trajectory satisfies a theoretical law would have to mean that it Sat_1s a sentence which, in turn, Sat_2s the law statement.

Second, the law statements when interpreted "explain" the presence of the trajectories, in the sense that the structures which satisfy the former are causally responsible for the latter. Hence, the equations (for laws) and the solutions (for trajectories) cannot be descriptions of the same relational structures (or models). The two types of models are certainly related, but one cannot be the models for the theory of the other.

Third, some theoretical law statements or equations do not have solutions except under special circumstances. Hence, the structures that satisfy the equations cannot be identified with the trajectories that satisfy the available solutions of the equations. But perhaps they can be identified with the possible solutions in the following sense. Whenever a law statement is given, all of its solutions are determined, whether we know them or not. Why cannot we identify those with the law statement? In principle we can, but practically such identification is meaningless. Suppose that we have a piece of crystal whose lawlike structure is given by a set of differential equations so complex that no solution is known. If we identify the set of trajectories that satisfy the solutions of those equations (about which we of course know nothing except their existence) as the equations' models, then we will not be able to know anything about the crystal, for instance, we cannot tell if it has certain symmetry under certain transformation groups, or whether some of its properties which are represented by the set of equations are reducible to some other properties which may be represented by another set of equations, and etc. That includes all the benefits that the semantic view is supposed to offer. In other words, if the set of solutions that 1-1 corresponds with the law statements is known, in general if not in detail, it is harmless (even though a mistake) to take their models to be the models of those laws. But if not, one must go back to the law statements' own models, namely, those structures that Sat_1 them, not the ones that Sat_1 their solutions.

Finally, microscopic models defined by postulates of theoretical entities and their behavior are widely used in science to explain macroscopic (or observable) phenomena represented in the form of sets of record experimental data (or the best-fit graphs resulting from those data). In this case, it is even more egregious to identify the structures that satisfy the best-fit graphs to be the same models that satisfy the laws governing the microscopic entities in whichever sense of "satisfaction". Moreover, although the causal laws of the microscopic entities are essential in our attempt to explain the macroscopic phenomena, the state-space structures of individual entities are rarely considered or even considerable. The most obvious example would be the causal explanation of thermodynamic behavior of ideal gas by evoking its molecular structures. Although the equation of motion for each molecule is important in the explanation, no one should dream of solving it for each molecule in the gas. In this case, the causal/structural models of the molecules in ideal gas are important but not the state-space models. The latter kind only comes into play at the macroscopic level.

I am of course not claiming that ontologically a system (e.g. an ideal gas) is not extensionally identical to its entire space-time instantiation. But ontology is not the issue here. At issue is what should be identified as *the theory* of a phenomenon. According to the semantic view, it is given by the relational structures that *satisfy* the law statements and some definitional sentences. If the above analysis is correct, those cannot always be identified with the structures that satisfy the statements' solutions in state-space.

A hybrid view

We must tidy up the terminology used in the above criticism before presenting a hybrid view whose content should be rather obvious by now. "Law statements", "differential equations", etc. have been used to denote those syntactic elements that the A-H models or structures satisfy, which is also in line with [V2]. Correspondingly, I used "trajectories", "solutions", "algebraic equations", etc. to refer to those syntactic elements that the Earman* models or structures satisfy, which is also in line with [V1]. Philosophically, these are clumsy if not misleading terms, but they have widespread currency among scientists and are therefore informative terms.

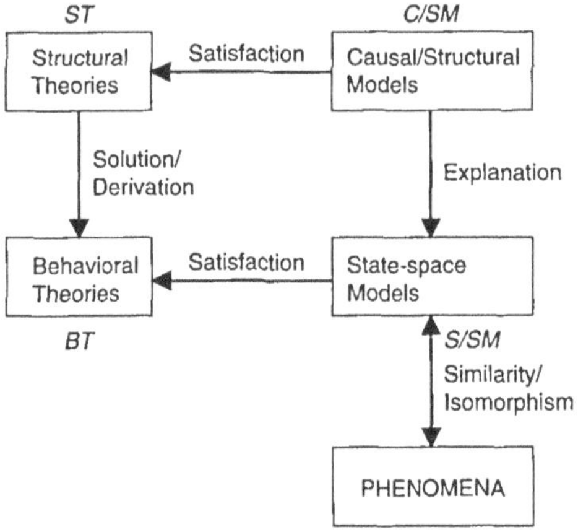

Figure 3. *A enhanced semantic view.*

To provide a clean and simple conceptual scheme for accomplishing the task at hand, some terms must be invented. The following seems true and simple enough. Most systems, actual or possible, have two kinds of properties: dispositional and categorical. Theories are conceptual devices that represent the structure and behavior of those systems. Each theory (in the broader sense) consists of at least two components: *T* (theories in the narrow sense) and *M* (models), where *T* contains linguistic items and *M* abstract structures. A theory that describe the dispositions of a system I call a *Structural Theory* (*ST*), and the relational structures that satisfy such a theory are *Causal/Structural Models* (*C/SM*). Similarly, we have a *Behaviorial Theory* (*BT*) and *State-space Models* (*S/SM*), for the categorical properties.

Except those low-level empirical generalizations, most lawlike statements describe dispositional properties of a system. This is especially true with differential equations which represent lawlike dispositional properties because they show with what rates or in which direction the system's properties change and what causes them to do so. Hence *ST* (or its members) can be used to refer to all those earlier called "law statements" or "differential equations" and *C/SM* (or its members) for those called the A-H models. When a system is given by its trajectory in the appropriate state-space, we typically say that it is known by its occurrent or categorical properties. So, *S/SM* (or its members) naturally refer to such trajectories, or in other words, the Earman* models, and the sentences or equations which are solutions of items in ST should be members of *BT*.

With these preliminary definitions of terms and classes, we have the flow-chart that represents a hybrid view (Figure 3). As in Figure 1, the upward direction marks the ascent of theoreticity.

Some observations of this view are in order.

While the *explanatory* relation between *C/SM* and *S/SM* can be seen as exceptionless on the level of their members, the *derivational* relation should only be seen as between the two classes.

Models in *C/SM* may also be called theoretical or explanatory models:[3] they are idealized systems whose structures or inner mechanisms, which are mostly lawlike and represent the essential properties of the real systems of which they are idealizations, are given by law—or lawlike—sentences (or differential equations). Suppose that $m \subseteq C/SM$ is the set of models for a $t \in ST$. m may contain structures of radically different kinds that are nonetheless isomorphic to one another with respect to t. Some of them may be well-known, others may not. This is where analogy among models comes into play in theory construction. Moreover, a realist may take some models to be real systems, while an empiricist heuristic devices.

Models in *S/SM* may also be called experimental, or idealized data, models. These are the state-space trajectories that satisfy either the solutions of those equations in *ST* or an extension of an actual data collection (e.g. the best-fit graph through the raw data points). Analogical reasoning in the business of theory discovery does not apply to these models.

As far as comparing a model with a real system is concerned, one can only use models in *S/SM*. That which may be isomorphic, according to van Fraassen, to an appearance must be expressible in a state-space. Empirically, that is the only way in which we have access to real systems; for it does not make much sense to compare an idealized system in *C/SM* with an actual system, except in the indirect sense of how much idealization has been introduced or how many symmetries it has. What corresponds to a model in *C/SM* are empirically shielded from us so that we can know them, if they exist, only through indirect means. Hence, Giere's theoretical hypothesis which says that a real pendulum, for instance, is similar to the one in our theory can only mean that a graph plotted by observing the movement of a real pendulum is similar to one of the models in its *S/SM*.

ST + *C/SM* and *BT* + *S/SM* may form a pair to serve either locally or globally on a theoretico-experimental landscape.[4] At the high end, say, we have mathematical physics whose theories are a long way from possible measurements. Still, one may find an axiomatized theory, *T* (in *ST*), which is satisfied by some highly idealized systems (in *C/SM*) (e.g. the Ising model of ferromagnetism), and also a theory, T^* (in *BT*), of some Gedanken experiments whose models (in *S/SM*) supply idealized data. Similarly, at the low end, we may have a theory of the experimental set-up (in *ST*) whose models are similar to the actual experimental apparatuses (in *C/SM*), and also we may have an extension of the collected data (in *BM*); the former is what Suppes calls "models of experiment" and the latter "models of data" (cf. Suppes, 1962, p. 259). The problem with the syntactic view is partly that its scheme (cf. Figure 1) can only be applied across the whole vertical landscape.

In combating the syntactic view, semanticists went overboard in denying the significant role language (especially a mathematical language) plays in scientific theories. After decades of "language" bashing, Suppes's original assessment of the situation still seems the best (see the quote from Suppes on p. 152). Therefore, I not only distinguish *T* and *M* (either dispositional or categorical) but put them on a par; and items in *T* could be called theories as long as one knows that *M* must be given to complete the story which is traditionally thought of a task of *T* alone. The real problem with the syntactic view is that models of highly theoretical sentences are never considered; theoretical reasoning is imagined to be carried out exclusively through syntactic rules until one obtains solutions that can be experimentally tested. So construed, the view is naturally plagued by the problem of underdetermination. What mitigates the problem is not that we should abandon language, but that we should take theoretical models (e.g. the A-H models) seriously, so that the relation between the highly theoretical and the experimental can be studied through that of their models.

The syntactic view makes realism a very implausible, although not impossible, position. If the calculus is not even interpreted until one reaches the level of experimentally obtainable results, it does not seem tenable to ask whether or to what the theoretical terms refer. The semantic view turns the table towards realism although not completely. Van Fraassen's constructive empiricism extols the utility, but denies the reality, of theoretical models, Suppe's quasi-realism takes them to be real but non-actual worlds, which may be viewed as realism along Cartwright's (1983) line, in which claims about the models are false but entities in them may nonetheless be real. Giere's (1988) is probably the strongest realist position stemming from the semantic view.

Other aspects of the semantic view

This concludes Part I of the paper, where the argument for the hybrid view is given in abstract. From the conceptions of the semantic view, of models in science, and of models in mathematical logic, I have constructed a detailed but still general view of what a scientific theory is. Obviously, many aspects of the theory needs to be examined in detail, which shall be the subjects of Part II. These include the following.

(1) What exactly is the relation between theory and phenomena in the semantic view? The concepts of "Similarity" and "Isomorphism" need to be fleshed out not in terms of specific examples but in terms of a general semantist theory of confirmation. The question comes down to how much of the current wisdom in confirmation theory, which has grown out of a syntactic tradition, is still compatible with the hybrid view, and how much revision is needed.

(2) Besides a loose usage of the term "model" in the praxis of science, there is an equally loose use of the word "interpretation", such as the interpretation of quantum mechanics. Does the semantic view in general and the hybrid view in particular shed any light on this notion? Is the Copenhagen interpretation one "model" of quantum theory and the Bohm interpretation another?[5]

(3) Does the hybrid view contribute to the current debate on scientific explanation? I believe it does and the currently promising notion of structural explanation[3] lends strong support to the hybrid view, and I hope to show that it is also true vice versa.

Acknowledgements

I would like to thank Giovanni Boniolo, Jim Brown, Wladyslaw Krajewski, Srdan Lelas, and James McAllister for constructive comments on an early version of this paper. I also want to thank the Dubrovnik IUC for inviting me to present that version of the paper.

Notes

1. At this stage I have not made the distinction between a system and its behaviors; therefore, it is intentionally left open as to whether the domain of the models in M^A consists of systems or the phase-space trajectories they causally produce.
2. "I claim that the concept of model in the sense of Tarski may he used without distortion and as a fundamental concept in all the disciplines ... In this sense I would assert that the meaning of the concept of model is the same in mathematics and the empirical sciences. The difference to be found in these disciplines is to be found in their use of the concept." (Suppes 1961, p. 165).
3. The scheme of this hybrid view—the division of the *ST + C/SM* and the *BT + S/SM* and the explication of their relations—is inspired in part by McMullin's theory of *structural explanation* (cf. Hughes, 1990; McMullin, 1978). Which I see as an improvement upon the dominant theory of scientific explanation which is largely formulated in terms of the syntactic relations in scientific theories.

4. Here I imagine a two-dimensional landscape with its horizontal axis denoting its breath, namely, the variety of disciplines, and the vertical axis its depth, namely, from highly theoretical and idealized to raw data.
5. I thank Jim Brown for raising this question to me during the Dubrovnik IUC conference where an earlier draft of this paper was read.

References

ACHINSTEIN, P. (1964) Models, analogies and theories, *Philosophy of Science*, 31, pp. 328–350.

ACHINSTEIN, P. (1965) Theoretical models, *British Journal for the Philosophy of Science*, 16, pp. 102–119.

ACHINSTEIN, P. (1968) *Concepts of Science: A Philosophical Analysis* (Baltimore, MD, The Johns Hopkins University Press).

APOSTEL, L. (1961) Towards the Formal Study of Models in the non-Formal Sciences, in: H. FREUDENTHAL (Ed.) *The Concept and the Role of the Model in Mathematics and Natural and Social Sciences* (Synthese Library) (Dordrecht, D. Reidel). pp. 1–37.

BALZER, W., MOULINES, C.U. & SNEED, J.D. (1987) *An Architectonic for Science: the Structuralist Program* (Dordrecht, D. Reidel).

BETH, E.W. (1949) Towards an up-to-date philosophy of the natural sciences, *Methodos*, 1, pp. 78–185.

BETH, E.W. (1961) Semantics of physical theories, in: H. FREUDENTHAL (Ed.) *The Concept and the Role of the Model in Mathematics and Natural and Social Sciences* (Dordrecht, D. Reidel), pp. 48–51.

BRAITHWAITE, R.B. (1962) Models in the empirical sciences, in: E. NAGEL, P. SUPPES & A. TARSKI (Eds) *Logic, Methodology and Philosophy of Science: Proceedings of the 1960 International Congress* (Stanford, Stanford University Press), pp. 224–231.

CARTWRIGHT, N. (1983) *How the Laws of Physics Lie* (Oxford, Clarendon Press).

EMCH, G. (1995) Models vs. models, *International Journal of Modern Physics B*, 9 (in press).

EARMAN, J. (1989) *World Enough and Space-Time* (Cambridge, MA, A Bradford Book, The MIT Press).

FREUDENTHAL, H. (Ed.) (1961) *The Concept and the Role of the Model in Mathematics and Natural and Social Sciences*, Synthese Library (Dordrecht, Reidel).

GIERE, R. (1988) *Explaining Science: A Cognitive Approach* (Chicago, University of Chicago Press).

GIERE, R. (1994) The cognitive structure of scientific theories, *Philosophy of Science*, 61, pp. 276–296.

HAUSMAN, D.M. (1992) *The Inexact and Separate Science of Economics* (Cambridge, Cambridge University Press).

HESSE, M. (1970) *Models and Analogies in Science* (Notre Dame, University of Notre Dame Press).

HILBERT, D. (1902) *Foundations of Geometry* (Chicago, University of Chicago Press).

HORGAN, J. (1994) Icon and bild: a note on the analogical structure of models—the role of models in experiment and theory, *The British Journal for the Philosophy of Science*, 45, pp. 599–604.

HUGHES, R.I.G. (1990) The Bohr atom, models, and realism, *Philosophical Topics*, 18, pp. 71–84.

KROES, P. (1989) Structural analogies between physical systems, *British Journal for Philosophy of Science*, 40, pp. 145–154.

LEATHERDALE, W.H. (1974) *The Role of Analogy, Model and Metaphor in Science* (Amsterdam, North-Holland).

LLOYD, E.A. (1988) *The Structure and Confirmation of Evolutionary Theory* (Westport, CT, Greenwood Press).

MALITZ, J. (1979) *Introduction to Mathematical Logic* (New York, Springer-Verlag).

MCMULLIN, E. (1968) What do physical models tell us? in: B. VAN ROOTSELAAR & J.F. STAAL (Eds) *Logic, Methodology and Philosophy of Science III* (Amsterdam, North Holland), p. 392.

MCMULLIN, E. (1978) Structural explanation, *American Philosophical Quarterly*, 15, pp. 139–147.

MONK, J.D. (1976) *Mathematical Logic* (New York, Springer-Verlag).

REDHEAD, M. (1980) Models in physics, *British Journal for the Philosophy of Science*, 31, 145–163.

SLOEP, P. & W. VAN DER STEEN (1987) The nature of evolutionary theory: the semantic challenge, *Biology and Philosophy*, 2, 1–15. (Also included in the same issue are responses from J. Beatty, E. Lloyd, and P. Thompson, and the rejoinder from Sloep & van der Steen, pp. 17–41.)

SNEED, J.D. (1971) *The Logical Structure of Mathematical Physics* (Dordrecht, Reidel).

SUPPE, F. (Ed.) (1977) *The Structure of Scientific Theories* (Urbana, IL, University of Illinois Press).

SUPPE, F. (1989) *The Semantic Conception of Theories and Scientific Realism* (Urbana, IL, University of Illinois Press).

SUPPES, P. (1961) A comparison of the meaning and uses of models in mathematics and the empirical sciences, in: H. FREUDENTHAL (Ed.) *The Concept and the Role of the Model in Mathematics and Natural and Social Sciences* (Dordrecht, Reidel), pp. 163–177.

SUPPES, P. (1962) Models of data, in: E. NAGEL, P. SUPPES & A. TARSKI (Eds) *Logic, Methodology and Philosophy of Science: Proceedings of the 1960 International Congress* (Stanford, Stanford University Press), pp. 252–261.

SUPPES, P. (1967) What is a scientific theory? in: S. MORGENBESSER (Ed.) *Philosophy of Science Today* (New York, Basic Books), pp. 55–67.

SUPPES, P. (1988) Representation theory and the analysis of structure, *Philosophia Naturalis*, 25, pp. 254–268.

SUPPES, P. (1993) *Models and Methods in Philosophy of Science: Selected Essays* (Dordrecht, Kluwer Academic Pub.).

VAN FRAASSEN, B. (1970) On the extension of Beth's semantics of physical theories, *Philosophy of Science*, 37, pp. 325–334.

VAN FRAASSEN, B. (1972) A formal approach to the philosophy of science, in: R. COLODNY (Ed.) *Paradigms and Paradoxes* (Pittsburgh, University of Pittsburgh Press), pp. 303–366.

VAN FRAASSEN, B. (1983) *The Scientific Image* (Oxford, Clarendon Press).

VAN FRAASSEN, B. (1989) *Laws and Symmetry* (Oxford, Oxford University Press).

WARTOFSKY, M.W. (1979) *Models: Representation and the Scientific Understanding* (Dordrecht, D. Reidel).

Part 4

Meaning of Spontaneous Symmetry Breaking

This part contains three previously published papers and a review; all of which deal with the philosophical issues regarding symmetries and symmetry breaking in classical and quantum systems and processes.

Spontaneous symmetry breaking (SSB) is one of the important notions in contemporary high-energy physics as well as in condensed matter physics. The existence of many elementary particles has been theoretically attributed to the results of SSB. One only has to recall the most famous such particles, the Higgs boson, whose recent confirmation made headlines and spawned multiple popular science books, to realize the importance of SSB.

Papers 15 and 16 trace the origin of the idea of SSB in classical mechanics and explore the connection of SSB with the classical notion of chance. One of the questions about SSB is: in what sense the breaking of the symmetry is "spontaneous"? The papers answer the question in all its relevant aspects. Paper 13 is a discussion of quantum SSB, in which an alternative to the standard formulation in high-energy physics is argued for.

For more recent discussions in this area see David John Baker and Hans Halvorson (2013) "How is spontaneous symmetry breaking possible? Understanding Wigner's theorem in light of unitary inequivalence," *History and Philosophy of Modern Physics* 44: 464-469; N. P. Landsman (2013) "Spontaneous Symmetry Breaking in Quantum Systems: Emergence or Reduction?" *Studies in History and Philosophy of Modern Physics* 44 (4): 379-394; Doreen Fraser (2012) "Spontaneous Symmetry Breaking: Quantum Statistical Mechanics Versus Quantum Field Theory." *Philosophy of Science*

79 (5): 905-916. And for issues of SSB relating to SSB of gauge symmetry in gauge field theories, see Simon Friederich (2013) "Gauge Symmetry Breaking in Gauge Theories — in Search of Clarification." *European Journal for Philosophy of Science* 3 (2): 157-182, and Holger Lyre (2008). "Does the Higgs Mechanism Exist?" *International Studies in the Philosophy of Science* 22 (2): 119-133.

ELSEVIER

Studies in History and Philosophy of Modern Physics 36 (2005) 137–163

Studies in History and Philosophy of Modern Physics

www.elsevier.com/locate/shpsb

Explaining quantum spontaneous symmetry breaking☆

Chuang Liu[a,*], Gérard G. Emch[b,c]

[a]*Department of Philosophy, University of Florida, Gainesville, FL 32611, USA*
[b]*All Souls College, University of Oxford, UK*
[c]*Department of Mathematics, University of Florida, Gainesville, FL 32611, USA*

Received 27 April 2004; accepted 28 September 2004

Abstract

Two accounts of quantum symmetry breaking (SSB) in the algebraic approach are compared: the representational and the decompositional account. The latter account is argued to be superior for understanding quantum SSB. Two exactly solvable models are given as applications of our account: the Weiss–Heisenberg model for ferromagnetism and the BCS model for superconductivity. Finally, the decompositional account is shown to be more conducive to the causal explanation of quantum SSB.

Keywords: Spontaneous symmetry breaking; Idealization; Thermodynamic limit; Foundations of physics

1. Introduction

The best known examples of spontaneous symmetry breaking (SSB) are found, many may think, in relativistic quantum field theory (QFT). For an up-to-date

☆GGE acknowledges the hospitality of All Souls College, Oxford, during the Hilary term of 2004, and in particular the stimulating conversations with Jeremy Butterfield. We thank Jeremy for reading a draft of this paper and giving us detailed comments. We would also like to thank Laura Ruetsche for her comments on parts of the paper.
*Corresponding author.
E-mail address: cliu@phil.ufl.edu (C. Liu).

1355-2198/$ - see front matter © 2005 Elsevier Ltd. All rights reserved.
doi:10.1016/j.shpsb.2004.12.003

account of the mathematical state of the art see, for instance, Ojima (2003) and the references therein. Then why, one may ask, do we choose to discuss SSB in non-relativistic quantum statistical mechanics (QSM)? While the problems of SSB in QFT may challenge physicists or mathematicians and fascinate philosophers of physics, they are not, as we shall argue, the best problems (i.e. offering the best models) for understanding the nature of quantum SSB. The tentative and controversial nature of some of the famous results of SSB in QFT is often an impediment to such an understanding, and the paucity of exactly solvable and experimentally realizable models does not help either. In contrast, SSBs in quantum thermo-systems with infinitely many degrees of freedom are well understood and have simple and realizable models. It is not at all accidental that when it comes to the discussion of SSB per se, authors in the QFT literature often resort to the analogous examples in QSM, examples such as ferromagnetism.

If so, one may wonder, why can we not search for the meaning of SSB in classical models, which would be an even simpler task? As shown in a detailed study (cf. Liu, 2003), classical SSB lacks several features that characterize *quantum* SSB. Therefore, part of the aim of this paper is to show how the essential features of SSB in QFT qua SSB manifest themselves clearly in systems of QSM, assuming that it is always preferable to study the simpler model, provided that all the essential features are captured, and none is left out.

Hence, in this paper we address the interpretative problems of explaining SSB in infinite quantum thermo-systems—the proper subjects of QSM. The main interpretative questions that are relevant to understanding how quantum SSB is understood include:

1. Why do quantum SSBs occur only in infinite systems? What justifies the use of such systems?
2. Why do quantum SSBs occur if, and only if, the symmetries in question are not unitarily implementable? What does this mean physically?
3. Why must the degenerate fundamental states[1] of an SSB system belong to unitarily inequivalent representations? What does this mean physically?
4. What is the physics of quantum SSB, in contrast to that of classical SSB?

To those readers who may worry that little philosophical interest can be generated from the above questions, we respond briefly as follows. (i) The nature of quantum SSB (markedly different from classical SSB) is itself of interest to metaphysics. An analogous case in this respect would be quantum measurement. (ii) Quantum SSB is closely related to the problem of quantum measurement. In a proposed solution to the problem (cf. Whitten-Wolfe & Emch, 1976), a measured quantum system triggers

[1]The phrase, 'fundamental states,' denote the 'lowest stable energy states' in classical mechanics, the 'equilibrium states' in statistical physics, and the 'vacuum states' of quantum field theory. In the literature, 'ground states' are often used for such a purpose, but the term can be misleading. In this paper we use this phrase to mean the lowest stable energy states in all contexts except in QSM, where a system's total energy is infinite, the role of such states are played by the KMS states.

a macroscopic measuring device originally in a symmetric state to fall into a symmetry-breaking state, thus registering the to-be-measured value. An understanding of quantum SSB may thus be prerequisite for solving the measurement problem. (iii) We address the question of what it means to explain quantum SSB as a natural phenomenon, which concerns the nature of explanation. (iv) We also explain what kind of idealization the macroscopic limit is and how we are justified to use it.

2. Two accounts of quantum SSB (the no-go results for finite systems)

In a recent paper broaching the concept of SSB for philosophers, Earman (2003a, b) brings to our attention two main features of quantum SSB, which we combine into the following.

[RA] A symmetry, α, of a system is *spontaneously broken* if and only if
(i) it is not unitarily implementable;[2]
or
(ii) some of the fundamental states (see footnote 1) of the system, related by α, generate (via the GNS construction) unitarily inequivalent representations of the algebra.

[RA] constitutes the first of two accounts of quantum SSB we examine in this paper, whose content will be explained step by step as we proceed and which we shall call the *representational account* (hence, [RA]). First, let us see why there is no possible SSB in *finite*, i.e. having finitely many degrees of freedom, quantum systems (a 'no-go' theorem) according to [RA] (cf. Emch, 1977). The theoretical framework for this account is the algebraic approach to quantum physics, which is suitable for both QFT and QSM.[3] For the algebraic account of a quantum system, we begin with an infinite n-dimensional Euclidean world, X. For the sake of simplicity, we shall study the algebras on a lattice world, $X = \mathbb{Z}^n$, unless explicitly stated otherwise. To each finite region, $\Lambda \subset X$, we associate a C*-algebra A_Λ, the self-adjoint elements of which are interpreted as the observables relative to Λ. On these algebras, a corresponding state ϕ_Λ is given that associates to any observable, $A_\Lambda = A_\Lambda^* \in A_\Lambda$, a real number, $\langle \phi_\Lambda; A_\Lambda \rangle$, which is its expectation value. These algebras are most instructively studied in their representations as algebras of operators acting on Hilbert spaces. To be effective, representations *must* be tailored to fit the specific physical situations at hand which are given by the states (more on this point when we discuss the GNS construction of representations in Section 4).

[2]In the algebraic language (on which more will be said later), we have, given A, an automorphism $\alpha \in \mathrm{Aut}(\mathrm{A})$ is *unitarily implementable* in a representation π (A) of A on H if there exists a unitary operator, $U: \mathrm{H} \to \mathrm{H}$ such that $\pi(\alpha[A]) = U\pi(A)U^{-1}\ \forall A \in \mathrm{A}$.

[3]For systematic treatments of the algebraic approach, see Emch (1972), Haag (1996), or Sewell (2002). For briefer and more philosophical discussions, see Clifton and Halvorson (2001), Arageorgis, Earman, and Ruetsche (2002) Arageorgis et al., 2002, Ruetsche (2002) and Emch and Liu (2002).

The ('no-go') argument may now be given as follows (some technical terms are explained later). First, for any finite region Λ, all irreducible representations of A_Λ are unitarily equivalent, i.e. for any two irreducible representations of A_Λ, π^k_Λ: $A_\Lambda \to B(H^k_\Lambda)$, $k = 1, 2$, (where $B(H_\Lambda)$ is the set of all bounded operators on the Hilbert space associated with Λ), there is a unitary transformation, $V : H^1_\Lambda \to H^2_\Lambda$, such that $\pi^2_\Lambda(A) = V\pi^1_\Lambda(A)V^{-1}$ for all $A \in A_\Lambda$. Then, for any two states, ϕ^k $(k = 1, 2)$, on A_Λ and their corresponding GNS representations, $(H_{\phi^1}, \pi_{\phi^1})$ and $(H_{\phi^2}, \pi_{\phi^2})$, the two states are said to be spatially equivalent whenever the two representations are unitarily equivalent. An automorphism, α, of A_Λ is said to be unitarily implementable with respect to a state ϕ if, and only if, $\phi \circ \alpha$ is spatially equivalent to ϕ, and similarly it is unitarily implementable with respect to a representation π whenever the representation $\pi \circ \alpha$ is unitarily equivalent to π. Since $A_\Lambda = B(H_\Lambda)$ for systems that can be modeled on finite regions of a lattice (with H_Λ finite or infinite dimensional), all automorphisms of A_Λ are inner and thus unitarily implementable in every representation. Now, given [RA], there cannot be SSB in finite quantum systems.

This argument remains opaque until one understands why and how the lack of unitary implementability of the relevant automorphism is a case of SSB. As we will see in Section 4, this first account of quantum SSB is not nearly as helpful in facilitating such an understanding as the second account we now announce (cf. Emch, 1977).

[DA] A symmetry $\alpha \in Aut(A)$ of a system S is *spontaneously broken* in a α-invariant state ϕ on A that is KMS with respect to a one-parameter group $\{\tau_t : t \in \mathbb{R}\}$ of automorphisms if and only if

(i) is *not extremal* KMS;
(ii) some of its extremal KMS components—in a *unique* convex decomposition of ϕ into extremal KMS states—are *not α-invariant*.

The details of [DA] will be discussed in the next two sections; for now one only needs to remember that a KMS state is one of the fundamental states mentioned in [RA] (roughly, it is the canonical equilibrium state in QSM); moreover, in order to *detect* the SSB in a system, one must find a *witness* among the macroscopic observables associated with ϕ (see [W] in Section 4). We shall call this account the *decompositional account* (hence, [DA]).[4]

In this account, there is also a 'no-go' theorem for finite systems, i.e. systems in Λ; it runs as follows. For finite systems, the Hilbert space formalism of quantum mechanics suffices. That means that the algebra of observables can be written as $A_\Lambda = B(H_\Lambda)$. Let the Hamiltonian H be a self-adjoint operator acting on H_Λ with,

[4]Some weaker conditions than the KMS may also give us the decompositional result. Among such weaker conditions are G-abelianness and η-abelianness (cf. Emch, 1972: Theorem II.2.8, and Corollary II.2, and 3 of Theorem. II.2.12). These conditions are always realized by space-translations but rarely satisfied by the group governing the time-evoluation. We decided here not to build our analysis on such exceptional conditions; moreover, defining a thermodynamic equilibrium state as just a time-invariant state would be grossly inadequate in general. Therefore, we opt for the KMS condition as the physically meaningful condition for, inter alia, the SSB in QSM.

C. Liu, G.G. Emch / Studies in History and Philosophy of Modern Physics 36 (2005) 137–163 141

for the sake of convenience, a discrete spectrum $Sp(H) = \{\varepsilon_i\}$ such that the partition function, $Z \equiv \sum_i \exp(-\beta\varepsilon_i)$, converges for all $0 < \beta = 1/kT < \infty$. The most general states in A_Λ are then positive, normalized, linear functionals on $B(H_\Lambda)$, which are of the form, $\phi = \phi_1 + \phi_2$, where ϕ_1 vanishes on the compact operators, and ϕ_2 is countably additive. Therefore, we only need to consider density matrices, dear to von Neumann and apt for the Hilbert space formalism. Let ρ be a density matrix, i.e. a positive operator of trace-class on H_Λ with $Tr(\rho) = 1$. It is proven in Appendix A that a state $\phi : A \in A_\Lambda \mapsto Tr(\rho A) \in \mathbb{C}$ is KMS if, and only if, its entropy, $S(\rho) = Tr(\rho \ln \rho)$ is maximal, given the constraint $Tr(\rho H) = E$ (the total energy). And then $\rho = \exp(-\beta H)/Tr[\exp(-\beta H)]$ is the Gibbsian distribution. From this expression for ρ we see immediately that for any symmetry, $\alpha[A] = VAV^{-1}$, such that $[V, H] = 0$, we have $[V, \rho] = 0$ as well. In other words, there is no possible (equilibrium) states of the system that can fail to commute with, and hence break, any symmetry of the Hamiltonian.

3. Infinite systems and KMS states

In theory, SSB becomes possible only in infinite systems. This fact holds for classical thermo-systems as well as quantum thermo- or field-systems. Both conceptually and technically, taking the macroscopic limit is by no means a trivial matter.[5] If one finds it conceptually difficult to stomach the notion of using infinite (model) systems to study the behavior of real systems that one would think are obviously finite, let us begin with the following observation. Taking a finite system to the limit of infinite size is certainly an act of idealization, but regarding a system as finite, which assumes either perfectly isolating walls or an infinite distance between systems, is also a drastic idealization. Unless specific surface or boundary effects are of interest, taking the infinite limit is (from a philosophical point of view) more sensible than not taking it. However, it is one thing to regard taking the macroscopic limit a desirable or acceptable idealization, but quite another to find out that it is necessary for the emergence of SSB. We assume here that the quantum thermo-systems to be considered are such that the macroscopic limit has been implemented, and we have a triple, $\{A, \phi, \tau(t)\}$, where A is a C*-algebra of observables, ϕ a state on A, and τ a one-parameter group of automorphisms of A representing the time evolution (for more on taking the macroscopic limit, see Liu, 2001; Batterman, 2002, 2004; Belot, 2004).

We can now proceed to define equilibrium states and thermodynamic phases of a quantum thermo-system. In the following, $\langle \phi; A \rangle$ denotes as usual the expectation value of the observable A in the state ϕ, and $\tau(t)[B]$ the time-evolute at time t of the

[5]The corresponding technical term in the physics literature is 'thermodynamic limit.' We adopt the more general term to denote the method by which the size of Λ is increased without bounds while the density is kept finite and *the ratio of the volume to the surface of Λ is normal.* The result of this procedure in QSM is the emergence of thermodynamic properties, but in this paper we want to emphasize the emergence of 'macroscopic' properties; hence our term, 'macroscopic limit.'

observable B under the time-evolution τ. Accordingly, $\langle\phi; A\tau(t)[B]\rangle$ and $\langle\phi; \tau(t)[B]A\rangle$ denote time correlations. The single most important concept for the following discussion is that of a KMS state, whose formal definition is given below.

[KMS] Given a C*-algebra A and a group of automorphisms $\tau : t \in \mathbb{R} \mapsto \tau(t) \in Aut(A)$, a state ϕ on A is said to be KMS with respect to $\tau(\mathbb{R})$ if:

(i) $\langle\phi; A\tau(t)[B]\rangle$ and $\langle\phi; \tau(t)[B]A\rangle$ are continuous in t $\forall A, B \in$ A;
(ii) $\forall A, B \in$ A, there exists a function, F_{AB}, defined on the closure of the strip $S_\beta = \{z \in C | Im(z) \in (0, \beta)\}$ and analytic inside it, such that $F_{AB}(t) = \langle\phi; A\tau(t)[B]\rangle$ and $F_{AB}(t + i\beta) = \langle\phi; \tau(t)[B]A\rangle \forall t \in \mathbb{R}$.

(Here, and through out this paper, the units are chosen such that $\hbar = 1$.)

Note that if ϕ is a KMS state on A with respect to $\tau(\mathbb{R})$, then $\phi \circ \tau(t) = \phi$ for all $t \in \mathbb{R}$, which is a sense of minimal stability of the system under time evolution. Specifically, a large body of evidence is known to support taking KMS as the definition of canonical equilibrium states in QSM, such as that a KMS state satisfies both local and global thermodynamic stability conditions against perturbations of state (cf. Sewell, 2002, pp. 118–126).[6] Further, since every KMS state is *uniquely* decomposable into *extremal* KMS states (i.e. states that are not decomposable to other KMS states), thermodynamic phases are then naturally defined as extremal KMS states.

4. Two accounts of quantum SSB (continued)

We now argue that the decompositional account, [DA], is better than the representational account, [RA], as far as understanding the nature of quantum SSB is concerned, and we show in some detail how the former explains quantum SSB.

There is now a rigorous formulation for a decompositional account for the SSB in QSM. There might also be such an account for the SSB in QFT, but we are not yet there; what we now have for that is [RA]. Below are some reasons for why [DA] in QSM is superior to [RA] in QFT in understanding the nature of quantum SSB. But before that let us see why a study of classical models of SSB would not do. There is a corresponding decompositional account of SSB for classical models (cf. Emch, 1977; Liu, 2003), but since notions such as representations and unitary implementability, which are crucial notions in quantum SSB, are absent in the classical regime, the account cannot be adequate for the understanding of quantum SSB.

[6]The first indication that KMS can be understood as more than a formal generalization to infinite systems of the notion of canonical equilibrium state appeared in Pucz and Woronowicz (1978), where it is argued from linear response theory. Sewell (2002) characterizes it in terms of local and global stability conditions, and he mentions in particular the relevance of allied notions to the decompositional view of SSB with applications to the theory of crystallization in Emch, Knops, and Verboven (1970).

QSM is a better arena than QFT for studying quantum SSB. *First*, there is a great deal of similarity between an infinite quantum thermo-system and a quantum field; so much so that arguments by analogy from the former to the latter are commonly used in the literature of QFT (cf. Anderson, 1963; Coleman, 1975). In a nutshell, to understand SSB in QFT one must deal with both (i) an infinite number of degrees of freedom, and (ii) special (or even general) relativity. Our view is that (i) is essential to the understanding of quantum SSB whereas (ii) is not; and the complexity introduced by (ii) tends to distract attention and/or obscure discussions of the SSB concept. Therefore, [DA] in QSM helps us by way of analogy to understand the nature of SSB in QFT. *Second*, there are simple and experimentally implementable models in QSM for [DA], while models in QFT for [RA] are either simple but unrealistic (i.e. not experimentally realizable) or realistic but complex (i.e. too complicated to be used to exhibit in controllable or simple terms what the account means).[7]

And (partly because of the above) [DA] is superior to [RA]. *First*, unlike the representational account in which the physical interpretations of the two features of quantum SSB—the lack of unitary implementability and unitarily inequivalent representations—are either obscure or not forthcoming, this account [DA] almost wears their physical meanings on its sleeves,[8] for the essence of SSB in general, for any system, is *the coexistence of solutions, each of which has less symmetry than the symmetry of the problem, while the symmetry of the problem acts transitively on the set of all solutions*; and that is exactly what [DA] describes. *Second*, adopting [DA] restitutes the correct conceptual order of explanation (or understanding): the two features in [RA] may be sure signs of a quantum SSB, but they should not be taken to define the concept. It is rather because a quantum system has SSB that it exhibits the two features of [RA], not the other way around (more on this in Section 6). *Third*, it can be shown (and we will show towards the end of this section) that all quantum SSB systems in the sense of [DA] are also SSB systems in the sense of [RA]. Hence, in studying quantum SSB in QSM one gets both [DA] and [RA] and their relationship, which one does not readily get, in studying SSB in QFT.

[7]As Earman (2003a, b) rightly pointed out in the conclusion of his papers, the quantum SSB as characterized by [RA] requires a great deal of interpretation before we can even make sense of it. The toy model of a simple quantum field Earman used to illustrate [RA] does not show how the lack of unitary implementability of some symmetry amounts to a case of SSB. The discussion of the Weyl algebra (a species of C*-algebra) gives us an illustration of how the lack of unitary implementability leads to having vacuum states in inequivalent representations; but again it does not shed any light on why the latter implies a SSB. Invoking the notion of a folium of states on a C*-algebra and of disjoint representations does not help either, since there are many cases of disjoint representations that are not connected to SSB. Earman is right in asserting that 'for pure algebraic states, spatial inequivalence equals disjointness,' but the connection between this fact and quantum SSB is still obscure. States of a system at different temperatures (above the critical temperature) are mutually disjoint, but this has nothing to do with SSB. Hence, inequivalence or disjointness per se does not necessarily tell us anything on whether or not systems having such states harbor quantum SSB.

[8]As Earman (2003a, b) again rightly pointed out, neither of the two features in [RA] explicitly say whether the symmetry in question is broken or in what manner it is broken. One wonders, from the outset, whether the two features, which seem to identify independently quantum SSB, are equivalent to the commonsense notion of it in classical SSB, and what explanation, causal or formal, these characterizations can offer to our understanding of quantum SSB (or SSB in general).

Let us assume next that the existence conditions for an infinite quantum thermo-system are met and we have $\{A, \phi, \tau\}$ with ϕ being KMS (see Section 3). As we said earlier, algebras as abstract objects are best studied and understood in a representation that is tailored to the physical situation in question summarized here by the state ϕ. For simplicity we restrict our attention to representations, (H, π), of A that comprise a separable, i.e. admitting a countable basis, Hilbert space H and a morphism, $\pi : \mathrm{A} \to \mathrm{B(H)}$, of A into the algebra of bounded operators B(H) of H. The meaning of tailoring a representation to the physical situation at hand can be seen in the result (proven by Gelfand, Naimark, and by Segal, hence GNS), which says that a state ϕ on A induces uniquely (up to unitary equivalence) a representation, $\pi_{\phi;}$, of A in a Hilbert space, $\mathrm{H}_{\phi;}$, with a cyclic vector, $\Phi_{\phi;}$, such that $\langle \phi; A\rangle = (\Phi_\phi, \pi_\phi(A)\Phi_\phi) \forall A \in \mathrm{A}$.[9] (To relieve notational congestion, we drop the sub-index $_{\phi;}$.) Given a GNS representation $\pi(\mathrm{A})$ of A, we consider the algebra, $\pi(\mathrm{A})'$, that comprises all the bounded operators in H that commute with all elements in $\pi(\mathrm{A})$, and we enlarge $\pi(\mathrm{A})$ to $\pi(\mathrm{A})''$, the bicommutant of $\pi(\mathrm{A})$, that comprises all the bounded operators that commute with all elements in $\pi(\mathrm{A})'$. The bicommutant set, $\pi(\mathrm{A})''$, which is known as a von Neumann algebra, is of special importance because it is the *weak-operator closure* of $\pi(\mathrm{A})$, which means that unlike the abstract A, it is anchored to a specific representation—hence to a specific physical situation at hand—and contains macroscopic observables while A only contains quasi-local ones, i.e. observables that can be approximated by local observables in the norm topology (whereas, the macroscopic observables require the much more encompassing weak-operator topology).[10]

Because of its central importance, we let $\mathrm{N} = \pi(\mathrm{A})''$. Among various subsets of N is its center, $\mathrm{Z} = \mathrm{N} \cap \mathrm{N}'$, which is important because it is the intersection of all complete sets of commuting operators—these sets contain *all the observables that can be measured simultaneously without dispersion.* In other words, it contains the *essential observables* of the system in question (cf. Wick, Wightman, & Wigner, 1952).

We are now ready to state the *central decomposition* Scholium, which is the core result for the decompositional account.

[CD] Let A, ϕ, α such that $\phi \circ \alpha = \phi$, H,π,N and Z be the objects as defined above. And suppose that for simplicity that the spectrum, $Sp(\mathrm{Z})$, is discrete, i.e. there is a partition $\{P_k\}$ of the identity in H by projection operators, P_k, such that $\mathrm{Z} = \{Z = \sum_k c_k P_k | c_k \in \mathbb{C}\}$.

Then, for $\mathrm{N}_k = P_k\mathrm{N}(= P_k\mathrm{N}P_k)$ and ϕ_k defined, $\forall N \in \mathrm{N}$, by $\langle \phi_k; N\rangle = \langle \phi; P_k N P_k\rangle / \langle \phi; P_k\rangle$ (with $\langle \phi; P_k\rangle \neq 0$), we have

1. $\mathrm{N} = \oplus \mathrm{N}_k$;
2. $\phi = \sum_k \lambda_k \phi_k$ with $\lambda_k = \langle \phi; P_k\rangle > 0$, and $\sum_k \lambda_k = 1$;
3. either $\phi_k \circ \alpha = \phi_k$ or $\phi_k \circ \alpha = \phi_j$, $k \neq j$.

[9] A vector Φ is cyclic for a representation π if $\pi(\mathrm{A})\Phi \equiv \{\pi(\mathrm{A})\Phi | A \in \mathrm{A}\}$ is dense in H.

[10] For the Double Commutant Theorem assumed in the above paragraph, see Kadison and Ringrose (1983, Theorem. 5.3.1, p. 326).

(For the proof and a more complete version of [CD], see Appendix B; in particular, [CD] can be straightforwardly tailored to accommodate the case where $Sp(Z)$ is not necessarily discrete.) [CD] is a scholium for a degenerate state that is uniquely decomposable into non-degenerate ones.[11]

From [CD], if we add that the system in question is not extremal KMS *and* it is the case that $\phi_k \circ \alpha = \phi_j$, $k \neq j$, we have the decompositional account of quantum SSB, i.e. [DA]. The first condition satisfies (i) in [DA], and the second satisfies (ii) in [DA] because condition (2) in [CD] above gives a unique convex decomposition of ϕ into extremal KMS states.

The notion of witness concerns the essential observables in the center of a von Neumann algebra. For QFT, they are usually various 'charges,' such as the electric charge or the baryon number. In a translation-invariant QSM, we have *macroscopic* observables, which are defined as space-averages of local observables. As they obtain only as weak-operator limits, they belong to $\mathrm{N} = \pi(\mathrm{A})''$, where π is the GNS representation associated with the invariant state ϕ in question. They also belong to $\mathrm{N}' = \pi(\mathrm{A})'$ as a consequence of locality, namely the fact that local observables that belong to spatially disjoint regions commute among themselves. These observables provide us with witnesses of SSB, which in general are defined as follows.

[W] An observable $W \in \mathrm{N}$ is a *witness* for SSB if it satisfies the following:

1. $W \in \mathrm{Z}$;
2. $\langle \phi_k; W \rangle \neq \langle \phi_j; W \rangle$, for some k different from j.

To be more specific, if the state is invariant under a translation group, say $\mathbb{Z}^n$ (for a lattice system) or $\mathbb{R}^n$ (for a continuous system), the locality condition (see above) requires that macroscopic observables (as space-averages of local observables) belong to the center of the GNS representation of ϕ, which means that they satisfy (1) above. And an example of (2) is the spontaneous magnetization of a lattice spin-$\frac{1}{2}$ system, namely, below the critical temperature, there are two values of magnetization, $\langle \phi_\pm; m \rangle \neq 0$ (corresponding to spin up or spin down), that show us—as witnesses ought to do—the presence of SSB.

Two questions arise from the above characterization of quantum SSB. (i) Does this also satisfy the two features given in [RA] so that it is also sufficient (though perhaps not necessary) for the alleged SSBs in QFT? (ii) In what exact sense is this a case of SSB for infinite quantum systems?

Because, as we mentioned earlier (see also Earman, 2003a, b), it is true for C*-algebras in general that an automorphism α of A, mapping one state to another, is

[11]The decomposition in (2) of [CD] is reminiscent of Lüders' version of the von Neumann description of the quantum measurement process $\phi \to \sum \lambda_k \phi_k$. Here, however, the process that starts from the LHS, namely ϕ, and ends with the QM-conditional expectation in the RHS, namely $\sum \lambda_k \phi_k$, reduces to the statement that LHS = RHS since we refer to a situation where the P_k's belong to the center; in contrast, recall that the center consists only of the scalar multiples of the identity operator in QSM of finite systems that predate the introduction of superselection rules (cf. Wick et al., 1952). Here, the possibility that the center be non-trivial comes from the consideration of systems with an infinite number of degrees of freedom, which appears in QSM involving the macroscopic limit. Furthermore, [CD] is akin to the decomposition in superselection sectors in QFT.

unitarily implementable if the GNS representations of the two states are unitarily equivalent, we only need to find out whether, given [DA], one or the other condition in [RA] holds. It turns out that the answer to the latter is affirmative. Assume [DA] and take the simplest case (where $k = +,-$): ϕ_+ and $\phi_{-;}$, where $\phi_\pm \circ \alpha = \phi_{\mp;}$. Since the two states are extremal KMS states, the corresponding GNS states are primary, which means their centers are trivial (multiples of identity). Suppose (for reductio) that the two representations were unitarily equivalent. We then would have a unitarily operator, U, such that $UW_+U^{-1} = W_-$, where the $W_\pm = \langle\phi_\pm; W\rangle P_\pm$ are the corresponding witnesses in the two centers; but that means $\langle\phi_+; W\rangle = \langle\phi_-; W\rangle$, which contradicts point (2) in [W]. This shows that *all quantum SSB systems in sense [DA] are also SSB systems in sense [RA].*

The demand for a witness in the [DA] account for SSB is a general feature pertinent to quantum systems. In a classical system, it is sufficient if the symmetry in question maps one lowest stable energy state to a different such state, since states of classical systems are determined by the values (precise or average) of the observables. For a quantum system, it is possible that the symmetry in question maps one extremal KMS state, e.g. ϕ_k, to another state, e.g. ϕ_j (where k is different from j), and yet there is no observable in the center that satisfies (2) of [W].

Further, the result of the decomposition is a mixture, not a coherent superposition; therefore, when a system crosses the critical value of the relevant parameter towards an SSB, the symmetry in question is *broken*, e.g. the system is in one of ϕ_k not in a coherent superposition of these states. The probabilities are of course given by e.g. $\lambda_k = \langle\phi; P_k\rangle > 0$.

For SSB in QSM, in addition to the possibility of a central decomposition, it is necessary for the systems to be in a KMS state; for otherwise the decomposition, though unique, is not necessarily one of extremal KMS states, which represent pure thermodynamical phases. One should also note that although the central decomposition of a KMS state is unique, a decomposition of such into primary states is not (nevertheless, see footnote 4). A different condition has yet to be found for quantum field-systems that would play a similar role. Whatever that condition is, it cannot be something such as the lack of unitary implementability or the inequivalent of representations, because none of them has been conceived as a condition for guaranteeing that the central decomposition of a state of quantum field results in states that spontaneously break the symmetry in question. Unlike the KMS condition, which can be proven to be equivalent to a condition of stable equilibrium states, the conditions of unitary implementability (or the lack of it) and unitary inequivalent of representations lack such direct physical meaning.

5. Models for the decompositional account

Among the various models of quantum SSB—which may be used to show the *consistency* of our account—the BCS model for superconductivity is probably the most famous; recently, it even caught the attention of philosophers of science (see, e.g. Cartwright, 1999). It occupies an important place in the history of physics,

C. Liu, G.G. Emch / Studies in History and Philosophy of Modern Physics 36 (2005) 137–163 147

and it is a case of the breaking of a U(1) gauge group (NB: the popular SSBs in QFT are also of gauge groups); however, since the symmetry group is continuous, which requires technical precautions (cf. Appendix B), we first discuss a simple discrete model: the Weiss–Heisenberg model for (anti)ferromagnetism (cf. Emch & Knops, 1970; Streater, 1967; Bianchi, Contucci, & Giardina, 2004).

To make it simple and yet still sufficiently general, we take the model to consist of a one-dimensional chain of quantum spins with a long-range, Ising-type interaction.[12] At each site k in the chain, $\mathbb{Z}$, sits a particle with a quantum half-spin, $\boldsymbol{\sigma}_k$, whose components are the three Pauli matrices. These components as observables generate an algebra A_k at each site, which is a copy of the algebra $M(2,\mathbb{C})$ of 2×2 matrices with complex entries:

$$M(2,\mathbb{C}) = \left\{ A = \begin{pmatrix} a & b \\ c & d \end{pmatrix} | a,b,c,d \in \mathbb{C} \right\}.$$

For any finite string Λ of sites along the chain, we have the algebra, $\mathrm{A}_\Lambda = \otimes_{k\in\Lambda}\mathrm{A}_k$, which is a copy of $M(2^{|\Lambda|},\mathbb{C})$; and an evolution described in the Heisenberg picture as

$$\tau_{\Lambda,B}(t) : A \in \mathrm{A}_\Lambda \mapsto \tau_\Lambda(t)[A] = \mathrm{e}^{\mathrm{i}H_{\Lambda,B}t} A \mathrm{e}^{-\mathrm{i}H_{\Lambda,B}t} \in \mathrm{A}_\Lambda, \tag{1}$$

where we choose the ferromagnetic Weiss–Ising Hamiltonian:

$$H_{\Lambda,B} = -\sum_k [B + B_{\Lambda,k}]\sigma_k^z, \quad \text{with } B_{\Lambda,k} = \frac{1}{2}\sum_j J_{\Lambda,jk}\sigma_j^z, \tag{2}$$

where

$$J_{\Lambda,jk} = \begin{cases} 0 & \text{if } j = k, \\ [2J/|\Lambda|] > 0 & \text{if } j \neq k. \end{cases}$$

Here, B is an external magnetic field in the z-direction and $J_{\Lambda,jk} \geqslant 0$ the strength of interaction between any two spins which favorizes their parallel, rather than anti-parallel, alignment. And we use $B_{\Lambda,k}$ to indicate the magnetic field (surrounding the k-site) that derives from the averaged effect of spins on other sites. The Hamiltonian not only determines the evolution of the system, it also determines the canonical equilibrium state, $\phi_{\Lambda,B,\beta}$, at natural (i.e. inverse) temperature $\beta = 1/kT > 0$:

$$\phi_{\Lambda,B,\beta} : A \in \mathrm{A}_\Lambda \mapsto \langle \phi_{\Lambda,B,\beta}; A\rangle = Tr(\rho_{\Lambda,B,\beta}A)$$

where

$$\rho_{\Lambda,B,\beta} = Z^{-1}\mathrm{e}^{-\beta H_{\Lambda,B}} \text{ and } Z = Tr(\mathrm{e}^{-\beta H_{\Lambda,B}}). \tag{3}$$

So far, the model only comprises finitely many spins, and hence (3) is the unique state on the local algebra, $\mathrm{A}_\Lambda = \otimes_{k\in\Lambda}\mathrm{A}_k$, that satisfies the KMS condition at natural temperature β with respect to the given time evolution $\tau_{\Lambda,B}$. In addition, when $B = 0$, the system is invariant under the symmetry, α, that is geometrically interpreted as a

[12]The long-rangeness of the interaction makes this model differ from the usual Ising models where nearest-neighbor interaction is assumed: the long-range interaction makes it possible for a 1-d model to harbor a phrase transition.

180° rotation around the y-axis. One can check that this rotational symmetry is unitarily implementable, with the matrix

$$V = \begin{pmatrix} 0 & 1 \\ -1 & 0 \end{pmatrix}$$

acting on each site. When $B = 0$ the model is also invariant under rotations around the other two axes, and even when $B \neq 0$ the model is invariant under the rotation around the z-axis. This means that the components of the magnetization per site of the string Λ, $\mathbf{m}_\Lambda = |\Lambda|^{-1}\sum_k \sigma_k$, have the following properties: $\langle \phi_{\Lambda,B,\beta}; m^x_\Lambda \rangle = 0 = \langle \phi_{\Lambda,B,\beta}; m^y_\Lambda \rangle$ and $\lim_{B\to 0}\langle \phi_{\Lambda,B,\beta}; m^z_\Lambda \rangle = 0$, which in turns means that there is no spontaneous magnetization (or no SSB), agreeing with the general 'no-go' theorem.

It is proven that taking this model to the macroscopic limit—$|\Lambda| \to \infty$—produces from (3) a well behaved state $\phi_{B,\beta}$ on A, where

$$\mathrm{A} = \overline{\bigcup_{\Lambda \subset \mathbb{Z}'} \mathrm{A}_\Lambda}$$

is a C*-algebra of quasi-local observables obtained as the C*-inductive limit of A_Λ. The evolution at the limit, $\tau(t)$, is defined on the von Neumann algebra $\mathrm{N}_{B,\beta} = \pi_{B,\beta}(\mathrm{A})''$ (not on A itself). It is then proven that the natural extension of $\phi_{B,\beta}$ to $\mathrm{N}_{B,\beta}$ satisfies the KMS condition with respect to $\tau(t)$. (For the sake of notational simplicity, we use $\phi_{B,\beta}$ to denote both the state on A and its extension to $\mathrm{N}_{B,\beta}$.) Moreover, finally, the magnetization per site at the limit whose components are $m^{x,y,z}_{B,\beta}$ is also well defined as follows. Because taking the limit $\Lambda \uparrow \mathbb{Z}$ for these components depends on the state of the chain, we have the following weak-limit:

$$w - \lim_{\Lambda \uparrow \mathbb{Z}} \pi_{B,\beta}(m_\Lambda)^{x,y,z} = m^{x,y,z}_{B,\beta},$$

which means that $\forall \varepsilon > 0$, and every pair (Φ, Ψ) of vectors in the Hilbert space of the representation, $\exists \Lambda_0$ with $|\Lambda_0| < \infty$, such that $\forall \Lambda \supseteq \Lambda_0$, $(\Psi, [m^{x,y,z}_{\Lambda,B,\beta} - m^{x,y,z}_{B,\beta}]\Phi) < \varepsilon$. Hence, $m^{x,y,z}_{B,\beta}$ is representation-dependent, and that is so because different states of such a system, which produce different GNS representations, give different expectation values for the magnetization. Further, because $m^{x,y,z}_{B,\beta}$ commute with all observables in the algebra $\mathrm{N}_{B,\beta}$, they are in its center: $\mathbb{Z}_{B,\beta} \equiv N_{B,\beta} \cap N'_{B,\beta}$.

We can now see how the general concept of SSB according to the decompositional account (i.e. [CD], [DA], and [W] above) applies to this model. For [CD], the automorphism under consideration in the model is α and the GNS-associated objects are all defined in terms of B and β because $\phi_{B,\beta}$ is. It is proven that (i) $\phi_{B,\beta}$ is extremal KMS with respect to the evolution $\tau(t)$ if, and only if, $\beta \leqslant J^{-1}$ (i.e. $T \geqslant T_c$), and (ii) if $\phi_{B,\beta}$ is extremal KMS, the three components $m^{x,y,z}_{B,\beta}$ are as follows:

$$m^x_{B,\beta} = 0 = m^y_{B,\beta}, \tag{4a}$$

$$m^z_{B,\beta} = \tanh\{\beta(B + Jm^z_{B,\beta})\}, \tag{4b}$$

where (4b) is a self-consistency equation, i.e. the same variable $m^z_{B,\beta}$ appearing on both sides of the equation. At fixed B, exactly one value of $m^z_{B,\beta}$ satisfies (4b); it is positive (negative) when $B>0$ ($B<0$), and

$$\lim_{B\to 0} m^z_{B,\beta} = 0.$$

In other words, there is no SSB when $\beta \leqslant J^{-1}$. However, when $\beta > J^{-1}$ (i.e. $T<T_c$) and $B=0$, we have

$$\phi_{0,\beta} = \tfrac{1}{2}(\phi^+_{0,\beta} + \phi^-_{0,\beta}) \tag{5}$$

(which is clearly a case of ϕ^k, with $k = +,-$), such that

1. $\phi^{\pm}_{B,\beta} \circ \alpha = \phi^{\mp}_{B,\beta}$
2. $\phi^{\pm}_{0,\beta}$ are extremal KMS and the corresponding GNS representations, $\pi^{\pm}_{0,\beta}$, are primary sub-representations of $\pi_{0,\beta}$.

And since $m^{\pm}_{\beta} = \langle \phi^{\pm}_{0,\beta}; m^z_{0,\beta}\rangle = P^{\pm}\mathbf{m}_{0,\beta}P^{\pm}$, and (from (4b)) $m^{\pm}_{\beta} = \tanh(\beta J m^{\pm}_{\beta})$, we have

3. $m^{\pm}_{\beta}$ are in the center, $\mathrm{Z}_{0,\beta} \equiv \mathrm{N}_{0,\beta} \cap \mathrm{N}'_{0,\beta}$;
4. $m^{\pm}_{\beta}$ has two non-zero solutions: $m^{\pm}_{\beta} = -m^{\mp}_{\beta}$; they serve as our witnesses for the SSB ('witness' as defined in [W]).

And hence,

5. the two representations, $\pi^{\pm}_{0,\beta}$, for the states, $\phi^{\pm}_{0,\beta}$, are unitarily inequivalent.

The last result, 5, follows from our earlier general discussion of the decompositional account of SSB. This model of the 'flip-flop' symmetry, α, *therefore* satisfies [DA]; moreover, it *also* satisfies [RA] because the corresponding GNS representations, $\pi^{\pm}_{0,\beta}$, are unitarily inequivalent (which implies that the symmetry is not unitarily implementable). And the conceptual advantage of [DA] is 4 above, which involves explicitly a macroscopic observable: the magnetization.

We now turn to a brief sketch of the BCS model, where the interaction between electrons and phonons in a metallic superconductor is mimicked by the interacting Cooper pairs, i.e. pairs of electrons, close to the Fermi surface, of opposite momenta and spin orientations (cf. Schrieffer, 1964). Its finite-volume Hamiltonian in a box Λ with volume V is

$$H_{\Lambda} = \sum_{p,s} \varepsilon(p) a_s(p) * a_s(p) + \sum_{p,q} b(p) * \breve{v}(p,q) b(q), \tag{6}$$

where $a_s(p)^*$ is the creation operator of electrons with momentum p and spin s, and $a_s(p)$ the correponding annihilation operator; and therefore the first term of H_{Λ} describes the free energy of the electrons. Similarly, $b(p)^* = a_{\uparrow}(p)^* a_{\downarrow}(-p)^*$ is the creation operator of Cooper pairs; and $b(p)^*\breve{v}(p,q)b(q)$ describes the interactional energy between two Cooper pairs.

In the *mean-field approximation*, the Hamiltonian (6) of the BCS model is diagonalized by a Bogoliubov–Valatin transformation,

$$\gamma_\uparrow(p)^* = u_\rho(p)^* a_\uparrow(p)^* - v_\rho(p)^* a_\downarrow(-p),$$
$$\gamma_\downarrow(-p) = v_\rho(p) a_\uparrow(p)^* + u_\rho(p) a_\downarrow(-p),$$

to give

$$H_{\Lambda,\rho} = \sum_{p,s} E_\rho(p) \gamma_{s,\rho}(p)^* \gamma_{s,\rho}(p), \tag{7}$$

which means that the system can now be viewed as having free elementary excitations created and annihilated by $\gamma_{s,\rho}(p)^*$ and $\gamma_{s,\rho}(p)$, respectively, with the energy

$$E_\rho(p) = [\varepsilon(p)^2 + \Delta_\rho(p)^* \Delta_\rho(p)]^{1/2}; \tag{8}$$

originally, u, v, and Δ were considered to be complex-valued functions. In particular, the observable 'energy gap' is obtainable from the self-consistency equation

$$\Delta_\rho(p) = -\sum_q \breve{v}(p,q) \frac{\Delta_\rho(p)}{2E_\rho(p)} \tanh\left(\frac{1}{2}\beta E_\rho(p)\right). \tag{9}$$

There is a critical tempreture, $T_c = (k\beta_c)^{-1}$, such that

(a) for $T \geqslant T_c$, Eq. (9) has no non-zero solution; and
(b) for $T < T_c$, a non-zero solution appears, which can be computed numerically and which agreed with the available data in 1957 well enough to be regarded as giving a satisfactory explanation to the then prominent and yet puzzling phenomenological features of superconductivity.

Moreover, one notices that the Bogoliubov–Valatin transformation, with the u_ρ and v_ρ complex-valued, leads to a *violation* of the gauge-invariance of the theory. Specifically, the action of the gauge group, $S^1 = \{e^{i\theta} | 0 \leqslant \theta < 2\pi\}$, is defined by

$$a_s(p) \to \alpha(\theta)[a_s(p)] = e^{+i\theta} a_s(p),$$
$$a_s(p)* \to \alpha(\theta)[a_s(p)^*] = e^{-i\theta} a_s(p)^*. \tag{10}$$

Clearly, the Hamiltonian (6) *is* invariant under this symmetry group, whereas the Hamiltonian (7) is *not*. This is indication that superconductivity is a SSB of gauge symmetry; it is however not properly accounted for in this mean-field approach for finite systems.

A flurry of researches in the 1960s reformulated the BCS model as an *infinite* system. The claims are that the mean-field approximation becomes *exact* in this limit, and that the system exhibits a controllable spontaneous breaking of a gauge symmetry (cf. Haag, 1962; Ezawa, 1964; Emch & Guenin, 1966; Thirring, 1969; Dubin, 1974, chapter 4). In this approach, the algebra appropriate for the BCS model is the algebra, A_ρ, associated with a representation, π_ρ, of the Fermi anti-commutation relations that the elementary excitations satisfy, i.e.

$$\{a_s(f), a_{s'}(g)\} = 0 \text{ and } \{a_s(f), a_{s'}(g)^*\} = (f, g)\delta_{s,s'}I, \tag{11}$$

where f and g run over $L^2(\mathbb{R}^3, \mathrm{d}x)$, and (f,g) is the scalar product in this space; in particular, $\forall p = n(2\pi/V^{1/2})$; $n \in \mathbb{Z}^3$ and $a_s(p) \equiv a_s(f_p)$ with

$$f_p(x) = \begin{cases} V^{-1/2}\mathrm{e}^{\mathrm{i}xp,} & x \in \Lambda, \\ 0 & \text{otherwise.} \end{cases}$$

The key to the understanding of how the mean-field Hamiltonian (7) becomes an exact diagonalization of the BCS Hamiltonian (6) in the limit $|\Lambda| \to \infty$ is to rewrite (6) as

$$H_\Lambda = \sum_{p,s} \varepsilon(p) a_s(p)^* a_s(p) + \frac{1}{2}\sum_p [b(p)^* \Delta_\Lambda(p)^* b(p)],$$

where $\Delta_\Lambda(p) = \sum_q \breve{\upsilon}_\Lambda(p,q) b(q)$ so that the canonical commutation relations (11) entail

$$[\Delta_\Lambda(p), a(q)] = 0 \text{ and } [\Delta_\Lambda(p), a(q)^*] = \breve{\upsilon}(p,q).$$

One needs now to recognize two mathematical facts. First, the physical constraints on the interaction between the Cooper pairs entail that there exists a constant υ such that

$$|\breve{\upsilon}(p,q)| < \frac{\upsilon}{|\Lambda|} \text{ so that } \lim_{|\Lambda| \to \infty} |\breve{\upsilon}(p,q)| = 0.$$

And second,

$$\Delta_\rho(p) = w - \lim_{|\Lambda| \to \infty} \Delta_\Lambda(p)$$

exists in the representation space associated with the canonical equilibrium state ρ of the infinite BCS model. As a consequence, $\Delta_\rho(p)$ belongs to the center of the von Neumann algebra obtained from this representation. Below the critical temperature T_c, this representation turns out to be a direct integral over primary representations that correspond to the pure thermodynamical phases of the model. In each of these phases, $\Delta_\rho(p)$ is a pure number (more precisely a scalar multiple of the identity operator) and satisfies the self-consistency Eq. (9).

Note further that for each $T \geqslant T_c$ the solution (9) is unique, *up to a complex number of modulus 1*. This ambiguity is essential; indeed, while the gauge symmetry is broken in each phase, the gauge group acts non-trivially on the center of the integral representation $\pi_{\rho;}$; in particular

$$\Delta_\rho(p) \to \alpha(\theta)[\Delta_\rho(p)] = \mathrm{e}^{2\mathrm{i}\theta}\Delta_\rho(p),$$

so that the gauge group acts transitively on the space of pure thermodynamical phases, as should happen in accordance with [DA]: one gets from any one of the components to any other by the action of an element of the gauge group S^1. Similarly, in connection with the Bogoliubov–Valatin elementary excitations, we have

$$\alpha(\theta)[u(p)] = u(p) \text{ and } \alpha(\theta)[v(p)] = e^{2i\theta}v(p)$$

so that

$$\alpha(\theta)[\gamma_s(p)] = e^{i\theta}\gamma_s(p) \text{ and } \alpha(\theta)[\gamma_s(p)^*] = e^{-i\theta}\gamma_s(p)^*$$

and the gauge symmetry of the evolution in the representation space of $\pi_{\rho;}$ is preserved. It is only the decomposition into pure thermodynamical phases that breaks the gauge symmetry.

6. Explaining quantum SSB

To further argue for the decompositional account [DA] we turn to the question of what constitutes an explanation of quantum SSB. We first briefly state the two senses of explanation that are involved in explaining classical SSB, and we then argue that the representational account [RA] can only be said to 'explain' quantum SSB in one of these two senses while the decompositional account explains it in both.

The essential features of classical SSB are culled from some simple mechanical systems in which some symmetry, say a rotational symmetry around the z-direction, is spontaneously broken because when the value of certain parameter passes a critical point, the original single fundamental state becomes unstable, and without any apparent asymmetrical causal antecedents, the multiple possible stable fundamental states individually break the symmetry (cf. Liu, 2003). The set of possible stable states is closed under the symmetry in that $\forall s \in S$ and $\forall g \in G, g[s] \in S$, but $g[s] \neq s$ for some g and s, where S is the set of stable fundamental states and G the symmetry group. Here we have the *'structural/kinematic'* aspect (or sense) of classical SSB. It refers to the conditions under which a SSB is possible. The *'causal/dynamic'* aspect (or sense) contains those features that characterize how (typically) actual SSBs take place. SSBs are mostly, though not necessarily, caused by arbitrarily small random fluctuations, and the conditions (or mechanisms) responsible for making SSBs possible necessarily differ from those actually producing them (otherwise, the breakings can no longer be regarded as spontaneous). To separate the two aspects, we will label as *structural* an explanation of SSB that tells us under what conditions an SSB is possible (and the obtaining of the conditions certainly has its own causal explanation, e.g. by lowering the temperature across the critical temperature); we label as *causal* an explanation that tells us the mechanisms for actual breakings: how systems harboring SSB end up in those symmetry-breaking fundamental states.

The causal aspect appears to be absent in both accounts (i.e. [RA] and [DA]) of quantum SSB; both seem to only tell us when a SSB is possible in an infinite quantum system (a magnet or a quantum field). However, we shall argue that the decompositional account is more conducive to providing a causal (or dynamical) explanation for quantum SSB than is the representational account. The causal explanation of classical SSBs already essentially involves classical SM because what actually causes mechanical SSB systems to go from an unstable fundamental state to a set of stable ones—some of which breaks the symmetry—cannot be explicitly stated as initial conditions in a mechanical (non-statistical) theory (detailed arguments for this point can be found in Liu, 2003). Moreover, what happens in the Weiss–Heisenberg model (and many other similar models[13]) can be roughly described as follows. Two opposing tendencies exist in a Weiss–Heisenberg system, namely, the one caused by the interactions to align the spins in the same direction and the other caused by thermal agitation to randomize the directions of individual spins. The strength of the latter obviously depends on the system's temperature and the strength of the former depends on how many spins are already aligned in parts of the system. When the temperature is above the critical value, the balance of the strengths is in favor of thermal agitation. Any large fluctuation of aligned segments of spins will quickly disappear rather than growing larger. But when the temperature drops below the critical value, the balance tilts in favor of interactional alignment of spins that will tend to grow larger and eventually result in spontaneous magnetization, i.e. having an average net magnetization even when no external field is present.

Since quantum SSB is a natural phenomenon (not merely a theoretical concept[14]), we deem it necessary for a theory of quantum SSB to provide a conception of how it is causally explained. In fact, authors who contemplate causal explanations for the SSBs in QFT models heavily depend on analogies from what we sketched above for the thermo-systems (cf. Anderson, 1963; Streater, 1965; Guralnik, Hagen, & Kibble, 1968; Coleman, 1975). For instance, attributing the causal mechanism of SSB in a gauge field to a correlation or coherence of phases within a vacuum state is helpful when one recalls what actually happens in a phase transition in a thermo-system, but not so helpful without such an analogy (cf. Moriyasu, 1983).[15] Therefore, although [DA] does not, and cannot, include the specific causal mechanism that spontaneously breaks a symmetry—for the mechanisms may differ in different systems—it better accommodates the inclusion of causal accounts by stating in explicit terms the physical consequence of all such mechanisms. It provides clues for, or constraints on, what such mechanisms may be like, whereas one does not see this in [RA]: neither not being unitarily implementable nor having fundamental states that generate

[13]There is a difference between the Onsager spontaneous magnetization (of a two-dimensional Ising model) (cf. Onsager, 1944; Schultz, Mattis, & Lieb, 1964) and the one in the Weiss–Heisenberg model we discuss here. In the former, the exact solution of the magnetization at B different from zero is not obtained explicitly, while such a solution for the Weiss–Heisenberg model is, as in (4b) in the previous section. Therefore, we can 'see' (in the solution) what happens with the magnetization when B approaches zero in the WH model, but we cannot directly do so in the Onsager solution, although the classical Peierls (1936) argument is an attempt towards the latter.

[14]For example, renormalization, first- or second- quantization, etc.

[15]One may argue that there are plenty of instances of causal explanations in the discussion of the Goldstone theorem (or zero-mass gauge particles) and Higgs mechanism in QFT. This is true; however, such causal stories are not about how the relevant gauge symmetries are spontaneously broken but rather about what happens as consequences of such breakings. The breakings are assumed from the outset of that discussion.

unitarily inequivalent representations seems to provide any clue as to what may have caused a system to break its symmetry spontaneously.

Even if we were to disregard the demand for a causal explanation, the representational account would still be weaker because it is not clear that it even provides a clear conception of a structural explanation. Knowing that an automorphism is not unitarily implementable does not by itself tell, even structurally or formally, why the associated symmetry is spontaneously broken. It itself needs explanation more than it can explain. Granted, we do gain some understanding of quantum SSB (at least in the sense it is possible) in the notion of having, say, vacuum states generate unitarily inequivalent representations. We shall explain how this is so in the next section; but it seems obvious that from the decompositional account we gain a better understanding. What we learn from [DA], together with [CD], is how under the KMS condition, the appearance of a non-trivial center, i.e.

$$\mathsf{Z} = \{Z = \sum_k c_k P_k | c_k \in \mathbb{C}\},$$

(together with having witnesses in it) makes a SSB possible. For since such KMS states decompose uniquely into sets of extremal KMS states, i.e. pure thermodynamical phases, and the representations associated with these extremal states are primary (i.e. having trivial centers), it is easy to see why these representations are unitarily inequivalent. If unitarily equivalent representations can be regarded as different mathematical descriptions of the same physical situation, it follows that two different extremal KMS states and their associated representations cannot be unitarily equivalent. Take the case of the Weiss–Heisenberg model. The two extremal states, spin-up, or spin-down, of the entire chain $\mathbb{Z}$, cannot be regarded as about the same physical situation. In other words, from the decompositional account one gets a structural explanation of how a SSB is possible, and *in addition*, one gets an account of why the states that spontaneously break the symmetry belong to unitarily inequivalent representations. As we said earlier, this is the right conceptual order in the explanation (even if only on the formal level) of SSB, which one does not see in [RA].

One may argue that we have misconstrued the representational account, especially when it concerns SSB in QFT. Is not the $P(\phi)_2$ field (quantum fields in a 2-d space-time) a concrete model in QFT, from which cases of SSB, strikingly analogous to the classical cases, can be derived (cf. Glimm & Jaffe, 1981, Simon, 1974)? To this we have two brief responses. First, this model provides a vivid picture of what an SSB looks like in a quantum-field model precisely because it bears a striking resemblance, as far as the geometry for the set of symmetry-breaking vacuum states are concerned, to the Weiss–Heisenberg model of ferromagnetism. Hence, one would think that it serves as a better illustration of [DA] than of [RA]. It is not even clear how the lack of unitary implementability and having vacuum states generating inequivalent representations play any crucial role in this QFT model of SSB. Second, unlike in the cases in which $\beta = 1/kT$ (or temperature T) is the control-parameter, the physical meaning of λ is not entirely clear, especially when $\sqrt{\lambda}$ is to be constantly interpreted as some kind of mass. Hence, even if we have in the $P(\phi)_2$ field model a control parameter, λ, that serves an analogous role as does $\beta = 1/kT$ in the Weiss–Heisenberg model and a double-dip curve when $\lambda < \lambda_c(\lambda_c = 0)$ that is characteristic of SSB, it is not clear what is physically responsible for the possibility of SSB in this case.

7. The nature of idealization

We have now seen that without taking the macroscopic limit a description of quantum SSB is impossible; but taking the limit is an act of idealization. What is idealization and what justifies the use of it?

Earman raises the issue of whether macroscopic limit is a justified act of idealization when he writes,

> a sound principle of interpretation would seem to be that no effect can be counted as a genuine physical effect if it disappears when the idealizations are removed. (Earman, 2003a, p. 21)

Since the removal of the macroscopic limit removes the (theoretical) possibility of SSB (see the no-go result in Section 2), we may have to disqualify such an idealization unless we can, *per impossibili*, deny the existence of SSB. What may have justified Earman's claim is perhaps that idealization has been characterized, not without initial plausibility, as an act of *neglecting the negligible*. Since removing or reintroducing the negligible should not affect the existence or absence of any genuine physical effect, the above principle appears sound.

However, idealization should not be understood as nothing more than an act of neglecting the negligible (cf. Liu, 1999). On the contrary, its more important function in theory construction is to help discover (or create)—via introducing appropriate predicates—*qualitatively* distinct properties or kinds out of ones that differ only quantitatively, more or less, prior to the idealization (cf. Batterman, 2002). And as such, the principle Earman invoked above does not seem to hold; for should we not regard qualitatively distinct properties or states as *genuinely* different *physically*, which will not be so if the idealizations are removed? One may argue for staying with finite systems but giving up on the idea that there is a qualitative difference between the symmetry-preserving state and the symmetry-breaking states, e.g. between the paramagnetic phase and the ferromagnetic phase. In other words, such qualitatively different properties are nothing more than theoretical artifices (cf. Callender, 2001).

We find this option unattractive. The difference between two phases, e.g. solid and liquid, is better captured by a singularity. Even before Gibbs, Maxwell had complemented the van der Waals model of real fluids by a construction, known as the 'Maxwell plateau,' the effect of which is to make the isotherms only piecewise analytic. Should we regard this idealization as no more than an ad hoc construction? The construction was not only supported later by entirely rigorous mathematical arguments, it was also, first by Andrews, observed by successively cleaner and more precise experiments that the isotherms are better described as piece-wise analytic rather than as having arbitrarily sharp corners.

However, one may think that the above option alone is consistent with a *minimalist* commitment in ontology, which says roughly that everything supervenes on the non-relational properties of the elementary parts and processes of the physical reality, and for these all differences are quantitative ones among their configurations.[16] But this is simply not true. Using such idealization procedures as taking the macroscopic limit to let physically distinct states or properties *emerge* from microscopic parts and processes is fully consistent with the same ontological commitment. It is perhaps not consistent with an *eliminativist* position regarding the emergent states or properties in the macroscopic realm, which holds that the results of such idealizations are predicates that do not refer to anything, that are fictitious, and that should be eliminated in the 'final' version of science. The broken symmetry in question is *not reducible* to the configurations of the microscopic parts of any *finite* systems; but it should *supervene* on them in the sense that for any two systems that have the exactly duplicates of parts and configurations, both will have the same SSB in them because both will behave identically in the limit. In other words, the result of the macroscopic limit is determined by the non-relational properties of parts of the finite system in question.

Moreover, taking the macroscopic limit is no more radical or implausible than taking space and/or time as continua; without these idealizations, the usual real analysis would not be applicable and hence many physical situations can neither be rigorously described nor inferred. It is only one of a great variety of methods physicists and other scientists use to deal with asymptotic behavior (see Batterman, 2002, pp. 122–123). Nevertheless, one may still ask whether we are justified to believe in them just because scientists have used them to obtain nice results so far. We obviously do not know how to convince a total skeptic on this matter; would Bishop Berkeley, who accused the Newtonians of using such monstrosities as differentials of the first, second, etc. orders as believing in 'the ghost of departed quantities' and the ghost of the ghost of ... etc., be convinced of the legitimacy of analysis, were he alive today (cf. Berkeley, 1734)? But we do have an answer to the question: why should we take the limit seriously when we know that the systems are finite and all actual experimental results are only approximate? First, we have the rigorous $\varepsilon - \delta_\varepsilon$ definition of limit, whose essence is not 'infinity' but 'asymptoticity' in the following sense:

$$f = \lim_{n \to \infty} f_n$$

means that for any $\varepsilon > 0$, there is a $\delta_\varepsilon > 0$ such that $|f - f_n| < \varepsilon$ for all $n > N_\varepsilon = \delta_\varepsilon^{-1}$. Then second, to say that f—a possible property of a large but finite system S—is instantiated in S is to say that for any neighborhood around f, a sequence of

[16]More precisely and realistically, one may say that only 'macroscopic properties' as values of a set of observables describing a system, e.g. a magnet or a beam of light, can be qualitatively distinct. Microscopic properties are not qualitatively distinct. For instance, with an SSB as a phase transition; only the macroscopic properties on the two sides of the transition are qualitatively different. The corresponding microscopic properties of the two sides differ only quantitatively and the transition may be regarded as continuous. And ontologically there really are no 'macroscopic properties.'

approximating measurements will eventually (after say the Nth in the sequence) produce results that all remain in the neighborhood; most importantly, this is so for neighborhoods of all sizes, i.e. for *all* $\varepsilon > 0$. Another system, S′, that is indistinguishable from S by finite sets of measurements may have radically different asymptotic behavior—e.g. take a smaller ε' and a corresponding N' some or all of the $f_{N'+1}$ ..., may lie outside the neighborhood—and therefore may be a radically different system. In most cases of this type, the properties at the ideal limit supervene on the corresponding ones of the actual systems, but there is no way of telling the difference (this is not merely an epistemic limitation) without taking the limit.

8. Conclusion

Even though a study of classical SSB systems is not sufficient for an understanding of SSB in general (for quantum SSB has its own unique features), it is neither necessary nor advisable to tackle the recondite cases in QFT; or so we have argued. Theories and models of SSB in QSM provide just the right material for the purpose. The discussion of two ‘no-go’ theorems for finite systems and Section 3 provides an answer to why quantum SSBs only occur in infinite systems (question 1 in Section 1). We discuss an account of what SSB is for quantum thermo-systems. We then give two concrete models to show that the account is indeed adequate. A connection is established between the decompositional account of SSB, which originates from QSM, and the representational account, which has been conceived with QFT in mind, and from this connection one can understand what is otherwise puzzling, namely, why SSB should be characterized as a symmetry that is not unitarily implementable or as having vacuum states (in QFT) that belong to unitarily inequivalent representations (an answer to questions 2 and 3). In Section 6 we give an account of what it means to physically explain a quantum SSB (question 4) and in Section 7 of how we should think about idealization so that taking macroscopic limit is fully justified.

Appendix A. KMS for the neophyte

Let H be a self-adjoint operator denoting the Hamiltonian, and assume that $\forall \beta > 0$ the partition function $Z = Tr \exp(-\beta H)$ is finite; let $\phi : A \in \mathrm{B(H)} \mapsto \langle \phi; A \rangle \in \mathbb{R}$ be the canonical equilibrium state defined by $\langle \phi; A \rangle = Tr \rho A$ with $\rho = Z^{-1} \exp(-\beta H)$. The evolution is defined $\forall t \in \mathbb{R}$ and $\forall A \in \mathrm{B(H)}$ by $\tau_t[A] = \exp(\mathrm{i}Ht) A \exp(-\mathrm{i}Ht)$. Note that the state ϕ is time invariant and faithful.

Alternatively, consider the following computation, which uses only the invariance of the trace under cyclic permutation, and that $\exp(X)\exp(Y) = \exp(X + Y)$, for any pair of $X, Y \in \mathrm{B(H)}$ that commute with each other:

$$\begin{aligned} &\mathrm{Z}^{-1}\, Tr \exp(-\beta H) \exp(\mathrm{i}Ht) X \exp(-\mathrm{i}Ht) Y \\ &\quad = \mathrm{Z}^{-1}\, Tr \exp(-\beta H)\, Y \exp\{\mathrm{i}(t + \mathrm{i}\beta)H\} X \exp\{-\mathrm{i}(t + \mathrm{i}\beta)H\} \end{aligned}$$

that is,

$$Tr\,\rho\tau_t[X]Y = Tr\,\rho Y\tau_{t+i\beta}[X]. \tag{A.1}$$

This is the KMS condition in its most naive form. It is essential to note that it characterizes ρ completely. Indeed, let v be any density matrix and suppose that the normal state $\psi : A \in \mathrm{B(H)} \mapsto Tr\,vA$ satisfies (A.1). In the case that $Y = I$, the condition reads: $Tr\,v\alpha_t[X] = Tr\,v\alpha_{t+i\beta}[X]$. As this relation holds for $X \in \mathrm{B(H)}$, it entails

$$\exp(-iHt)v\exp(iHt) = \exp\{-iH(t+i\beta)\}v\exp\{iH(t+i\beta)\}$$

i.e. $\exp(-\beta H)v = v\exp(-\beta H)$ and thus v commutes with H. Consequently ψ is a stationary state, i.e. $\forall t \in \mathbb{R} : \psi \circ \tau_t = \psi$; more generally, this can be seen as a consequence of the Liouville theorem which says that any analytic function bounded over $\mathbb{C}$ must be constant.

Now let $t = 0$ and ρ be substituted by v in (1), we have,

$$Tr\,vXY = Tr\,v\{Y\exp(-\beta H)X\exp(\beta H)\}.$$

Since $Y \in \mathrm{B(H)}$ is arbitrarily chosen, the above entails: $\{\exp(\beta H)v\}X = X\{\exp(\beta H)v\}$, i.e. every $X \in \mathrm{B(H)}$ commutes with $\{\exp(\beta H)v\}$, which in turn entails that this operator is a multiple of the identity, or equivalently that $v = a\exp(-\beta H)$, where a is a constant that, when normalized ($Tr\,v = 1$) is Z^{-1}. Hence, the density matrix v (w.r.t. ψ) is none other than the canonical equilibrium density matrix, just as ρ is (w.r.t. ϕ). Hence, for finite systems, to say that a state satisfies the KMS condition for $\beta > 0$ is equivalent to saying that it is the canonical equilibrium state for β.

Further remarks:

1. The KMS condition can be naturally generalized to infinitely extended systems (see Section 3).
2. For infinite systems, the KMS condition can be viewed unambiguously as a stability condition (cf. e.g. Sewell, 2002 or Emch & Liu, 2002).
3. For infinite systems, there may be more than one state that satisfy the KMS condition.
4. In any infinite systems the convex set of all states satisfying the KMS condition for the same temperature is a simplex, i.e. the KMS states decompose uniquely into a set of extremal KMS components. Physically, that KMS states are taken as equilibrium states supports the pragmatic expectation that equilibrium states decompose uniquely into a set of pure thermodynamical phases (see Appendix B below).

Appendix B. Central decomposition

To illustrate the idea at the core of the general theorem on central decomposition, we consider the following simpler result, which in itself is all that is needed for much of the arguments in the main body of the paper.

Scholium 1. *Suppose* $\alpha \in Aut(\mathrm{A})$ *and* $\phi \circ \alpha = \phi$, *which implies that* α *is unitarily implementable in the GNS representation* $\pi(\mathrm{A})$ *associated with* ϕ. *Then,*

1. $\forall N \in \mathrm{N} \equiv \pi(A)''$, $\alpha[N] \equiv U^*\pi(N)U$ *extends* α *to an automorphism on* N, *and* $\langle \phi; N \rangle = (\Phi, N\Phi)$ *extends* ϕ *to a normal state on* N.
2. $\mathrm{Z} = \mathrm{N} \cap \mathrm{N}'$ *is stable under* α (i.e. $\forall \mathrm{Z} \in \mathrm{Z}, \alpha[\mathrm{Z}] \in \mathrm{Z}$), *and the restriction of* α *on* Z, *is also an automorphism.*
3. *If* ϕ *is KMS w.r.t. evolution* τ, *then the extention to* N *is again KMS w.r.t. the extention of* τ.

Scholium 2. *Let* $\mathrm{N} \subset \mathrm{B(H)}$ *be a von Neumann algebra admitting a cyclic and separating vector* Φ *such that its norm* $|\Phi| = 1$. *Assume further that the spectrum SpZ of the center* $\mathrm{Z} = \mathrm{N} \cap \mathrm{N}'$ *is discrete. Then the normal state,* $\phi : N \in \mathrm{N} \mapsto (\Phi, N\Phi)$, *can be written as a convex combination,* $\phi = \sum_k \lambda_k \phi_k$, *of primary states on N.*

Proof. The discreteness of the spectrum of the center of the algebra means that there exists a partition of the identity, $\{P_k\}$, into mutually orthogonal projectors such that $\mathrm{Z} = \{\mathrm{Z} = \sum_k z_k P_k | z_k \in \mathbb{C}\}$; in particular $\sum_k P_k = 1$. Correspondingly $\lambda_k \equiv \langle \phi; P_k \rangle \mathrm{H} = \oplus \mathrm{H}_k$, where $\mathrm{H}_k = P_k \mathrm{H}$ is the closed subspace $\overline{\mathrm{N}P_k\Phi}$. Let $\mathrm{N}_k \equiv \pi_k(N)$ be the von Neumann algebra obtained as the restriction of N to the N-stable subspace H_k. Let now; and note that since Φ is separating for N and therefore ϕ is faithful, $\lambda_k \neq 0$. Then

$$\phi = \sum_k \lambda_k \phi_k \quad \text{with } \lambda_k > 0, \quad \sum_k \lambda_k = 1$$

and

$$\phi_k : N \in \mathrm{N} \mapsto \langle \phi_k; N \rangle = \frac{\langle \phi; P_k N P_k \rangle}{\langle \phi; P_k \rangle} \in \mathbb{C}. \tag{B.1}$$

ϕ_k are positive, linear complex valued function on N and are normalized by $\langle \phi_k; I \rangle = 1$; i.e. they are states on N. Moreover, the above equality is a particular case of the following equality:

$$\forall N \in N \text{ and } \forall Z = \sum_j z_j P_j \in Z : \langle \phi; ZN \rangle = \sum_k z_k \lambda_k \langle \phi_k; N \rangle. \tag{B.2}$$

Upon noticing that $\{\pi_k, \mathrm{H}_k, \Phi_k \equiv P_k \Phi\}$ is the GNS triple associated with ϕ_k, the fact that $\forall Z = \sum_j z_j P_j \in Z : \pi_k(Z) = z_k P_k$ entails that the states ϕ_k are primary. □

Scholium 3. *If* ϕ *is KMS, the central decomposition in Scholium 2 provides a unique decomposition into extremal KMS states.*

For the extension of these results to not necessarily discrete spectrum of the center, see Emch, 1972, p. 215, and this requires replacing direct sums by direct integrals and the concomitant measure-theoretical trimmings.

160 C. Liu, G.G. Emch / Studies in History and Philosophy of Modern Physics 36 (2005) 137–163

Appendix C. Equivalence of representations

Three notions of equivalence between representations of a C*-algebra are pertinent to this paper: (1) unitary equivalence, (2) quasi-equivalence, and (3) weak equivalence, which we briefly explain below; for details see Emch (1972) or Kadison and Ringrose (1986). These notions satisfy the following:

$$(1) \models (2) \models (3).$$

The notion of quasi-equivalence was discussed already in Mackey (1953) and Haag and Kastler (1964) argue, on the basis of a theorem by Fell (1960), that weak-equivalence ought to be identified as 'physical' equivalence. In a broader philosophical context, this interpretation was re-examined and endorsed by Clifton and Halvorson (2001); see also Ruetsche (2002).

Let $\pi_i : \mathrm{A} \to \mathrm{B(H)}_i$ $(i = 1, 2)$ be any two non-zero representations of a C*-algebra A; π_i is said to be primary whenever $\pi_i(\mathrm{A})'' \cap \pi_i(\mathrm{A})' = CI$, and it is said to be irreducible whenever $\pi_i(\mathrm{A})' = CI$. $\mathrm{N}_i \equiv \pi_i(\mathrm{A})''$ is the von Neumann algebra obtained as the weak-operator closure of $\pi_i(\mathrm{A})$ in $\mathrm{B(H)_i}$. $Ker\,\pi_i \equiv \{A \in A | \pi_i(A) = 0\}$ denotes the kernel of the representation π_i. A C*-algebra is said to be simple whenever it admits no closed two-sided proper ideals; and then, in particular, $Ker\,\pi_i = \{0\}$.

Definition 1. π_1 and π_2 are said to be weakly equivalent whenever $Ker\,\pi_1 = Ker\,\pi_2$.

This condition can be variously reformulated in terms of the states associated with π_i, see e.g. Emch (1972) [Theorem II.1.7]; for the purpose of QSM, the most directly relevant reformulation may be the following result.

Scholium 1. *π_1 and π_2 are weakly equivalent whenever every density matrix associated with one representation can be approximated, point-wise on* A, *but arbitrarily closely, by a net of density matrices associated with the other representation.*

Definition 2. π_1 and π_2 are said to be quasi-equivalent whenever there exists a *-isomorphism $\alpha : \mathrm{N}_1 \to \mathrm{N}_2$ such that $\alpha[\pi_1(A)] = \pi_2(A)\ \forall A \in \mathrm{A}$.

Scholium 2. *π_1 and π_2 are quasi-equivalent whenever both of the following two conditions are satisfied*: (i) *π_1 and π_2 are weakly equivalent [and thus π_2 there exist a natural *-isomorphism $\alpha_0 : \pi_1(\mathrm{A}) \to \pi_2(\mathrm{A})$]; and* (ii) *$\alpha_0$ extends to a *-isomorphism $\alpha : \mathrm{N}_1 \to \mathrm{N}_2$.*

Definition 3. π_1 and are said to be unitarily equivalent whenever there exists a unitary operator $U : \mathrm{H}_1 \to \mathrm{H}_2$ such that $U\pi_2(A)U^{-1} = \pi_1(A), \forall A \in \mathrm{A}$.

At the opposite extreme sits the following notion.

Definition 4. π_1 and π_2 are said to be disjoint whenever every subrepresentation of π_1 is not unitarily equivalent to any subrepresentation of π_2, and vice versa.

Scholium 3. *π_1 and π_2 are disjoint whenever they have no quasi-equivalent subrepresentations; conversely π_1 and π_2 are quasi-equivalent iff π_1 has no subrepresentation disjoint from π_2 and vice versa. Furthermore, two primary representations π_1 and π_2 are either quasi-equivalent or disjoint. Moreover, two irreducible representations are weakly equivalent iff they are unitarily equivalent, and thus iff they are quasi-equivalent. Finally, all representations of a simple* C**-algebra are weakly equivalent.*

Scholium 4. (Takesaki, 1970). *Two representations π_1 and π_2 that correspond to different temperatures are disjoint.*

References

Anderson, P. W. (1963). Plasmas, gauge invariance, and mass. *Physical Review, 130*, 439–442.

Arageorgis, A., Earman, J., & Ruetsche, L. (2002). Weyling the time away: the non-unitary implementability of quantum field dynamics on curved spacetime. *Studies in History and Philosophy of Modern Physics, 33B*, 151–184.

Batterman, R. W. (2002). *The devil in the details: asymptotic reasoning in explanation, reduction, and emergence*. Oxford: Oxford University Press.

Batterman, R.W. (2004). Response to Belot's 'Whose devil? Which details?' (preprint).

Belot, G. (2004). Whose devil? Which details? (preprint).

Berkeley, G. (1734). *The analyst (or a discourse addressed to an infidel mathematician)*. Reprinted In G. Berkeley, *The Works*, III, B. Bell & Sons, 1898 (pp. 53–89).

Bianchi, A., Contucci, P., & Giardina, C. (2004). Thermodynamical limit for mean-field spin models. *Mathematical Physics Electronic Journal, 9*(6) 15pp. arXiv: math-ph/0311017 v2 8 Jan. 2004.

Callender, C. (2001). Taking thermodynamics (too) seriously. *Studies in History and Philosophy of Modern Physics, 32B*, 539–553.

Cartwright, N. (1999). Models and the limits of theory: quantum Hamiltonians and the BCS models of superconductivity. In M. Morgan, & M. Morrison (Eds.), *Models as mediators* (pp. 241–281). Cambridge: Cambridge University Press.

Clifton, R., & Halvorson, H. (2001). Are Rindler quanta real? Inequivalent particle concepts in quantum field theory. *British Journal for the Philosophy of Science, 52*, 417–470.

Coleman, S. (1975). An introduction to spontaneous symmetry breakdown and gauge fields. In A. Zichichi (Ed.), *Laws of hadronic matter* (pp. 138–215). New York: Academic Press.

Dubin, D. A. (1974). *Solvable models in algebraic statistical mechanics*. Oxford: Clarendon Press.

Earman, J. (2003a). Spontaneous symmetry breaking for philosophers (preprint).

Earman, J. (2003b). Rough guide to spontaneous symmetry breaking. In K. Brading, & E. Castellani (Eds.), *Symmetries in physics: philosophical reflections* (pp. 335–346). Cambridge: Cambridge University Press.

Emch, G. G. (1972). *Algebraic methods in statistical mechanics and quantum field theory*. New York: Wiley-Interscience.

Emch, G. G. (1977). An algebraic approach for spontaneous symmetry breaking in quantum statistical mechanics. In P. Kramer, & M. D. Cin (Eds.), *Groups, systems and many-body physics* (pp. 246–284). Braunschweig/Wiesbaden: Friedr. Vieweg & Son.

Emch, G. G., & Guenin, M. (1966). Gauge invariant formulation of the BCS model. *Journal of Mathematical Physics, 7*, 915–921.

Emch, G. G., & Liu, C. (2002). *The logic of thermostatistical physics*. Berlin: Springer.

Emch, G. G., & Knops, H. J. F. (1970). Pure thermodynamical phases as extremal KMS states. *Journal of Mathematical Physics, 11*, 3008–3018.
Emch, G. G., Knops, H. J. F., & Verboven, E. J. (1970). Breaking of Euclidean symmetry with an application to the theory of crystallization. *Journal of Mathematical Physics, 11*, 1655–1668.
Ezawa, H. (1964). The representation of canonical variables as the limit of infinite space volume: the case of the BCS model. *Journal of Mathematical Physics, 5*, 1078–1090.
Fell, J. M. J. (1960). The dual spaces of C*-algebras. *Transactions of the American Mathematical Society, 94*, 365–403.
Glimm, J., & Jaffe, A. (1981). *Quantum physics: a functional integral point of view*. New York: Springer.
Guralnik, G. S., Hagen, C. R., & Kibble, T. W. B. (1968). Broken symmetries and the Goldstone Theorem. In R. L. Cool, & R. E. Marshak (Eds.), *Advances in particle physics* (pp. 567–708). New York: Wiley Interscience Publishers.
Haag, R. (1962). The mathematical structure of the Bardee–Cooper–Schrieffer Model. *Nuovo Cimento, 25*, 287–299.
Haag, R. (1996). *Local quantum physics: fields, particles, algebras* (2nd ed). Berlin: Springer.
Haag, R., & Kastler, D. (1964). An algebraic approach to quantum field theory. *Journal of Mathematical Physics, 5*, 848–861.
Kadison, R. V., & Ringrose, J. (1983). *Fundamentals of the theory of operator algebras*, Vol. I. New York: Academic Press.
Kadison, R. V., & Ringrose, J. (1986). *Fundamentals of the theory of operator algebras, Vol. II: advanced theory*. New York: Academic Press.
Liu, C. (1999). Approximation, idealization, and the laws of nature. *Synthese, 118*, 229–256.
Liu, C. (2001). Infinite systems in SM explanations: thermodynamic limit, renormalization (semi-) groups, and irreversibility. *Philosophy of Science, 68*(S), S325–S344.
Liu, C. (2003). Classical spontaneous symmetry breaking. *Philosophy of Science, 70*(S), S1219–S1232.
Mackey, G. W. (1953). Induced representations of local compact groups II. *Annals in Mathematics, 58*, 193–221; see also, The theory of unitary group representations. Mimeographed lecture notes, 1955; later published with an Appendix by University of Chicago Press, 1976.
Moriyasu, K. (1983). *An elementary primer for gauge theory*. Singapore: World Scientific.
Ojima, I. (2003). A unified scheme for generalized sectors based on selection criteria: order parameters of symmetries and of thermality and physical meaning of disjointness. *Open Systems & Information Dynamics, 10*, 235–279.
Onsager, L. (1944). Crystal statistics: I. A two-dimensional model with an order–disorder transition. *Physical Review, 65*, 117–149.
Peierls, R. E. (1936). On Ising's model of ferromagnetism. *Proceedings of the Cambridge Philosophical Society, 32*, 477–481.
Pucz, W., & Woronowicz, S. L. (1978). Passive states and KMS states for general quantum systems. *Communications in Mathematical Physics, 58*, 273–290.
Ruetsche, L. (2002). Interpreting quantum field theory. *Philosophy of Science, 69*, 348–378.
Schrieffer, J. R. (1964). *Superconductivity*. New York: Benjamin.
Schultz, T. D., Mattis, D. C., & Lieb, E. H. (1964). Two-dimensional Ising model as a soluble problem of many fermions. *Reviews of Modern Physics, 36*, 856–871.
Sewell, G. L. (2002). *Quantum mechanics and its emergent macrophysics*. Princeton: Princeton University Press.
Simon, B. (1974). *The $P(\phi)_2$ Euclidean quantum theory*. Princeton: Princeton University Press.
Streater, R. F. (1965). Spontaneous breakdown of symmetry in axiomatic theory. *Proceedings of the Royal Society of London, Series A, 287*, 510–518.
Streater, R. F. (1967). The Heisenberg ferromagnet as a quantum field theory. *Communications in Mathematical Physics, 6*, 233–247.
Takesaki, M. (1970). Disjointness of the KMS states of different temperatures. *Communications in Mathematical Physics, 17*, 33–41.

Thirring, W. (1969). The mathematical structure of the BCS model and related models. In A. Crus, & T. W. Peist (Eds.), *The many-body problem* (pp. 125–160). London: Plenum.

Whitten-Wolfe, B., & Emch, G. G. (1976). A mechanical quantum measuring process. *Helvetica Physica Acta*, *49*, 45–55.

Wick, G. C., Wightman, A. S., & Wigner, E. P. (1952). Intrinsic parity of elementary particles. *Physical Review*, *85*, 101–105.

International Studies in the Philosophy of Science
Vol. 19, No. 1, March 2005, pp. 85–100

Book Reviews

Symmetries in Physics: Philosophical Reflections
KATHERINE BRADING and ELENA CASTELLANI (Eds.)
Cambridge, Cambridge University Press, 2003
445 pp., ISBN 0521821371 £ 65.00, US$ 100.00 (hardback)

This expert collection of articles on many aspects of symmetries in physics is much needed and long overdue. The last collection of this kind was published 16 years ago and contains contributions from physicists and historians only. Things have changed: in this volume; all but a couple of original contributions are given by philosophers of physics. The collection is divided into four parts: Part I: continuous symmetries, Part II: discrete symmetries, Part III: symmetry breaking, and Part IV: general interpretive issues. Each part begins with a few excerpts from the classic texts, such as from Weyl, Wigner, Leibniz, Kant and Curie. The excerpts are parsimonious; some excessively so – had I not been already familiar with the originals, I would not have understood what as being said there. Since the excerpts are indeed from the classics that readers of this volume should know, I suggest that the originals be carefully read before one reads the articles.

The central philosophical issue in Part I regards the meaning and significance of the gauge principle and the 'gauge argument' that stems from it. I am pleased to see an exerted effort to figure out the connection between the covariance principle in the general theory of relativity (GTR) and the gauge principle in quantum gauge field theory (GFT). Ismael and van Fraassen's article in Part IV should be read with these articles. Here is what I take to be the key challenge: how can we reconcile the apparent disparity between gauge symmetries as expressions of surplus mathematical structures in a theory and gauge symmetries as responsible for the presence of gauge fields? Part II is less focused. Its intended focus, as the editors explain in the introduction, is on the permutational and the reflectional symmetry among elementary particles. The former creates the issue of particle identity and the latter – in the form of parity violation – of whether space-time is relational or substantival. However, the reader is left to figure out what the connection is between these arguments and the ones by Leibniz and Kant that are excerpted in front. The subject of Part III – symmetry breaking – has come to philosophers' attention only recently. Of the three contributions from philosophers, two are devoted exclusively to the discussion of spontaneous symmetry breaking. The contributions are mostly aimed at clarifying concepts and casting about, trying to catch

ISSN 0269–8595 (print)/ISSN 1469–9281 (online) © 2005 Inter-University Foundation
DOI: 10.1080/02698590500051183

meaty philosophical issues. The focus of Part IV, suggested by the excerpts from Wigner, appears to be on the principle of invariance of laws of nature, the idea that a lawlike thing would not be a law if it did not obey the principle; but it is not. It is mostly about the relationship between symmetry and structural redundancy in theories. (For instance, none of the contributions in this part confronts the philosophical question why laws of physics must be Lorentz invariant.)

The best place to begin in Part I, in my opinion, is Michael Redhead's article, which lays out in simple and clear terms what gauge symmetries are, how they are related to surplus mathematical structures, and what the interpretive challenges are. If – a big 'if' that is investigated in detail by Ismael and van Fraassen in Part IV – a distinction between the physical and the mathematical structure (P and M) of a system can be made, there is a way of construing the surplus structure in M – the distinct states in M that correspond to the same state in P – as related to a gauge symmetry. A clear sense of why GTR is a GFT is concisely given there, but for a thorough discussion of it see the next article, by John Earman. Here lies a problem that requires some philosophical reflections. GTR can be counted as a GFT only if it is put in a canonical formulation. In this (aptly called (3+1)) formulation, the configuration variables are 3-geometries on a spatial slice at a given coordinate time. This formulation, however, renders the 'time evolution' of a general relativistic system a matter of gauge transformations, which *prima facie* means that 'no genuine physical magnitude (= gauge-invariant quantity) changes its value with time' (Earman, 152). Notice that while the indeterminism problem from the 'hole argument' may be resolved if there exists a True Gauge, this problem cannot. It seems to me that before any effort is – and a great deal has been – spent in finding a way out, we should ask whether or not the problem is created by the canonical formulation of GTR. Such an apparently disastrous result may serve, in many other circumstances, as the premise of a *reductio* against the formulation. Unless the same problem is demonstrated in the original non-canonical formulation – the formulation used by physicists until the demand for quantising the gravitational field arose – the burden of proof rests on those who espouse the canonical formulation.

The central interpretive challenge arises mainly on two fronts, both covered by excellent discussions in Part I. It arises in the interpretation of Noether's theorems, which is given in a clear and careful treatment by Brading and Brown (for in-depth discussions on many aspects of the theorems see other works produced by them, jointly or separately). The challenge is this: how can conservation laws be derivable from symmetries, some if not all of which are supposed to represent the mathematical surplus of the theory in question? Given the tacit assumption no 'physics' in, then no 'physics' out, there must be something physically real or significant about the symmetries in order for them to 'produce' the conservation laws, which are supposed to be among the most physically real. B and B answer as follows:

> *The very fact that a global transformation does not lead to empirically distinct predictions is itself non-trivial.* In other words, the freedom in our descriptions is no 'mere' mathematical freedom – it is a consequence of a physically significant structural feature of the theory. (Brading and Brown, 98)

What kind of metaphysical commitment does this imply? If the elements of the 'space' that the symmetry renders as mathematical surplus are not real, then whatever non-trivial facts one can report about them, the symmetry itself has no physical 'oomph' (Martin, 53), at least not enough oomph to give rise to, say, the conservation of charge. I see two options here. One can be a substantivalist about such elements, or one can be a 'regularity' relationist about them (this comes to me from reading Huggett's second article in Part II, 285–8). Actual phase differences are physically real enough (shifts of interference patterns are accounted for by such differences), and therefore relationist spaces are metaphysically less suspect, and also one should regard conservation laws as regularities. The problem is that I am not sure that such a relationist position of any space can work out (more on this later).

The same challenge arises between GTR and GFT: namely, on the one hand, history tells us that the latter is fashioned almost exactly out of the former, but on the other hand, if the general covariance principle (which is a gauge principle) in the former lacks physical significance (the Kretschmann thesis), how could the gauge principle in GFT introduce any gauge field? Christopher Martin (53–7) is pessimistic about a positive solution; but isn't such a solution pointed to, if not explicitly given, by John Norton in Part I? Norton's insight is that, very roughly, even if general covariance – the existence of a covariant (= gauge-invariant) formulation of a theory – is non-contingent, the formal properties of the theory in a given covariant formulation may still be highly contingent. To put this in concrete terms, the answer to our challenge seems to be this: it is the 'gauging' of a globally gauge-invariant theory that 'introduces a field'. By gauging I mean localising the gauge transformations, and by introducing a field I mean deriving in a given theoretical framework statements containing variables that range over values of the strength of that field, where *the derived has the same metaphysical status as the deriving*. What is physically real about the gauge symmetries – even if they represent mathematical redundancies – is that such redundancies exist in different spatio-temporal locations independently of one another. *In the right gauge-invariant formulation*, this fact is accountable only by the existence of a field.

I am quickly running out of space, so let me be selective and brief. Antigone Nounou's proposal for an alternative explanation of the AB effect is interesting. However, I am not yet clear what type of explanation she wants hers to be. Using the topology of a fibre bundle to explain the AB effect can be seen as giving a causal explanation or a non-causal one. To opt for one or the other entails different metaphysical commitments to the relevant potentials and bundles. If she regards hers as a causal explanation, I do not see how it differs from the 'holonomy' explanation; but if it is meant to be non-causal, more needs to be said: e.g., should there also be a causal explanation? And in what sense is hers an explanation alternative to the previous three she criticises, since they are clearly meant to be causal (local or non-local)? Perhaps she thinks that no causal explanation is tenable, but that is not something easily demonstrated.

Regarding Nick Huggett's ingenious solution to the challenge of coming up with a relationist account of orientability, I have the following concern. Huggett's idea is to transplant the best-system regularity account for laws of nature (à la, e.g., Lewis) to this case, which results in a 'regularity' relationist account. The analogy in some sense is indeed a close one: on one side it is the regularity vs. realist account of laws, and on the other, the relationist vs. substantivalist (or realist) account of space(-time). While a law-regularitist regards a law as a special kind of regularity among actual events, a relationist regards a geometrical property as a relational property of actual objects or events. However, *prima facie*, there is this disanalogy: when criteria such as simplicity and empirical strength (or overall fit) are used to single out laws, they are used on the totality of actual regularities, but when analogous criteria are used to single out an embedding space, they are used not on the totality of spatial relations among actual objects and events but on alternative totalities, some of which contain spatial relations among unoccupied points. I do not yet know how important this disanalogy is; but the following has me worried. The key virtue of the best-system regularity theory of laws is that one need not go beyond the actual to determine which regularity is or is not a law. This has to be kept, I think, in any transplantation in order for the result to be called a 'regularity' account. Huggett's regularity account of relational space(-time) loses this virtue because one has to assume and look at alternative possible relational structures in order to apply, for instance, his simplicity criterion, to pick out the desirable one. If Lewis had to use alternative possible sets of regularities (one set for a different possible world, let us say), his wouldn't be properly called a regularity account of laws of nature, would it? However, I am not yet clear about what kind of consequences such a disanalogy has and whether some of them are incompatible with relationism.

There are many more worthy points in this volume than I have space to discuss. In general, I found the collection timely, informative and stimulating. Many of the contributions are deft and elegant in technical treatment and clear and thoughtful in philosophical analysis. I recommend this volume to anyone who is seriously interested in philosophical issues on symmetries in physics.

CHUANG LIU
Department of Philosophy, University of Florida
330 Griffin-Floyd Hall, P.O. Box 118545
Gainesville, FL 32611-8545
email: cliu@phil.ufl.edu

Classical Spontaneous Symmetry Breaking

Chuang Liu†‡

This paper aims at answering the simple question, "What is spontaneous symmetry breaking (SSB) in classical systems?" I attempt to do this by analyzing from a philosophical perspective a simple classical model which exhibits some of the main features of SSB. Related questions include: What does it mean to say that a symmetry is spontaneously broken? Is it broken without any causes, or is the symmetry not broken but merely hidden? Is the principle, "no asymmetry in, no asymmetry out," violated by SSB? What really distinguishes SSB from the usual types of symmetry breaking?

1. Introduction. When people today use the phrase "spontaneous symmetry breaking" or the initialism "SSB," they usually refer to the technical concept in condensed matter physics (cf. Anderson 1997) and high-energy physics (cf. Coleman 1975, Weinberg 1996). The theories of SSB in those fields are complex and recondite, and so it could be difficult to see clearly what SSB really means. By reading articles on the subject in the major journals of physics and mathematical physics, one does not usually come away with a clear understanding of such simple questions as why the symmetry breakings are *spontaneous* or even what the word "spontaneous" really means in those contexts. What is the difference between SSB and the usual types of symmetry breaking? We sometimes hear in the literature the claim that "SSB" is a misnomer (cf. Coleman 1975, 142), and the correct phrase should be "hidden symmetry." What does the latter mean and is the claim true? We also hear talks about the breakings of symmetry in SSB being "arbitrary" or "a matter of a free choice" or "a

†To contact the author, please write: Department of Philosophy, University of Florida, 330 Griffin-Floyd Hall, P. O. Box 118545, Gainesville, FL 32611-8545; e-mail: cliu@phil.ufl.edu.

‡I would like to thank Gérard Emch for comments on an earlier draft of this paper. I would also like to thank Jeremy Butterfield, Craig Callender, Nick Huggett, John Stachel, and Sang Wook Yi for questions and comments at the PSA 2002 session at which I read a short version of this paper.

Philosophy of Science, 70 (December 2003) pp. 1219–1232. 0031-8248/2003/7005-0030$10.00

matter of pure chance"(cf. Moriyasu 1983, 99; Brout and Englert 1998, for Salam's metaphor). Whether given as literal or metaphorical, how should such talks be interpreted? Moreover, we often hear the philosophically inclined announce that in SSB all doubts about the existence of emergent properties can finally be put to rest (cf. Anderson 1997, 263–266). Is this justified?

Fortunately we can answer most of the above questions concerning the nature of SSB by examining some simple mechanical models (see *inter alias,* Greenberger 1978, Sivardière 1983, Drugowich de Felício and Hipólito 1985). The SSB in such models can be shown to have the main features of SSB.[1]

2. The Model. Imagine a circular metal hoop vertically suspended as shown in Figure 1. It rotates freely, and on it is frictionlessly threaded a bead. The only forces acting on the bead are gravity, *mg,* and the normal force, *N,* the bead receives from the hoop.

Question: when is the bead stationary on the hoop? In other words, what is (are) the equilibrium state(s) of this rotating system? The bead may be stationary whenever the forces are balanced out, which means,

In z-direction:

$$N_z - mg = N\cos\theta - mg = 0, \tag{1}$$

In y-direction:

$$N_y - ma = (N - m\omega^2 R)\sin\theta = 0 \tag{2}$$

where, a is the linear acceleration of the bead, which is

$$a = \omega^2 r = \omega^2 R \sin\theta.$$

When (1) is true, $N = mg/\cos\theta$. Substitute this into (2) to eliminate N, we have

$$m(g/\cos\theta - \omega^2 R)\sin\theta = 0$$

1. For other related mechanical models in which spatial symmetries are spontaneously broken, see Alben 1972, Aravind 1987, Johnson 1997, Racz and Rujan 1975, Sivardière 1985, Sivardière 1997. Much more widely mentioned, but seldom explicitly demonstrated cases of classical SSB include Euler's rod, which buckles under a perfectly vertical load, and Poincaré's rotating "nebula" in self-gravitating equilibrium, which separates into a two-body system when a critical value of its angular velocity is passed.

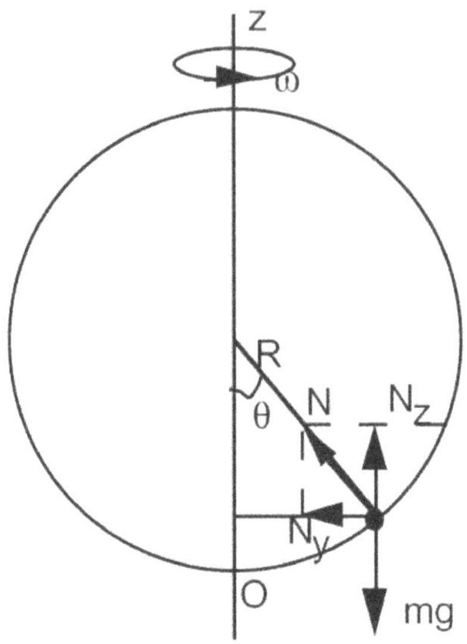

Figure 1. A bead with mass, $m \neq 0$, is free to move frictionless on a circular hoop of radius, R, which rotates around the z-axis with a variable w.

or

$$m\sin\theta(1 - \beta\cos\theta) = 0 \qquad (3)$$

where $\beta = \omega^2 R/g$.

There are obviously two solutions to this equation:

i. $\theta = \theta_0 = 0$ such that $\sin\theta_0 = 0$, which therefore satisfies (3).
ii. $\theta = \pm\,\theta_1 \neq 0$ such that $\cos\theta_1 = 1/\beta$.

Solution (**ii**), θ_1, has a further property: $\cos\theta_1$ increases when ω decreases (with R and g held constant), and yet $\cos\theta_1$ has an upper bound of 1. Hence, there is a *critical value,* $\omega = \omega_1$, where $\cos\theta_1 = 1 = g/\omega_1^2 R$, such that below this value solution (**ii**) makes no sense in physics. The standard way of dealing with such a situation in physics is to say that then solution (**i**) takes over, so that for all ω such that $0 \leq \omega \leq \omega_1$, $\theta = \theta_0 = 0$ is the solution at which the system is in equilibrium; and for those ω such that $\omega > \omega_1$, solution (**ii**)—$\cos\theta_1 = 1/\beta$—takes over so that all angles θ_1, where $0 < \theta_1 < \pi/2$, are possible stationary positions for the bead on the hoop.

Or alternatively, though perhaps less transparently, the same results can be reached from the Lagrangian (or Hamiltonian) formulation of this problem. Besides obtaining the two solutions, it also gives us the potential energy (or a dimensionless function of it) as a function of the angle, θ. Without going into the actual derivation (see Greenberger 1978 for details), let me summarize the main steps of the argument. The Lagrangian of the bead is of the usual form: L = (kinetic energy) − (potential energy). Since we are only concerned with the stationary situations, we only need to look at the potential energy term, which has two components: one due to the bead's weight and the other its centrifugal tendency. Writing everything out explicitly, we have the potential energy in the following form:

$$V = mgRU(\theta), \qquad (4)$$

where $U(\theta)$ is the dimensionless potential and it has the following form.

$$U = 2\sin^2(\theta/2)\left[1 - \beta\cos^2(\theta/2)\right], \qquad \beta = \omega^2 R/g.$$

Now, for the bead to be in equilibrium on the hoop at a give ω its potential must be an extremum, which means that

$$\partial U/\partial\theta = \sin\theta(1 - \beta\cos\theta) = 0. \qquad (5)$$

This equation, one should realize, is essentially the same as (3); and so we can derive the same two solutions as given earlier. There are two further benefits from this approach: an exact knowledge of the solutions' stability and a diagram—in which $U(\theta)$ is plotted against θ—that vividly shows the SSB in this model.

First, the solutions from (5) are stable if $\partial^2U/\partial^2\theta > 0$; and now $\partial^2U/\partial^2\theta = \cos\theta - \beta\cos^2\theta$. So,

iii. for $\theta = \theta_0 = 0$, we have $\partial^2U/\partial^2\theta = 1 - \beta$, which means that the bead is stable at this position for $\beta < 1$; and
iv. for $\theta = \pm\theta_1$ or $\cos\theta_1 = 1/\beta = g/\omega^2R$, we have $\partial^2U/\partial^2\theta = \beta - 1/\beta$, which means that the bead is stable at these positions for $\beta > 1$.

In other words, $\beta_c = 1$ or $\omega_1 = \sqrt{g/R}$ is a critical value where the stable equilibrium solution switches from $\theta = \theta_0 = 0$ to $\theta = \theta_1 > 0$, as shown in Figure 2; and the latter no longer preserves the reflectional symmetry[2] across the z-axis.

The structural properties of SSB in this model, as highlighted above, are present in all the better known models in previously mentioned fields. In condensed matter physics, models of ferromagnetism (e.g., the Heisenberg model), for instance, have temperature (analogous to β) as the controllable parameter, the crossing of whose critical value (the Curie temperature) brings about a breaking of the rotational symmetry that the systems'

2. One may argue that what is broken here is the rotational symmetry around the z-axis. To be safe, let us say that it is a breaking of a reflectional symmetry to an observer who is tied to any of the moving part of the ring.

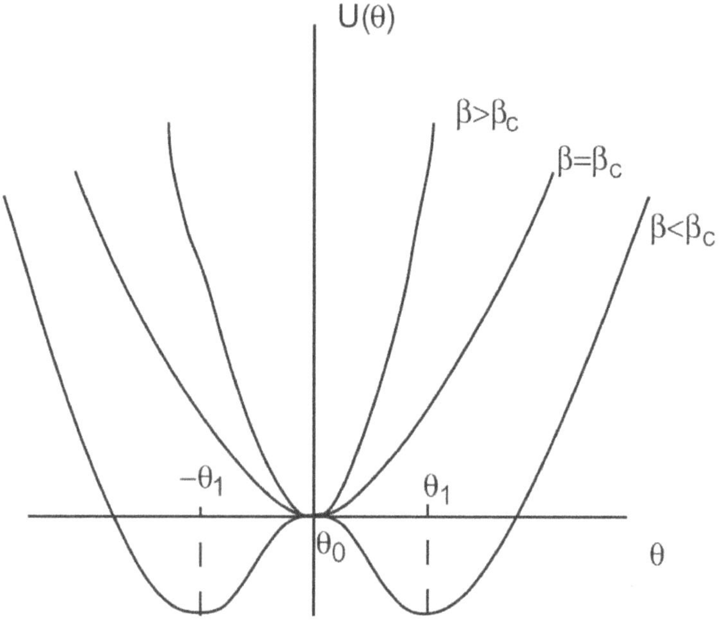

Figure 2. The dimensionless potential, $U(\theta)$, against θ. Note, $\beta = \omega^2 R/g$ and $\beta_c = \omega_1^2 R/g = 1$.

Lagrangians preserve, which results in ground states—the lowest energy states—of spontaneous magnetization (analogous to the departure of the bead from O). And in high-energy physics, the models of gauge fields (e.g., the SU(2) gauge field) have an effective mass term, μ^2, as their parameter, which has a critical value that separates the symmetrical from the symmetry-breaking region. The symmetries broken in these cases are the respective gauge groups (e.g., SU(2)) that their Lagrangians preserve. In fact, Figure 2 can be used to illustrate the SSB of a scalar gauge field with a simple replacement: $\theta \rightarrow \phi$ and $\beta \rightarrow \mu^2$, where ϕ represents the field and $\mu_c^2 = 0$. There are even analogs of equation (5) (or (3)) in gauge fields whose solutions produce the bifurcations (cf. Anderson 1997, Strocchi 1985, and Quigg 1983).

3. The Meaning of SSB. I shall begin the dissuasion with the simplest question:

1. Why is the symmetry breaking in the model not a symmetry breaking of the usual type?

A common answer in the literature goes as follows: An SSB occurs in a system when its Lagrangian or dynamical law has certain symmetries which the ground state solutions do not preserve (cf. e.g., Anderson 1997, 266; Coleman 1975, 141–142). Let's see what this claim means in our model. The independent variables in the model which determine the bead's states are, in polar coordinates, $\theta, \dot{\theta}$; $\varphi, \dot{\varphi} = \omega$, which are, respectively, the vertical angle from the z-axis, the velocity; the horizontal angle of rotation, the angular velocity. Now φ does not explicitly appear in the Lagrangian, which means that the Lagrangian has the rotational symmetry (i.e., the transformations $\varphi \rightarrow \varphi' = \varphi + \alpha$, where α is an arbitrary angle, do not affect the Lagrangian). The Lagrangian is such a function of θ and $\dot{\theta}$ that it is also reflectionally symmetrical (i.e., invariant under $\theta \rightarrow -\theta, \dot{\theta} \rightarrow \dot{\theta}$).[3] But from Figure 2 we see that $U(\theta_1)$—a clear example of a ground state—has no reflectional symmetry.

This is clearly different from the breaking of a symmetry when the Lagrangian in question contains a term that does not preserve the symmetry.

2. Do all cases of nonspontaneous symmetry breakings have some asymmetrical terms in their Lagrangians?

What the Lagrangian represents may be regarded as the lawlike properties of a system, which include both its intrinsic properties and nonaccidental interactions with its environment. And a particular state of a system should be the result (or evolution) of its lawlike properties with a given set of initial and/or boundary (i/b) conditions (cf. Wigner 1979). It is certainly possible that a symmetry is broken not because of any asymmetrical lawlike properties—that would show up in the Lagrangian, but because of some special i/b conditions—that would not (Wigner 1979). Then it is a natural question to ask:

3. Can a symmetry breaking caused by i/b conditions be an SSB?

It seems that the answer would have to be "no," for the SSB in our model does not appear to be a case of symmetry breaking by any special i/b conditions. To see this point, just imagine that we begin with $\omega = 0$ and let ω very slowly (e.g., adiabatically) increase. The argument in the previous section is taken to imply that once ω passes the critical value, the bead will depart from O and ascend the hoop, thus breaking the symmetries; nowhere in this argument are any asymmetrical i/b conditions introduced.

Note that it is not a legitimate objection to simply say that because the model is idealized it does not tell us what happens in a real hoop-bead system where any number of asymmetrical causal factors may exist to actually break the symmetries. The bifurcation of the ground state beyond the

3. The Lagrangian for our model is: $L = T - V = (1/2)mR^2\dot{\theta}^2 + (1/2)m\omega^2R^2\sin^2\theta - mgR(1 - \cos\theta)$ (Greenberger 1978). The sine and cosine functions are symmetrical with respect to θ and $-\theta$.

critical point is a result for the model as it is given, idealized and totally symmetrical; therefore no asymmetrical factors of a real system are relevant.

We now ask the question:

4. Is a symmetry broken in SSB without any cause?

Since neither a lawlike property nor any i/b conditions of a system may be regarded as responsible for a true SSB, the answer may obviously appear to be affirmative. But the matter is not so simple, so let us approach it more carefully. First, van Fraassen (1989, chs. 10 and 11) discussed two general forms of what he calls "the symmetry arguments":

> There are two forms of argument which reach their conclusion "on the basis of considerations of symmetry." One, the symmetry argument proper, relies on the metaprinciple: that structurally similar problems must receive corresponding similar solutions. A solution must "respect the symmetries" of the problem. The second form, rather less important, assumes a symmetry in its subject, or assumes that an asymmetry can only come from a preceding asymmetry. (1989, 233)

Since the second form is rather a strategy in theory constructions than a principle (and one may argue that it follows, or is strongly suggested by, the contrapositive of the metaprinciple), I shall only consider the first form in connection with our question above. Let us then look at the following question:

4.1. Is the metaprinciple—structurally similar problems must receive corresponding similar solutions—true among deterministic systems?

Here is another way of putting the metaprinciple: problems that can be transformed into one another by a symmetry must receive solutions that are related by the same symmetry. A symmetry here means a group of transformations that leaves what is transformed "essentially the same" or "the same in all essential properties" (van Fraassen 1989, 235–236). And we only consider deterministic systems because (i) our model is a deterministic system, and (ii) there is a sense in which the principle is already violated by any indeterministic systems.[4]

Prima Facie, the answer is "no" and the SSB in our simple model is sufficient for such an answer. The "problems" in our model would be the

4. See the discussion in van Fraassen 1989, 239–240; also see pp. 250–257 for his discussion of the connection between the symmetry principle and determinism. Ismael (1997) argued against van Fraassen and concluded that Curie's principle holds in the indeterministic contexts as well (1997, 176–178). However, such a conclusion derives from Ismael's particular reading of the principle, which is more restrictive than van Fraassen's. I regard Ismael's reading of the principle a bit too narrow, but I will not argue the point in this paper.

different equations or Lagrangians connected by the reflectional and/or rotational transformations mentioned earlier (i.e., the symmetries for the problem). If the principle holds, the ground-state solutions should also preserve the symmetries; but they do not preserve the reflectional symmetry in the region of $\omega > \omega_1$. Hence, the metaprinciple fails.

Second, as mentioned in the introduction, some people argue that SSB is a misnomer; the phenomena it refers to should properly be called "hidden symmetries." The laws which govern the behavior of a system has one set of symmetries and the behavior another, usually lesser, set. Since what we observe is behavior, the symmetries of the laws are therefore *hidden* from us. Now the question:

4.2. Is there a breaking of symmetry in the case of hidden symmetries?

Some people (e.g., Coleman 1975, 141) argued that there is nothing unusual about a symmetry being hidden in the aforesaid manner, and therefore no symmetry needs be regarded as "broken" in SSB; and all the worry about an uncaused "breaking" of something is a misguided exercise caused by an inappropriate name! If so, the answer to our previous question—whether the metaprinciple holds—is still a "no" since the symmetries are still not preserved; however, it seem to suggest that "there is no reason why" such a principle should hold or why anybody should regard it a kind of violation to have symmetries in a problem (e.g., a Lagrangian) which are absent in its ground-state solutions.

In a limited sense this is just right for our model as I explained earlier. However, if you look at the model and consider the movement of the bead as ω increases, something extraordinary does happen and hence a symmetry breaking really seems to be present. To repeat what I said earlier in response to the third question, as the hoop begins to rotate and very slowly increases its angular velocity, at first the bead sits at the bottom of the hoop, motionless; but as ω passes its critical value ω_1 the bead starts climbing the hoop and thus breaks the reflectional symmetry. However one regards the symmetries of the Lagrangian or the laws in this model (e.g., as "hidden" or otherwise), there is presumably a physical process during which the system goes from preserving them ($0 \leq \omega \leq \omega_1$) to not preserving one of them ($\omega > \omega_1$). Therefore, calling the symmetries "hidden" rather than "broken" does not make the legitimate quest for an explanation of why the bead suddenly departs from O goes away.

Third, if there is a breaking of a symmetry, it might not be too far-fetched to ask: what then is its dynamics? So here we have,

4.3. Is there a dynamic process for SSB and, if there is, do we, or are we likely to, have a theory for it?

What we said above seems to indicate that there is a dynamic process through which the bifurcation occurs (or the symmetry is broken); and as we saw earlier that this is true for all models of SSB (a bifurcation which breaks the symmetry occurs when the values of some (controllable) parameter crosses its critical value). But such parameterized dynamic processes are not the ones that are responsible for the actual breakings of any symmetries, and if there are such dynamic processes, they may never be accurately accounted for by any physical theory.

First, one should not think that if there is ever a real system exactly like our model, then the bead, initially sitting motionlessly at the bottom of the hoop, would ever depart from that point however fast the hoop rotates. This is because it is one thing for that position (i.e., O in Figure 1) to no longer be the stable equilibrium state for the bead when $\omega > \omega_1$ but quite another for the bead to actually move out of it (and to be transferred into one of the new stable equilibrium states). One can easily see by looking closely at Figure 2 that O is still an equilibrium state when $\omega > \omega_1$; only it is no longer a *stable* one. The difference between these two kinds of states is essentially that a system in the latter will return to it when subject to perturbations,[5] while a system in the former will not. In other words, if there are perturbations present in our model, the bead will not stay at O when $\omega > \omega_1$ but will stay at θ_1 or $-\theta_1$. Since, strictly speaking, there are no perturbations of any kind in our idealized model, there should not be any movement of the bead which causes the *actual breaking* of any symmetries of its Lagrangian or its laws.[6]

Therefore, second, our model should best be seen as telling us the *possibility* of SSB, not what it actually is. In other words, it does not tell us what the dynamic process is by which the bead actually breaks the reflectional symmetry, namely, it's making the transition from a symmetrical state to an asymmetrical one. The actual breaking can happen only if some asymmetrical causal factors are assumed to exist in or around the system. Because of the instability of position O (or $\theta_0 = 0$), neither a sustained force nor a disturbance of measurable magnitude is necessary; small, and usually random, perturbations (or fluctuations) are sufficient to do the job. Because of the ubiquitous presence of such perturbations, people often neglect to mention them and talk as if an SSB may occur even in an idealized model as ours.[7]

5. I shall use the term "perturbation" as a technical term which simply means any small deviations of a system from a given state, the cause of which may be endogenous or exogenous. It specifically does *not* refer to the "perturbers" of a system in question, such as the air around the bead in our model.

6. Poincaré (1952, 167) makes the same point on a model of a stationary cone balanced on its point.

7. To my best knowledge, there are two exceptions: Radicati 1987 and Ismael 1997, both

Therefore, the dynamic process in our model which is parameterized by the angular velocity of the hoop and which tells us a complete story of how the ground state solution of the system bifurcates from a single symmetrical state to two asymmetrical states is not the process by which an SSB occurs, if it actually occurs at all. The "real" (or efficient) cause is something for which the model has no account.

Now, there must be dynamic processes that link the perturbations and asymmetrical ground states if SSB are regarded as deterministic phenomena. However, in no case of SSB in any fields of physics can one hope to know, or to write down explicit accounts for, the actual processes in terms of specific i/b conditions. Hence, it is better to characterize perturbations as random elements and to study the actual transitions in SSB in terms of probabilistic distributions.[8]

Hence, there are two ways—one may think of it as SSB_1 and SSB_2—of giving the notion of SSB a precise meaning: (i) it means that there are stable ground states that do not preserve some of the symmetries of the Lagrangian (or the laws); or (ii) it means that any actual breaking of a system's symmetry can be achieved by random perturbations. Of course, were the symmetrical ground state not unstable, perturbations would not be capable of breaking the symmetry. Whether SSB proper only means (i) is a question that requires careful considerations (discussed elsewhere, also see Kosso 2000 for a rather different view). Whatever the answer to it, the following is true of SSB: *The difference between SSB and other types of symmetry-breakings is that in the latter both the possibilities and the actual breakings are produced (or caused) by the same agent, while in the former they are not. That which produces the possibility of SSB is usually completely different from that which causes the actual breakings.*

Finally, with this much said, the answer to our main question (i.e., question 4)—whether SSB is uncaused—must be negative. Someone may object to this answer as follows. Is it not true according to my own account of SSB that even without perturbations, the new stable ground state is already asymmetrical, and therefore the metaprinciple is violated in that the symmetry in the problem is no longer a symmetry in the solution?

---

are discussions of Curie's principle of symmetry and each contains a brief section on SSB. Both in broad outlines give the right analysis of SSB but reach opposite conclusions as to whether Curie's principle is violated by, inter alia, SSB. Radicati: yes; while Ismael: no.

8. This point about the relationship between SSB and chance (or indeterminism) is discussed in another paper. The upshot is this: SSB is a case of instability that provides rigorous justification for treating as a matter of chance which state among the degenerate ground states the system in question will end up in, despite the deterministic contexts in which SSBs are considered. Cf. Poincaré (1952, 64ff); van Fraassen (1989, 233ff); Ismael (1997, 176–178).

The point would be justified if an interesting feature of the ground states which are suppose to comprise the "symmetry breaking" solution is absent. From Figure 2 one can clearly see that the ground states when $\omega > \omega_1$ always come in pairs: $(\theta_1, -\theta_1)$, which taken together is reflectionally symmetrical, and if one imagine a 3-d extension of the figure, i.e., with $0 \leq \varphi < 2\pi$, one can see that the solution is also rotationally symmetrical. This is what physicists call a (continuously) degenerate solution, meaning that from a formal point of view any of these infinite number of ground states—$\langle\theta_1, \varphi\rangle$—is a possible stable ground state. Of course, each of these states is not reflectionally symmetrical, but taken together the solution, which is what our model offers, still has a symmetrical "shape." Hence the formal results in this sense preserve the symmetries of the Lagrangian.

This is a remarkable feature of SSB: it shows that even with the possibility of symmetry breaking, SSB respects at some level the symmetries in the problems. And it is shared by all the better known cases of SSB in condensed matter physics and high-energy physics. In fact, this ought to be so, for if the ground states do not form a degenerate set whose members are transformable from one to another by the same symmetry that each is supposed to break, it would be a genuine puzzle as to how random perturbations could produce a set of nonrandomly distributed possibilities. For instance, if in our model, state $U(\theta_1)$ is possible but $U(-\theta_1)$ is not, then it would be a puzzle as to why the perturbations in the direction of θ_1 can cause the bead to ascend the hoop while the perturbations in the direction of $-\theta_1$ can not; or it would be equally puzzling why there are perturbations in the direction of θ_1 but none in the direction of $-\theta_1$, given the symmetries assumed in the model. On the other hand, if it ever turns out that there are such SSB, then the metaphysical landscape of physics would be radically changed. So far, we do not see a need for such a change.

In connection with van Fraassen's metaprinciple or to visit question 4.1 again, we may say the following. "No asymmetry in, no asymmetry out" holds in spite of the existence of SSB. If we banish the random perturbations (which we do for all idealized models), the "broken symmetrical" solution comprise a set of degenerate ground states, each of which breaks the symmetry but all of which together preserve it. If we count the random perturbations, then "no asymmetry in" is no longer true.

And in connection with our question (3) of whether SSB may be a case of symmetry breaking by i/b conditions, we now see that the answer will depend on our conception of random perturbations or fluctuations. According to Poincaré (1952, 64ff), being the results of perturbations is what it means to be the results of chance. Along this line, we should not count in i/b conditions such things as perturbations or fluctuations, for otherwise, we are clearly not following the schematic division of dynamical laws and i/b conditions that Wigner (1979) has emphasized. And hence, SSB is not a case of symmetry breaking by i/b conditions. However, this short remark does not do justice to this rich subject about chance and perturbations, which is discussed elsewhere.

Finally, we come to the question: Why should the increase of the angular velocity in our model, nowhere during which is any asymmetry introduced, "cause" a transition from a symmetry preserving ground state, $\langle\theta_0, 0\rangle$, to a set of symmetrical breaking ground states, $\langle\theta_1, \varphi\rangle$, $0 < \varphi < 2\pi$? The answer for this case turns out to be very simple. The centrifugal tendency of the bead is proportional to ω^2, and where the ground state of the bead lies depends on the balance between the weight and this tendency of the bead. When $\omega < \omega_1$, the balance is in favor of the weight, and hence the ground state is at O; but when $\omega > \omega_1$, the balance tips the other way, and hence the ground state "climbs" up the hoop. But the centrifugal tendency is only a potential. So without an actual initial deviation, the symmetry will not be broken. An analogous but more interesting answer exists for a similar question in condensed matter physics. Certain directions of phase transitions in a thermo-system are examples of SSB: for instance, a transition from liquid to a solid or from a paramagnet to a ferromagnet. What happens in these transitions is the formation of "long-range" (or large-scale) orders in systems that are not the result of any external influcences and that break the (characteristic) symmetry in the previous phase. Now, why should the decrease of temperature of such systems across a critical value spantenously breaks the symmetry of the system? The balance in this case is the local tendency to form an order and the global thermo-agitation that prevents such a tendency to achieve a global "alignment." When the temperature is high, the balance is in favor of the thermo-agitation, and the phase has the symmetry of being "without order" as in the case of a liquid in equilibrium; but when the temperature is lowered across the critical value, the balance tips over to the other side, where the tendency of forming long-range order becomes dominant. This is when a local fluctuation towards an order may easily grow into a long-range order and thus spontaneously breaks the symmetry.

4. Conclusion. From a simple classical model we are able to see that what people usually refer to as SSB has in fact two different meanings: SSB_1 means having stable and degenerate ground states each of which taken singly breaks the symmetries of the Lagrangian (or the laws); and SSB_2 means that the breakings of symmetries can be done by random perturbations or fluctuations; and also how in such phenomena, a symmetry in question is and is not broken, depending on which level one is considering it. Thus, we can give simple answers to most of the questions I sampled in Section 1. When people argue that SSB is a misnomer and that it should be called "hidden symmetry," they are thinking of SSB_1 but not SSB_2, for it is not proper to call a symmetry hidden if SSB_2 is part of its meaning. The metaphorical image of objects "choosing" to break a symmetry without apparently having a sufficient reason can now be understood as alluding to the fact that, according to SSB_2, the symmetry may not be broken by anything that can be counted in physics as a given cause. Of course no object really makes a free choice in an SSB; but analogous to Buridan's ass which goes for one of the stacks of hay, the arbitrariness with which an object moves into one of the symmetry-breaking ground states reflects the fact that all such states are of equal chance. Other questions, such as whether SSB has demonstrated the existence of emergent properties, will have to be discussed elsewhere.

REFERENCES

Alben, R. (1972), "An Exactly Solvable Model Exhibiting a Landau Phase Transition", *American Journal of Physics* 40: 3–8.

Anderson, Philip W. (1997), *Basic Notions of Condensed Matter Physics*. Reading, MA: Addison-Wesley.

Aravind, P. K. (1987), "A Simple Geometrical Model of Spontaneous Symmetry Breaking", *American Journal of Physics* 55: 437–439.

Brout, R., and F. Englert (1998), "Spontaneous Symmetry Breaking in Gauge Theories: A Historical Survey", http://arXiv.org/hep-th/9802142 v2 (18 May): 1–9.

Coleman, Sidney (1975), "An Introduction to Spontaneous Symmetry Breakdown and Gauge Fields", in A. Zichini (ed.), *Laws of Hadronic Matter*. New York: Academic, 138–215.

Drugowich de Felicio, J. R., and O. Hipolíto (1985), "Spontaneous Symmetry Breaking in a Simple Mechanical Model", *American Journal of Physics* 53: 690–693.

Greenberger, Danial M. (1978), "Esoteric Elementary Particle Phenomena in Undergraduate Physics—Spontaneous Symmetry Breaking and Scale Invariance", *American Journal of Physics* 46: 394–398.

Ismael, Jenann (1997), "Curie's Principle", *Synthese* 110: 167–190.

Johnson, R. C. (1997), "Floating Shells: The Breaking and Restoration of Symmetry", *American Journal of Physics* 65: 296–300.

Kosso, Peter (2000), "The Epistemology of Spontaneous Broken Symmetries", *Synthese* 122: 359–376.

Moriyasu, K. (1983), *An Elementary Primer for Gauge Theory*. Singapore: World Scientific.

Poincaré, Henri (1952), *Science and Method*. New York: Dover.

Quigg, Chris (1983), *Gauge Theories of Strong, Weak, and Electromagnetic Interactions*. Reading, MA: Benjamin/Cummings.

Racz, Z., and P. Rujan (1975), "Dynamics of a Simple Model Exhibiting a Landau Phase Transition", *American Journal of Physics* 43: 105–107.

Radicati, Luigi A. (1987), "Remarks on the Early Developments of the Notion of Symmetry Breaking", in M. G. Doncel, A. Hermann, L. Michel, and A. Pais (eds.), *Symmetries in Physics (1600–1980)*. Barcelona: Universitat Autonoma de Barcelona, 195–207.

Sivardière, Jean (1983), "A Simple Mechanical Model Exhibiting a Spontaneous Symmetry Breaking", *American Journal of Physics* 51: 1016–1018.

——— (1985), "Mechanical Model for a First-Order Phase Transition", *American Journal of Physics* 53: 363–365.

Sivardière, Jean (1997), "Spontaneous Symmetry Breaking in a Cavendish Experiment", *American Journal of Physics* 65: 567–568.
Strocchi, F. (1985), *Elements of Quantum Mechanics of Infinite Systems*. Singapore: World Scientific.
van Fraassen, Bas (1989), *Laws and Symmetry*. Oxford: Clarendon Press.
Weinberg, Steven (1996), *The Quantum Theory of Fields*, vol. 2, *Modern Applications*. Cambridge: Cambridge University Press.
Wigner, Eugene P. (1979), *Symmetries and Reflections: Scientific Essays of Eugene P. Wigner*. Woodbridge, CN: OX Bow Press.

Spontaneous Symmetry Breaking and Chance in a Classical World*

Chuang Liu†‡

This essay explores the nature of spontaneous symmetry breaking (SSB) in connection with a cluster of interrelated concepts such as Curie's symmetry principle, ergodicity, and chance and stability in classical systems. First, a clarification of the two existing senses of SSB is provided and an argument developed for a proposal for SSB, in which not only the possibilities but also the actual breakings are referred to. Second, a detailed analysis is given of classical SSB that answers the questions: (i) how we are justified in regarding it as a matter of chance, and (ii) why the breakings in it are equally probable. The answer provides some support to the applicability of ergodicity in special systems (such as ours).

1. Spontaneous Symmetry Breaking. Even though spontaneous symmetry breaking (SSB) as a technical concept first gained its currency in the studies of superconductivity, superfluidity, and quantum fields (Coleman 1975), it is now recognized that the phenomena it refers to are quite common in physics. Many classical systems also exhibit SSBs; the better known of which include Euler's rod, which buckles in an arbitrary direction when the vertical load on it crosses a threshold weight, and Poincaré's rotating "nebula" in self-gravitating equilibrium, which separates into a two-body system when a critical value of its angular velocity is passed. There are even simpler classical systems whose SSBs can be studied thoroughly with no more than high-school mathematics. Elsewhere (Liu 2003), I have given

*Received April 2002; revised Febraury 2003.

†To contact the author write to Department of Philosophy, University of Florida, 330 Griffin-Floyd Hall, P.O. Box 118545, Gainesville, FL 32611–8545; e-mail: cliu@phil.ufl.edu.

‡I would like to thank Craig Callender for reading an earlier draft of the paper and commenting on a shorter version of the paper at the APA Eastern Division Meeting in Philadelphia on December 30, 2002. I also thank two anonymous referees for their constructive suggestions.

Philosophy of Science, 70 (July 2003) pp. 590–608. 0031-8248/2003/7003-0008$10.00

a detailed study of a simple mechanical model (the ring-bead model), from which the metaphysical nature of classical SSB is laid bare. I now give in this section a sketch of the same model with enough details for the main tasks of this essay: the senses of SSB and the connection between SSB and deterministic chance.

To those who are skeptical about the scope and significance of such a simple mechanical model for the study of SSB in physics, let us hope the following brief remarks suffice. Indeed, not only does SSB occur in radically different systems, mainly in condensed-matter physics and high-energy physics, but there are also different types of SSB: the symmetries that may be spontaneously broken are either external (i.e. spatial and/or temporal) or internal symmetries, and for each, either global or local symmetries, and again for each, either discrete or continuous; and the symmetries may be applied to a classical or a quantum system; and the systems in which SSBs occur may be of finite or infinite dimension and/or number of degrees of freedom. What our simple model harbors is a spontaneous breaking of either a global external discrete symmetry or—by a simple extension of the model—a global external continuous symmetry in a finite classical system. For various rather obvious reasons that I will not enumerate here, a comprehensive treatment of all the varieties of SSB in this short essay is not feasible; a division of labor is thus inevitable: the quantum SSBs—the more recondite ones—will have to be studied separately from classical SSBs; and what else could be better than starting from a simple and typical classical model? I doubt anyone would say that a study of classical SSB is irrelevant to our general understanding of SSB (could a study of classical physics be irrelevant to our understanding of physics in general)? And if one doubts that there is anything philosophically interesting about classical SSB, then let the rest of this essay show how wrong s/he is.

That said, let me give a sketch of the model (Greenberger 1978; Sivardiere 1983).[1]

The model comprises a metal ring vertically suspended and free to rotate (without friction) and a bead frictionlessly threaded on it (see Figure 1). Imagine that we set the ring into rotation and very slowly (i.e. at equilibrium every step of the way) increase its angular velocity ω. A complete story of the system's mechanical behavior can be told in classical mechanics. At first when ω is low, the ground state of the system—the state at

1. It may be helpful to point out that in the physics literature where this and similar models are given, what one finds are complete mathematical analyses whose results are then shown to closely resemble those obtained in various areas in condensed-matter physics and high-energy physics. If one does not already understand what SSBs are, reading these studies will not help. To me, this is getting things upside down. These simple models give us a much better initial understanding of the nature of SSB.

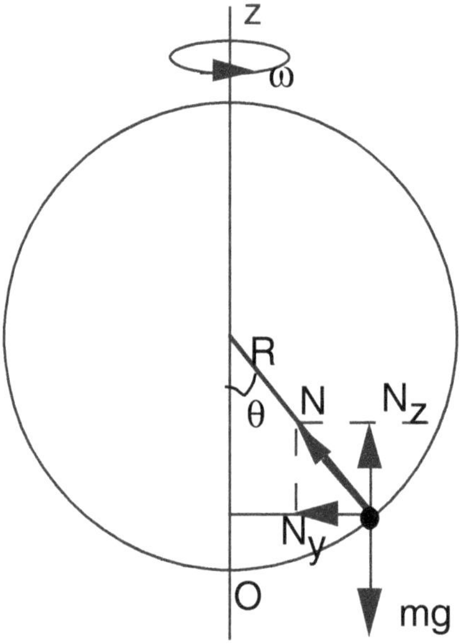

Figure 1. A bead with mass, $m \neq 0$, is free to move frictionlessly on a circular wire of radius, R, which rotates around the z-axis with variable ω.

which the bead has the lowest energy—is the one with the bead resting at the lowest point of the ring (i.e. at O or $\theta = 0$). This situation remains until ω passes a certain value—henceforth the "critical value"—when O stops being the ground state of the system. Instead, the ground state begins to "split" and "climb" the ring as ω continues to increase. From Figure 1 it means that the bead will be in its ground state at successive values of θ or $-\theta$, where $|\theta| \neq 0$, with corresponding values of ω. The position of the bead's ground state will become higher and higher on the ring (i.e. the angle θ will become larger and larger as ω increases, and it approaches 90° (or $-90°$) when ω approaches infinity) (see Figure 2).

Therefore, the bead, originally resting at O, will, in seeking the new ground state, depart from O and ascend one or the other side of the ring; and because this rotating system initially possesses the reflectional symmetry between the two sides of the ring with respect to the z-axis (to an observer who is co-moving with the ring[2]), this symmetry is said to be broken by its ground states after ω passes its critical value (Wigner 1979, pp. 3–50). The symmetry appears to be spontaneously broken because there are no apparent causes (asymmetries) in the model that are responsible for the breaking. The results can be easily transplanted to a similar

2. I thank one of the referees for pointing this out for me.

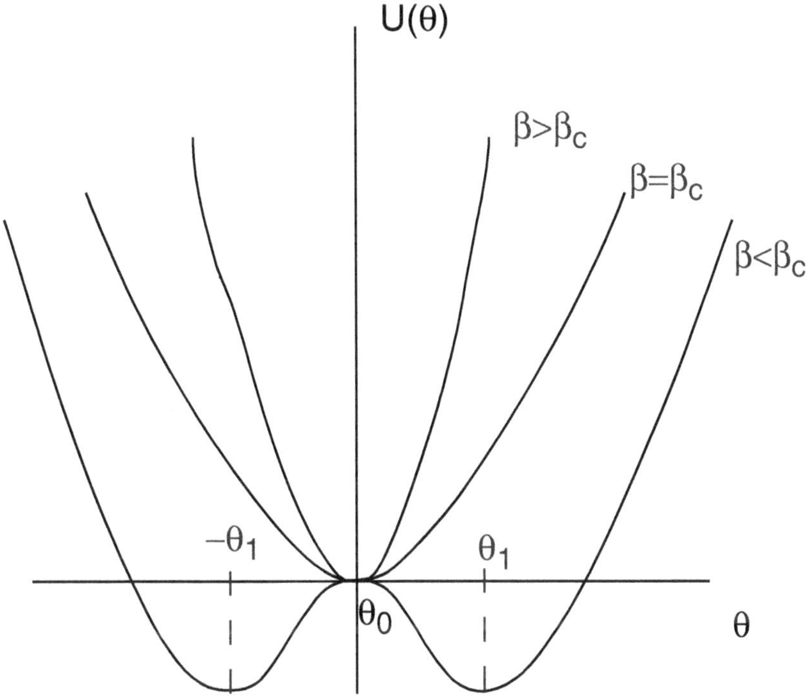

Figure 2. The dimensionless potential, $U(\theta)$, against θ. Note, $\beta = \omega^2 R/g$ and $\beta_c = \omega_1^2 R/g = 1$.

model that spontaneously breaks a continuous symmetry of rotation (Liu 2002; see also the case of Poincaré's top in Section 4 below).

2. The Two Senses of SSB. Are the symmetries truly broken *spontaneously* in SSB—i.e. without any causes in the form of asymmetrical antecedents? The answer is no, for if one thinks that the bead will inevitably climb the ring from O in our simple model—exactly as it is described—then one is mistaken. If there were a physical system in reality that was exactly like our model, we would have no reason to believe that the bead would budge at all however fast the ring rotates. This is because even though the point O is no longer the ground state of the system beyond the critical point, it is still an equilibrium point in the following sense: there is nothing in the mathematical argument which indicates that O is not a state in which the bead will remain forever unless bumped or disturbed, however slightly, toward another state. Therefore, it seems that the actual breaking of a symmetry is only attributable to some impact, external or internal to the system, large or small. In our model, the bead won't actually ascend the ring unless something at O causes it to do so (Poincaré 1952, 64–90; Ismael 1997, 179–180).

However, there is a sense of SSB in which the symmetry-breaking is obviously spontaneous (as uncaused). In our model in which no asymmetrical elements are considered, the symmetrical ground state at O is "broken up"—i.e. rendered unstable—when the angular velocity passes the critical value. The bifurcation of the ground state which provides the *possibility* for a breaking of the symmetry is present without any asymmetrical antecedents. This is connected to the following general feature of SSB. What the initial unique ground state breaks into after the parameter crosses its critical value is always a *set* of more than one ground state—in our model they are those at θ and $-\theta$—which are *degenerate,* meaning that they have the same value (e.g. energy) so that any transformation—of the same symmetry they are supposed to have broken—from one state to another within the set does not change the value (see Figure 2). Hence one may say that the ground states together as a set still preserve the symmetry which each breaks.

Therefore, two senses of SSB should be distinguished:

> SSB_1 which gives us its formal/mathematical aspect and which is therefore about the *possibility* of SSB;

and

> SSB_2 which gives us its causal/physical aspect—e.g. symmetry-breakings caused by perturbations[3]—and which is therefore about the *actual* breakings.

Which of the above should one take for SSB? SSB_1 seems right because some of the salient formal features of SSB, as described above, are captured by it: the control parameter, its critical value, the bifurcation (or the spread) of the ground state, etc.; but should one not adopt SSB_2 instead? Some may argue that one should not because all SSB means, and could mean, in physics is SSB_1.

There indeed seems to be some support in the physics literature on SSB for this view. All those who insist that "spontaneous symmetry-breaking" is a misnomer and a better alternative would be "hidden symmetry" would presumably be taking such a position (Coleman 1975; Aitchison 1982). For them, no symmetry is broken in these cases; rather the system in question is such that the symmetry of its Lagrangian is different from (usually more general than) the one of its ground states and is thus hidden

3. I will use the word "perturbation" as it is used in the physics literature, namely, to mean any kind of small deviations of the configuration of a system from a certain understood state—the unperturbed state. They may be exogenous or endogenous.

by the ground states. Or, one may argue that only in SSB_1 is the use of 'spontaneous' appropriate, because in SSB_2 the actual breakings turn out to be caused and therefore can no longer be regarded as spontaneous. Or, most importantly, one may argue that as long as a case satisfies SSB_1, it is SSB regardless of how the symmetry is actually broken. There are indeed cases regarded as SSBs in the physics literature in which the systems satisfy SSB_1 but the breakings are done not by perturbations but by explicit initial/boundary (i/b) conditions. So it is incorrect to single out perturbations as relevant to the meaning of SSB.

Despite its initial plausibility, this—SSB_1 being sufficient for the meaning—is not a defensible position. For a starter, it runs into the following general philosophical problem. To say a symmetry is spontaneously broken just because the system has the potential of breaking it is like saying a person is spontaneously dead whenever he or she has the potential of dying (and therefore, everyone is spontaneously dead). (Remember that the above view says " . . . regardless of how the symmetry is actually broken.") Or in general: x is spontaneously ϕ if x has the potential of being (or becoming) ϕ, which does not seem an appropriate way of stretching the meaning of "being spontaneously ϕ."

However, more specifically from the physics literature, it is clear that 'SSB' is used to refer to a *physical phenomenon,* such as a phase transition, rather than to some theoretical aspect of it; and the phenomenon it refers to does involve the actual breakings of some symmetries. Moreover, many physicists prefer to use instead the term "spontaneously broken symmetry" (Weinberg 1996), which cannot be understood as only referring to the possibility (but not the actuality) of the symmetry breaking. It seems that the physics literature also contains strong evidence to support the adoption of SSB_2.

However, it would not do to simply identify the meaning of SSB with SSB_2 either. If one only includes the perturbations as the legitimate causes of SSB, then SSB_2 is too restrictive, as mentioned above, for there are SSBs that are caused by non-perturbative influences. If one includes non-perturbative elements, then would one not have to admit all symmetry breakings as SSBs? What then does "spontaneous" refer to in the notion of SSB? Hence, SSB_2 may be sufficient for SSB but it is not necessary.

At this point, it seems that SSB_1 is necessary but not sufficient for the notion of SSB, while SSB_2 is sufficient by not necessary. The real meaning of SSB must lie in an extension of SSB_1 in the direction of SSB_2. We may start from the following observation. The difference between SSB and other types of symmetry breakings seems to be that in the latter both the possibilities of and the actual breakings are produced (or caused) by the same agent, while in the former they are not. That which produces the possibility of SSB (and also maintains the broken symmetry) is different from that which causes the actual breakings. In our model, the possibility—the bifurcation of the ground state—is caused by the crossing of the critical value of , while an actual breaking is the result of (mostly) perturbations at O. This suggests that a symmetry breaking is an SSB if, and only if, the system in which the symmetry is broken satisfies SSB_1 and the causal agent responsible for that breaking does not override whatever creates the possibility of, and maintains, it. For an illustration, let us look at the case of a modified ring-bead model. The model can be modified by either or both of the following conditions:

A: having a device on the right lower quarter of the ring that can catch and hold the bead (a boundary condition)
B: a hammer striking the bead at O to the right at a certain time and imparts to it a speed sufficient to send it to the location of the device even if the ring is stationary (an initial condition)

In one situation, only B is introduced at an greater than its critical value (i.e. the hammer strikes but no device is attached to catch the bead). In this case, we would regard the result as a case of SSB, because even though the hammer's strike is non-perturbative, it does not override the system's potential to have a broken symmetry. In other words, had the ring been rotating below the critical , the symmetry would not have been broken at all. Now, in another situation where both B and A are introduced at an greater than its critical value, we would not regard the result a case of SSB, because whether the system is ready for SSB (i.e. whether it satisfies SSB_1) the symmetry will be broken when the device catches the bead. In this case, that which actually breaks the symmetry does override the system's potential to have a broken symmetry.

If the meaning of SSB should refer to actual breakings and the latter are not without asymmetrical antecedents, why should such a symmetry breaking be regarded as spontaneous? Two reasons make up the answer to this question. First, the perturbations—the main contributors to the actual SSBs—are, contrary to other types of asymmetrical antecedents, cannot be theoretically accounted for either as extra terms in a system's Lagrangian (or Hamiltonian) or as i/b conditions. Second, and more importantly, the new ground states of an SSB system always form a set that is closed under the symmetry each of its members breaks and the chance of the system falling into any one of these states, as I will show later, must be equal. (Systems with such features are said to be *unstable*—an important technical term one usually runs into in the study of chaotic systems (Smith 1998); and indeed another way of defining the instability of a system is to say that it is in such a state in which perturbations alone are sufficient to make it leave the state and (practically) never return.) As a consequence, it is not possible to predict either when—i.e. at what instant of time—such a breaking will take place or into which of a set of degenerate ground states the system will fall when an SSB happens. Hence, it does not seem inappropriate to call such a breaking of symmetry spontaneous.

3. Curie's Principle and SSB. One may wonder whether Curie's principle of symmetry—which in slogan form reads: "asymmetry out only if asymmetry in"—holds in SSB? From the previous section it seems that the answer is yes; despite the sense of spontaneity, the breakings are caused by asymmetrical antecedents. However, if the existence of a possibility for symmetry-breaking is already symmetry-breaking enough, i.e. if SSB_1 is the meaning of SSB, one may say that Curie's principle is violated. There is, of course, a question of whether Curie's principle even applies to SSB in this case. If one is justified in regarding a theoretical bifurcation (or spread) of a ground state into a set of symmetry-breaking states as an asymmetrical "effect", one can indeed say that the principle no longer holds. I take this to be another reason against identifying SSB with SSB_1.

What if SSB should be treated as an indeterministic phenomenon? Does Curie's principle hold in indeterministic processes? The answer, I will argue here, is yes and no, depending on which aspects of an indeterministic process one is concerned with. The upshot of the argument is that the principle is violated in indeterministic processes as far as the individual effects are concerned, while it holds among the probabilistic regularities—some of which are natural laws—in the processes.

The idea that Curie's principle does not hold in indeterministic processes is briefly explained in van Fraassen 1989 (239–243) and 1991 (23–24). Here is an argument in the spirit of van Fraassen's. Suppose that we have a system that radiates particles one at a time along a single spatial dimension—in the positive or the negative direction with respect to the emitter—and the chance of one particle in either direction is 1/2. And suppose that the state of the system before any emission is reflectionally symmetrical. The resulting system after any emission—the emitter and the emitted—would be a state that is not reflectionally symmetrical. Given the emissions are indeterministic, namely, there is no hidden mechanism in the system that determines which particle is emitted in which direction, the Curie's principle is violated by this model: no asymmetry in but some asymmetry out.[4]

4. Even though most of the known types of SSB are regarded as happening in deterministic systems, there may be indeterministic cases of SSB. In fact, one of the most interesting and difficult questions about SSB is whether the processes of quantum measurement—the heart of indeterminism—are of SSB. However, one must note that the SSB cases in condensed-matter physics and in high-energy physics are not examples of indeterministic SSB. The symmetries are spontaneously broken in those cases at the level of probabilistic distributions rather than at the level of purely chancy events.

However, at the level of chance distribution, the system before and after any such emissions is always reflectionally symmetrical. Indeed, were the chances in the two directions not equal, we would be entirely justified to assume that some asymmetrical elements exist in the emitter which cause the unequal chances. I know of no instance of a scientific theory in which an asymmetrical probabilistic distribution is not accounted for by any asymmetrical antecedents or, failing that, is regarded as an entirely satisfactory result. But such a request must be distinguished from the request for hidden variables in quantum mechanics. The former is at the level of determining chance distributions and the latter individual measurement results. In fact, at least in quantum mechanics, chance distributions evolve deterministically because the evolution, of either state functions (in the Schrödinger representation) or observables (in the Heisenberg representation), which determines such distributions at any instant, given an initial distribution, is entirely deterministic. Indeed, one may say, vulgarly as it were, the principle holds in any processes that are treated deterministically, whether they are in fact classical or quantum processes.

The above should be sufficient to resolve an apparent dispute between van Fraassen and Ismael, where van Fraassen (1989, 1991) claims that Curie's principle fails, while Ismael (1997) claims that it holds, in the contexts of indeterminism. They ask different questions and, not surprisingly, get different answers. Van Fraassen's question is whether the principle holds in any indeterministic processes, and the answer, as one can see from the above, is clearly no. Ismael's question is "does Curie's principle have any application where the laws in question are *in*deterministic . . . " (176) and the answer, also in accord with what is just said, is yes. The application of the principle, as Ismael explains in detail, is indeed on the level of probabilistic distributions.[5]

4. Chance and SSB. Given that at least many SSBs happen in deterministic systems, e.g. our model and most of the thermo-systems in which phase transitions take place, we now ask first whether and in what sense we are justified to call the actual breakings a matter of chance; and secondly, why the breakings in SSB are equally probable?

The first question is, as we will see, a special case of a more general question, namely, how is chance compatible at all with determinism? David Lewis and Karl Popper are famously known to have rejected the com-

5. Ismael also realized that the principle does not apply among the purely chancy processes, but she does not regard such as a violation of the principle in the indeterministic contexts, for her interpretation of Curie's principle is in fact such that the purely chancy processes in the indeterministic contexts are not in the proper domain of the principle's application.

patibility of the two, but more recent works (Clark 1987; Loewer 2001) have argued for the opposite from two different directions. While Clark (1987) argues for how ergodicity can ground chance in a deterministic world, Loewer (2001) suggests an alternative solution from a Best System account (á la Lewis) of laws and the Past Hypothesis (á la Albert 2000). Since it is neither the aim nor within the scope of this essay to continue that discussion, it suffices to mention that I am more sympathetic to Clark's approach and take what follows to be in a sense providing further support to it. Very roughly put, Clark poses two connected questions for the compatibility of chance and determinism: (i) whether chance can be regarded as "real", and (ii) if it can, whether it can be made consistent with the underlying deterministic processes. Clark's answer to (i) is that chance is real even though its probability measures are ultimately reducible to 1 or 0; or " . . . reducibility does not imply unreality or non-objectivity" (Clark 1987, 204). And the ergodic theory provides an affirmative answer to question (ii). Loewer's complaints about this solution evokes both technical and conceptual objections to ergodicity: few systems, if any real ones, are proven ergodic and some are proven non-ergodic, and even for those proven ones, how does ergodicity gives us any reason to believe the probability it gives for finding a single outcome at a certain moment? I hope the following discussion will show how the notion of ergodicity can help to answer my two questions even though SSB systems are obviously non-ergodic.

Moreover, I want to mention before our in-depth analysis what might be the first case in which a connection between chance and SSB is made in a deterministic context, although to my best knowledge it was not recognized as such. To my surprise (and delight) the first example in Poincaré's now famous discussion of chance (Poincaré 1952) is a special case of SSB: a stationary cone balanced on its point on a flat surface. It will topple towards an unpredictable direction since rotational symmetry is already "broken" in that the upright position of the cone in balance is an unstable equilibrium state and the stable one (i.e. the ground state) is one of the infinite number of states, $[0,2\pi)$, in which the cone lies on its side. The toppling of the cone towards any particular direction may be caused by a particular "very slight trepidation, or a breath of air" (67).[6] Again, why is the toppling of the cone in any direction a matter of (equal) chance, even though it is determined by a specific perturbation?

6. One should note that the difference between our ring-bead system and Poincaré's cone with respect to SSB is almost trivial. The cone case can be easily modified so that it also has a one-parameter controlled process of SSB. For instance, one can think of a rotationally symmetrical system whose lower part continuously change its shape (which is the parameter) from spherical to an inverted cone shape.

Those who are familiar with the literature of the foundations of statistical mechanics (SM) (Sklar 1993; Guttmann 1999) may think that the answers to our questions are, details aside, simple and straightforward. To put it roughly, if the use of objective probabilities (i.e. chances) among observable (macro-)states (i.e. the coarse-grained states) is consistent with the assumed underlying determinism among dynamical (micro-)states (i.e. the fine-grained states) in SM, then one can simply say the same (again details aside) for the chance in SSB. That is however not true, even if we assume that the antecedent of the above conditional is unproblematic (e.g. assuming that we are dealing with ergodic systems; more on this later). The case of SSB is similar to, but not the same as, those in standard SM. First, some important differences (i.e. the ones that are relevant to the treatment of chance in deterministic contexts), and then, a direct answer to our questions.

To begin with, SSB systems are not the ones in thermo-equilibrium; instead they are in transitions from one equilibrium state to another (i.e. from the old ground state to one of the new ground states). Therefore, the appearance of chance in them cannot be justified in the same way as is the appearance of chance in systems of equilibrium. Even if ergodicity is what we now know justifies with rigor the consistency of chance in the observable states with determinism in the dynamical states in equilibrium SM, it would not be directly applicable to SSB systems.

That leaves us with the possibility of regarding SSB as non-equilibrium phenomena, as processes that eventually approach equilibrium. We may think of the new ground states (the states obtained when the parameter is beyond the critical value) as equilibrium states and the transfer from the unstable state to one of them as analogous to what one encounters in the transport phenomena in the kinetic theory. Again, that is not the case.

The kinetic theory—at whose core is the Boltzmann equation of the one-particle density function—does not apply straightforwardly to SSB systems either. First of all, it is one of the main challenges of the theory to show that for a thermo-system the equilibrium state at which the Boltzmann distribution—the most probable distribution—holds is indeed the state to which all systems in other (non-equilibrium) states will eventually approach and from which they, once there, will never permanently leave. The other challenge is to show that the approach to equilibrium is irreversible regarding observable states despite the fact that the laws that govern the dynamical states are completely time-reversible. But none of these is a problem for the SSB systems. An SSB system when regarded as a thermo-system—considering the ring-bead system and its immediate environment as a collection of molecules strictly obeying the dynamical laws of classical mechanics—is *'attracted'* towards one of its new ground states, θ or $-\theta$, not because of its tendency of becoming "one of the most probable state," as in the kinetic theory case, but rather because—in a simpler sense of attraction—its tendency of moving to a lower energy state. This is true at both the observable and the dynamical level, because even if we look exclusively at the individual trajectories in the phase space of the ring-bead system, all of them should be "deterministically" attracted towards the subspaces defined for θ or $-\theta$ and the corresponding potential energy.[7] In other words, the transition of the ring-bead system from the former ground state at the bottom of the ring to one of the latter ground state higher up is not really a probabilistic phenomenon; it is a mechanical one with an energy gap. Essentially the same holds for the case of Poincaré's cone. It is not a matter of "becoming more probable" that the upright cone topples, but rather it is mechanically determined to do so.[8] Because of this feature, there is no good sense in which one may talk about the physical possibility that a large fluctuation, for instance, may return the system, the bead on the ring or the Poincaré cone, to its old ground state, while this would certainly be possible if the system were a case of the kinetic theory. In other words, while it is only an overwhelming probability that prevents the cream from suddenly separating on its own from the coffee once the two are well mixed, there is an energy gap—a difference at the observable level—that prevents the system in question from going back to the previous (symmetrical) ground state in SSB.

Therefore, we cannot directly borrow from SM to answer our questions, and yet a combination of some notions there, perhaps slightly modified, may just get us what we wanted. The rest of the section provides a conception of such an attempt. I first give the answer in intuitive but imprecise terms, and then try to precisify it by connecting it, wherever possible, to what is well established in the foundations of SM.

The breaking of a symmetry in a case of SSB is a matter of chance partly because they may be caused by perturbations that are randomly distributed in the neighborhood of the unstable equilibrium point, and

7. One must note that in this paper, and especially in the rest of it following this note, there are two kinds of spaces: the 3-dimensional space and the 6N-dimensional phase space (N being the number of particles). In principle one should always make it clear which space one is talking about when one describes a situation and avoid mixing them in such descriptions. But there are occasions where a mixed used is convenient and free of the risk of confusion. So, I will occasionally say things such as "the phase states—or states—of the ring-bead system around O," which simply means "the phase states . . . around the phase point that corresponds to the bead's being at rest in the 3-d space at O."

8. Another, more precise, way of putting it is that the Boltzmann transport to the most probable state does not involve work, while the transport to the lower energy state by SSB does. Hence, while large fluctuations may send a Boltzmann system to a non-equilibrium state without costing any work, restoring a broken symmetry does cost work.

partly because the stable equilibrium states are symmetrically distributed around (but not close to) it. Our ring-bead model may be conceived as a system composed of a large number of molecules and surrounded by air. When the angular velocity goes beyond the critical value, the state at O (see Figure 1) becomes unstable, which when translated into the language of ensembles means that of all the possible systems in the ensemble in the phase-space neighborhood of that state, only those with configurations that make them occupy exactly the position O will not depart permanently from O and make the transition to one of the new ground states, θ or $-\theta$. Any system whose configuration makes it deviate from O will, because of the energy gap, make the transition. And since the possible states in which such departures will end up are all degenerate, meaning that they all have the same value for the macroscopic quantity, e.g. the energy, there is no reason why the system should not end up in any of them with equal probability.

Let me now make this answer to our two questions more accurate, first in the structure of the answer.

1. An SSB system acquires a chancy character when its symmetrical ground state becomes unstable. The values of the chance—the probabilities—are determined by how the perturbations, if present, are distributed before the state becomes unstable, i.e. in the system's pre-SSB stage.
2. The characters of the breakings are entirely determined by such individual perturbations via individual dynamical processes.
3. The symmetrical arrangement of the asymmetrical ground states and the degeneracy of the values of the SSB-causing property, such as the energy gap, ensure that the actual breakings have the same probability distribution as the distribution of the perturbations.

To fill out the structure, we begin with the question of how the perturbations are distributed. We can simply say as I did earlier that the perturbations are *randomly* distributed and then go on to give a precise sense of randomness, if the latter turns out to be good enough to justify the equal-chance character of the symmetry breakings. It is beyond the scope of this essay to even touch on the various approaches to, and controversies over, a precise definition of randomness (Earman 1986, chs. 8–9; Emch and Liu 2001, 188–215); it suffices to predict that no satisfactory answer is forthcoming if one seeks a direct connection between randomness and the equal chances of perturbations in all directions from O in the ring-bead model. Further, even if such a connection can be figured out accurately, we are still left with the more difficult question of how we know that the perturbations are indeed random.

We do not have such a (double) problem with the notion of ergodicity, for we know exactly what ergodicity means; the question—and a difficult one—is how we know whether or not a certain system is ergodic. We know that before an SSB becomes possible, the system in question is in a stable ground state. If it can be suitably modeled as a thermo-system—systems with a large number of degrees of freedom—there is a possibility that we can employ the notion of ergodicity, or comparable notions such as quasi-ergodicity, to justify the uniform distribution of the perturbations we intuitively suspect.

First, can we model SSB systems as thermo-systems? Some of them obviously can readily be so modeled, e.g. the systems in which such SSB as ferromagnetism, superconductivity, and superfluidity take place. One may worry that the SSB systems in quantum field theories are not thermo-systems in any straightforward sense. However, there is so many similarities between the SSB in quantum SM and in quantum field theory (Aitchison 1984, ch.6), and from a certain ontological perspective quantum fields are many-body systems of infinite degrees of freedom (Strocchi 1985), that there is no reason to doubt that what accounts for the chance of SSB in the former case cannot be applied to the latter.

More importantly, the question is whether or not such SSB systems as our ring-bead system or Poincaré's cone can be modeled as thermo-systems. Prima facie, they are modeled as rigid-body systems, which are not amenable to thermo-statistical laws. But that is only true when no "very slight trepidation, or a breath of air" is considered. Once we include perturbations as the chief causes of the SSBs in these systems, they should indeed be treated as thermo-systems.

Second, how do we know that such pre-SSB systems are ergodic? For those who know the story of the struggle with ergodicity, it should be obvious that it is practically impossible to give a rigorous proof here that the SSB systems in question are or are not ergodic. And for those who share Loewer's (2001) skepticism of ergodicity, the question may seem a conundrum. But let us not give up so easily.

Theorists only know how to construct such a proof for physical systems that do not remotely look like thermo-systems, and yet many simple thermo-systems seem obviously ergodic from an intuitive point of view (Emch and Liu 2001, chs. 8–9, especially 317ff.; Earman and Rédei 1996; Toda, Kuba, and Saitô 1995, ch. 5). At the intuitive level what would prevent a system from being ergodic is for a possible trajectory of it to wander into a region of its phase space and stay there (practically) forever. The reason for this is quite simple. For ergodicity to hold for a system, the time average of a function of its variables (as $t \to \infty$) must equal its ensemble average, so that it justifies the idea that the probability that the system will be found in a certain region of its phase space is proportional to the phase area of that region (in Lebesgue measure). But for this to be true, it is obviously necessary that the system's trajectory should not stay in a certain region for a disproportionally long time. Such islands in which trajectories may wander in and never leave are called KAM tori in the literature, and to be an ergodic system is not to have KAM tori in the system's phase space. It is usually very difficult to prove the absence of KAM tori even if it is rather obvious from a direct inspection.

Our models when in the symmetrical state of stable equilibrium seem indeed to be ergodic[9] from an intuitive point of view in the same way that many simple thermo-systems are. It appears rather obvious that it is not possible to have islands or subspaces in the neighborhood of the state (except the state itself) into which any phase trajectory of the system—either the ring-bead or the cone—would wonder and stay forever. Because of the energy constraint, any trajectory starting away from (or leading out of) the ground state would eventually, if not quickly, come (return) to that state. This analysis shows a couple of possibilities that are of interest to the problem of chance and determinism. First, this system may well be ergodic even though we cannot produce a rigorous proof of it; second, even if it is not ergodic, it may be close enough to be ergodic (e.g. quasi-ergodicity in the sense of Vranas 1998), and that is good enough to make chance compatible with determinism.

Given that the systems can be modeled as thermo-systems and they are intuitively ergodic, we are able to conclude that the probability of having a particular perturbation occur in a certain region of the phase-space neighborhood of the unstable equilibrium state should be proportional to the area of that region. This implies that perturbations are uniformly distributed around that state, namely, a small deviation from it in one direction is equally likely as one in another direction. (To put it more accu-

9. One may wonder whether this is in conflict with the idea that SSB is a form of *ergodicity breaking* (Goldenfeld 1992, 55ff). But there is no such conflict. The SSB we have been discussing is precisely a process that breaks ergodicity of the whole system, because once the system gets into any one of the asymmetrical ground states, it will be trapped there and thus breaks—analogous to having a KAM torus—the ergodicity the system may have before. In quantum statistical mechanics, SSB is identified as the appearance of a decomposition in a state, which may be ergodic to begin with, into extremal KMS states, each of which is identifiable with a pure phase and has a lower symmetry than the original state (Emch 1977). See also Huang (1998, 295), with respect to the case of ferromagnetism. (However, Huang wrote, "[w]hat causes spontaneous symmetry breaking, then, is a breakdown of ergodicity[,]" which reads like a categorical mistake. It is as if we can say what causes mental disorder is insanity. Ergodicity is a form of symmetry (just as sanity a form of mental health), and hence the breakdown of it is a form, not a cause, of SSB. Here we may have an example of confusing the formal from the causal sense of a physical concept.)

What I am considering here is whether or not the system before such breakings may be regarded as ergodic.

rately, think of a phase-space ring around the unstable equilibrium state. Given that the pre-SSB system is ergodic, the probability that a perturbation occurs in one finite section of the ring must be the same as the probability of its occurring in any other section of the same size. The result is the same even if we let the size of the section approach zero. Hence a perturbation from the equilibrium state is equally likely in any directions.)

The second element (number 2 above) in our answer is the transitions from the unstable equilibrium state to the stable ones, which are caused by the perturbations. When the systems in question reach a state in which their SSB-parameter is beyond the critical value, the state becomes unstable—which means that the systems will not return to it when they acquire however small a perturbation from it—and yet the distribution of the perturbations for these systems should not be affected. It is so because, as we have seen earlier, making a system capable of SSB introduces no causal asymmetry with respect to the equilibrium state although it renders it unstable. Hence, when symmetry breaking becomes possible for a system, the system will be disposed to make a transition from one state to another with the presence of any perturbation. But since the distribution of the perturbations is the same, the probability of the system departing in one direction should be the same as its departing in any other possible direction.

Furthermore, how the symmetry in question is actually broken also depends on what the ground states (the new stable equilibrium states) are and how they are situated (number 3 above). If they do not form a set of degenerate states which together has the same symmetry—i.e. one being transformable into another by the same symmetry group, then the distribution of actual breakings may not agree with the system's probabilistic disposition for SSB. For example, if in our ring-bead system the perturbations from O are equally distributed between the left and the right direction and yet only one of the ground states, say, θ (but not $-\theta$), is allowed (never mind how this can actually be the case, but it is certainly physically possible), then the probability of the bead to deviate from O would be 0.5 in either direction while the probability of it settling into the ground state is unity; and therefore, the two probabilities do not match. (Here I assume that every perturbation deterministically leads to a transition of the system to its (or one of its) ground state(s).)[10]

In sum, the three elements together do seem to provide an adequate—

10. If one finds this case difficult to imagine (because switching directions inevitably requires the passing of the point O), then think of the case of Poincaré's cone, where all directions of perturbations from the upright position are possible and yet not all directions along which the cone lies on its side are allowed; and the set of these disallowed directions is not of measure zero. The same argument, mutatis mutandis, should go through.

though short of a proof—and unified answer to the two questions I posed in the beginning of this section, namely, they provide reasons for us to believe that SSB is a chancy process and actual breakings are equally probable. Moreover, seeing how the three elements fit together to provide a simple and realistic answer to our questions should tell us something about the adequacy of each element, the most controversial being the first, namely, the ergodicity of the pre-SSB ring-bead system in the manifold of O's phase-space neighborhood. I will not speculate on how much the above argument contributes to the status of ergodic theory in general, but it ought not be dismissed as insignificant when it provides one of its crucial elements. I also think that it goes some way in calming the qualm about ergodic theory caused, for example, by Loewer's questions, as mentioned above. For instance, it does show, once we are convinced that the pre-SSB system is ergodic or quasi-ergodic, why the chance of SSB is proportional to the phase area in the phase neighborhood of O.

As a final point, let me speculate on the idea that SSB may have something to do with making chance work in a deterministic world. Even though complexity or a large number of degrees of freedom is not necessary for chance in a deterministic world (Sinai 1970), experience tells us that the more complex a system is the more likely it will harbor chancy observable processes. There are many different ways for a system (or a universe) to become complex (or more complex); but they all have the general feature of bringing about stable (or more stable) observable states that "cover"[11] more and more dynamical states. One of the simplest examples of such a process is the free expansion of a gas from one volume to another bigger volume. Imagine a gas first confined in one of the two halves of a container, which is separated by a door in the middle. When the door is suddenly lifted, the gas molecules will rush into the empty chamber, and by that is the number of their possible positions and momenta increased. When the gas finally permeates the whole container and reaches the equilibrium, the system has become more complex in exactly the sense I have given above.

Now, SSB is not unlike the free expansion of gas. The crossing of the critical value of the parameter is analogous to the lifting of the door, for when either happens, the space of possible observably distinct states is widened (in the ring-bead model, having the bead at rest at positions other than O); and with the parts of the system in question occupying the possible states, the system becomes more complex. Of course, the free-expansion case of gas is not a case of SSB (because, inter alia, it is a Boltzmann

11. Whenever I say that an observable or macroscopic state covers or contains a set of dynamical or microscopic states, I mean no more than that no permutations among the members of the latter can change the value of the former.

transport phenomenon), so SSB is only one of the ways of making the world more complex. One may say that an SSB is a process by which new observable states are produced, which have distinct symmetries, each being different from the one of the initial observable state.

5. Conclusion. To summarize, the justification of the use of (equal) chance in SSB—a unified answer to two separate questions—turns out to have a three-element structure: (1) the uniform distribution of the perturbations that holds even when the ground state becomes unstable; (2) the deterministic transition from the unstable state to the stable states; and (3) the symmetrical (of the same symmetry) arrangement of the symmetry-breaking states. The first is again justified when we realize that the system, including the perturbations, before the SSB becomes possible can be regarded as an ergodic (or quasi-ergodic) thermo-system; and the second when we see that the SSBs are transitions that break that ergodicity. We then see how SSB may be a type of instability which is responsible for producing diverse chancy processes in our universe that contributes to its complexity.

REFERENCES

Aitchison, Ian J. R. (1984), *An Informal Introduction to Gauge Field Theories.* Cambridge: Cambridge University Press.

Albert, David (2000), *Time and Chance.* Cambridge, MA: Harvard University Press.

Clark, Peter (1987), "Determinism and Probability in Physics", *Proceedings of the Aristotelian Society,* Suppl. LXII: 185–210.

Coleman, Sidney (1975), "An Introduction to Spontaneous Symmetry Breakdown and Gauge Fields", in Antonio Zichichi (ed.), *Laws of Hadronic Matter.* New York: Academic Press, 138–215.

Earman, John (1986), *A Primer on Determinism.* Dordrecht: D. Reidel.

Earman, John, and Miklós Rédei (1996), "Why Ergodic Theory Does Not Explain the Success of Equilibrium Statistical Mechanics", *British Journal for Philosophy of Science* 47: 63–78.

Emch, Gérard G. (1977), "An Algebraic Approach for Spontaneous Symmetry Breaking in Quantum Statistical Mechanics", in Peter Kramer and Mario Dal Cin (eds.), *Groups, Systems and Many-Body Physics.* Braunschweig: Friedr. Vieweg & Sohn, 246–284.

Emch, Gérard G., and Chuang Liu. (2001), *The Logic of Thermostatistical Physics.* Berlin, Springer-Verlag.

Goldenfeld, Nigel (1992), *Lectures on Phase Transitions and the Renormalization Group.* Reading, MA: Addison-Wesley.

Greenberger, D. M. (1978), "Esoteric Elementary Particle Phenomena in Undergraduate Physics—Spontaneous Symmetry Breaking and Scale Invariance", *American Journal of Physics* 46: 394–398.

Guttmann, Yair M. (1999), *The Concept of Probability in Statistical Physics.* Cambridge: Cambridge University Press.

Ismael, Jenann (1997), "Curie's Principle", *Synthese* 110: 167–190.

Liu, Chuang (2003), "Classical Spontaneous Symmetry Breaking", *Philosophy of Science* (Proceedings), forthcoming.

Loewer, Barry (2001), "Determinism and Chance", *Studies in History and Philosophy of Modern Physics* 32: 609–620.

Poincaré, Henri (1952), *Science and Method.* New York: Dover.

Sinai, Ya. G. (1970), "Dynamical Systems with Elastic Reflections", *Russian Mathematical Surveys* 25: 137–189.
Sivardiere, J. (1983), "A Simple Mechanical Model Exhibiting a Spontaneous Symmetry Breaking", *American Journal of Physics* 51: 1016–1018.
Sklar, Lawrence (1993), *Physics and Chance.* Cambridge: Cambridge University Press.
Smith, Peter (1998), *Explaining Chaos.* Cambridge: Cambridge University Press.
Strocchi, Franco (1985), *Elements of Quantum Mechanics on Infinite Systems.* Singapore: World Scientific.
Toda, Morikazu, R. Kubo, and N. Saitô (1995), *Statistical Physics I.* 2d ed. Berlin: Springer-Verlag.
van Fraassen, Bas (1989), *Laws and Symmetry.* Oxford: Clarendon Press.
——— (1991), *Quantum Mechanics: An Empirical View.* Oxford: Clarendon Press.
Vranas, Peter (1998), "Epsilon-Ergodicity and the Success of Equilibrium Statistical Mechanics", *Philosophy of Science* 65: 688–708.
Weinberg, Steven (1996), *The Quantum Theory of Fields, vol. II: Modern Applications.* Cambridge: Cambridge University Press.
Wigner, Eugene P. (1979), Symmetries and Reflections: Scientific Essays of Eugene P. Wigner. Woodbridge, CT: Ox Bow Press.

Part 5

Infinity Systems and Emergent Phenomena

This part contains three previously published papers, which are loosely connected by the theme of an interpretational study of the emergent phenomena in infinite quantum systems.

Phase transitions and critical phenomena are found among recurrent events in daily life. The boiling of water is a phase transition from water's liquid phase to gaseous phase, and so is the freezing of water from liquid to solid. During the freeze at the border region of the two phases, a critical phenomenon occurs when water becomes opaque before turning into ice. Such common phenomena possess a theoretical challenge to mathematical physicists when they have to come up with a *rigorous* explanation of when and how they happen. From a quantum statistical mechanical point of view, a no-go theorem has been proven such that it is impossible for any finite systems (all natural systems we study are finite) to have such phenomena taken place in them. It turns out that a specific type of idealization known as "taking the thermodynamic limit" is to be used in order to circumvent the no-go theorem and produce the rigorous explanation. And taking the thermodynamic limit is to imagine that the size of the system in question goes to infinity while its density remains finite.

Papers 18 and 19 explore different aspects of the mathematical techniques that are used to deal with the no-go result for phase transition and critical phenomena, and paper 17 is a discussion of the nature of idealization as a scientific methodology by using such method as taking the thermodynamic limit or using the renormalization group. The issues concerning the so-called emergence or emergentism are also discussed in the papers. An important work that further this line of research is Robert W. Batterman (2002) *The Devil in the Details: Asymptotic Reasoning in Explanation, Reduction, and Emergence*. Oxford University Press. For a critical discussion of this work, see Gordon Belot (2005) "Whose Devil? Which Details?" *Philosophy of Science* 72 (1): 128-153. See also Sorin Bangu (2009). "Understanding Thermodynamic Singularities: Phase Transitions, Data, and Phenomena." *Philosophy of Science* 76 (4): 488-505, and Robert Batterman (2005) "Critical Phenomena and Breaking Drops: Infinite Idealizations in Physics." *Studies in History and Philosophy of Science Part B* 36 (2): 225-244.

CHUANG LIU

APPROXIMATIONS, IDEALIZATIONS, AND MODELS IN STATISTICAL MECHANICS

ABSTRACT. In this paper, a criticism of the traditional theories of approximation and idealization is given as a summary of previous works. After identifying the real purpose and measure of idealization in the practice of science, it is argued that the best way to characterize idealization is not to formulate a logical model – something analogous to Hempel's D-N model for explanation – but to study its different guises in the praxis of science. A case study of it is then made in thermostatistical physics. After a brief sketch of the theories for phase transitions and critical phenomena, I examine the various idealizations that go into the making of models at three difference levels. The intended result is to induce a *deeper appreciation* of the *complexity* and *fruitfulness* of idealization in the praxis of model-building, not to give an abstract theory of it.

1. INTRODUCTION

The overall structure of the paper's main argument goes as follows. The existing theories of approximation and idealization are found to be inadequate because they oversimplify the measure of approximation and misidentify the purpose (and therefore the measure) of idealization, respectively. It is then argued that to idealize is not necessarily to postulate conditions which are themselves approximate (or realistic) or whose gradual removal guarantees asymptotically more approximate results; it is also – mainly (I argue) – to set up theoretical models so that approximate as well as true laws of nature can be either applied (if they are already known) or discovered (if they are not yet known). These points have been argued extensively in Liu (1999a, 2004), therefore, they will only be *summarized* here with a new example. Because of this close connection with model-building in science and because of the extreme scope and complexity of the activity, it does not seem plausible that there is a logical model for idealization – something analogous to Hempel's D-N model for explanation. Rather, I contend that to understand the rich content of this notion, one must conduct detailed analysis of its different guises in the praxis of science. Thermostatistical physics is chosen here because it contains the oldest and best understood microscopic accounts of macroscopic behavior;

Erkenntnis **60:** 235–263, 2004.

it is where a clear divide between the phenomenological and the fundamental laws shows up in its simplest form (see Cartwright 1983, 1989, for the importance of this point); and it is fundamental in that without it, we cannot even understand why there is 'anything' (i.e., observable macro-objects) rather than 'nothing' (i.e., merely unobservable micro-objects). My hope is to unearth some important aspects of idealization through a detailed analysis of thermostatistical models.

2. IDEALIZATION AND APPROXIMATION (CF. LIU 1999A, 2004)

According to the traditional characterization, an idealized law (sentence)[1] or theory is intuitively understood as one that is true under the ideal conditions which differ from the actual conditions either slightly – so that the law is approximately true – or significantly but, when the differences are systematically reduced, the law becomes more and more approximate. And a sentence is approximately true if it is in fact false but very close to being true (cf. Barr 1971, 1974). There are mainly two kinds of theories for approximate truth: the state-space and the possible-world theory. According to the most comprehensive state-space theory (Niiniluoto 1984, Chap. 7, 1986,[2] see also Marquis 1991), the approximation to truth is measured by the metric distance between two cells (discrete space) or points (continuous space) on an Euclidean space of the states of the world. Let S be a predicate space which fully describes all the possible states of the world by having n families of one-place, mutually orthogonal predicates (i.e., S is spanned by n axes). The distance between any two cells in an S-space is the usual Euclidean measure: $D = \sqrt{\sum d_i^2}$, where d_1 is the difference between the two cells on the ith axis. (This scheme can be extended to the continuous case and to the case for generalizations or laws (cf. Niiniluoto 1986).) And then, one needs to specify the limit of approximation (a function of D) within which all sentences are considered approximately true. One should note that this is a theory of what approximation is, not one of how we come to know what is, and what is not, approximately true.

The possible-world theory (Hilpinen 1976; Weston 1987, 1992) runs roughly as follows. Let us think of a possible world, w, (or a possible way the world might be) as a possible model. Then we have the following:

> A sentence ϕ is approximately true (under Σ) in w if, and only if, there is another world, w', such that it is 'α-close' to w and satisfies ϕ.

The central element of this theory is the measure of similarity or closeness – Σ – between possible models (or worlds). Hilpinen (1976) follows Lewis in trying to articulate the notion of similarity among possible worlds, but comes short of offering a quantitative measure for it – the difficulty is carefully explained by Lewis (1986), while Weston's (1987, 1992) construal of Σ is very similar to D in the state-space theory.

Then, the notion of idealization can also be made precise, and there are mainly two traditional theories for it. For one, we mainly follow Laymon's (1980, 1985, 1989) comprehensive theory of confirmation, of which idealization is a central component. Given a generic theory T, and a set of n ideal-condition claims: $I = \{I_i | i = 1, \ldots, n\}$, such that I_j is less idealized (or more realistic) than I_i, where $j > i$, we have that for all $i \in \{1, \ldots, n\}$, $I_i : T \mapsto T_i$ (i.e., the ideal-condition claim maps the generic theory to an idealized one), and $T_i| = P_i$, where $P = \{P_i | i = 1, \ldots, n\}$ is a set of predictions. Suppose that the non-idealized theory, T^U, predicts the true P^U; and let $\Delta^- I = |I_j - I_i|$ and $\Delta^+ P = |P_j - P_i|$ (where $j > i$) denote respectively the factual distance between two ideal conditions and predicted states of affairs, we then have,

[A] T is a good idealized theory (or the idealization is justified) if it is always true that $T : \Delta^- I \Rightarrow \Delta^+ P$; otherwise, it is not (or the idealization is not justified).

In other words, when $I \to 0$, we have $T \to T^U$ and $P \to P^U$, if T is a good idealized theory. I suppose that one can likewise define a rate of convergence for idealizations of the same kind – such as $\gamma = -\Delta^+ P / \Delta^- I$, when $\Delta^- I \to 0$ – and claim that one idealization in the set is better than another if its rate of convergence is higher than that of the other. The other theory is Nowak's (1972, 1980), which, as shown for instance in Liu (1999a), can be understood as a special case of Laymon's theory, if some slight modifications are made.[3]

2.1. *The Main Problems*

Suppose that we follow the possible-world theory of approximation, which is equipped with a clearly formulated sense of approximation, Σ. The space of possible worlds (or models) is supposed to be Euclidean. This implies that when one says, for instance, that sentence A is more approximate than sentence B, given that they are comparable (e.g., they are about a person's height), one means that the total factual distance between A and the truth is smaller than that between B and the truth. The *total* factual distance depends not only on the magnitude of each individual distance but also on the number of such distances; and a *factual* distance is a magnitude that is easy to define between certain types of states but difficult for other types. It is safe to assume that as long as one is dealing with quantitative theories in which possible states of affairs are determined by the values of the main variables (i.e., degrees of freedom), a factual distance is simply defined as the Euclidean distance between two sets of such values.

However, this is not true for many such spaces. The point can be easily understood by imagining a space partitioned by laws of nature which prohibit the crossing from one region to another. For example, a phase space for particles may be divided into two regions: in one, particles move with speeds less than c (the speed of light) – i.e., ordinary objects – and in the other, with speeds greater than c – i.e., tachyons. Suppose that a particle moves at a speed very close to, but less than, c. To attribute any speed, $\nu < c$, to it would be a better approximation than giving it a speed slightly over c. But according to the Euclidean metric, the latter is factually closer to the truth than most of the former, and so the latter should be more approximate, which is clearly the wrong judgment. One may argue that authors such as Niiniluoto, Hilpinen and Weston could certainly rebut this challenge by noting the fact that the particle is forbidden by a physical law to move faster than the speed of light. However, there may be particles, such as tachyons, which do move faster than the speed of light; although no particle can be accelerated from a sub-light-speed state to a super-light-speed one, but that is beside the point. Unless these authors want to put in their theory of approximation different stipulations on how factual closeness is to be used on different species of objects, the challenge stands. I assume that the general measures of closeness proposed above are independent of the specific dispositions of objects and the laws associated with them.

One may perhaps hope that the problem can be avoided by identifying fault lines in each case and construct multiple Δ's accordingly for the space. However, the matter does not seem so simple, even if it is taken as a metaphysical – rather than an epistemological – problem. The space of either possible worlds or possible states of affairs is very much unlike the pVT (pressure-volume-temperature) phase space of thermal systems, where one has mostly homogeneous pure phases except a few lines of phase transitions near which the systems' behavior is singular (more on this later). To see this point, let us consider again our example with the speed of light. With regard to particles that are not tachyons, suppose that it is true that a particle, α, is moving with a speed ν that is less than c but quite close. There are two statements, one says that α moves with a speed ν' and the other says that it moves with a speed ν'', where $\nu' \in (c, \nu)$ and $\nu'' \in (\nu, 0)$, both measured in the same reference frame of ν; and moreover, $|\nu - \nu'| < |\nu - \nu''|$. Which statement is more approximate then?

If the space of possible states of α below c is Euclidean, the statement with ν' should be regarded as more approximate than the one with ν''; but the opposite seems more reasonable because a particle *is more likely* to be moving at ν'' than at ν', considering the difficulty with which a particle has to be accelerated to a speed of ν'.

These examples seem to point to a qualitative approach, as a solution, for the problem of approximation (cf. Balzer et al. 1987; Kuipers 2000). This is certainly the right direction in general; and in an integrated approach to approximation and idealization, there might be a way in which a qualitative treatment of idealization yields what one may regard as a semi-qualitative treatment of approximation. However, the commonsense notion of approximation is a quantitative one; it may be vague but it is certainly quantitative. Therefore, whatever the final theory of approximation and idealization is, there must be a place for the numerical measures of some kind of closeness that accounts for approximation. My integrated approach, which is only roughly sketched in Liu (2004), has the idea that the qualitative limits of approximation is only supplied by idealization, which is a tool for carving up nature at its joints in that models are built to discover and test laws of nature, and these in turn supply the qualitative boundaries for the quantitative evaluations of approximation claims.[4]

Now, a problem for the traditional theory of idealization. It concerns idealized laws, which are false because of being idealized but which still approximate the true laws. Let us represent an idealized law hypothesis as: L given I, where, I is a set of ideal conditions, L is in the form of '$N(F, G)$',[5] where N denotes natural necessity.[6] According to the theory of idealization given above, we should say that a law is approximately true if, and only if, either all of L's factual consequences (e.g., instantiations) are approximately true under I or their approximation is asymptotically improved when I is gradually removed. But here a problem arises. Without any lost of generality and in connection with our later discussion of thermostatistical models, let us suppose that our true law is the isotherm, L, of a system in which a phase transition takes place (see Figure 1), and that L_1 and L_2 are two laws which are true under ideal conditions, I_1 and I_2.

Which of L_1 and L_2 is more approximate, given L obtains? If we, by following the traditional theory, compare their factual consequences (i.e., point-wise distances), L_1 is closer to L than L_2, but we can clearly see that *as a matter of law*, it is the opposite: L_2 at least has singularities in it (i.e., the points at the two ends of the Maxwell plateau), whereas L_1 is all smooth – which as we will see in a moment means that there is no phase transition in L_1! And, worse still, part of L_1 implies that $\partial p/\partial V > 0$, namely, the pressure increases while the system expands at a constant T!

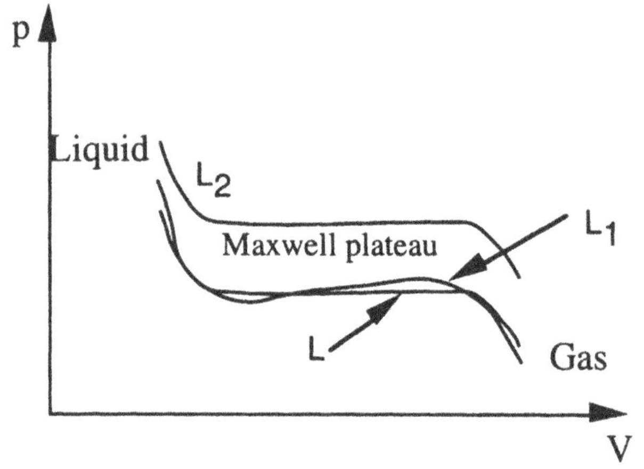

Figure 1.

Similarly, we would say that holding I_2 is a better idealization than holding I_1, and yet I_2 defines a factually closer world (or state) to L than I_2 does. L_1 is what Pauli is said to have called a 'not even false' law hypothesis, which in our context must mean that the hypothesis totally misses the *form* of the intended law. One could always improve the approximation – the empirical fit – of such a hypothesis, but should one? But why shouldn't one, if the traditional theory of idealization is correct?

Nor would it help – for very similar reasons – by requiring asymptotic improvement of approximation when the ideal conditions are gradually removed.

We now see that there are two dimensions for evaluating idealized laws: factual closeness and *form closeness* (resemblance). The traditional theory only has the first dimension, which is why it fails.

2.2. *A Theory of Idealization*

The reason for the failure of the traditional theories is that there are different types of idealizations, of which approximation production is only one type – e.g., those idealizations one depends on in finding generalizations in a collection of experimental data. But this is not the same kind of reasoning as the one that assumes that the universe contains only the sun and the earth or is without electromagnetic field. Nor are both the same type as assuming the ratio of an object's speed to that of light is zero or taking the volume of a thermo-system to infinity (i.e., taking the thermodynamic limit, as will be explained later). Nor are they the same as the construction of lattice models for bulk matter and for quantum fields. From the perspective of approximation production, we simply do not see anything in common in these different acts of idealizations. We must therefore begin afresh.

Within the scope of science it seems that acts of idealization serve the purpose of 'inventing' or 'isolating' models in which laws of nature can be found (if not known) and/or applied (if known) (see McMullin 1985; Cartwright 1989). The complexity and variability of natural phenomena and our desire to explain them with a small number of simple and invariant laws demand that we *disentangle* them. The best way of achieving that is to conceptually imagine or design a setting (not obtained in the actual world) in which systems operate in simple and uniform manners so that approximate or true laws can be discovered and (in some cases) tested; and this is what idealizations are primarily used for. (One should not therefore miss the close connection between idealization and *thought experiments*.) It is only when the actual world with its complex phenomena is thus disentangled into separate theoretical models by idealization is it hopeful that genuine laws of nature may be discovered or applied. Some laws so discovered and tested in one context (or discipline) are not directly applicable in another, even if they are universally true. Again, phenomena in the latter context (or discipline) need to be disentangled, and idealized models need to be set up for the detached components, before the laws become applicable again. Can mechanical laws of motion for Newtonian particles be used to study the phenomena of heat if idealized models are not allowed?

A particular form of this idea is perhaps first clearly articulated by J. S. Mill in Book VI (1874), namely, genuine laws of moral science can never be discovered by looking for regularities in the observed social events, for though the regularities may be close to the observed facts (i.e., approximately true), they can never be universal enough to qualify for laws. Instead, one should think of social phenomena – similar to chemical phenomena – as results of complex combinations of simple systems or processes which obey simple and universal laws – à la chemistry (see also Cartwright 1997; Wimsatt 1987, 1994; Morrison 1997). The more complex and/or variable the phenomena and the deeper the posited mechanisms of which the laws are to be formulated, the wider the gap between the laws and the observable facts. And the gap should not affect the truth of the laws or theories in question. If one can conceptually (and to some extent even experimentally) cut the world apart at its joints, one can better understand how it actually operates – i.e., giving explanations – and predict how not yet existing phenomena may occur (out of a composition of causes which obey the laws).

The same point can be made from a different perspective. We humans are situated in nature in such a way that some of its component strands – e.g., regular macroscopic processes – are made obvious to us. Little art of idealization was required for the ancients to formulate a precise and mathematically sophisticated science of astronomy. That is because the relevant characteristics were clearly revealed to the astronomers by nature – i.e., by making the sizes of the heavenly bodies and their observers much smaller than the distances that separate them. No such luck existed, for example, for the science of mechanics: few if any surfaces at Galileo's time could be made smooth enough to allow an easy recognition of the separation between motion and friction. It required Galileo's ingenious idealization (cf. McMullin 1985; Suàrez 1999) – which made such a separation – to set the foundation for the discovery of the law of motion. To include, for instance, friction – which de facto accompanies all macroscopic motions in nature – in the description of a body's state of motion would be to entirely miss a mechanical joint of nature.

What are joints of nature? It is not a question that admits a simple general answer. First, they are supposed to be places in natural processes where component systems or processes meet to produce observable phenomena. Hence, the important things are not the joints but what are, at least conceptually, *detachable* from the joints. What are these? Are they the same as natural kinds? I do not think they are, but the terms that refer to such detachable parts are similar to natural kind terms. As natural kind terms, they are not absolute but may come and go in the history of science. (For several illustrative examples, see Liu 2004.) One of the results of a major scientific revolution is certainly the disappearance of some old terms for detachable parts and the introduction of some new ones. But the concept of such parts – the most important of them are the fundamental ones – only makes sense if one accepts the metaphysical view that nature is somewhat like a machine whose working is the result of the combined working of its intricate parts; and different combinations of the same, or of different, parts usually result in different natural processes. They are obviously incompatible with a radically holistic view of nature. Again as natural kind terms, it is up to science to find out which are detachable components in nature and which are not. If there are detachable parts, there must be joints; and the combined actions of the parts through the joints produce the variety of phenomena we observe. If one is a realist about the joints and the parts they connect, one believes that the progress of science is in part a search for the real joints and, hence, parts of nature, while if one is an empiricist, one may believe that the joints and the parts they connect may be purely theoretical constructions, whose justification lies in how well they save the phenomena and satisfy the various methodological criteria of theory selection.

What exactly is the relation between idealized models and laws? Idealized models are needed for *discovering* or *applying* laws in systems or situations other than those in which they are supposed or known to hold.[7] These are true especially for fundamental laws at the microscopic level. It is no exaggeration to say that laws of nature often express exact and true relations of physical necessities among properties of systems which themselves do not exist in the actual world, e.g., frictionless planes. In the context of law discoveries, the acts of detachment, or disentanglement, I mentioned before are often needed to separate the different elementary components or causal processes; and in the context of law applications, one needs to detach the different components of a complex system so that laws which obviously apply to them separately can be tried out on the system to see whether the results of their combined application matches the system's observed (or observable) behavior. The model of a frictionless region, for instance, is a state of the world in which the region of space in question has nothing inside (except the testing object); and a process in which both electromagnetic and gravitational forces are present can be disentangled into two model processes each of which has one force alone present, other aspects of the processes being the same (cf. Liu 2001b). We will see in a moment when we discuss thermostatistical models some subtle complications in this way of seeing model-building in science (see also Herfel et al. 1995). But as will be shown, the general conception given here is still sound.

What exactly are models in our approach? Are they of the same type as those logicians use? Opinions are divided on this issue in the literature on scientific theories. People such as van Fraassen (1989) and Suppe (1989) (see also Giere 1988) believe that they are the same, while people such as Cartwright (1999) and Morrison (1999) think that they are not.[8] The structuralist conception of models (cf. Sneed 1971; Stegmüller 1973; Balzer et al. 1987) is closer to the former, namely, it takes a model for a theory to be the set-theoretic object that satisfies the theory. The task of scientific theorization is, in a nutshell according to the structuralists, to discover the logical structure of the theory by constructing set-theoretic predicates that describes the models. To have a theory of thermodynamics, for instance, is to find out the structure of the predicate, 'being a thermodynamic system', and construct abstract models that are in the extension of the predicate. Then, to know whether such statements as 'the air in this room is a thermodynamic system' is true, one can run tests to obtain results for the observables of the air and compare them to the corresponding results given by the models. The primary task the structuralists set for themselves is making explicit the logical structure of any scientific predicates, or any predicates that deserve such a name.

However, according to the latter, models are mediators between theories and reality (or phenomena). Here I suggest that the two views can be reconciled. A model can be defined by all the sentences that are true in it or it can be defined independently of these sentences. The set of all real numbers, for instance, can be defined independently of many arithmetic statements that are true of it; and yet it is a model for all such statements and more. The same can be said – though not as precisely – of a scientific model that either van Fraassen or Cartwright may have in mind. The ideal gas model is a relational structure in the general sense that it defines a possible world – one in which the molecules are sizeless and forceless – and it can be, and is, specified independently of any laws that are true in it. Cartwright et al. are right to say that such models are not the theories themselves since a theory consists in part of law-statements, e.g., Boyle's law for the ideal gas model, but van Fraassen et al. are also correct since models such as the ideal gas model are indeed a possible structure (or a family of possible structures) in which laws such as Boyle's law are true.

How can laws that are true in an idealized model also be true simpliciter? Cartwright (1983, 1989) argued that they are not, but I disagree.[9] Here is a rough sketch of the main reason. Given there are laws, $N(F, G)$, where F and G refer to dispositional properties (or propensities), such laws may hold universally even though one needs an idealized model to display their operations in their pure form. To know how exactly electromagnetic force operates in its pure form, one has to have a system in which only the electromagnetic force acts; but there are in fact no such systems – hence it is an idealized model. But it does not mean that when there are other forces operating – what actually happens – the law of electromagnetism fails. As a law relating dispositional properties, it holds at every instant of its presence anywhere and anytime. In the philosophical literature of laws of nature (Dretske 1977; Armstrong 1983, 1978; Lewis 1973, 1983; Carroll 1994), dispositional properties are suspect and therefore not included among the legitimate properties relatable by laws. However, if one has any knowledge of physical sciences, one cannot avoid noticing that there the majority of laws are laws relating dispositional properties. For instance, all laws expressed in the form of differential equations are laws of this kind. The law of motion in e.g., Newtonian mechanics is represented by the differential equation relating the second-order time-derivative of a position function and the first-order space-derivative of a potential, not the infinite number and varieties of trajectories as the solutions of the equation under different initial/boundary conditions. The categorists (see e.g., references cited above) cannot even claim in this case that the law of motion comprises the set of all possible solutions of the equation of motion, because then the law is neither definable nor knowable.

However, not all laws are like these. There are plenty of laws in physics or elsewhere that are false but true only in their tailor-made models. Boyle's law of ideal gas is a good example; and so are the mean-field models that we shall see in a moment. And all those laws that are regarded as 'limiting cases' of the more general true laws – e.g., all the laws in non-quantum and/or non-relativistic theories – are also of this kind. We may call them either 'approximate' or 'idealized' laws (cf. Kyburg 1990, Chap. 10). My point above was simply that it is not necessarily the case that idealized models only produce idealized laws.[10] And the examples I used above are all true laws in idealized models.[11]

By now it seems apparent that whatever an alternative theory of idealization may be, it will not comprise a set of adequacy conditions and a simple measure for its degrees. However, two concepts (or dimensions) seem crucial to the measure of idealizations: *detachability* and levels of *stability*. Both are indispensable for the correct and effective description (and understanding) of nature. What we want to achieve by idealizations is really to create models with the right amount of detachment and/or separation of levels. For the former, we see ample examples in the discoveries or applications of simple and universal causal laws in the face of complex phenomena, which by hypothesis are the result of the coinstantiation of more than one of such causal laws. And for the latter, examples in thermostatistical physics or material science provide excellent illustration.

In general terms, what are levels of stability in natural phenomena? What is the essence of this concept? In a sense, the phrase 'of stability' is superfluous, for something may be counted as a level only if it is stable. The notion in question is mainly an epistemic notion, having to do with the extent to which phenomena are investigated and learned. Although the notion of a level is an epistemic one, the stability associated with it is physical, namely, the stability is a set of physical conditions that guarantees the irrelevance of details that go beyond the level. For instance, the level of molecules and the level of atoms are distinguishable epistemically because one can learn about a whole category of properties of systems by regarding the systems as comprising molecules as fundamental units, despite the fact that many interactions involve the exchange of atoms not whole molecules. As we will see in the next section, whether a level is acceptable depends on how well a set of properties that belong solely to the level can be use to formulate laws or lawlike statements that admit acceptable empirical tests.

Note, sometimes we talk of a hypothesized level, which turns out not to work; that simply means that it seems a well defined level from theoretical considerations, but it disagrees with experimental results. We should also keep in mind that known disagreements with experimental results may not be always sufficient to condemn a seemingly well defined model in theory.

3. IDEALIZATIONS IN THEORIES OF PHASE TRANSITIONS (PT) AND CRITICAL PHENOMENA (CP)

In this section I offer a case-study of an area in physics, which I argue will show us how idealization as the essential tool for model-building is at work. Since idealization works mainly in two ways – the detachment of casual agents and the separation of levels of phenomena – and I have offered extensive study of the former elsewhere (cf. Liu 2004), the following will be an attempt to take a closer look at the latter. The account will be given with as little technical details as is possible not to compromise the accuracy of the description. The theory that describes and explains PT and CP is part of thermostatistical physics, which is usually divided into thermodynamics and statistical mechanics. The theory for PT and CP consists of three different levels, on which different types of models are used. On the first (phenomenological) level, PT and CP are described in pure thermodynamic terms (Stanley 1971; Toda, et al. 1983). The most discussed PTs and CPs are two: those of fluid, such as the liquid-gas transition and its critical region, and those of magnet, such as the paramagnetic-ferromagnetic transition and its critical region. Although being entirely different phenomena, they are remarkably similar with respect to PT and CP. It turns out that at the transition from one phase to another below the critical temperature, T_c, the isotherms develop singularities (i.e., non-differentiable points);[12] and as $T \to T_c$, all relevant thermodynamic functions diverge in the form of a power law: $F(t) \propto |t|^{-y}$, where $t = (T - T_c)/T_c$ and $y = \alpha\ or\ \beta\ or\ \delta$ is the critical exponent. The components are, respectively, the component of specific heat at constant volume, of density difference at constant pressure and temperature, of correlation length, of compressibility at constant temperature, and of pressure-density ratio (Fisher 1974). The respective critical exponents of fluid or magnet have the same values, which is known as the universality of CP that has no explanation on this first level.

To explain the first-level phenomena, two types of models have been developed. On the second (semi-microscopic) level, we have the mean-field models, which include the van de Waals (vdW) model for fluid and the Weiss model for ferromagnet (cf. Stanley 1971; Goldenfeld 1992). The vdW law (or equation of state) for the 'real' gas, $[p+a/v^2][v-b]=kT$, is an improvement of the ideal gas law, where a and b are introduced to capture for each substance the effect of (i) the attractive interaction between molecules and (ii) the hard-core of each molecule. From this law one is able to derive the critical values: T_c, p_c, and V_c, and thus the law of corresponding states: $[\tilde{p}+3/\tilde{v}^2][\tilde{v}-1]=8\tilde{T}$ (where for any x, $\tilde{x}=x/x_c$), an equation free of the explicit presence of a and b and so *universally* valid for all substances. The 'wiggles' in the isotherms at $T<T_c$ are made by Maxwell (1875) into the Maxwell plateaus (see Figure 1), at the two ends of each of which are the needed singularities that signal a PT (cf. Emch and Liu 2002, Section 11.3). Despite being a crude model from a microscopic perspective, the vdW model – and the law that is true in it – can already do a great deal. As mentioned above, the law of corresponding states shows already the universal nature of thermodynamic properties. For PT and CP, we have mentioned that with the Maxwell plateaus, the PT for fluid are accurately accounted for, and one can use the law to obtain critical exponents. The exponents however disagree with the experimental values, e.g., while the experimental value of β, the density-difference component, is 1/3, the vdW law predicts that $\beta=1/2$; but they preserve some experimentally established relations among these exponents.

The Weiss model is designed to explain ferromagnetism. According to this model, the magnet consists of dipoles as tiny compasses which are not only influenced by the external field, B, but also by one another, the latter of which is expressed by I – the 'mean-field' created by the rest of the dipoles in the magnet (or part of it). With such a model, one can do a simple substitution: $B\rightarrow B+\alpha I$, in the equations – more precisely, in the Hamiltonian – for a paramagnet (i.e., magnets with no PT) and derive ferromagnetism. The Weiss model is a microscopic model without the details of the micro-structure of the magnet. It succeeds in explaining the PT but fails drastically with the estimates of the critical exponents. Moreover, both the vdW model and the Weiss model predict that there are PTs even if the system is only one-dimensional, which is deemed incorrect when the Ising-type models are used.

The simplest and genuinely microscopic models (at the third level) are the Ising-type models (Emch 1971; Toda et al. 1983; Goldenfeld 1992). All models of this type have the following ingredients:

(i) a lattice of N fixed sites of equal distance, and each site is occupied by a spin-1/2 particle that can be in one of two possible states: spin-up ($\sigma_i=+1$) and spin-down ($\sigma_i=-1$);

(ii) an interaction between any two sites is given by the term, $J_{ij}\sigma_i\sigma_j$, where, i *or* $j = 1, 2, \ldots, N$, and $J_{ij} > 0$ for specified sites (e.g., nearest neighboring sites) and $j_{ij} = 0$ otherwise.

The Hamiltonian of the system is therefore written as

$$(1) \qquad H_N = -\sum_{<i,j>} J_{ij}\sigma_i\sigma_j - B\sum_i \sigma_i,$$

where B is the external field and $< i, j >$ means pairs of sites with specified relations only. According to statistical mechanics (cf. e.g., Sklar 1993, Chap. 2), the partition function,

$$Z_N = \sum_{\sigma_1} \cdots \sum_{\sigma_N} \exp(-H_N/kT),$$

turns out to be sufficient for recovering all thermodynamic variables with well-established rules and therefore for obtaining the isotherms. However, it can be shown (Yang and Lee 1952; Griffiths 1972), without solving the Z_N, that no isotherms can possibly harbor singularities, as are required for showing PT and CP (i.e., a general no-go theorem). The problem is resolved when it is realized that when taking to the thermodynamic limit (TL), i.e., $N \to \infty$ and $V \to \infty$ while N/V remains finite, a system in thermo-equilibrium is well-behaved, namely, F_∞ (the free energy under TL) and p_∞ (the pressure under TL), etc. not only exist but are also smooth except at those points where PTs take place. Exact solutions (Onsager 1944; see also Schultz et al. 1964) of the one- and two-dimensional Ising systems show that PTs only occur at systems of at least two dimensions, and for such systems the derived critical exponents reasonably agree with the experimental values. But for more complex systems a method of renormalization group (RG) must be used instead (Kadanoff 1966; Wilson and Kogut 1974; Fisher 1983, 1998; Batterman 1998). RG provides a recursive technique of rescaling the system so that one can repeatedly collapse blocks of degrees of freedom so that in the end the macroscopic features of PT and CP appear.[13]

3.1. *The Macroscopic Model*

As described above, PT and CP are observable phenomena whose patterns can be seen on plotted isotherms of data points, and the isotherms can also be predicted by empirical laws such as the vdW law. Already, the isotherms are idealized in thermodynamics so that PT and CP points are shown as singularities, but the idealizations are minimal. In this sense, there is some similarity between the mechanics of celestial bodies and thermodynamics.

As I mentioned earlier, one of the reasons that a relatively precise science of motion for the celestial bodies came long before such a science could even be conceived for the terrestrial bodies is that nature appears to have done the idealization for those who studied the celestial movements. The same seems to be true for thermodynamics in contrast to statistical mechanics. One may nonetheless ask the following question: is it not the case that the truth of all thermodynamic laws depends on a drastic idealization that ignores the molecular nature of bulk matter? Should we not therefore say that all thermodynamic models are among the most (unrealistically) idealized models in science? Comparing to this, whatever idealization one takes for a mechanical model should be far more realistic since it at least recognizes the molecular nature of matter.

This question brings up an important, and so far neglected, point about idealization. If one is asked to compare a thermodynamic model with an Ising-type model, which should one say is more realistic – i.e., less idealized? One may say that since ignoring the molecular structure and constructing laws as if fluid is continuous violate our basic metaphysical belief of the molecular nature of bulk matter, thermodynamic models are far less realistic (or far more idealized). But one may also say that since thermodynamic models are far closer to the systems from which empirical generalizations are discovered and/or tested than the Ising-type models are (in fact no empirical generalizations can be drawn from an Ising model), the Ising-type models are far less realistic (or far more idealized).

This analysis illustrates the point about the relevance of levels in the phenomena to the evaluation of idealizations. Even though it is one of our fundamental beliefs that matter is made of small particles, the degrees of idealization should not be measured on an absolute scale dictated by this ontological picture. Models of celestial mechanics in which great and complex bodies such as stars are regarded as nearly point masses and models of thermodynamics in which fluid and magnet are regarded as continuous matter are not to be regarded as highly idealized models because they are so close to being the way such systems are presented to us by nature. To us humans, the mid-size macroscopic level is one of the most stable levels of natural phenomena on which a whole host of idealized models should be evaluated independently. There are other levels, from the cosmological to that of the elementary particles, on which exist stable phenomena, from which idealized models can be made and evaluated. We should not think that there is an absolute scale. Hence, idealizations are level dependent; and the ones belonging to different levels are not comparable (those in thermodynamics are not any better or worse than the ones in statistical mechanics). One can only speak of degrees of idealization and make comparison among idealized models within the same level.

3.2. *The Mean-Field Models*

Serious efforts of idealization arise from the desire to explain why the singularities, among other thermodynamic features of bulk matter, are there. For the PT and CP in fluid, we have (as described above) the vdW model, accompanied with the Maxwell construction (see the Maxwell plateau in Figure 1), and for ferromagnetism, the Weiss model. Once bulk matter is regarded as being made of corpuscular parts, the most extremely idealized model for gas is the ideal gas model, from which Boyle's law can be derived (and thus explained): not only the size of, but also the non-collisional interactions among, the particles are ignored. The model is so impoverished that it cannot represent anything except diluted gases at $T \gg T_c$. Moreover, the model is so idealized that some more drastic idealizations can be added to derive the same Boyle's law, (dramatically simplifying the derivation). To assume that the molecules in a diluted gas all move in straight lines, with the same constant speed, and in directions perpendicular to the walls of the container would be a very unrealistic idealization by anyone's standard, and yet if one subjects the ideal gas model to this additional assumption, one still recovers Boyle's law.[14] This assumption can of course be dropped and replaced by the famous Maxwell distribution of the molecular velocities, and yet we recover the same Boyle's law.

Here we see yet another interesting aspect of the notion of idealization that is not noticed before: given a law or a theory, the choice of idealized models for it may be underdetermined, which implies that a choice among such models – i.e., a choice of the idealizations – cannot be made by comparing their effects on the law or theory alone. In the case of the Maxwell vs. the Bernoulli model of ideal gas, it is clear that the former is chosen because it is more realistic. It is not unlikely that this is one of the two ways by which such cases are resolved. Here is how I see it. Whenever a underdetermined case as described appears, there are two possible ways of making a determination, if one is genuinely called for: either one extends the theory in question to include more phenomena (if it can be done) and the extension breaks the underdetermination, or one makes no extension but evaluates the models against a network of theories and empirical findings.

In the vdW model, the two essential ideal conditions in the ideal gas model is removed. However, neither the exact size nor the exact form of the attraction are specifiable in the model; rather their effects are phenomenologically estimated by the two constants, b and a, such that the assumption of the hard-core produces a correction: $v \to v-b$, and that of the attractive force a correction: $p \to p + a/v^2$, in the equation of state; and the values of the constants are ultimately determined by experiments. From this one can see that the vdW model is not yet a fully microscopic model, although it is microscopic by conception.

As briefly explained earlier, the Weiss model of ferromagnetism consists of dipoles in a solid whose average total spin orientation determines its magnetization. When $T < T_c$, the agitation of the dipoles is subdued so much so that a self-induced mean-field of the aligned spins appears with the external field and does not disappear when the external field vanishes. Similar to the vdW model, there is no precise argument for how the model gives rise to ferromagnetism; rather the effect of the model is given phenomenologically by the constant, α, in the added term, αI, which is in turn determined experimentally. One can see a striking parallelism between the vdW and the Weiss model in the 'ideal gas model → vdW model' transition and the 'paramagnetic model → Weiss model' transition; both involve microscopic but inexact models and the simple insertion (in the equations resulting from the first models) of experimentally determined constants for whatever effects such models may produce. Because the ideal conditions are qualitatively correct and quantitative inexact, there should be plenty of room for further fine-tuning. This type of idealizations agrees well with the traditional theory (see, [A] above): the evaluation of the mean-field models should depend on whether the fine-tuning eventually works out. To my best knowledge, no such fine turning were taken seriously.

The first fully microscopic model from which the vdW law is recovered with an explicit microscopic definition of the two constants is perhaps Ornstein's given in 1908.[15] The Hamiltonian of an N-particle system is the sum of two terms: the kinetic and the potential term, which also splits the total partition function of the system accordingly. The kinetic part is the same as the one in the Maxwell-Boltzmann theory of ideal gas, and the potential part contains the two-body inter-molecular force term:

$$(2) \qquad \phi(\bar{x}_{ij}) = \begin{cases} \infty & \bar{x} < \delta \\ \phi^a(\bar{x}_{ij}) & \text{otherwise} \end{cases}$$

where $\bar{x}_{ij} = |x_i = x_j|$ is the distance between the two molecules and $\phi^a(\bar{x})$ the attractive part of the force which is weak and very long-range. With such a force, the molecules have a hard-core with a radius of $\delta/2$, and outside the core they attract one another with $\phi^a(\bar{x})$. The full derivation should not concern us here (see Emch and Liu 2002, Section 11.3); suffice it to point out that the crucial idealization is taken when one replaces the contribution of $\phi^a(\bar{x})$ to the partition function by its average:

$$(3) \qquad a = (1/V)\int dx\phi^a(x),$$

where the integration is taken over the whole system, and a is one of the constants in the vdW law (and b is of course identified with δ, the diameter of the hard-core). Another major model (or a series of models) is the one-dimensional Kac–Baker model in the early 1960s (cf. Kac 1959; Baker, Jr. 1961). In this model, the idealization (3) is removed and replaced with a stipulation of the exact function of $\phi^a(\bar{x})$. This model is much more sophisticated than the Ornstein model: it can derive not only the vdW law with more rigor but also the Maxwell plateaus at $T < T_c$, which the Ornstein model cannot.[16]

One may argue that the Kac–Baker model is obviously better than the Ornstein model, so I am wrong in saying earlier that the micro-models for the vdW law are underdetermined. This is not quite right. As far as recovering the vdW law is concerned, the Kac–Baker model does manage to avoid idealization (3), but it introduces such things as the thermodynamic limit. But is it not true that one can derive the Maxwell plateaus from the Kac–Baker model but not from the Ornstein model? The correct response to this should not be that the Kac–Baker model is a better vdW model, but rather it is probably a better model than any candidate for the vdW model.

This point raises the question of which model should be taken as a vdW model. We see from our discussion that the answer to this question is not at all trivial. There is one definite vdW law, which by conception takes the molecular structures of bulk matter seriously. But a model in which the law is true can be a purely thermodynamic model (a rather far-fetched possibility), a model which leaves the microscopic details of the two constants unspecified (the most fitting model, I think, to be called a vdW model), a model in which the vdW law is derivable with a critical idealization that replaces the attractive part of the inter-molecular force with its average value (the idealization explains why the vdW is a mean-field model), or a fully microscopic model with some highly unusual conditions of idealization (which seems to have gone beyond the vdW model).

The fully microscopic model from which the Weiss equation of ferromagnetism can be derived is a special case of the Ising-type models, to which I now turn.

3.3. *The Ising-Type Models*

If the traditional theory of idealization were right, the development of thermostatistical explanations of PT and CP should have followed the direction of vdW and Weiss models. However, what happened in history was not a pursuit of less idealized laws (or models) along the line of mean-field models – better values for a, b, and α or more constants for more refinement – but an invention of models to which well-established laws in other fields of physics, such as the laws of (classical or quantum) mechanics and electrodynamics, apply. We have already seen above some fully microscopic models for fluid in the efforts to derive the vdW law. The Ising-type models, as briefly described above, are mostly for ferromagnetism, even though there is a lattice model for fluid, as we shall see in a moment.

First, let us look at the recovery of the Weiss model from an Ising model. If we follow Weiss in thinking of ferromagnetism as the result of each, say the *ith*, spin feeling the collective influence of all the other spins and the external field, we can rewrite the Hamiltonian in (1) as follows:

$$(4) \qquad H_N = -\sum_i \left(\sum_{j \neq i} J_{ij} < \sigma_j > +B\right)\sigma_i,$$

(where the fluctuation term, $\sum_{j \neq i} J_{ij}(\sigma_j - < \sigma_j >) \approx 0$), which differs from a paramagnet's Hamiltonian, $H_N = -\sum_i B\sigma_i$, by a simple substitution of $B \to B + \sum_{j \neq i} J_{ij} < \sigma_i >$ (where, $J_0 \sum_{j \neq i} J_{ij}$). This is the Weiss model, with its substitution scheme: $B \to B + \alpha I$, and here we see the meaning of the coefficient, α, which is the sum of the coupling constants between any two sites. Similar to the Ornstein model for recovering the vdW equation of state, the Ising model shows how exactly the mean-field appears in the Weiss model; and also like the Ornstein model, there is no precise way of calculating $I =< \sigma_j >$ or $\alpha = J_0$ either rigorously or in a controlled approximation for all Ising-type models.

We now turn to the discussion of the idealizations involved. The variety of the Ising-type models is generated by the number of dimensions, the type of site occupancy, and the type of interactions. Some models are simple to use but unrealistic, while the more realistic ones are notoriously difficult to handle: the simplest model is perhaps a one-dimensional string of finite length on which spin sites are located at equal distance; and each site takes only one of two spin states – up or down – and interacts only with its nearest neighbors. And the most complex is perhaps a three-dimensional honeycomb shape lattice of equal distanced sites on which two spin values are possible, and the interactions are not only among the nearest neighbors.[17] When I mentioned the derivation of the Weiss model above, I deliberately avoided specifying which among these Ising-type models must be used. To recover the equation and leave the value of $J_0 = \sum_{j \neq i} J_{ij}$ and $< \sigma_j >$ in (4) to the experiments, one can use any of the Ising models.

It should be obvious by now that from the mean-field models to the Ising-type ones a fundamental shift has taken place in terms of the quality or the degrees of idealization. A sense of this may be had if we ask the question: which type of models is a better *idealized* one? If this question is put to us concerning instead the ideal gas model and the vdW model, we would have no difficulty in giving a definite answer. The ideal conditions for the latter is not only more realistic than the former but also produces via the vdW law a more approximate description of thermo-systems than the former can produce via Boyle's law. The same cannot be said unambiguously for the previous pair – between the mean-field and the Ising-type models.

For the ferromagnet, we have the standard Ising-type models, as described earlier; but for fluid, the Ising-type model is the lattice gas (cf. Yang 1952), which is built out of an Ising magnet by a simple transformation. For this, any Ising lattice will do: instead of having on each site a spin of up or down, we have the site as 'occupied (by a molecule)' or 'not occupied'. Hence, we switch from a configuration of spin orientation pattern on a lattice to that of occupation pattern on a lattice. In terms of random variables, where σ is for spin and n for the occupation number of molecules, the switch is as follows:

$$\sigma : i \in \Lambda \mapsto \sigma_i \in \{+1, -1\} \Rightarrow n : i \in \Lambda \mapsto n_i \in \{0, 1\}.$$

Despite its rather artificial appearance, the lattice gas model can do a great deal (as briefly noted above). And with some modification, such as in the Kac–Baker model, one can even use it to rigorously recover the vdW law. It is especially a suitable model for solutions and alloys.

To truly answer the question of which type of idealized models are better, we need to consider several aspects of this comparison. First, the mean-field models are more for the discovery of laws than for the application, whereas the Ising-type models are the opposite. I say 'more for' rather than 'definitely for' because some known laws are used – even if implicitly – in the mean-field models. However, the construction of mean-field models, such as the vdW and the Weiss model, is mainly for the discovery of semi-empirical laws which account for the phenomena. The laws are derivable neither from purely thermodynamic principles and/or laws nor from statistical mechanical laws within the models; therefore, they are new laws discovered in the corresponding mean-field models. The Ising-type models are also constructed to account for the phenomena – especially PT and CP, but the aim is to see to it that the account follows from the (heavily tailored) application of laws in mechanics and electrodynamics in the models.

Second, idealizations that make the mean-field models are certainly not as *exact* as the ones that make the Ising-type models, but they are by no means less *realistic*. A judgment on which one is more realistic is difficult because the difference is subtle. The key idea of mean-field models is precisely to use the 'mean-field', namely, to use the averages of what are in reality so complex as to defy any precise knowledge. One of the key assumptions behind this use of average is of course the well-supported belief that the deviations are negligibly small. On the other hand, the key idea of the Ising-type model is precisely to use the 'lattice', a rigid structure with all the particles separated by equal, or multiples of equal, distances. Such models (or their idealizations) are clearly less realistic than the mean-field models (or their idealizations). Perhaps one can say that the mean-field models are as *inexact* as the Ising-type models are *unrealistic*; but this is not quite true, and a simple discrepancy of the two shows why. As I mentioned earlier, while the Ising-type models predict the right critical exponents, the mean-field models get the wrong values. This must mean that taking the averages in the ways in which the mean-field models take them is too crude for characterizing the fluid or the ferromagnet near the critical point. Hence, the mean-field models are simply not acceptable models regardless of whether they are more realistic than the Ising-type models.

Third, despite their theoretical power, the Ising-type models are simply not realistic. We can see this by a simple argument. There is a no-go theorem for all Ising-type models which says that there is no possibility for any PT (and hence no CP) in any such model systems with finite extensions. To recover PT, one must take the thermodynamic limit, as briefly described earlier. Once that is taken, it is found that there is no PT in a one-dimensional string of nearest-neighbor interactions or interactions of similar short-ranges, and one needs at least a two-dimensional system for that to emerge. Now, taken literally, the no-go theorem should be a simple refutation of the Ising-type models, which should naturally lead to a rejection of the idealizations (as unrealistic). Literally speaking, most, if not all, PTs take place in finite systems, and yet when modeled by Ising models as their microstructures, none of them can harbor PTs. The best interpretation of this situation is I believe the following. A ferromagnet, for instance, is neither an Ising lattice nor an infinite system; but there are good theoretical reasons to idealize it into a model of infinite Ising-type lattice, and among the reasons, I am certain, is that this is a, if not the, model in which a relatively simple adaptation of laws in mechanics and electromagnetism can be carried out to rigorously derive PT and CP. As is argued elsewhere (cf. Liu 1999b), the thermodynamic limit is in fact a species of idealization that, in different guises, is widely used in physical sciences; the most common variation of which is to assume an otherwise porous and malleable body solid and rigid ('filling in' the gaps by an infinite number of parts). It seems that by introducing two radically unrealistic idealizations – the Ising lattice and the thermodynamic limit – one is able to do better than what one can with the less idealized mean-field models.

4. CONCLUSION

Let me now summarize the main points of arguments and observations in this paper and point out some remaining issues for future investigations. Approximations in general cannot be evaluated by a single (a few) simple measure(s). This is because the simple measures cannot take into account of the roles laws of nature play in assigning different dispositions to different species of objects. Further, a single (a few) simple measure(s) for idealization, such as approximation approbation, is (are) also unlikely to work because there are not one but two dimensions in the measure of 'closeness' (to a true law) of an idealized (or approximate) law: how much it resembles the facts and how much it resembles the form (or 'shape') of the regularity (which also applies to counterfactual cases). I then argued that it seems rather hopeless to try to come up with a single measure of the closeness of forms in approximate laws; and even if such a measure can be had, one has to deal with the *balance* between it and the closeness to facts. Moreover, idealizations – and idealized models as their products – are not just for idealized laws. If some acts of idealizations are used to detach component systems or processes from joints of nature, the result may well be the discovery or application of true laws. Indeed, none of the theoretical laws at the microscopic level can be discovered or applied without idealized models, and yet there is no reason to believe that they are therefore not true laws of nature. The second part (Section 3) of the paper is intended to offer a detailed study of a case in which acts of idealization are carried out and laws and lawlike hypotheses are 'discovered' as a result. It is not intended to help formulate a general theory of idealization, nor should we look for one there.

We see three different levels of idealized models in the theory of thermostatistical physics. The thermodynamic level is the closest to direct experiments and therefore needs the least idealization, except the level itself is sealed off by an almost metaphysical idealization which ignores the molecular nature of bulk matter. The result of this idealization is thermodynamic laws and state equations that describe fluctuationless processes. The semi-microscopic level sees models that acknowledge the molecular structures but pack the information into average quantities that appear in the phenomenological laws with empirical constants whose values are determined experimentally. At the fully microscopic level we find models with their microstructures fully – if somewhat artificially – specified. Just when one might think that the idealizations which make the Ising-type models may deprive of them the possibility of explaining PT and CP (i.e., the no-go theorem), another idealization – taking the thermodynamic limit – comes to the rescue; and together not only PT is recovered but with the help of renormalization group technique one can also make accurate calculations to reveal the details of CP regions.

This paper mainly discussed those idealized models that are intended for the discovery or the application or both of laws of nature. Some such laws are true but some are only true in the models – which I call idealized or approximate. Now, are thermodynamic laws on the first level true or approximate laws? This seemingly innocent question is actually difficult to answer. Judging from the enormous effort by the physicists and mathematicians in trying to 'recover' those laws from statistical mechanics, one would conclude that they are true laws; but the very effort clearly reveals that from a microscopic perspective those laws must admit exceptions, however unlikely such excepts are.[18] It is fairly obvious that the laws in the mean-field models (at the second level) are meant to be taken as improvable approximate laws. But the true laws for thermo-phenomena turn out not to be greatly improved mean-field laws, but true laws from other areas of physics which apply in highly unrealistically idealized models, such as the Ising-type models.

This paper has barely scratched the surface of the notions of idealization and approximation. Many deeper issues or possibilities remain. Must ideal conditions be physically possible? The answer seems to be 'Yes', but what do we do then with infinite volumes and continuous edges? I argued for the conclusion that some idealized models give true laws but some only approximate ones.[19] It seems that going down one level always gives one cases of exceptions for the laws obtained as universal at the upper level. But this is a very preliminary speculation. One may simply ask the following question: why should Boyle's law be false – though somewhat approximate – while the law of inertia be true? Very roughly put, Boyle's law is true for a diluted tank of sizeless and forceless particles while the law of inertia is true for a forceless region. Shouldn't they have the same truth value?

ACKNOWLEDGEMENTS

I would like to thank Gérard Emch for many discussions on issues related to this paper. I also want to thank five anonymous reviewers from NSF for providing constructive criticisms on an earlier version of this paper. This paper is supported in part by a NSF Grant No. SES-9910889.

NOTES

1 Since there is no danger of mistaking a law-sentence for a law in the context of our discussion here, I shall often use the term 'law' for both and let the contexts tell which is meant. This saves me from writing such odd phrases as 'Boyle's law-sentence' rather than 'Boyle's law', which in some context refers to the statement of the law and in others the law itself.

2 It is far beyond the scope of this paper to give an adequate account of Niiniluoto's theory of approximation and verisimilitude. For a recent survey, see Niiniluoto (1998).

3 For other similar approaches, see Krajewski (1977); also Schwartz (1978), Rosenkrantz (1980), Niiniluoto (1987), Cartwright (1989, Chap. 5) and Kyburg (1990, Chap. 10).

4 For other criticisms of the traditional theory, see Ramsey (1992). It is not the intention of this paper to work out a satisfactory theory of approximation by carefully engaging the literature in this area. For this job, another paper is required.

5 One should note that most laws in physics and many laws in science are such that the F's and G's are property variables – magnitudes or quantities – which are really classes or bundles of properties. For such laws, the form may be better given as, $N[P, Q|Q = \mathbf{f}(P)]$, which means that for every value of P and Q, it is naturally necessary that all P's are Q's; or more accurately, given D_p and R_q as the domain for P's values and the range for Q's values, for every value of $P^i \in D_p$, there is a $Q^j \in R_q$, such that $\forall x N[P^i(x), Q^j(x)]$.

6 The natural necessity is the same as physical necessity for a physicalist; but for a non-physicalist, the natural is broader than the physical. Psychophysical laws, if exist, are natural but not physical.

7 Idealized models may also be used for testing a law or a theory, but this role is quite different from the other two roles – discovery and application – because if by 'testing' we mean checking the law in question whether it is consistent with a set of other laws to which it should belong, some models certainly serve this logical function; but if by that we mean checking – i.e., doing experiments and interpreting data – whether the law is actually true, it must be possible for the idealized models to be approximately realized. This is a severe demand that most highly theoretical models, e.g., in the microscopic or cosmological realms, cannot meet. Therefore, many theoretical laws are not directly tested in their own models but rather indirectly tested in models that are close to the phenomena.

8 For early discussions of models, see Apostel (1961), Braithwaite (1962), McMullin (1968) and Hesse (1970); for alternative views, see Redhead (1980) and Horgan (1994); and for a review, see Liu (1997).

9 For the details of this argument, see Liu (2004).

10 For more on this, see also Creary (1981), Kline and Matheson (1986) and Laymon (1989).

11 One may worry about the following possibility: we have not discovered a single law in science and we may never discover any since all of our known 'laws' and all those we will in fact discover are only approximate ones. If so, the distinction between approximate laws and true laws in idealized models would be of no practical meaning. This is however not quite true. Perhaps none of the laws we regard true is in fact true, but those we do have reasons to believe as true will not be false because of the idealized models we use to either discover or apply them. The current law of gravitation may only be approximately true, but it will not turn out to be false because we studied it in separation from other force fields, such as electromagnetism. This is why I can use Newton's theory of gravity and Maxwell's theory of electromagnetism to illustrate how laws that hold in idealized models may hold in general. Both are false theories but not false because we have studied them separately.

12 For instance, as shown in Figure 1, the liquid-gas PT of a fluid below the critical temperature is bounded by two (singular) points in the pV isotherm, across which 'rates' such as $[\partial p/\partial V]_T$ are not continuous.

13 See Liu (1999b, 2001) for more detailed discussion.

14 D. Bernoulli in fact did the calculation under such a model with a few more ideal conditions in 1738, see Emch and Liu (2002, Section 3.2).

15 The original derivation is given by L. S. Ornstein in his dissertation (in Dutch) in Leiden with a title whose English translation reads: Application of the statistical mechanics of Gibbs to molecular theoretical questions. For its contemporary rendering, see Emch and Liu (2002, Section 11.3).

16 The derivation is more rigorous at least in the sense that by replacing the attractive force, $\phi^a(\bar{x})$, with its average, (3) is not an approximation that one can estimate in a controllable manner, i.e., being able to produce a numerical estimate of whatever degree of approximation. However, it is only a one-dimensional model and involves some mathematical maneuvers which seems to be extreme measures of idealization: one needs to take the thermodynamic limit in order to derive the vdW law in this model, and one has to put the quantity, $A^{-1}\int_0^\infty x\phi^a(x)dx$ to zero while maintaining that $A = \int_0^\infty \phi^a(x)dx$ be finite. I shall discuss in detail the nature of thermodynamic limit later. The extension to higher dimensions seems to be conceptually transparent, but technically it is by no means straightforward. In this aspect, the Kac–Baker model is very much like the Ising-type models.

17 There are more complex models. One must note that the Ising-type models are really classical models because the spin variable on each site is not treated as a quantum operator as it should in quantum theory but rather as a two-valued classical variable. The corresponding quantum models are know as the Heisenberg models, on whose sites sit particles of quantum spins, namely they take all the eigenvalues of a quantum spin as possible values.

18 For an in-depth discussion of the whole issue, see Callender (2001).

19 More accurately I should say that I argued, mostly elsewhere (Liu 2004), that idealized models can produce universally true laws. I have not argued that idealized models can also produce approximate laws, which in view of laws such as Boyle's and Weiss's I take to be fairly obvious.

REFERENCES

Apostel, L.: 1961, 'Towards the Formal Study of Models in the non-Formal Sciences', in H. Freudenthal (ed.), *The Concept and the Role of the Model in Mathematics and Natural and Social Sciences*, Synthese Library, D. Reidel, Dordrecht, pp. 1–37.

Armstrong, D. M.: 1978, *Universals and Scientific Realism*, Vol. 2: *A Theory of Universals*, Cambridge University Press, Cambridge.

Armstrong, D. M.: 1983, *What Is a Law of Nature?* Cambridge University Press, Cambridge.

Baker, Jr. G. A.: 1961, 'One-dimensional Order-Disorder Model Which Approaches Phase Transition', *Physical Reviews* **122**, 1477–1484.

Balzer, W., C. U. Moulines, and J. D. Sneed: 1987, *An Archetectonic for Science: The Structuralist Approach*, D. Reidel, Dordrecht.

Barr, W. F.: 1971, 'A Syntactic and Semantic Analysis of Idealizations in Science', *Philosophy of Science* **38**, 258–272.

Barr, W. F.: 1974, 'A Pragmatic Analysis of Idealizations in Physics', *Philosophy of Science* **41**, 48–64.

Batterman, R. W.: 1998, 'Why Equilibrium Statistical Mechanics Works: Universality and the Renormalization Group', *Philosophy of Science* **65**, 183–208.

Braithwaite, R. B.: 1962, 'Models in the Empirical Sciences', in E. Nagel, P. Suppes, and A. Tarski (eds.), *Logic, Methodology and Philosophy of Science: Proceedings of the 1960 International Congress*, Stanford University Press, Stanford, pp. 224–231.

Callender, C.: 2001, 'Taking Thermodynamics (Too) Seriously', forthcoming in *Studies in History and Philosophy of Modern Physics*.

Carroll, J. W.: 1994, *Laws of Nature*, Cambridge University Press, Cambridge.

Cartwright, N.: 1983, *How the Laws of Physics Lie*, Clarendon Press, Oxford.

Cartwright, N.: 1989, *Nature's Capacities and Their Measurement*, Clarendon Press, Oxford.

Cartwright, N.: 1997, 'Models: The Blueprints for Laws', *Philosophy of Science* **64**, S292–S303.

Cartwright, N.: 1999, 'Models and the Limits of Theory: Quantum Hamiltonians and the BCS Models of Superconductivity', in Morgan and Morrison (1999), pp. 241–281.

Creary, L.: 1981, 'Causal Explanations and the Reality of Natural Component Forces', *Pacific Philosophical Quarterly* **62**, 148–157.

Dretske, F.: 1977, 'Laws of Nature', *Philosophy of Science* **44**, 248–268.

Emch, G. G.: 1971, *Advanced Statistical Mechanics*, Lecture Notes, preprint, Rochester, NY.

Emch, G. G. and C. Liu: 2002, *The Logic of Thermostatistical Physics*, Springer-Verlag, Berlin.

Fisher, M.: 1974, 'The Renormalization Group in the Theory of Critical Behavior', *Review of Modern Physics* **46**, 597–616.

Fisher, M.: 1983, 'Scaling, Universality and Renormalization Group Theory', in F. J. W. Hahne (ed.), *Critical Phenomena: Proceedings of the Summer School, the University of Stellenbosch, South Africa, January 18–29, 1982*, Springer-Verlag, Berlin, pp. 1–139.

Fisher, M.: 1998, 'Renormalization Group Theory: Its Basis and Formulation in Statistical Physics', *Review of Modern Physics* **70**, 653–681.

Giere, R.: 1988, *Explaining Science: A Cognitive Approach*, University of Chicago Press, Chicago.

Goldenfeld, N.: 1992, *Lectures on Phase Transitions and the Renormalization Group*, Addison-Wesley, Reading, MA.

Griffiths, R. B.: 1972, 'Rigorous Results and Theorems', in C. Domb and M. E. Green (eds.), *Phase Transitions and Critical Phenomena*, New York, Academic Press, pp. 8–109.

Herfel, W., W. Krajewski, I. Niiniluoto, and R. Wójcicki (eds.): 1995, *Theories and Models in Scientific Processes* (*Poznan Studies*, Vol. 44), Rodopi, Amsterdam.

Hesse, M.: 1970, *Models and Analogies in Science*, University of Notre Dame Press, Notre Dame.

Hilpinen, R.: 1976, 'Approximate Truth and Truthlikeness', in M. Przlecki, K. Szanianwski and R. Wójicki (eds.), *Formal Methods in the Methodology of Empirical Sciences*, D. Reidel, Dordrecht, pp. 19–42.

Horgan, J.: 1994, 'Icon and Bild: A Note on the Analogical Structure of Models – The Role of Models in Experiment and Theory', *The British Journal for the Philosophy of Science* **45**, 599–604.

Kac, M.: 1959, 'On the Partition Function of a One-Dimensional Gas', *Physics of Fluids* **2**, 8–12.

Kadanoff, L. P.: 1966, 'Scaling Laws for Ising Models near Tc', *Physics* **2**, 263–272.

Kline, A. and C. Matheson: 1986, 'How the Laws of Physics Don't Even Fib?', in P. D. Asquith et al. (eds.), *PSA 1986*, East Lansing, *Philosophy of Science Association* **1**, 336–350

Krajewski, W.: 1977, *Correspondence Principle and the Growth of Knowledge*, D. Reidel, Dordrecht.

Kuipers, T. A. F.: 2000, *From Instrumentalism to Constructive Realism*, Kluwer Academic Publishers, Dordrecht.

Kyburg Jr., H. E.: 1990, *Science and Reason*, Oxford University Press, Oxford.

Laymon, R.: 1980, 'Idealization, Explanation, and Confirmation', in P. D. Asquith and R. N. Giere (eds.), *PSA 1980*, East Lansing, *Philosophy of Science Association* **1**, 336–350.

Laymon, R.: 1985, 'Idealizations and the Testing of Theories by Experimentation', in P. Achinstein and O. Hannaway (eds.), *Observation, Experiment and Hypothesis in Modern Physical Science*, MIT Press, Cambridge, MA, pp. 147–173.

Laymon, R.: 1989, 'Cartwright and the Lying Law of Physics', *Journal of Philosophy* **86**, 353–372.

Lewis, D.: 1973, *Counterfactuals*, Harvard University Press, Cambridge, MA.

Lewis, D.: 1983, 'New Works for a Theory of Universals', *Australasian Journal of Philosophy* **61**, 343–371.

Lewis, D.: 1989, *On the Plurality of Worlds*, Blackwell, Oxford.

Liu, C.: 1997, 'Theories and Models I: the Semantic View Revisited', *International Studies in the Philosophy of Science* **11**, 147–164.

Liu, C.: 1999a, 'Approximation, Idealization, and the Laws of Nature', *Synthese* **118**, 229–256.

Liu, C.: 1999b, 'Explaining the Emergence of Cooperative Phenomena', *Philosophy of Science* **66(S)**, S92–S106.

Liu, C.: 2001, 'Infinite Systems in SM Explanations: Thermodynamic Limit, Renormalization (semi-) Groups, and Irreversibility', *Philosophy of Science* **68(S)**, S352–S344.

Liu, C.: 2004, 'Laws and Models in a Theory of Idealization', *Synthese* (forthcoming).

McMullin, E.: 1968, 'What Do Physical Models Tell Us?', in B. van Rootselaar and J. F. Staal (eds.), *Logic, Methodology and Philosophy of Science III*, North Holland, Amsterdam, p. 392.

McMullin, E.: 1985, 'Galilean Idealization', *Studies in the History and Philosophy of Science* **16**, 247–273.

Marquis, J.-P.: 1991, 'Approximations and Truth Space', *Journal of Philosophical Logic* **20**, 375–401.

Maxwell, J. C.: 1875, 'On the Dynamical Evidence of the Molecular Constitution of Bodies', *Nature* **11**, 357, 374; or *The Scientific Papers of James Clark Maxwell*, Vol. 2, Dover (1965), p. 418.

Mill, J. S.: 1874, *A System of Logic*, 8th edn., Harper and Brothers, New York.

Morgan, M. S. and M. Morrison (eds.): 1999, *Models as Mediators: Perspectives on Natural and Social Science*, Cambridge University Press, Cambridge.

Morrison, M.: 1997, 'Physical Models and Biological Contexts', *Philosophy of Science* **64**, S315–S324.

Morrison, M.: 1999, 'Models as Autonomous Agents', in Morgan and Morrison (1999), pp. 38–65.

Niiniluoto, I.: 1984, *Is Science Progressive*, D. Reidel, Dordrecht.

Niiniluoto, I.: 1986, 'Theories, Approximations and Idealizations', in R. B. Marcus et al. (eds.), *Logic, Methodology and Philosophy of Science*, North Holland, Amsterdam, pp. 255–289.

Niiniluoto, I.: 1987, *Truthlikeness*, D. Reidel, Dordrecht.

Niiniluoto, I.: 1998, 'Verisimilitude: The Third Period', *The British Journal for the Philosophy of Science* **49**, 1–29.

Nowak, L.: 1972, 'Laws of Science, Theories, Measurement', *Philosophy of Science* **39**, 533–548.

Onsager, L.: 1944, 'Crystal Statistics. I. A Two-Dimensional Model with an Order-Disorder Transition', *Physical Review* **65**, 117–149.

Ramsey, J. L.: 1992, 'Towards an Expanded Epistemology for Approximations', in D. Hull, M. Forbes, and K. Okruhlik (eds.), *PSA 1992*, Vol. 1, *Philosophy of Science Association*, East Lansing, Michigan, pp. 154–164.

Redhead, M.: 1980, 'Models in Physics', *The British Journal for the Philosophy of Science* **31**, 145–163.

Rosenkrantz, R.: 1980, 'Measuring Truthlikeness', *Synthese* **45**, 463–488.

Schultz, T. D., D. C. Mattis, and E. H. Lieb: 1964, 'Two-Dimensional Ising Model as a Soluble Problem of Many Fermions', *Review of Modern Physics* **36**, 856–871.

Schwartz, R. J.: 1978, 'Idealization and Approximations in Physics', *Philosophy of Science* **45**, 595–603.

Sklar, L.: 1993, *Physics and Chance*, Cambridge University Press, Cambridge.

Sneed, D. L.: 1971, *The Logical Structure of Mathematical Physics*, D. Reidel, Dordrecht.

Stanley, H. E.: 1971, *Introduction to Phase Transitions and Critical Phenomena*, Clarendon Press, Oxford.

Stegmüller, W.: 1976, *The Structure and Dynamics of Theories*, Springer Verlag, New York.

Suàrez, M.: 1999, 'The Role of Models in the Application of Scientific Theories: Epistemological Implications', in Morgan and Morrison (1999), pp. 168–196.

Suppe, F.: 1989, *The Semantic Conception of Theories and Scientific Realism*, University of Illinois Press, Urbana, IL.

Toda, M., R. Kubo and N. Saitô: 1983, *Statistical Physics I: Equilibrium Statistical Mechanics*, Springer-Verlag, Berlin.

van Fraassen, B.: 1989, Laws and Symmetry. Oxford, Clarendon Press.

Weston, T.: 1987, 'Approximate Truth', *Journal of Philosophical Logic* **16**, 203–227.

Weston, T.: 1992, 'Approximate Truth and Scientific Realism', *Philosophy of Science* **59**, 53–74.
Wilson, Kenneth G. and J. Kogut: 1974, 'The Renormalization Group and the ϵ Expansion', *Physics Report* **12C**, 75–197.
Wimsatt, W. C.: 1987, 'False Models as Means to Truer Theories", in M. Nitecki and A. Hoffman (eds.), *Neutral Models in Biology*, Oxford University Press, Oxford, pp. 23–55.
Wimsatt, W. C.: 1994, 'The Ontology of Complex Systems: Levels, Perspectives, and Causal Thickets", in M. Matthen and R. X. Ware (eds.), *Biology and Society: Reflections on Methodology*, Canadian Journal of Philosophy, Supplementary Vol. 20, pp. 207–274.
Yang, C. N. and T. D. Lee: 1952, 'Statistical Theory of Equations of State and Phase Transitions I. Theory of Condensation', *Physical Review* **87**, 404–409.
Yang, C. N.: 1952, 'The Spontaneous Magnetization of a Two-Dimensional Ising Model', *Physical Review* **85**, 808.

Infinite Systems in SM Explanations: Thermodynamic Limit, Renormalization (semi-) Groups, and Irreversibility

Chuang Liu†‡
University of Florida

This paper examines the justifications for using infinite systems to 'recover' thermodynamic properties, such as phase transitions (PT), critical phenomena (CP), and irreversibility, from the micro-structure of matter in bulk. Section 2 is a summary of such rigorous methods as in taking the thermodynamic limit (TL) to recover PT and in using renormalization (semi-) group approach (RG) to explain the universality of critical exponents. Section 3 examines various possible justifications for taking TL on physically finite systems. Section 4 discusses the legitimacy of applying TL to the problem of irreversibility and assesses the repercussions for its legitimacy on its home turf.

1. Introduction. To do thermo-statistical physics (TSP) is in a large part to devise microscopic models for bulk matter and to use them to reconstruct, and thus to explain, its macroscopic behavior. When the bulk system in question is in a single phase and in equilibrium and of classical parts,[1] the task is rather simple and straightforward. Complications arise if these three conditions—single phase, equilibrium, and classical parts—

†Send requests for reprints to the author, Department of Philosophy, University of Florida, 330 Griffin-Floyd Hall, P.O. Box 118545, Gainesville, FL 32611–8545; e-mail: cliu@phil.ufl.edu.

‡I would like to thank Gérard Emch for discussions on topics I deal with in this paper. I would also like to thank Bob Batterman and Craig Callender for questions and comments on an earlier draft. This paper is supported in part by a NSF grant (SES-9910889).

1. I shall in this paper use adjectives such as 'classical', 'quantum', 'purely thermodynamical', etc., to mean the ways in which properties or objects are regarded in the appropriate theories. In this sense, the same object or property can be called classical or quantum or thermodynamical or mechanical, depending on which theory it is regarded in.

Philosophy of Science, 68 (Proceedings) pp. S325–S344. 0031-8248/2001/68supp-0027$0.00

no longer hold. If only one of the conditions fails, the complication is usually manageable, and in this paper I examine some foundational problems in one of these cases—multiple-phase systems in equilibrium with classical (or essentially classical) parts. The theoretical difficulties lie in the microscopic description and explanation of the transitions among, and transitional regions of, phases.

Another phenomenon which has long caused philosophical puzzlement and conceptual difficulties in TSP is the irreversibility of thermal processes. I shall show in this paper that there is a close connection between the treatment of phase transitions and that of irreversibility, and how the difference between the two sheds light on the justification for such treatments. Besides this methodological connection of the two phenomena, one should note that they are the two major and basic *cooperative phenomena* in nature. They are cooperative because they involve processes that are more than a collection of random mechanical processes of microscopic parts and are not accountable simply by mechanics and statistics; we shall see how statistical mechanics of finite systems are incapable in principle of rigorously accounting for either phenomenon and that radical measures have to be introduced to provide such an account. Further, they are basic because they involve the simplest *ordered* states in nature. Analogous but much more complex are such phenomena as the transition from organic to inorganic matter and from life to death of an organism. If one is serious about reductionism in the belief that all phenomena are eventually reducible to the physics of microscopic particles, one should learn one's lesson well with the reduction of phase transitions/critical phenomena and irreversibility.

In Section 2, I give a brief summary of the chief methods by which a rigorous micro-reconstruction of phase transitions (PT) and critical phenomena (CP) is given by the mathematical physicists. Such methods include taking the thermodynamic limit (TL) to recover PT and using the renormalization (semi-) group approach (RG) to explain the universality of critical exponents. Then, I examine in Section 3 various possible justifications for taking TL on physically finite systems. Section 4 discusses the legitimacy of applying TL to the problem of irreversibility and assesses the repercussion of its legitimacy on its home turf.

2. Phase Transitions and the Renormalization Group. The experimental observations and the purely thermodynamic (TD)—i.e., phenomenological—account of PT are unproblematic. For a fluid system, which may be in the phase of solid, liquid, or gas (cf. Figure 1), its thermodynamic states are determined by three variables, p (pressure), V (volume), and T (temperature). The equation, $f(p,V,T) = 0$, defines a surface in the phase space whose projections onto two-dimensional spaces are shown in Figure 1.

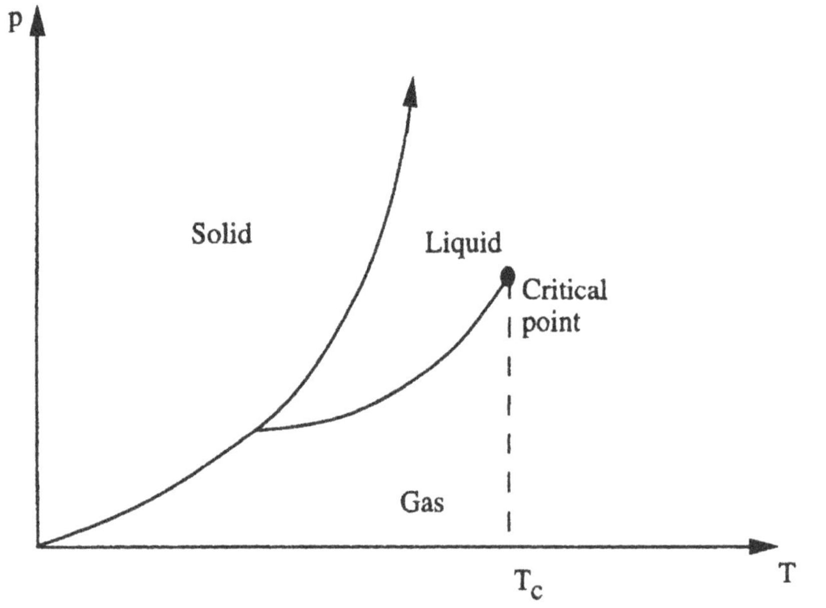

(a) pVT --> pT

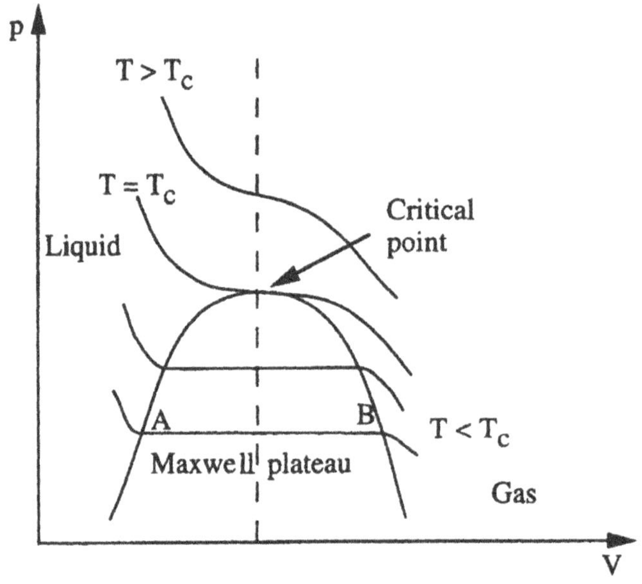

(b) pVT --> pV

Figure 1. Phase Transitions in Fluids. Projections of the pVT surface in two different phase planes.

Very similarly, a magnet system's thermodynamic states are determined by three analogous variables: B (external field), M (magnetization), and T (temperature). A similar equation defines a similar surface in its phase space (cf. Liu 1999). From Figure 1 (a) and (b), one can read off PTs as regions in the phase space which are bounded by singular points, such as A and B, and a critical point as the terminating singular point of a phase boundary line. The points are singular in the sense that derivatives, such as $\partial p/\partial V$, are discontinuous functions there. For instance, when a system reaches the region bounded by A and B, an increase of volume at constant temperature will not correspond to a change of pressure, which appears to be a singular behavior in TSP.

Idealizations are involved both in plotting the data into mostly smooth isotherms and in deriving them from the laws of TD, the most important of which is that since bulk matter is dense, no fluctuation occurs (hence the fact that the isotherms are mostly smooth). Such an idealization is necessary because no experiments, no matter how finely tuned, can ever determine whether the 'corners' which bound PT regions, such as A and B in Figure 1(b), are sharp or round. The theoretical reason for preferring sharp corners (and hence singular points) is quite simple: were sharp corners not preferred, the isotherm would have to first pass A and go down below the AB line and then curve back up above the AB line before curving back down again towards B, thus forming a horizontal 'S' shape. This entails such absurd conclusions as $\partial p/\partial V > 0$ at some points, i.e., that it is possible to increase a system's pressure by expanding its volume at a constant temperature.

Behavior around the critical point constitutes CP and is thermodynamically accounted for by a set of power laws, which show how various bulk variables of a system in the critical region vary with $t = (T - T_c)/T_c$. For fluids, they can be summarized as follows:

When $(t \to 0^{\pm})$,

a. compressibility at constant T	$\propto \begin{matrix} t^{-\gamma} \\ (-t)^{-\gamma'} \end{matrix}$	$(t > 0)$ $(t < 0)$
b. density difference at PT	$\propto (-t)^{\beta}$	$(t < 0)$
c. pressure-density ratio	$\lvert p - p_c \rvert \propto \lvert \rho - \rho_c \rvert^{\delta}$	$(t = 0)$
d. specific heat at constant V	$\propto \begin{matrix} t^{-\alpha} \\ (-t)^{-\alpha'} \end{matrix}$	$(t > 0)$ $(t < 0)$
e. correlation length	$\propto \begin{matrix} \lvert t \rvert^{-\nu} \\ \lvert t \rvert^{-\nu'} \end{matrix}$	$(t > 0)$ $(t < 0)$

where $f(t) \propto t^x$ means $x = \lim_{t \to 0}(\ln f(t)/\ln t)$. There is a similar set of power laws for ferromagnets with the same set of exponents (cf. Stanley 1971).

These exponents, $\alpha(\approx\alpha')$, β, δ, $\gamma(\approx\gamma')$, $\nu(\approx\nu)$, are called *critical exponents*, and the fact that whole classes of radically different systems and phenomena share the same critical exponents is known as the *universality* of the critical exponents, a feature which has no explanation in TD. The systems and phenomena can vary from fluids, for the transition from the one-phase state of gas to the two-phase state of gas and liquid, to magnets, for the transition from the paramagnetic to the ferromagnetic phase. Therefore, the universality seems to say not only that radically different physical systems behave similarly near criticality but also that liquidation is no different from spontaneous magnetization! (Try to imagine being told that there is a certain state of the earth at which not only all human beings necessarily walk (or run, etc.) at roughly the same pace and all dolphins necessarily swim at roughly the same speed, but the pace of a human's walking (or running, etc.) is also very likely to be necessarily the same as the speed of a dolphin's swimming.)

To recover, and improve upon, the above results in statistical mechanics (SM) via Hamiltonian mechanics—plus some statistical assumptions—a bulk system must be modeled to comprise parts (or subsystems) of a single kind or a few different kinds that possess their properties independently of where they are in the system—i.e., they are *translationally invariant*. Typical examples include the van der Waals models for gas and the Ising-type models for ferromagnets and lattice gases (cf. Goldenfeld 1992). This is an essential idealization without which almost nothing can be done in SM, as I shall now briefly show. Given such a model, all thermodynamical magnitudes, except V,N,T, are obtainable through a reductional scheme of which the following is a typical example.[2]

- *The reconstruction* $H \Rightarrow \langle H \rangle$: from the microscopic total energy to the average energy.

1. From mechanics: $H(p,q,s)$,
2. via the distribution: $\rho = (1/Z)e^{-\beta H}$
   (or $\rho = C$ or $\rho = (z^N/Z_G)e^{-\beta H}$), $(\beta = 1/kT)$,
3. to thermodynamics: (i) $Z = Tr[e^{-\beta H}]$
   (ii) let, $Z \equiv e^{-\beta F}$
   (iii) $-\partial_\beta Z = Fe^{-\beta F} = Tr[He^{-\beta H}]$
   (iv) $F = Z^{-1}Tr[He^{-\beta H}] = \langle H \rangle$.

In step 1, we have mechanics—classical, semi-classical, or quantum—which gives the total Hamiltonian of the microscopic parts and the Hamiltonian laws of motion that the parts obey. Step 2 introduces the statistical

2. I thank Gérard Emch for suggesting this example to me.

elements of SM in the form of three distribution functions: canonical distribution for systems which exchange only energy with their environments, microcanonical distribution for isolated systems, and grand canonical systems for systems which exchange energy and particles with their environments, respectively. In step 3, one calculates the partition function with the *Trace* ($Tr\{\bullet\}$) operation over all parts of the system. Here, $Tr\{\bullet\}$ = $Tr\{\bullet\}$ (for a quantum system), or = $\Sigma(\bullet)$ (for a discrete system), or = $\int(\bullet)ds$ (for a continuous system). The rest is a simple derivation. This derivational scheme shows why the idealization is essential; for otherwise—if each part has a distinct Hamiltonian—neither H nor the Trace operation would be manageable.

This scheme of SM works exceedingly well, within the acceptable limit of approximation, for a single-phased system at equilibrium. One may reasonably expect that it also works well for multiple phases at equilibrium, but it does not! To explain this let us begin with a system in the far right region of Figure 1(a) (or far top of Figure 1(b)) and suppose we have written out its partition function, which, as shown above, is a function of T. Let T drop past the critical temperature. If our theory is right, the partition function should have in it the phase boundary between liquid and gas and, in particular, it should contain the two singular points, such as A and B in Figure 1(b)—the two ports, if you like, of a PT region. Similarly, we should in principle be able to reproduce, with some negligible give and take, from the partition function the power laws of CP and an estimation of the critical exponents.

Here comes the dramatic twist: one can tell with or without carrying out the Trace operation that the partition function of a finite system—where the interaction of its parts is stable and strongly tempered—can never harbor any singularities! One can see this either in the algebraic properties of any such partition functions or in the solution of a partition function for a special system, i.e., a normal 2-dimensional Ising system (the latter is an anachronism, see the paragraph after the next). In other words, if a partition function accurately describes all isotherms in Figure 1(b), then whatever shape it gives to an isotherm below T_c, the isotherm will not have the two 'kinks', such as the ones at A and B; nor does it show that when the temperature changes across the critical point there is a discontinuous change from a one-phased system to a two-phased one (or vice versa).

To give a very brief sketch of this simple argument (for more details, see Liu 1999), let us suppose that we have a system modeled as follows: it comprises a finite number N of identical parts in a finite volume V and satisfies two conditions for the interactions among its parts: it is *stable*, which means that only a finite number of parts can be crammed into V, and it is *strongly tempered*, which roughly means that a system of large N

won't explode.[3] The partition function of such a system is obtained by two steps: first we formulate a partition function for each number N of parts and then we sum over all possible numbers of parts, from zero to the maximum N_{max}. The first function is $Z_N = Tr_N\{e^{-\beta H}\}$, (i.e., the canonical partition function) and the second $Z_G = \sum_{N=0}^{max} z^N Z_N$, where $z = exp(\mu/kT)$ is the term caused by the exchange of parts with the system's environment. Since z is positive, Z_G is a polynomial with a positive coefficient for every term. According to SM, the only way for there to be singularities is if $Z_G = 0$; and the roots, if there are any, of this equation, $z_c = exp(\mu_c/kT_c)$, should give us the critical temperature. But the theory of polynomials tells us that there exist no real positive roots for $Z_G = 0$, which implies that there is no temperature at which multiple phases appear or disappear.[4] In other words, we are stopped at the gate and no simplifications or shortcuts along this line can even get us off the ground.

One may find this a perfect case for anti-reductionism or emergent properties. The impossibility of rigorously reconstructing PT and CP from micro-models may well be taken by some as the nail that could finally seal the coffin of reductionism. For it appears to have deprived reductionists of their prized possession—a clear-cut case of reduction of TD to SM. Indeed, this seems to involve a failure of type-type identification across theories (Fodor 1974) that renders impossible a reduction as domain replacement (Nickles 1973). There is no distinct type of micro-state in SM which can be identified with the type that TD singles out as PT or critical points (i.e., singularities).[5]

However, mathematical physicists thought differently, and some of them pursued and eventually proved that PT as a type can be rigorously accounted for by SM (see Yang 1972; Griffiths 1972, I-VI). The key idea is both simple and radical: take the volume and number of parts of a bulk system to infinity, while keeping the density finite—a maneuver now known as 'taking the thermodynamic limit (TL)'. If the limits for the major bulk variables exist—they exist for all systems whose internal interactions are stable and strongly tempered—singularities are proven to emerge again from partition functions (see Liu 1999 for more details). For a simple

3. Conceptually, stability against implosion (stability) and stability against explosion (strong temperedness) implies thermodynamic stability (existence of TL). Mathematically, stability = homogeneity: for certain reasonable functions, $h, h(ax) = ah(x)$; strong temperedness = subadditivity: $h(x + y) \leq h(x) + h(y)$; and thermodynamic stability = convexity: $h(ax + (1 - a)y) \leq ah(x) + (1 - a)h(y)$ (cf. Wightman 1979).

4. Although $z = 0$ is a root, it is not physically meaningful.

5. For an extended discussion of issues in multiple realizability and universality which touches on Fodor and reduction, see Batterman n.d.

model—the 2-d Ising model—whose partition function can be exactly solved, one obtains both its value of T_c and its critical point under TL (Onsager 1944; Schultz et al. 1964). TL is generally justified as an *idealization* of bulk systems whose size and number of parts are so large that compared to their individual parts they are 'approximately' infinite.

The next natural step is to come up with a micro-explanation of CP: to derive the power laws and recover the values of the critical exponents. This is accomplished by the methods of *scaling* and *renormalization*. By scaling, a 'master equation' for the free energy density is obtained, from which various equality relations among different critical exponents are derivable, and such relations effectively reduce the number of independent exponents from half a dozen to only three. Then the *renormalization (semi-) group* (RG) theory is discovered to derive the master equation—or the *scaling hypothesis*—from the microscopic models and to determine the values of the independent exponents. The RG theory is perhaps better understood as an approach or method, of which only the most general scheme and its application to special cases are worked out. The central idea of RG is coarse-graining by lumping a large number of degrees of freedom together and 'decimating' them (cf. Fisher 1982, 1998; Goldenfeld 1992). Let us see, on a conceptual level, first why this is needed and then how it can be done.

For the success of SM, which uses Trace operations to bridge the macro and the micro, the micro-models must be made such that the description of a bulk systems is insensitive to the level or scale of description; or more figuratively speaking, when one descends (or ascends) from the macro- (or the micro-) scale of a system, the general configuration in one's 'view-finder', so to speak, stays very much the same. For the single-phased system, this requirement is already satisfied by the standard models, such as the ideal gas and the Ising models—the random distribution of values for its micro-variables is the same (to a good approximation) at whichever level one focuses one's 'view-finder'. This is why the straightforward application of Trace operations over the Hamiltonians of micro-parts gives us the right macro-properties. However, this is no longer true when a system is near its critical point, where *long-range orders*—stable correlations—among parts form; and what a region of such a system 'looks' like at a micro-level usually does not resemble what it appears at, say, a mesoscopic level. In other words, if we insist on trying to carry out the normal Trace, $Z_N = Tr_N\{e^{-\beta H}\}$, as if values are still randomly distributed all the way down, we will get a distorted picture of the system's thermodynamic behavior near the criticality.

What is needed is a systematic—e.g., a reiterative—set of transformations which *coarse-grain out* the degrees of 'freedom' that are tied up by the long-range order. For an Ising-type lattice model, blocks of spin sites are clumped together and considered as single sites, and the old interaction parameters among the original sites are *transformed* into new ones for the coarser set of sites. Then, the lattice is rescaled—which may be seen as switching levels of description—so that it 'looks' the same as the original system except with fewer degrees of freedom (cf. Figure 2). Its end result in terms of physics is to preserve all the thermodynamic properties of the

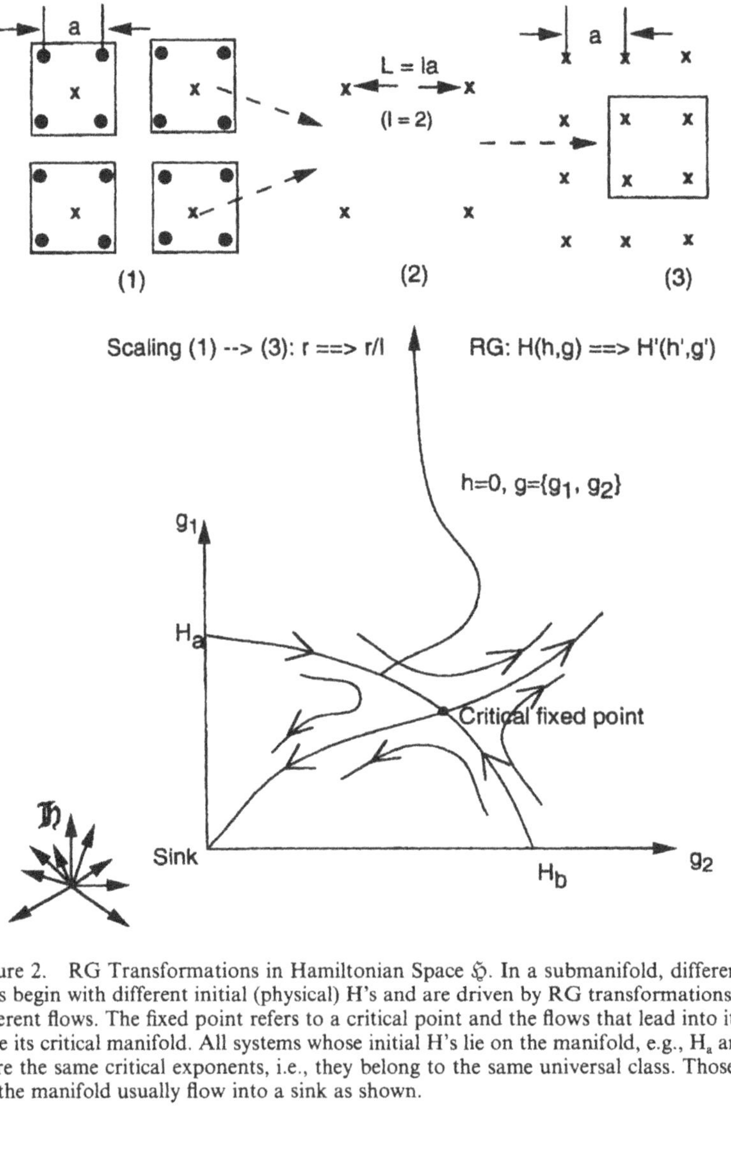

Figure 2. RG Transformations in Hamiltonian Space $\mathfrak{H}$. In a submanifold, different systems begin with different initial (physical) H's and are driven by RG transformations along different flows. The fixed point refers to a critical point and the flows that lead into it comprise its critical manifold. All systems whose initial H's lie on the manifold, e.g., H_a and H_b, share the same critical exponents, i.e., they belong to the same universal class. Those lying off the manifold usually flow into a sink as shown.

system as one descends to the micro-levels. Such are the *renormalization transformations* which form a *semi-group* and which decimate the 'fake' degrees of freedom in $\{s^f\}$. Hence, the original set of micro-variables is divided into two: $\{s\} = \{s^t\} + \{s^f\}$, where $\{s^t\}$ contains the remaining degrees of (true) freedom after the decimation. Once the degrees in $\{s^f\}$ are collapsed, we only have those normal variables as if there has never been any long-range order. To these variables the Trace operation is applicable again, and we shall have, instead of the original partition function,

$$Z_{N'} = Tr_{N'}\{exp\tilde{H}'(s^t)\},$$

where for the sake of convenience we have $\tilde{H} = -\beta H$. Now, because of the decimation, we have $N' = N/l^d$ and $\tilde{H}'(s') = \mathbf{R}_l\tilde{H}(s)$, where ($\mathbf{R}_l$ ($l >$ 1) is an element of RG, l the number of sites that $\mathbf{R}_l$ decimates, and d the number of dimensions of the system. RG thus allows us to partially trace out the expression, $exp\ \tilde{H}$, so that we have $exp\ \tilde{H}' = Tr_{(N-N')}[exp\ \tilde{H}]$, where $N - N'$ is the number of 'fake' degrees. By 'shrinking' the system $N \rightarrow N/l^d$, one also shortens the correlation length, $\xi \rightarrow \xi/l$, so that when one descends toward the critical point, one may keep the correlation length in one's 'view-finder' manageable by the reiterations of $\mathbf{R}_l$. All these coarse-graining transformations must guarantee the unitarity of $\mathbf{R}_l$, namely, the partition function must be invariant under the transformations $Z_{N'}[\tilde{H}'] = Z_N[\tilde{H}]$, which simply means that the TD of the system in question remains invariant under $\mathbf{R}_l$. Moreover, we have $f(\tilde{H}') = l^d f(\tilde{H})$.[6]

Besides a few highly simplified cases, such as the 1-d Ising model, the RG transformations are usually too complex to obtain exactly. But is it usually not too difficult to estimate the global features of RG and to obtain linearized local behavior at some special regions. Figure 2 gives an illustration of what the global features of the RG actions may look like. It shows a submanifold of the total Hamiltonian space, **H** (usually infinitely dimensional), spanned by thermodynamic fields: h (= 0), g_1,g_2. Different types of systems with different Hamiltonians initially occupy different regions in this space and are then driven, as *flows*, by RG transformations to other regions, the most interesting of which are the *fixed points* and their *critical manifolds*. The former are defined as those points which satisfy the condition:

$$\tilde{H}^* = R_l\tilde{H}^* \text{ (for any } l > 1)).$$

6. $\tilde{H}$ is usually written as $\tilde{H}(\{s\},h,g,\ldots)$, where $\{s\}$ are all the micro-variables and h,g, ... are the macro-variables, often known as the 'thermodynamic fields' because their values determine thermodynamic states; and therefore, they determine where the system begins in $\mathfrak{H}$ and where it flows to under RG (see Figure 2). Such fields include coupling strength/temperature, external fields, and any other relevant external influences.

These are singular points—somewhat like the blackholes in Einsteinian spacetime—into which once a system 'flows' it can never flow out. The latter consist of all those initial points in $\mathfrak{H}$ which eventually flow into the respective fixed points: $\tilde{H}^* = \lim_{l\to\infty} R_l \tilde{H}$ (see Figure 2). Fixed points represent different thermodynamical states of a system, some stable and some unstable. *Sinks* are one type of fixed points and represent the state when $T \to 0$ or $T \to \infty$, where a system once there not only will never flow out on its own but also cannot be brought out by any manipulations of other thermodynamic fields. Another type are the *discontinuity fixed points*, which represent the PT regions such as those at $T < T_c$ in Figure 1(b); and most importantly, there are the *critical fixed points*, which separate the single-multiple phase regions, such as the one at T_c in Figure 1(a). These global features of $\mathfrak{H}$ tell us several important things: (i) once a system is found or put on the critical manifold of, say, a critical fixed point, then the values of its thermodynamic fields mark the critical point in the physical phase-diagram; (ii) the CP of that system at that critical point can be obtained by investigating the properties of the system's $\tilde{H}$ in a linear approximation on the critical manifold.[7] Lastly and most importantly, (iii) the universality of critical exponents can be read off $\mathfrak{H}$ and its flows as the fact that a whole class of many different submanifolds flow into the same fixed point.

3. Interpreting TL and RG. Even though TL grounds the entire rigorous approach to SM, we all know that actual systems in which PT and CP take place are finite. Hence, what justifies the taking of TL?

There is a seemingly straightforward answer (for details, see Liu 1999): PT appears in TD as singularities because of a special idealization I emphasized earlier: bulk systems are considered dense with their parts (or without distinct parts) and so lack fluctuations. The idealization is a good approximation because molecules in any bulk system are so small and so numerous that the system is practically continuous. This idealization is inapplicable in SM because given that bulk systems are stable, as I mentioned before, only a finite number of molecules can be crammed inside a finite volume. To compensate, we take the size and the number of molecules to infinity while keeping the density finite.[8] TL effectively kills the

7. In a nutshell, we Taylor expand the effective Hamiltonian near the fixed point as $\mathbf{R}_l(\tilde{H}^* + \varepsilon h) = \tilde{H}^* + \varepsilon \mathbf{L}_l h + O(\varepsilon)$, as $\varepsilon \to 0$. In other words, we linearize $\mathbf{R}_l$ by $\mathbf{L}_l$, and from $\mathbf{L}_l$ we can set up an eigenvalue equation from which we can obtain some eigenvalues which turn out to be directly related to the critical exponents. Further, from $f(\tilde{H}') = l^d f(\tilde{H})$ we can recover the scaling hypothesis.

8. To be more precise, in TD the density of a system is given as either 1/V or M/V, a

random fluctuations in a system so that the isotherms may again appear mostly smooth and PT as singularities.

Therefore, the singularities which represent PT and CP should be regarded as mere artifacts or fictions of TD that do not exist in reality. PT are just places where rather dramatic changes of thermo-variables take place so that their derivatives make sharp, but not discontinuous, changes. If this is right, one should retire both idealizations and treat PT and CP as what they really are. Hence, there shouldn't have been any 'crisis' of the 'impossibility of PT' or any 'no-go' theorems, and the effort to introduce TL seems a consequence of the failure to notice the simple truth that a crucial idealization in TD should no longer be honored in SM.

Even though I believe the account of the two idealizations given above is right, the conclusion may sound outrageous. It implies that the rigorous methods (including RG) are a waste of time, which has the implication that PT and CP do not form distinct kinds qualitatively different from other states of a bulk system. Should we not then extend the conclusion to such conjectures as that none of the other kinds, such organic vs. inorganic or life vs. death, are distinct kinds, but are rather merely quantitatively different configurations of molecules in a system? One can think of a number of ways to get around this conclusion, if one wants to; and here I can only discuss a few. One approach is not to deny the main thrust of the conclusion but to appeal to the pragmatic virtues of TL for its defense. Another is to fully embrace special sciences, such as thermodynamics and biology, and regard their kind predicates as of genuine natural kinds despite the micro-structures' showing otherwise. Yet another is to separate mathematical predicates from the physical properties they are used to pick out. The predicates may be exactly defined and related, but in their applications a certain range of approximation is introduced so that the predicates can pick out those objects whose properties they refer to within the range. I shall discuss them briefly in turn.

First, we should not forget that there are 'approximate' models, such as the mean field models, which reproduce PT and CP with finite systems (cf. Wannier 1966; Goldenfeld 1992). The fact that they make wrong predictions does not mean that they are not improvable. The 'no-go' theorem does not necessarily apply to mean field models because they do not assume the stability and strong tempered conditions for the interactions. Onsager's success with a 2-d Ising lattice at TL and $B = 0$ certainly lends strong support for TL, but it is such a special case which is unlikely extendible to any other more realistic situations (not even for the same model

continuous function where M is the system's mass and V its volume, while in SM it is N/V, a discontinuous function where N is the total number of parts. When $N \to \infty$ and $V \to \infty$, N/V converges to a continuous function if it converges.

at $B \neq 0$). The strongest practical virtue of TL is that it makes the RG approach possible, which not only works for CP but also has applications in quantum field theories.

In SM, RG is prized for its alleged success in giving an *explanation* of the *universality* of critical exponents, which in TD is merely a curious coincidence (Fisher 1998; Batterman 1998). From Figure 2 we see that all systems whose physical criticalities are eventually driven by an infinite iteration of renormalizations into a fixed point belong to the same universal class and share the same set of critical exponents. From the critical manifold we also learn which thermodynamic fields are relevant and which are not to the calculation of the exponents. But what kind of explanation is this? The question is: why should phenomena as different as a fluid system's approaching opalescence and a ferromagnet's approaching spontaneous magnetization behave the same near their own critical points? The answer from renormalization is: because their critical points flow into the same fixed point in the space of Hamiltonians! It is true that universality is no longer a shear coincidence, but do we have an explanation of it? If we are asked to explain why different planets in the solar system move in elliptical orbits around the sun, and we say, "because all orbits of the sun's planets have the geometric shapes of conic sections," have we given an explanation? I believe not; for that we need a causal law or story, such as the law of gravity.[9]

If one is looking for a causal explanation for the critical phenomena, i.e., why we have the same asymptotic behavior (cf. Balashov 1997) near criticalities, the renormalization group does not deliver it. It does not tell us which compositional or structural features are relevant or irrelevant for the asymptotic behavior. What it does tell us is, at a critical point, which features are relevant or irrelevant to the estimation of the critical exponents. We should note that the space of Hamiltonians is not a configuration space, and the flows on Hamiltonian space are not physical processes. Therefore, *the relevancy of the flows to a fixed point determined by the renormalization group* should not be confused with *the relevancy to the asymptotic processes of systems in the same universal class approaching their respective critical points*. The fact that the latter may provide an explanation for the former shows why the former alone is not yet an explanation for the universal asymptoticity of critical phenomena.

For the latter, it seems to me that theoreticians know almost nothing

9. We must note that there is a difference between an explanation of universality via the micro-structures of thermo-systems and a reductional explanation of some TD laws, such as the van der Waals law of gases. In the former case, there is no TD explanation of universality. Therefore, if RG explains, it cannot do so by only giving a precise enough estimation of the exponents. It must explain why diverse thermo-phenomena all approach their own criticalities in the same manner and with the same exponents.

beyond the fact that correlation lengths diverge towards criticalities and therefore certain kinds of micro-degrees of freedom become asymptotically frozen and irrelevant. And this I believe is not sufficient for a full causal explanation of universality. However, if this is sufficient, then RG is dispensable for the explanation, but if this is not, RG is not the kind of theory that can fill out the gap in the causal story. Remember, universality is not just that the density differential of different kinds of fluids share the same critical exponent at criticality and the spontaneous magnetization of different kinds of ferromagnets share the same exponent, but that these two types of exponents are the same. The irrelevance of micro-structures due to divergent correlation lengths points towards an explanation but is not one yet.

From a slightly different angle, the situation is this. Suppose it is logically possible to calculate any partition function if the Hamiltonian is given and there is a computer powerful enough to do the calculation in finite time. Then we can precisely calculate the partition functions of systems near their critical points to obtain the exponents; and it turns out that all of them are numerically equal. Do we have an explanation of universality by this result alone? I think not. We know that universality is not a shear coincidence, but we have no idea what is physically responsible for such 'miracles'. Using RG gives us a better idea because if it were not for the divergence of correlation lengths and the irrelevance of certain kinds of degrees of freedom, RG would not have worked. Therefore, rather than saying that RG explains universality, one should more accurately say that what makes RG work in deriving the universal critical exponents is also responsible for universality. To explain universality is to know fully what that is. To say that the RG has given us a full description of what that is is premature, if not misleading. We know what is responsible for the manner in which the planets revolve around the sun by knowing the law of gravity, but not by knowing that the orbits of all the planets in the solar system have the geometric shapes of conic sections.[10]

Even if the pragmatic answer succeeds, one may still find the main

10. After the completion of this paper I received a copy of Bob Batterman's most recent paper (Batterman n.d.), which explains in detail in what sense the RG explanation of universality is an explanation (or what kind of an explanation it is). What is not clear to me is whether Batterman takes such an explanation to be a causal explanation—in the broadest sense of the phrase, which means, roughly, giving sufficient conditions under which the fact, i.e., the universality, obtains. It may well be that the kind of explanation RG provides, according to Batterman, takes only some mystery out of the macro-micro relations between kinds in special sciences and their microscopic realizations, but it does not entirely eliminate it. I think my example of the planets is still a good analogy in this regard. I leave a detailed discussion of this subject to another occasion.

thrust of the conclusion unacceptable. By accepting it one would have in fact abolished PT and CP—so far as they are distinct thermodynamic kinds—as genuine physical states or properties, and all our talk of PT and CP in terms of TL and RG is talk about artifacts or fictions. Moreover, this will have to apply to all the other special sciences, such as chemistry and biology. Therefore, the second response begins with TD and takes points of PT and CP as singularities (or whatever distinct kinds) to be about genuine properties of bulk matter. Then, whatever our hypothesis of the micro-structure of reality is, it will have to be made compatible with the above by recovering at least PT and CP. Therefore, the 'idealization' in TD—if it is at all an idealization—is not to be removed simply because we realize that bulk matter consists of discrete parts. If this is right, when we realize that finite systems with the posited micro-structure, such as the Ising-type models, forbid PT and critical points, either we come up with a good idealization or the models are thrown out. TL appears to be the only idealization by which all microscopic models are saved with regard to PT and CP. Therefore, TL is not an idealization to compensate or 'cancel out' the 'idealization' in TD concerning PT and CP; rather it is one to compensate for the artificiality of such micro-models as the Ising-type models or of such conditions as the stability and strong temperedness of inter-molecular interactions. To argue for this response, one may invoke the spirit, albeit not the actual theory, of logical-positivism and first affirm the primacy of TD, since it is closer to the observables. One then looks at the lattice structures in the Ising-type models and the two stipulations for the interactions between lattice sites as equally theoretical as the theoretical size of a system which may harbor PT and CP. Then, TL may be seen as the necessary theoretical element in a consistent and optimal theoretical reconstruction of PT and CP. All the elements—at least the ones mentioned above—in this theory are equally idealized and may be subject to the same kind of philosophical criticism as TL is. But the fact that together they recover the thermodynamic description of PT and CP gives them justification as an integrated whole. It would be inappropriate and unfruitful if we single out TL for such a criticism.

This would be acceptable if TL is a normal kind of idealization. Idealizations are acts in theory construction by which conditions are said to obtain when they actually do not. Normal idealizations specify at least physically possible conditions; but TL seems to be flat out contradictory: a finite system is of infinite size![11] Even though it is, I believe, a problem

11. There are many examples of idealizations in physics which assume non-actual but physically possible conditions. Complete isolation is one example, having all lattice sites of equal distance is another, and even such mechanisms as the famous Cooper-pairs, which initially provided an explanation for superconductivity, are of this kind.

for this approach, one may have a way out, as some discussions in the mathematics literature seem to suggest (Griffiths 1972). Even though PT and CP do in fact happen in finite systems, what really takes place there is entirely independent of the systems' sizes (given that the densities remain the same). It means that where the boundary of the system is has no relevance to the behavior in question.[12] This can be seen in the fact that PT and CP are described entirely by intensive quantities, such as the free energy density, when the TL of these quantities exist. Therefore, even though finite and infinite systems are logically incompatible, it doesn't really matter because the physical properties of PT and CP are identical in both kinds of systems. But there is a subtle shift in this argument which may ultimately defeat it: the thermodynamic phenomena of PT and CP in a finite system and the SM account of them in a corresponding infinite system are the same. What we do not and cannot know is whether some (or any) SM account of PT and CP in a finite system is the same as in a corresponding infinite system. If we know that is true, then we are indeed justified to use the results in an infinite system to account for the thermodynamic phenomena of PT in a finite system. But if we know that, we wouldn't be taking the trouble of going TL.

The third response is probably the closest to the truth for our case. Predicates we used to describe PT and CP in TD or infinite systems in SM are mathematical ones that result from an accentuation or exaggeration of the corresponding physical properties by neglecting or filling out negligible differences. With such predicates, scientists must demand strict exactness among their relations, whether they are lawlike or reductional statements. But when such predicates are applied to actual physical systems, estimates of approximation are brought in so that the right kind of systems are picked out by the predicates. For instance, being a critical point as a mathematical predicate is defined by a singular point on an isotherm; but when using it to pick out a physical critical point a certain range of approximation to the singular point should be understood so that it picks out the right set of systems. This is similar to our use of most exact magnitudes, such as 'weighing 100 kg'. We are justified to use it to pick out objects whose weight is not exactly 100 kg but very close to it. Here, unlike in the first response, most mathematical predicates we use in physics or other scientific disciplines whose precision cannot always be assumed in their applications do pick out genuine properties, just as 'weighing

12. Note that there is a big difference between **where** the boundary is and **what** it is. For a single bulk phase, it doesn't matter where or what the boundary is; but for systems in multiple phases, what the boundary condition is could sometimes be crucial, but it doesn't matter where it is, since the long-range order is supposed to hold even if the boundary is infinitely far away.

100 kg' does, even though a certain degree of approximation is implied in their meanings. Unless there is a danger that approximation of any kind, when introduced to the SM predicates of PT and CP, may compromise them so much so that it is impossible to use it to pick out the right systems (which are picked out by the corresponding predicates in TD), we do not seem to have a problem here. If this is right, infinite systems provide the extension for a mathematical predicate which when applied to real physical systems picks out finite, but very large systems. It is not difficult to imagine that this is how all mathematical predicates which depend on limits or convergences, such as being differentiable, being continuous, and being smooth, should be understood. One should note that this response may not apply to all kind-predicates in special sciences; it may only apply to those quantitative ones which are related by precise mathematical relations.

4. TL and Irreversibility. If TL is not just a mathematical trick for handling the problems of PT and CP, it ought to apply to other problems concerning bulk matter; it ought to work, for instance, for a system approaching equilibrium from a non-equilibrium initial state. But does it and should it? Here I briefly explore this question which, to my best knowledge, is broached in the philosophy literature by Sklar (1993, 244ff, 277ff).

Boltzmann's H-theorem (without its probabilistic interpretation) says roughly that the H-function of a system can never increase through time, given that it satisfies the simple assumptions of exclusively binary collisions (a pure mechanical assumption) and of Stosszahlansatz (a statistical assumption). This produces two problems. The *reversibility problem* (Loschmit) says that since all mechanical processes taking place at the microscopic level are reversible, for any process that leads the system towards an equilibrium state there may be an exact reverse of it which leads the system away. And the *recurrence problem* (Zermelo) says that given Poincaré's theorem of recurrence—which says that an isolated system of finite size, number of parts, and energy will return in finite time to an arbitrarily small neighborhood of almost any state from which it begins—one should expect that a system will return in finite time to the neighborhood of a non-equilibrium state if it begins there. From either problem, the approach to equilibrium is not monotonic and the equilibrium state is not stable.

The similarity with the case of PT and CP is striking: the H-theorem is intended to recover the second law of TD in SM, just as the evaluation of the partition function is for recovering the thermodynamic description of PT and CP; and the reversibility or the recurrence problem gives a kind of 'no-go' result to monotonicity as the 'no-singularity' result does to PT and CP. Several questions arise at this juncture, of which I can only discuss two: Is TL used to solve the problems? And should it be so used?

Introducing TL makes, prima facie, a big difference. Poincaré's theorem of recurrence in its current form is no longer true, for it assumes a finite system with finite energy. Taking TL throws out at least one of these conditions. Can the theorem be extended to infinite systems? That is not clear at all. If the energy is made unbounded and the size and number of parts infinite, the theorem clearly cannot hold. Even if the energy is finite, the theorem still seems unlikely to hold. If Poincaré's theorem fails, the recurrence objection can no longer be made. However, Loschmit's objection still seems to stand since whether or not the mechanical processes of micro-constituents are reversible has nothing to do with whether or not the system they comprise is finite. It is not even clear whether taking TL helps Boltzmann's probabilistic interpretation of his H-theorem, especially if it is assumed that all micro-states are equally probable, i.e., each 'towards-equilibrium' state is as probable as its reversal, an 'away-from-equilibrium' state (cf. Sklar 1993, 42). If it helps at all, it may help in getting monotonicity back into a *local evolution* of H from a non-equilibrium state. If for a finite system, the overwhelming majority of states are at or close to equilibrium, for an infinite system the set of states *not* at or close to equilibrium will be of zero measure. But do we really want strict monotonicity? I shall come back to this point.

Returning to the recurrence problem we should realize that TL has been introduced not to extend Poincaré's theorem but to directly derive the H-theorem from certain models. Kac (1956; also Wannier 1966) devised a 1-dimensional ring model in which taking TL produces a monotonic approach to equilibrium (i.e., no recurrence). But the model is too abstract to be useful even for idealized gases. TL is applied to a model of gas by Lanford (1983) and monotonicity (non-strict) is proven for a short time after any initial states (1/5 of the mean-free-time). We begin with a Boltzmann gas model and apply to it the Boltzmann-Grad limit, which sends the volume and the number of particles to infinity, while the diameter of the particles goes to zero in such a way that Nd^2 is finite but $Nd^3 \to 0$ (so the density goes to zero); and thus the gas becomes infinitely diluted with exclusively binary collisions. Even with such a dramatic TL, what is proven for 1/5 of the mean-free-time is *not* that for any system if it begins ($t = 0$) at a state χ near f, $T^t\chi$ ($t > 0$, and small) is near f_t, (where t: $f \Rightarrow f_t\chi$ is due to the Boltzmann equation and T: $\chi \Rightarrow T^t\chi$ to mechanics of particles), but rather this: the overwhelming majority of those systems at a state χ near f end up via T near f_t.[13] No extension of this result to a

13. One should be careful in interpreting Lanford's result in connection with Boltzmann's probabilistic interpretation of H-theorem. What Lanford proves is an H-

larger time scale has been produced so far, although reasonable speculations point to a successful extension (and the mean-free-time of such an infinitely diluted gas is not as short as one might assume).

Although Lanford's model does not 'recover' Boltzmann's H-theorem and the second law of TD, it (together with Kac's) points to the necessity of TL in such endeavors. But now we must confront a more serious question, namely, whether the H-theorem should be recovered in its strict sense. If it is indeed true that what the H-theorem represents is not a law of nature but the result of some widespread initial/boundary conditions of the actual universe, then the reversibility and recurrence problems should no longer be problems and Boltzmann's probabilistic interpretation should be the right answer. Judging from the latest investigations of these issues (cf. Callender 1998), the more sensible attitude towards them seems to be that we should fully accept the statistical nature of irreversibility and reconcile the fact that our current cosmological regime appears to phenomenologically support the strict H-theorem with the set of special initial conditions (low entropy states) of the regime and the smallness of the recurrences within the time span of the regime. Such an attitude seems to reinforce one of the earlier views of TL in PT and CP, namely, that we should fully accept the 'non-singular' nature of PT and reconcile the fact that PT and critical points appear to be singularities with the smallness of fluctuations in bulk matter.

However, one may still ask why there is such disparity in the attitudes towards the use of TL in the community of physicists and mathematicians—why TL was taken so seriously in dealing with the recovery of PT and CP and yet not nearly as seriously in dealing with Boltzmann's H-theorem. There is a dissimilarity in the two cases which may contribute towards an explanation: PT and CP need to be qualitatively distinguished from other states of a bulk system in TD (as distinct kinds) while no such distinction is needed for processes from non-equilibrium states to equilibrium states. All such processes should obey the H-theorem, however it is interpreted, so one does not have the problem that under one interpretation the 'H-processes' are threatened to collapse into another qualitatively distinct thermodynamic kind while in another interpretation they are not.

---

monotonicity in a short time for the mechanical evolution of the majority of systems satisfying the Boltzmann-Grad limit, while Boltzmann's probabilistic interpretation gives no monotonicity to any mechanical evolution but only high probabilities for those systems starting at non-equilibrium states to end up at the equilibrium state.

REFERENCES

Balashov, Yuri (1997), "Comments on R. Batterman's 'Why Statistical Mechanics Works: Universality and the Renormalization Group' ", preprint.

Batterman, Robert M. (1998), "Why Equilibrium Statistical Mechanics Works: Universality and the Renormalization Group", *Philosophy of Science* 65: 183–208.

——— (n.d.), "Multiple Realizability and Universality", *British Journal for Philosophy of Science*, forthcoming.

Callender, Craig (1998), "The View from No-when", *British Journal for Philosophy of Science* 49: 135–159.

Fisher, Michael E. (1982), "Scaling, Universality and Renormalization Group Theory", in F.J.W. Hahne (ed.), *Critical Phenomena: Proceedings of the Summer School, The University of Stellenbosch, South Africa, January 18–29, 1982*. Berlin: Springer-Verlag, 1–139.

——— (1998), "Renormalization Group Theory: Its Basis and Formulation in Statistical Physics", *Reviews of Modern Physics* 70: 653–681.

Fodor, Jerry (1974), "Special Sciences", *Synthese* 28: 77–115.

Goldenfeld, Nigel (1992), *Lectures on Phase Transitions and the Renormalization Group*. Reading, MA: Addison-Wesley.

Griffiths, Robert B. (1972), "Rigorous Results and Theorems", in C. Domb and M. E. Green (eds.), *Phase Transitions and Critical Phenomena*. New York: Academic Press, 8–109.

Kac, Mark (1956), "Some Remarks on the Use of Probability in Classical Statistical Mechanics", *Bull Acad Roy Belg* 42: 356.

Lanford III, Oscar E. (1983), "On the Derivation of the Boltzmann Equation", in J. L. Lebowitz and E. W. Montroll (eds.), *Nonequilibrium Phenomena I: The Boltzmann Equation*. Amsterdam: North-Holland, 1–17.

Liu, Chuang (1999), "Explaining the Emergence of Cooperative Phenomena", *Philosophy of Science* 66 (Supplement): S92-S106.

Nickles, Thomas (1973), "Two Concepts of Intertheoretic Reduction", *Journal of Philosophy* 70: 181–201.

Onsager, Lars (1944), "Crystal Statistics. I. A Two-Dimensional Model with an Order-Disorder Transition", *Physical Review* 65: 117–149.

Schultz, T. D., D. C. Mattis, and E. H. Lieb (1964), "Two-dimensional Ising Model as a Soluble Problem of Many Fermions", *Review of Modern Physics* 36: 856–871.

Sklar, Lawrence (1993), *Physics and Chance: Philosophical Issues in the Foundations of Statistical Mechanics*. Cambridge: Cambridge University Press.

Stanley, H. Eugene (1971), *Introduction to Phase Transitions and Critical Phenomena*. Oxford: Clarendon Press.

Wannier, Gregory H. (1966), *Statistical Physics*. New York: Dover.

Wightman, Arthur S. (1979), "Introduction: Convexity and the Notion of Equilibrium State in Thermodynamics and Statistical Mechanics", in R. B. Israel (ed.), *Convexity in the Theory of Lattice Gas*. Princeton: Princeton University Press, ix–lxxxv.

Yang, C. N. (1972), "Introductory Note on Phase Transitions and Critical Phenomena", in C. Domb and M. E. Green (eds.), *Phase Transitions and Critical Phenomena*. New York: Academic Press, 1–5.

Explaining the Emergence of Cooperative Phenomena

Chuang Liu†‡
University of Florida

Phase transitions are well-understood phenomena in thermodynamics (TD), but it turns out that they are mathematically impossible in finite SM systems. Hence, phase transitions are truly emergent properties. They appear again at the thermodynamic limit (TL), i.e., in infinite systems. However, most, if not all, systems in which they occur are finite, so whence comes the justification for taking TL? The problem is then traced back to the TD characterization of phase transitions, and it turns out that the characterization is the result of serious idealizations which under suitable circumstances approximate actual conditions.

1. Introduction. Emerging properties as a philosophical category are at once fascinating and suspect. They pose a serious threat to reductionism of any stripe, but few people seem to be able to prove that such properties exist. While emergentists see emerging properties everywhere, anti-emergentists see none; and few seem to be able to give a precise enough characterization of such properties and to prove that a given property under such a characterization 'emerges' from other properties but is not reducible to them (cf. Teller 1992, Humphreys 1997).

The property of *phase transitions*, as I shall argue in this paper, is just such a property, for there is, as we shall see, a precise notion of its being emergent and a rigorous argument for its not being reducible to properties in statistical mechanics (SM). In fact, it is one of the two

†Department of Philosophy, 330 Griffin-Floyd Hall, P.O. Box 118545, Gainesville, FL 32611–8545; e-mail: cliu@phil.ufl.edu.

‡I would like to thank Gérard Emch for countless discussions and suggestions related to this paper. I would also like to thank those who attended the PSA 1998 session at which I read an earlier version of this paper, asked questions, and/or gave comments.

Philosophy of Science, 66 (Proceedings) pp. S92–S106. 0031-8248/99/66supp-0008$0.00

main *cooperative phenomena* in SM (cf. Stanley 1971, Emch 1977, Toda et al. 1983)—the other being *irreversibility*—that are of this kind.

One should first distinguish *collective* phenomena from *cooperative* ones. The main feature of the former (see a similar notion in Wimsatt 1997) is the invariance of phenomenal properties under "mereological manipulations" of nearly identical parts (or a limited number of species of such parts)—operations of *random* interchange, rearrangement, addition, or subtraction.[1] Exhibiting the usual thermodynamic (TD) properties, such as bulk, temperature, pressure, and entropy, is a collective phenomenon. This explains why reducing these properties in SM, though complicated in some cases, is a solvable problem, and why they are not emergent properties. The main feature of the latter is the *order* or *cooperation* of the parts; for instance, the correlation among parts is qualitatively different in a solid than in a liquid phase. But why are cooperative phenomena, such as phase transitions, also emergent?

In a nutshell, the answer is as follows. Phase transitions are well-understood phenomena in thermodynamics (TD), but it turns out that they are mathematically impossible in finite SM systems. Hence, phase transitions are truly emergent properties. They appear again at the thermodynamic limit (TL), i.e., in infinite systems. However, most, if not all, systems in which they occur are finite, so whence comes the justification for taking TL? The problem is then traced back to the TD characterization of phase transitions, and it turns out that the characterization is the result of serious idealizations which under suitable circumstances approximate actual conditions.

2. The Problem of Phase Transitions. Giving a rigorous account of phase transitions in purely TD terms encounters no apparent conceptual problems (Stanley 1971, 1–9). It is illustrated in Figures 1 and 2.

For fluid systems, we see (in Figure 1a) three phases—gas, liquid, and solid—separated by surfaces in pVT space on which any isotherms (representing quasi-static changes of the systems), otherwise smooth, develop singularities, i.e., discontinuities in their tangent curves (Figure 1b). A similar case appears for ferromagnetic systems—systems which show spontaneous magnetization in the absence of external magnetic field. Note the correspondence: $p \leftrightarrow B$, $V \leftrightarrow M$, $T \leftrightarrow T$. In general, phase transitions as TD phenomena are shown as *singularities* (or *non-*

1. Wimsatt's notion, "aggregativity," is different from mine and is perhaps not intended to capture collective phenomena. Wimsatt requires invariance under (actual?) decomposition of a system for an aggregative property. Among his examples are total mass, charge and energy of a system—all of which are conserved quantities in physics. We regard the distinction between collective and cooperative more germane to the discussion of microreduction than the one between aggregativity and non-aggregativity.

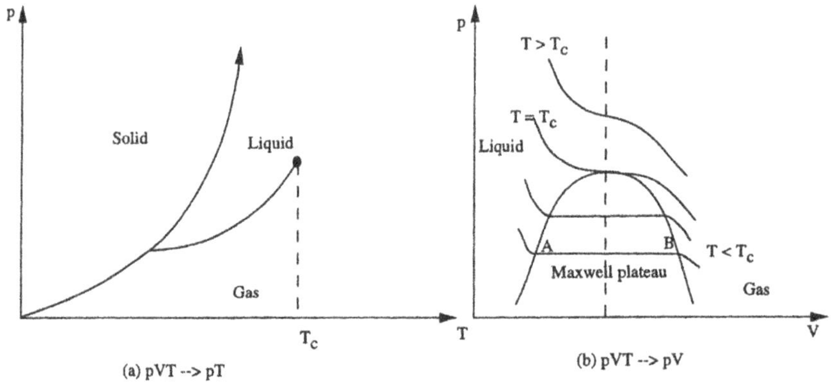

Figure 1. Phase Transitions in Fluids: projections of the pVT surface in two different phase planes.

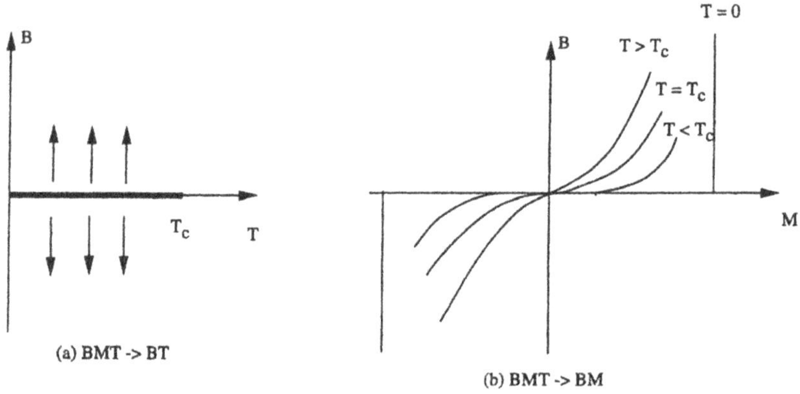

Figure 2. Phase Transitions in Magnets: projections of the BMT surface in two different phase planes.

analyticities) of a macrosystem's isotherms, a feature that any micro-explanations (or SM explanations) of phase transitions must recover.

For the SM explanation of phase transitions, we begin with a brief sketch of the most general and shared features of our model (Toda et al. 1983, 27–68). A dynamical system, S, is represented by a phase space, Γ, and a parameter family of homomorphisms or diffeomorphisms (the flow): ϕ_t: $\Gamma \rightarrow \Gamma$ $(t \in \Re)$, where each $\gamma \in \Gamma$ denotes a state of S and ϕ_t dictates how γ evolves in Γ (a dynamics). As a phase space, Γ is equipped with a σ-algebra of subsets, $\mathfrak{A}$, and a normalized measure, ρ, on $(\Gamma, \mathfrak{A})$ such that $\rho(\Gamma) = 1$ and $\forall t, \forall\, A \subseteq \mathfrak{A}$, $\rho(\phi_t(A)) =$

$\rho(A)$—i.e., it is invariant under the flow. For states of equilibrium, S is fully determined by (Γ, ρ), where ρ—known as the equilibrium probabilistic measure (e.p.m.)—is different for different ensembles. For a microcanonical ensemble which represents isolated macrosystems, the e.p.m. is $\rho_{mic} = \delta(H - E_0)$ (where H is the system's Hamiltonian and E_0 its total energy); for a canonical ensemble which represents systems in which an exchange of energy, but not of molecules, with outside takes place, we have $\rho_{can} = exp(-\beta H)$, where $\beta = 1/kT$ (k = the Boltzmann constant); and for a grand canonical ensemble—a canonical ensemble where an exchange of molecules with outside is also permitted, the e.p.m. is $\rho_{gran} = exp[-\beta(H - \mu N)]$, where μ is the chemical potential and N the total particle number. All TD quantities are, in this scheme, weighted (by the proper e.p.m) averages of the corresponding micro-quantities. For any system, S, we select an e.p.m. of the ensemble which best describes S, and then for any TD function, f, of S—i.e., any integrable functions whose averages correspond to known TD quantities—we have,

$$\langle f \rangle = (\int_\Gamma \rho ds)^{-1} \int_\Gamma f \rho ds,$$

$$\text{or } \langle f \rangle = (\sum_i \rho_i)^{-1} \sum_i f_i \rho_i,$$

$$\text{or } \langle \hat{f} \rangle = (Tr\hat{\rho})^{-1} Tr(\hat{f}\hat{\rho}),$$

depending on whether S is continuously classical or discretely classical or quantum. Here lies the complete scheme for a micro-explanation of TD phenomena.

It turns out that most TD functions, such as pressure, entropy, and energy, are partial derivatives of the partition functions: Z (or Ξ) = $(\int \rho ds)$ or $(\Sigma \rho_i)$ or $Tr(\rho\hat{})$, (where Z corresponds to ρ_{can} and Ξ to ρ_{gran}). For instance, for the canonical ensemble, if we define free energy, F, as $F = -lnZ/\beta$, then we have, the pressure: $p = -[\partial F/\partial V]_T$, and the entropy: $S = -[\partial F/\partial T]_V$.[2] And in the grand canonical ensemble, we have the pressure as

$$p = (\beta V)^{-1} ln\Xi \qquad \text{and} \qquad \sigma = N/V = z(\partial ln\Xi/\partial z) \tag{1}$$

Therefore, in principle, the TD properties of a system is determined once its partition function is known, and the partition function is known once the Hamiltonian of the system is given and the ensemble

2. Since the free energy is clearly an ensemble-average, all the other functions obtained from it via partial differentiations are certainly also ensemble-averages. Therefore, I shall not use the bracket for such functions unless the context does not make it clear.

chosen. However, what is in principle the case is far from so in practice. The entire enterprise of thermo-statistical physics is built on the ingenuity of finding the right model for certain macrosystems such that the Hamiltonian of the model is complex enough to approximate them but not so complex as to render the calculation of the partition function infeasible.

We are now ready to appreciate the problem of phase transitions in its most general form. Phase transitions such as given in Figure 1 and 2 are singularities in the isotherms below the critical temperature T_c. In the fluid case, if we can calculate $p(V,T)$ from the partition function (from equation 1), we only need to see if and where a 'kink' appears in $p(V,T)$ below T_c.

But here's the rub: Suppose (cf. Yang and Lee 1952, Griffiths 1972, Penrose 1979) that we have a classical system, S, which occupies a finite region $\Lambda \subseteq \Gamma$ with a volume V (determined by Λ). Further, we suppose that the intermolecular interaction has the potential, $\Phi(\mathbf{r})$, such that it is *stable* (i.e., the total potential $U(\mathbf{r}_1 \ldots \mathbf{r}_N) \geq -cN$, where $c > 0$ is a constant) and *strongly tempered* (i.e., $\Phi(\mathbf{r}) \leq 0$ for large $\mathbf{r}$).[3] It follows from stability that inside V, only a finite number (say N_{max}) of molecules can coexist. The grand canonical partition function can be written as follows:

$$\Xi_\Lambda(V,z,T) = \sum_{N=0}^{N_{max}} \sum_{i=1}^{N} exp[-\beta(H_i - \mu N)]$$

$$= \sum_{N=0}^{N_{max}} exp(\mu\beta N) \sum_{i=1}^{N} exp(-\beta H_i) = \sum_{N=0}^{N_{max}} z^N Z_N,$$

where $z = exp(\beta\mu)$ and $Z_N = \sum_{i=1}^{N} exp(-\beta H_i)$ is the partition function for an canonical ensemble of N particles. But the last expression in the equation is a simple polynomial of z, i.e.,

$$\Xi_\Lambda(V,z,T) = 1 + zZ_1 + z^2 Z_2 + \ldots + z^{N_{max}} Z_{N_{max}}, \qquad (2)$$

which, from the above expression of Z_N, has only non-negative coefficients. This means that there exists no positive real root, z_i, for the equation, $\Xi_\Lambda(V,z,T) = 0$. In other words (from equation 1), neither

3. Conceptually, stability against implosion (stability) and stability against explosion (strong temperedness) implies thermodynamic stability (existence of TL). Mathematically, stability = homogeneity: for certain reasonable functions, h, $h(ax) = ah(x)$; strong temperedness = subadditivity: $h(x+y) \leq h(x) + h(y)$; and thermodynamic stability = convexity: $h(ax + (1-a)y) \leq ah(x) + (1-a)h(y)$ (cf. Wightman 1979).

$p_\Lambda(z)$ nor $\sigma_\Lambda(z)$ can have singularities for any positive real z because singularities only appear when $\Xi_\Lambda(V,z,T) = 0$.

One can also construct a similar argument for magnetic systems by using the correspondence of $p_\Lambda \leftrightarrow B_\Lambda$ and $s_\Lambda \leftrightarrow m_\Lambda$ (where $m_\Lambda = M_\Lambda/N_{max}$) which reaches the same conclusion: no phase transition (e.g., spontaneous magnetization) takes place in ferromagnets as long as they are of finite size (more on this in detail in Section 3).

It now appears that a rigorous SM explanation of some TD phenomena of finite systems is impossible. This is not true for some approximate methods, e.g., the mean-field method (cf. Wannier 1966; Goldenfeld 1992, 85–116), but all of such methods introduce considerations not justifiable solely on grounds of SM.

3. The Rigorous Solution: Thermodynamic Limit. There exist in the literature of mathematical physics two ways of giving a rigorous solution to the problem, in both of which the idea of TL is essential. The first considers infinite systems from the beginning (cf. Emch 1977, Penrose 1979, 1957–1972) while the second begins with a finite system and then considers what happens when the system's volume goes to infinity while keeping its density finite. The first general solution in the latter approach is given in Yang and Lee 1952. Its main idea, whose justification we shall discuss in detail later, is to show (i) that the limits of TD functions exist, and (ii) that singularities may appear in those functions at the limit. At this level, the proofs are strictly for claims of existence, not for what the limits are and where the singularities lie, which can only be obtained in special models, as we shall see later in this section.

For our system S as defined above, Yang and Lee proved two theorems, which can be simplified and combined into one as follows.

[T1] In a region in the complex z-plane which contains all real $z > 0$, if $V_\Lambda \rightarrow \infty$ while $\sigma = N/V$ remain finite, then $p_\infty = \lim_{\Lambda\to\infty} \frac{1}{\beta V} \ln\Xi_\Lambda = \lim_{\Lambda\to\infty} p_\Lambda$, $\sigma_\infty = \lim_{\Lambda\to\infty} \frac{\partial}{\partial z}(\frac{z}{V} \ln\Xi_\Lambda) = \lim_{\Lambda\to\infty} \sigma_\Lambda$, . . . exist ($p_\infty$ is also monotonically increasing along z) and are analytic for all positive real z *except where* $\Xi_\Lambda(V,z,T) = 0$ *(i.e., the roots* z_i*).*

[T1] essentially says that any systems like S will be in a pure phase except possibly at those z_i's which are roots. But non-real and non-positive (recall that $z = exp(\beta\mu)$) z_i's are not physically meaningful. So, if no roots lie on the positive z-axis, there are no multiple phases and hence no phase transitions. One must note the generality of this theorem: besides the density being finite and the interaction stable and strongly tempered, little else is assumed.[4]

We know from the previous section that for a finite system, $\Xi_A(z) = 0$ can never have positive real roots. The trick is that taking TL makes the number of terms of the polynomial infinite and so is the number of roots. Thus, we have the following.

[T2] When $V_A \to \infty$, the number of roots of $\Xi_A(V,z,T) = 0$ becomes infinite which may force some z_i onto the positive real axis of z.

For fluids, $p_\infty(z)$ is still continuous at those points but $\sigma_\infty(z) \approx \partial p_\infty/\partial z$ is not and for ferromagnets, $F_\infty(B)$ is continuous but $m_\infty(B) \approx \partial F_\infty/\partial B$ is not. Hence, there is a z_i at TL where it is possible that $p_\infty(z)$ or $F_\infty(B)$ become nonanalytic. These theorems give the possibility of a rigorous SM explanation of phase transitions; however, one needs specific models for certain systems to know where a phase transition occurs in the systems.

The simplest systems in the literature are classical *lattice systems*, the majority of which can be studied under the *Ising model* which has one form to represent a species of magnets (i.e., ferro-(or anti-ferro-)magnets) and another a species of fluid (i.e., lattice gas). The model was not invented for deriving rigorous results of TLs, but it turned out to be ideal for such purposes, and the lattice gas model allows us to apply rigorous results obtained in magnets to fluids or vice versa.

The Ising model of a magnet of finite size describes it as a macro-system satisfying the following conditions.

i. It consists of N fixed sites each of which is occupied by an spin-1/2 particle which can be in one of two possible states: spin-up ($s_i = +1$) and spin-down ($s_i = -1$). It therefore occupies a finite region, Λ, in the configuration space with a volume, V_Λ.
ii. The interaction between two sites is given exclusively by the terms: $J_{ij}s_i s_j$, for every site i or $j = 1, \ldots, N$, where $J_{ij} > 0$ if sites i and j are nearest neighbors and $J_{ij} = 0$, otherwise.[5]

4. Another technical requirement for [T1] is that the area of the surface of V increases no faster than $V^{2/3}$ as V tends to infinity. This simply means that S should not be enclosed in a surface whose area is proportional to R^s, where R is the one-dimensional size of the surface (e.g., its radius), and $s > 2$. Since $s = 2$ is the number of dimensions of most surfaces, it must take a great deal of twists, such as in a sponge, to make a surface's area go beyond that.

5. This condition can be relaxed, but it should not concern us here.

iii. The Hamiltonian of the system, therefore, is given by:

$$H_A = -\sum_{\langle i,j\rangle} J_{ij} s_i s_j - B \sum_i s_i \tag{3}$$

where B is the external magnetic field and $\langle i,j\rangle$ means pairs of nearest neighbors only.

Since the total number of the spins in an Ising system is constant, we use the partition function of the canonical ensemble, i.e., $Z_A = \sum_{s_1} \cdots \sum_{s_N} exp(-\beta H_A)$, for sites, $1, 2, \ldots, N \in A$, and we then have the fee energy, $F_A = -(\beta)^{-1} lnZ_A$ and the magnetization,

$$m_A = \frac{1}{N} \sum_{i=1}^{N} \frac{1}{2} \langle s_i \rangle = -\frac{1}{N} \frac{\partial F_A}{\partial B}. \tag{4}$$

To get a model of lattice gas, imagine that all the sites on the lattice is occupied either by one gas particle or by none, and instead of the spin, $s_i = 1$ or -1, we have the occupation number of a site, $n_i = 1$ or 0. If we assume, as in the magnet model, nearest-neighbor interactions only, i.e., $\Phi_{ij} < 0$ if sites i and j are nearest neighbors and $\Phi_{ij} = 0$, otherwise, where $\Phi_{ij} n_i n_j$, $(i$ or $j = 1, \ldots, N)$ is the interaction Hamiltonian, then the two models are isomorphic with the map: $s_i = 2n_i - 1$. Because of this isomorphic relation, one can prove a result in the magnet case and transfer it directly to lattice gas, and then if some restrictions of the latter can be relaxed (so as to make the model closer to the more realistic gas models) without forfeiting the result, we would have proven it in general.

For Ising systems, Lee and Yang (1952) again proves the following theorem.

[T3] Given an Ising system of magnet or lattice gas with $J_{ij} \geq 0$ or $\Phi_{ij} \leq 0$, $Z^*_A = 0$, only if $|z| = 1$, namely, all the roots of the partition function lie on a circle in the complex z-plane whose radius is of unit length. (Here, $Z^*_A = \sum_{N=0}^{N_{max}} exp(-\beta BN) \sum_{s_1} \cdots \sum_{s_N} exp(-\beta H_A)$ $= \sum_{N=0}^{N_{max}} z^N Z_N$.)

The phase transition of a ferromagnet known as the *spontaneous magnetization* is described thermodynamically in Figure 2. Just as in the case of fluids, SM says that it is not possible for a finite magnet to exhibit such a phenomenon at $0 < T \leq T_c$. But as $V_A \rightarrow \infty$, the collection of roots on the unit circle, $|z| = 1$, becomes denser and denser until eventually one of them 'pinches' the positive real axis, whereby a singularity at $z = 1$ occurs. From [T2] and because $z = exp(-\beta B)$, we have that a phase transition at a temperature $T > 0$ is possible in an Ising system at $B = 0$ only at TL. In other words, at T, $m_\infty (B) \approx -\partial F_\infty / \partial B$ is not equal to zero when $B \to 0^+$, nor is it equal to $m_\infty (B)$ when $B \to 0^-$. This result obtained at TL is usually seen as that of a *broken symmetry*; the symmetry being $m_A(B)\ (B \to 0^+) = m_A(B)\ (B \to 0^-) = 0$. (See Figure 3 for an illustration of the symmetry breaking.) Note that the theorem does not say that a phase transition (or symmetry breaking) must occur at $B = 0$ (when $0 < T \leq T_c$), and thus could not explain why it actually occurs. The z_i's on the circle of $|z| = 1$ becoming dense as $V \to \infty$ does not implies that every point on the circle eventually becomes a root; and therefore the perforation of the positive z-axis is not guaranteed by taking TL. In fact to have a phase transition at $T \leq T_c$, an Ising system must have no less than 2 dimensions. In the case of a 2-dimensional system, the partition function can be solved by the method of transfer matrix so that we know exactly how the symmetry breaks (i.e., $\partial F_\infty / \partial B|_{(B=0^+)} \neq \partial F_\infty / \partial B|_{(B=0^-)}$) at $T \leq T_c$ (cf. Onsager 1944, Yang 1952). And to know the detailed behavior of a system near the critical point—the *critical phenomenon*—the method of renormalization is by

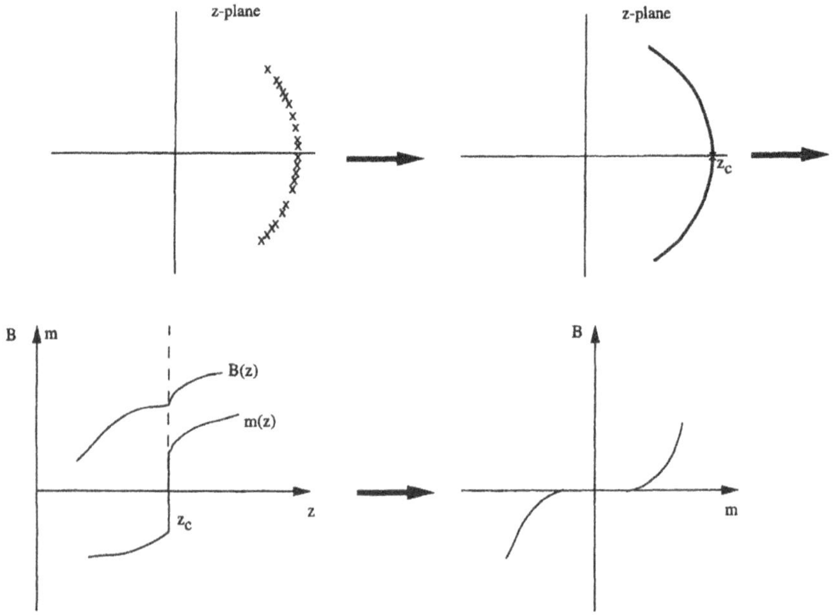

Figure 3. When the positive real z-axis is 'pinched', B(z) develops a kink and m(z) a break at B = 0 and z_c = 1. Eliminating *z* in B(z) and m(B), m(B) shows a singularity at B = 0.

far the best available (cf. Kadanoff et al. 1967, Wilson 1974). But whatever method one uses, a necessary condition for phase transitions is to take TL.

4. The Nature of Thermodynamic Limit. Why, one may ask, is the kind of rigorous solutions, as delineated above, solutions to the problem of phase transitions? Is the problem not really this: *while most if not all actual macrosystems are finite, thermostatistics predicts rigorously that no such systems could possibly exhibit the phenomenon of phase transitions*? And further, it is indeed finite systems which show phase transitions in TD. Taking TL, one may argue, appears to be avoiding rather than solving the problem.

The prevailing response to such an objection is that TL serves as a good *approximation* because actual macrosystems contain so many particles that they are practically indistinguishable from infinite systems. One may also mention that such a method of approximation is not infrequently used in physics. Are we not using the same kind of approximation when we regard rigid bodies in mechanics and fluids in hydrodynamics as differential systems?

But there are two problems with this response. First, the number of parts of a finite system does not change whether it is studied by TD or by SM. If it is treated as finite in TD, why should it be treated as infinite in SM? Second, it would be the right answer if the question were about the justification of using limit quantities to represent TD quantities; for instance, the justification of using p_∞ and m_∞ rather than using p_Λ and m_Λ. The limits, $m_\infty = lim_{\Lambda\to\infty}\, m_\Lambda$, etc., are indeed the same kind of limits as, say, the mass density of a solid: $\sigma_M = lim_{\Delta V\to\infty} \Delta M/\Delta V = dM/dV$. In other words, imagine a series of ΔV's, in which each later ΔV is smaller than the earlier one. At each ΔV_n, the average mass density, $\Delta M_n/\Delta V_n$, is well-defined and as $\Delta V_n \to 0$ (i.e., $n \to \infty$): $\forall\, \varepsilon > 0,\ \exists N(\varepsilon) > 0$, such that $\forall n > N(\varepsilon), |(\Delta M_n/\Delta V_n) - \sigma_M| < \varepsilon$. And since the molecules in the solid are so densely populated, the difference between the actual average mass density and the limit density is so small as to make the latter a good approximation for the former. It is essentially the same story for p_∞, m_∞, etc. However, this does not seem to apply to phase transitions. The singularities in p_∞ and m_∞ are not reached asymptotically towards TL, for (i) at no stage of the process in which $V, N \to \infty$, is a singularity of a system even roughly or approximately defined, (ii) nor is the singularity approached or approximated in any proper sense of approach or approximation. In [T3], we see that unless a root z_i assumes the value of 1, it is not even a meaningful physical quantity (or solution) regardless of how 'close' it is to that. So, there is no sense in which a complex z_i, close to but $\neq 1$, could give an approximation of a phase transition in either a ferromagnet or a system of lattice gas. (Note: it is not that no sense of limit applies in this case, but that the limit has nothing to do with approximation. As the circle of $|z_i| = 1$ becomes dense, it is true that $\forall \varepsilon > 0, \exists N(\varepsilon) > 0, \ni \forall n > N(\varepsilon), |z_n - z^*| < \varepsilon$, where $z^* = 1$. But no z_n is approximate to z^*.)

Most people do not give reasons to justify the use of TL, but a few did (e.g., Griffiths 1972, Toda et al. 1983, Penrose 1979, Sklar 1993), and their reasons can be summarized as follows. TL is needed for (i) the equivalence of all three ensembles for TD functions, (ii) an exact solution for the problem of phase transitions, (iii) an elimination of any effects of a system's boundary, and (iv) an elimination of fluctuations in a system. But these cite only the beneficial consequences of using TL; they do not explain why it can be used in the first place: (ii) begs the question; (i) means that SM is consistent, for if the three ensembles give different values in TL for TD functions, SM would be inconsistent (because an infinite system is both 'isolated' and without 'boundary'—that which prevents the exchange of energy and particles); (iii) and (iv) are a matter of pragmatics in that most inter-molecular interactions are short-ranged[6] such that unless one studies the behavior of a system near its boundary, the boundary's influence should be negligible, and the number of molecules in any finite macrosystem is so large that fluctuations are virtually undetectable. None of these answers our question, except that (iv) points to the right direction along which an answer can be sketched as follows.

Phase transitions are characterized as singularities in TD because there is no fluctuation in TD systems, and this is so because TD systems are considered as of continuous matter. In other words, systems are *idealized* in TD to be dense with their parts in any given V; and the idealization is warranted by the large number of molecules any macro-volume contains. However, from an SM perspective, no volume of matter can be dense with particles, which means that there are always fluctuations in a finite system and hence no possibility for singularities. Taking TL is the only resort (i.e., idealization) in SM by which the TD account of phase transitions (as singularities) can be reproduced. Actual systems are finite and phase transitions in them are never real singularities (due to *fluctuations*). Both TD and SM with TL idealize their systems, but they idealize them in opposite directions: in TD, systems remain finite

6. Electromagnetic and gravitational interactions are not short-ranged, so they need separate treatments. For the former, an assumption of shielding is made so that TL is guaranteed, while for the latter, we assume that it is so weak in most terrestrial systems that it does not affect the TL. This is certainly not true with celestial bodies and that is where gravitational collapse is possible.

but are idealized to be continuous, while in SM, they remain discontinuous but are idealized to be infinite (because having infinite number of particles in a finite volume is not an option). And the idealization in SM, which does not seem an approximation of any actual condition, is required by the idealization in TD, which does appear to be a good approximation—in practical terms, a macrosystem is dense with its parts. Therefore, the need for TL is really the result of an artifact in TD. The dramatic claim that it is impossible in SM to have phase transitions in finite systems is in fact an exaggeration if not totally false. Finite systems do harbor phase transitions, only they are not singularities.

What exactly is the role of fluctuations? Suppose that we have a finite system of liquid, S (with finite N and V). We increase its volume while keeping its temperature, $T < T_c$, constant and measuring its pressure. We repeat the same experiment many times and record the isotherm for each. If we then look closely at one of the extreme points of the Maxwell plateau of all the isotherms (e.g., points A or B in Figure 1b), we will find that they are not at the same spot. This is one of the effects of fluctuations. If we were to have a system with relatively small N, the fluctuation would be large and the average isotherm would not resemble the one for any actual fluid system. But the larger the N, the smaller the fluctuation; and eventually, when $N \rightarrow \infty$, the fluctuation disappears and the isotherm becomes one of those shown in Figure 1b. The isotherm for an actual system undergoing a phase transition at T will be close to the ideal ones in Figure 1b, but not showing the singularities. Or in other words, with the existence of fluctuations, the transition is neither 'smooth' nor 'singular' (see Figure 4 for a mere illustration).

It will be interesting, at this juncture, to compare the case of phase transitions with that of the second law of TD. The question in the latter is whether there is a SM justification for the monotonic non-increase of entropy of an isolated macrosystem. Since the physical processes occurring at the level of the system's micro-constituents are all reversible, the irreversibility represented by the non-increase of entropy can at most be a statistical—i.e., non-monotonic—property. Would taking TL help to restore the strict irreversibility and reëstablish the unqualified truth of the second law? Here, I can only offer a few conjectural remarks. In so far as eliminating fluctuations is concerned, taking TL is sufficient to kill any recurrence in a process which begins at a non-equilibrium state. It should be possible to prove a theorem to the effect that for any recurrence level, there is an N such that no fluctuation can exceed that level. However, I am not sure what can be said about this from the perspective of reversing initial/boundary conditions. How does one deny the possibility of a Loschmidt process (cf. Sklar 1993, 36)? How does one even define a non-quasi-static transition from one state to another under TL?

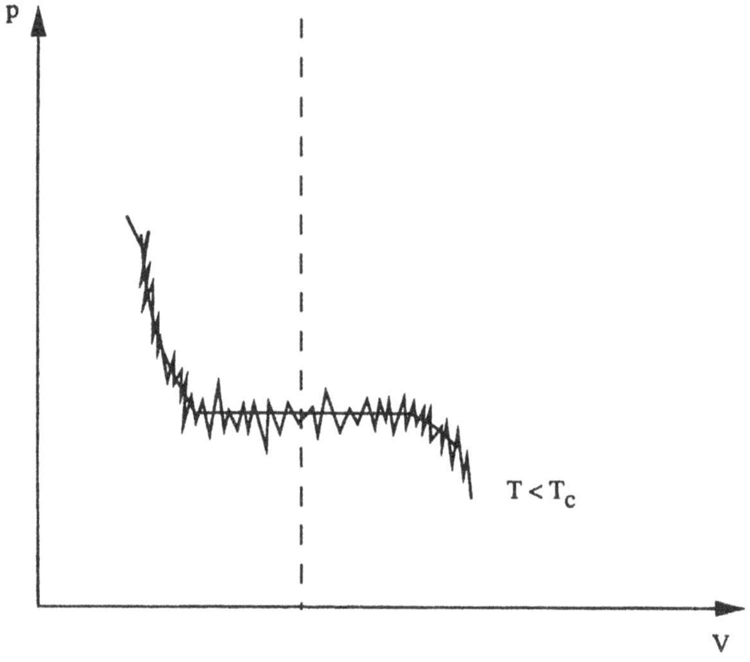

Figure 4. Possible fluctuations along an isotherm at $T < T_c$.

Would it make sense to talk about reversing, say, the momenta of all the particles at a certain state while taking N to infinity? These are interesting questions that need further investigations.

Be that as it may, the following seems right in the light of this comparison. Phase transitions as singularities are no more and no less an artifact of TD as the transport phenomena (in isolated systems) are as processes of a monotonic non-increase of entropy. If one needs a rigorous SM explanation, the other should need one as well. This, surprisingly, does not seem to be the attitude of most physicists and mathematicians. For those who find the impossibility of recovering TD singularities in finite SM systems unacceptable do not seem to be bothered by the similar impossibility of upholding the strict irreversibility in those systems.

5. Concluding Remarks. So far, I have tried to explain in what sense a phase transition, like irreversibility, is an emergent property—i.e., emerging from, but not reducible to, properties of parts in finite systems. It is emergent precisely because it is a property of finite systems and yet only reducible to micro-properties of infinite systems. It would not be so if there were infinite systems and it were a property of them alone. But does TL has any explanatory value? Does it help us to understand phase transitions as a cooperative phenomenon? We know from the above discussion that what explains, say, the phase transitions in Ising systems is the postulated structure and interaction of the model. It turns out that a system modeled as such only has phase transitions when it goes to TL. There certainly are other possible types of interactions among, and/or arrangements of, a given system's molecules so that singularities arise even if the system is finite, but the two conditions, the strong temperedness and stability, are so natural and any other additional constraints so implausible from a microscopic point of view that it seems unreasonable to consider those types. This is why TL becomes a much better option. Therefore, TL provides a *necessary condition* for an approximate causal explanation of phase transitions. Further, should one accept the SM theory of phase transitions under TL? And what would be one's justification if one did? Does such a theory save any phenomena? What it saves is not a set of phenomena but another theory—the idealized TD of phase transitions (including the critical phenomena[7]). It does not try to go beyond or correct TD but to save it instead. I wonder whether it would be better to give up the standard TD description of phase transitions (as singularities) and provide an account of them in terms of finite SM systems. This seems, as I mentioned earlier, to be the attitude towards the issue of irreversibility or recurrence in the physics community. So, why could we not say that since we now know that there are fluctuations in finite macrosystems, phase transitions in them should no longer be seen as singularities? It is quite possible that if such a conceptual shift takes place among theoreticians, the isotherms of systems with multiple phases coming back from the laboratories will no longer have singularities in them.

REFERENCES

Emch, Gérard G. (1977), "An Algebraic Approach for Spontaneous Symmetry Breaking in Quantum Statistical Mechanics", in P. Kramers and M. Dal Cin (eds.), *Groups, Systems and Many-Body Physics*. Braunschweig: Friedr, Vieweg & Son, 246–284.

Goldenfeld, Nigel (1992), *Lectures on Phase Transitions and the Renormalization Group*. Reading, MA: Addison-Wesley.

Griffiths, Robert B. (1972), "Rigorous Results and Theorems", in C. Domb and M. E. Green (eds.), *Phase Transitions and Critical Phenomena*. New York: Academic Press, 8–109.

Humphreys, Paul (1997), "Emergence, Not Supervenience", *Philosophy of Science* 64 (Proceedings): S334-S245.

Kadanoff, L., W. Götze, D. Hamblen, R. Hecht, E. Lewis, V. Palciauskas, M. Rayl, and

7. Unfortunately, I do not have space to discuss the subject of critical phenomena in this paper. It gives a better illustration to the necessity of taking phase transitions as singularities in TD, but it does not affect my discussion of the justifications of TL.

J. Swift (1967), "Static Phenomena near Critical Points: Theory and Experiment", *Reviews of Modern Physics* 39: 395–431.
Lee, T. D. and C. N. Yang (1952), "Statistical Theory of Equations of State and Phase Transitions II. Lattice Gas and Ising Model", *Physical Review* 87: 410–419.
Onsager, Lars (1944), "Crystal Statistics. I. A Two-Dimensional Model with an Order-Disorder Transition", *Physical Review* 65: 117–149.
Penrose, Oliver (1979), "Foundations of Statistical Mechanics", *Reports on Progress in Physics* 42: 1938–2006.
Sklar, Lawrence (1993), *Physics and Chance*. Cambridge: Cambridge University Press.
Stanley, H. Eugene (1971), *Introduction to Phase Transitions and Critical Phenomena*. Oxford: Clarendon Press.
Teller, Paul (1992), "A Contemporary Look at Emergence", in A. Beckermann, H. Flohr, and J. Kim (eds.), *Emergence or Reduction*. Berlin: Walter de Gruyter, 139–153.
Toda, M., R. Kubo, and N. Saitô (1983), *Statistical Physics I: Equilibrium Statistical Mechanics*. Berlin: Springer-Verlag.
Wannier, Gregory (1966), *Statistical Physics*. Dover: New York.
Wightman, Arthur S. (1979), "Introduction: Convexity and the Notion of Equilibrium State in Thermodynamics and Statistical Mechanics", in R. B. Israel (ed.), *Convexity in the Theory of Lattice Gas*. Princeton: Princeton University Press, ix-lxxxv.
Wilson, Kenneth G. and J. Kogut (1974), "The Renormalization Group and the ε Expansion", *Physics Report* C 12: 75–197.
Wimsatt, William C. (1997), "Aggregativity: Reduction Heuristics for Finding Emergence", *Philosophy of Science* 64 (Proceedings): S372-S384.
Yang, C. N. (1952), "The Spontaneous Magnetization of a Two-Dimensional Ising Model", *Physical Review* 85: 808.
Yang, C. N. and T. D. Lee (1952), "Statistical Theory of Equations of State and Phase Transitions I. Theory of Condensation", *Physical Review* 87: 404–409.

Part 6

Quantum Phenomena and Realism

This part contains four previously published papers, which are loosely connected by the theme of interpretive understanding of quantum objects (i.e. entities that are described by quantum theory, which include such things as the wave pockets).

The two papers, "The Aharonov-Bohm Effect and the Reality of Wave Packets," and "Potential, Propensity and Categorical Realism," explore the implications of the Aharonov-Bohm Effect with respect to coming up with a largely realist interpretation of the wave packet (or that which the wave function represents). And this is taken as a metaphysical or ontological issue in foundations of physics. The first paper is an in-depth philosophical study of the Aharonov-Bohm Effect, while the second spells out its wider implications in philosophy of quantum mechanics. There have been many studies of the Effect in the philosophy of science literature since then, which cannot all be listed here. The journal, *Foundations of Physics*, has published the majority of papers concerning the Aharonov-Bohm Effect, and a simple search on *PhilPapers* with the key phrase "Aharonov-Bohm Effect" leads one to most of the papers. Here I mention a few that are most relevant to the line of inquiry pursued in above-mentioned papers. Richard Healey (1997) "Nonlocality and the Aharonov-Bohm Effect," *Philosophy of Science* 64 (1): 18-41, focuses on the nonlocality aspect of the wave packet. Stephen Leeds (1999) "Gauges: Aharonov, Bohm, Yang, Healey," *Philosophy of Science* 66 (4): 606-627, makes a connection of the Aharonov-Bohm Effect to the wider realm of quantum gauge field theories. Antigone. M. Nounou (2003)

"A Fourth Way to the Aharonov-Bohm Effect," in Katherine Brading & Elena Castellani (eds.), *Symmetries In Physics: Philosophical Reflections*. Cambridge University Press, provides an alternative way to understand the philosophical implications of the Effect. James Mattingly (2007) "Classical Fields and Quantum Time-Evolution in the Aharonov–Bohm Effect," *Studies in History and Philosophy of Science Part B* 38 (4): 888-905, provides a novel construction of the Effect that involves only locally definable quantum electrodynamic field strength; and argues that such a construction shows "gauge fields in general make no metaphysical demands, and commit us to no novel ontology."

The other two papers, "Particularism and Holism Revisited," and "A Possible Unified Understanding of Probabilistic Objects," try to tease out a sense in which quantum entities are no different from classical ones, i.e., in some precisely definable sense, electrons are no different from coins; and why particularism may still be regarded as true in quantum phenomena. The sense in which that is true is of course very limited, and yet it is unmistakenly there, if my arguments are sound.

A Possible Unified Understanding Of Probabilistic Objects

By comparing two simple cases, the toss of fair coins (classical objects) and the two-slit experiment of electrons (quantum objects), I suggest a way of understanding the so-call wave-particle duality of quantum objects, which deprives it of its paradoxical appearance. I argue that it is paradoxical because we are misled into thinking that quantum objects ought to have categorical properties as classical corpuscles do, whereas many classical objects, if characterized by their dispositional properties, can be just as puzzling. In the end, I show how the distinction between categorical and dispositional properties may help to dispel the wave-particle paradox of quantum objects. And my analysis, if correct, lends support for the position which takes disposition (and propensity) seriously.

I

Even though quantum theory (relativistic or not) is now indispensable for physical sciences, when one asks what a quantum (or microscopic) object, e.g. an electron, really is, it gives no clear answer. An electron is known to be partly a corpuscle and partly a wave, but what does that mean? It cannot mean that an electron consists of physically separate parts, one of which acts like a particle and the other like a wave. What it is usually taken to mean is that it ***sometimes*** acts like a particle and ***sometimes*** like a wave. By acting like a particle or a wave, I mean no more than that the object of interest possesses properties which are usually associated with either particles or waves in classical physics. The properties in question are, for particles, positions, momenta, energy, angular momenta (if they are of finite extension), & etc., and for waves (especially for wave packets), extensions, amplitudes (intensities), wave lengths (or frequencies), group velocities (if they are packets), & etc. The properties do not have to be observable; ***they only have to be ones that an object can possibly possess***.

As to when quantum objects are waves and when they are particles, the following brief account should suffice. When an electron is measured, either by a Geiger counter or through a cloud chamber, the recorded effect is such that whatever produces it must resemble a small particle, namely, that its total energy and momentum are delivered almost instantaneous to the measuring device over a very small region. A classical wave can never deliver its energy and momentum in such a manner; the total energy of a wave packet must be transferred to the device which receives it through the time that takes the packet to pass and over the region which matches the packet's size. Therefore, we ***infer*** that the electron must be a particle when measured. However, when not subject to measurement, it behaves like a wave (or wave packet). However, this statement needs careful explanation lest it is misunderstood (while misunderstandings of the wavelike nature of quantum objects abound).

As explained above, singular measurement outcomes do not indicate that electrons are like waves. However, if one looks at the collective outcome of a large number of measurements done on identical electrons and in identical environments, one sees it as consistent with each electron being wavelike.[1] In other words, 'being wavelike', in this context, is a phrase that refers to ***what*** is capable of producing certain results in multi-particle experiments. The ***inference*** goes essentially as follows. The experiments are set up as such that if they were carried out on classical (or macroscopic) particles, they would result in identical outcomes. With identical electrons in such experiments, the outcomes are not identical but form a distribution of different individual outcomes whose total effect looks (when the details are ignored) exactly like what a classical wave would have left if measured in the same time interval. Since each single outcome is corpuscular, the wavelike behavior which is represented mathematically by the electrons' wave functions is no more than a property that produces probabilistic outcomes in an ensemble of identical electrons. Of course, identical electrons may not in fact be identical, but they are prepared in identical observable conditions and put through processes of identical observable properties and measured by identical apparatus and

[1] Here, the word 'identical' should be understood properly because, strictly speaking, no two electrons are identical. It means 'being identical in all relevant aspects', where what is or is not relevant is determined by physical laws and the requirements of the experiments involved.

observers.

Electrons would not be considered puzzling if they can be viewed as corpuscles whose behavior is irreducibly probabilistic. The balls in a game of bagatelle board are objects of this kind[2], and so are fair coins and dice. But the kind of probabilistic laws the electrons obey are fundamentally different from those obeyed by the classical objects, and they might be the result of some puzzling causal processes at the quantum level. I shall explain this point in the next section. The main challenge for this paper is to find a way of seeing this difference so that quantum objects, such as electrons, are not fundamentally different from classical ones, such as coins and dice. That they are usually seen as such is the result, I shall argue, of looking at the wrong set of properties when comparing them. In other words, it will be shown that if one insists on having an electron's states determined uniquely by the values of its observables, one gets the unsavory, or some even say paradoxical, conclusion: no electron (nor any other quantum object) has any definite property unless or until some measurement shows that it does.[3]

II

Quantum objects are, nevertheless, not paradoxical in this respect, or so shall I argue in this paper: an electron is metaphysically no more mysterious or paradoxical than any ordinary object such as a fair coin or a die, provided that one accepts the idea that some classical objects are also irreducibly probabilistic. Electrons, for instance, appear puzzling only because we are misled by looking at them the wrong way: thinking that they ***ought to be*** like small charged particles simply because when they are measured (such as being put through a cloud chamber), they appear as such.

Let us now compare the tossing of fair coins with the two-slit experiment of electrons. Suppose that we have a fair coin which stays fair no matter how many times it is tossed and toss it repeatedly onto a hard and smooth

[2] We suppose that the game is played out under ideal conditions, so that the distribution at the bottom of the board is irreducibly probabilistic.

[3] The same claim can be made with respect to local hidden-variables. For this, see the articles in Cushing and McMullin (1989).

surface (which stays hard and smooth no matter how many times it is hit by the coin) upon which it rests with either heads up or tails up (the only two possible outcomes of our experiment). Suppose again that the coin is tossed fairly or properly, meaning that the direction and speed by which it is tossed every time does not favor its landing on either heads or tails.[4] The probability of this coin's landing on heads under such tossing conditions is 0.5 and so is that of its landing on tails.

Let us assume that the probabilistic behavior of a fair coin when properly tossed is fundamental, namely, that no further specification of the physical conditions before and during the toss can change (i.e. increase or decrease) the probability of heads or tails. In other words, with a complete description of the states of the coin and its environment, the chance of its landing on heads or on tails are 50/50. One can of course change the probability by changing the set of tossing conditions: there must exist conditions under which all tosses of the coin result in heads up and there must be conditions under which all tosses result in tails up. What we have assumed here is that there is at least one set of conditions (i.e. fair or proper tossings of the coin) such that no description of which can improve on the estimation of the chance of the coin's landing heads or tails. By expressing the case in this way, I have certainly assumed that the coin's chance is objective under the given conditions. Beyond this objectivist prejudice on how the probabilities in this case should be interpreted, nothing more is assumed. Especially, I pass no ruling between the frequency and the propensity interpretation of probability. One can view the chance of the fair coin in question as the limiting frequency of an infinite ***random*** series defined as having maximal complexity (cf. Martin-Löf 1966, 1969, 1971), or one can view it as propensity, i.e. some physical property possessed by the coin and the given set of conditions (cf. Mellor 1971, ch.4). To my best knowledge, both interpretations, which can also be subdivided into several different interpretations, have problems but are the best candidates for the objectivists (cf. Hawson 1995).

However, one may reject the above assumption and insist on taking the coin-tossing experiment as deterministic in principle. In other words, if we

---

[4] By properly tossing a fair coin, I mean the standard, if not too idealized, conditions under which the coin is tossed. Whether there is such a thing as the proper tossing of a fair coin and the probability of landing on heads or on tails in such tosses is 0.5 are issues I cannot discuss in this paper. Here they are taken for granted.

toss the coin once and it falls on the surface heads up, then repeating the toss under exactly the same conditions will always make the coin falling heads up; and the same is true, *mutatis mutandis*, if the coin falls on the surface tails up. In this view, we have the distribution of 50/50 because we are not exactly repeating the same experiment, but doing different ones in which conditions producing either outcomes are randomly realized. In that case, we have no other choice but to abort the discussion. There is no point of arguing that a non-probabilistic object, if a fair coin is one, is not fundamentally different from a probabilistic one, such as an electron. Therefore, the reader may want to take the major claim of this paper to be a conditional: ***if coins and electrons are both probabilistic objects***, they are not fundamentally different. And further I might add that there are good reasons for thinking that classical processes, such as the tossing of a fair coin or the throwing of a die, are irreducibly probabilistic (cf. Earman 1984, ch.3).

We now turn to the case of an electron, a quantum object, and explain why it ***appears*** to be fundamentally different from a fair coin. In a two-slit experiment, a beam of electrons is produced by an oven such that one electron at a time is projected towards a measuring device: either a photo-sensitive surface or an array of Geiger counters. A thin wall is placed in between upon which two slits are cut in parallel, one directly above the other, so as to let the electrons through from two well separated openings. The outcomes of this experiment are dots or clicks that after a certain time interval form a distribution on the plate or among the Geiger counters. It is well known that the distribution is peculiar from a classical point of view. The probability of any of the electrons landing at a certain place on the photo-plate when the two slits are open simultaneously is not equal to, as it should be if classical probability calculus is applicable, the sum of the probabilities of its landing there when only the upper or only the lower slit is open respectively. In short: $P_{OO} \neq P_{OC} + P_{CO}$, where the symbol P_{OC} denotes the probability when the upper slit is **o**pen and the lower slit **c**losed, P_{CO} for the opposite situation, and P_{OO} when both slits are open.

To see why this is problematic or even mysterious, let us consider the following. Suppose that the electron in question is a particle, which means that it has certain position in space (and perhaps some other properties as well) at every instant, whether we know it or not. This implies that it passes through either the upper or the lower opening if it passes through the wall at all. In other words, if it has its position at certain instant right in the middle of the upper opening, then it cannot be at the middle of the lower one at that instant, and vice versa. But if this is true, then it follows that $P_{OO} = P_{OC} + P_{CO}$, which contradicts the above result. This seems to imply that the opening of a slit which the given electron does not pass through has influence over where that electron may land. There is no apparent causal explanation for such an influence of an open slit upon an electron some distance away.[5] It is this kind of quantum phenomena that forces some pioneers of quantum theory to give up the idea that quantum objects possess, say, definite locations before they are observed to do so. Therefore, the alleged fundamental difference between a quantum object and a classically probabilistic one is that while it can be said that the latter possesses definite values at any instant of those properties which determine its state at that instant, albeit we may not or never know what they are, it cannot be said so for the former.

Now, the conclusion reached above is only partially correct. To see how a different perspective should be given to this discussion which may dispel the mystery that quantum objects enjoys (they don't have definite properties until they are measured), let us take a closer look at the two cases again.

The two-slit experiment is only puzzling if we reason as follows. We see an electron impinging on the photo-plate with a precise position, and so we infer that it must or ought to have definite positions at every instant on its way to the plate. And when we find out that such inference breaks down because the mere assumption that it does have those positions (which may not be knowable to us) yields inconsistent statements, we conclude that it has no properties unless it is measured. It should not take a genius to see that the argument is fallacious. To be more clear and precise, observe the following. Isn't this the same as the experiment of the fair coin? We see the

[5] One may think that the bit of material that covers the lower slit exerts some kind of causal influence upon the electron passing nearby so that the absence of which could alter the electron's later chance of landing somewhere on the plate. But quantum theory has no account of such a causal interaction. So, if it exists and is strong enough to alter the behavior of the electron, it must be some interaction that quantum theory fails to describe. Given quantum theory is complete, this is not possible, and if this is possible, quantum theory is radically incomplete. Bohmian mechanics is a theory that attempts to complete quantum theory. According to Bohmian mechanics, there is an interaction between the electron which passes through the upper slit and the physical condition at the lower slit, and it provides an account for such interactions. I can't discuss the status of Bohmian mechanics here, for that see Cushing *et al.* 1996.

coin landing on the given surface with its heads facing up or down, and from that we should infer, by making the same kind of inference as in the electron case, that at every instant during its flight before it hits the surface, the coin has one of the two properties (heads up or heads down). But such an assumption would also lead us into inconsistent claims about the coin's behavior. Having such a property at each instant during its flight would contradict the predicted results of having the coin finally landing on the surface with a 50/50 chance of heads up or down. In this respect, the two cases — one of quantum, and the other of classical, objects — are indeed the same. If we are sure that in the case of tossing the fair coin, the coin is not having heads up or heads down at any instant of its movement towards the given surface, then we should be equally sure that the electron does not have any definite position at any instant of its movement towards the photo-plate. If nothing seems paradoxical or mysterious about the former, why should it be so for the latter?

One may argue that having a definite position (or energy or momentum) is intrinsic to an object, quantum or classical, while having heads up or down on a surface is not; therefore, the inference in the electron case is justified while the one in the case of a fair coin is not. We shall see later in section III that this idea about which properties are intrinsic and which not is far from as simple as is assumed here. One may also argue that the coin is different from the electron in that we can calculate as well as 'see' (with a quick enough video camera) coin's trajectory after it is tossed and before it hits the surface. That is certainly true, and we cannot say the same for the electron. However, having definite trajectories does not address the questions in the coin-tossing experiment, just as having a definite electric charge (and we know it) before the electron hits the measuring device does not help answer the questions in the two-slit experiment; (and nor does having a definite wave function of the electron at each instant).

Since the inference goes from the values of a certain property determined after a measurement to those of the same property before that, switching properties does not seem the right way to resolve the problem. Perhaps an argument can be made to relate different properties before and after a measurement, but this point cannot be clarified until we say more about different types of properties, which is the focus of section III. At present, the point is this. Since what is measured are the positions where the electrons hit the plate, having definite values for other properties is not relevant. The same is true with the coins. The outcomes of the toss are not the positions and the orientations of the coin, but whether its heads or tails face up when it settles on the surface. Nowhere in the process of the toss is a fair coin lying on a surface, so we must say that the coin does not possess definite values of that which is to be observed in the experiment. Furthermore, it would not help if one were to let a similar surface intersect the coin somewhere during the toss and find out whether or not it lands on its heads (and repeat this many times), because (a) if the experiment stops there, we still have the same problem; and (b) if one lets the experiment continue by releasing the coin again every time and letting it fall on the final surface, the probability distribution will not be the same (i.e. it is no longer a toss of a fair coin which has 50/50 chance of landing on its heads or tails).[6] Similarly, we may intersect an electron somewhere in the two-slit experiment with a Geiger counter (perhaps immediately behind one of the openings) and find out its definite position; and we either end the experiment there and face the same problem or let the electron go through and reach the photo-plate (or the Geiger counters), on which we will eventually (i.e. when we complete many runs) see a different probability distribution. In short, since what is relevant in both experiments is that which is to be observed in the end, neither kind of objects possesses precise or definite values for such properties in the middle of their respective experiment.

To summarize, if a fair coin's landing on heads or tails is a fundamentally probabilistic event and a coin is characterized by these two properties — heads up and tails up, a fair coin is just as mysterious (or non-mysterious) as an electron. The fact that we often choose to characterize a coin in some other way, for instance, by the trajectory of its center of mass and the orientations of its two faces in the air, either is beside the point or should prompt us to do the same with electrons. We automatically assume, because of our knowledge of classical objects, that electrons, when not subject to measurement, should possess those observables whose values are in fact only obtainable after the measurements. We now see that if we make the same assumption to a fair coin, we would get an equally puzzling object.

However, this is not to argue that we should characterize a tossed fair coin only by the properties of its landing on heads or tails. On the contrary,

[6] The distribution may incidentally end up the same as when the tosses are not intersected by a surface. But since it can only be incidental, the scenario does not affect the argument.

we should not do that. A tossed fair coin is better characterized by its mass and trajectory in the air. But if, again, the probabilities are fundamental, the trajectory plus the environment does not determine a unique outcome of the toss. Is this not the same with an electron? It is better characterized by its charge and wave function; but so characterized we cannot determine which of its observable outcomes will obtain (we can only determine each outcome's probability, just as in the case of the coin). The only real difference between a fair coin and an electron in this respect is that the trajectory of the former is itself an observable (we can 'see' it) or characterizable by observables, whereas what determines an electron's states between measurements does not seem to be so characterizable. The wave function has its range in the set of complex numbers and what it represents (which physicists call a 'wave packet'), if anything at all, is not a wave in the four dimensional spacetime which one can set up an experiment to measure. The 'movement' represented by a wave function of an electron, for instance, does not describe the electron's movement in any straightforward sense; it mathematically determines, rather, (in a complex space) the change of the probabilistic distribution of the possible values of a given set of its observables. Therefore, what is observable over time is the distribution not the wave.

III

Even though physical objects are not identical to the collection of their properties (as well as relations among them), the best way to settle questions regarding them is to examine those. An object may have one or both of the two kinds of properties: ***categorical*** properties and ***dispositional*** properties (for a discussion of categorical v. dispositional properties, see, inter alia, Armstrong 1969; Mellor 1974; Pargetter & Prior 1982). One may further divide the latter into ***dispositions*** and ***propensities***, where the latter are probabilistic versions of the former.[7] Categorical properties are those occurrent or actual properties of an object whose values at a certain instant

[7] I shall use 'disposition' for deterministic dispositional properties and 'propensity' for indeterministic dispositional properties. Similarly, a dispositional (propensional) object is an object which is primarily characterized by dispositional (propensional) properties.

determine in part the object's state at that instant.[8] The values of such properties at one instant may be connected to their values at another instant by physical laws, but they are by no means determined by, or definable in terms of, those laws.[9] For many, a world, e.g. the actual one we inhibit, is determined by the totality of its categorical properties.[10] Dispositional properties are not occurrent; they are defined, rather, in terms of certain conditions or procedures that could produce certain occurrent properties. Suppose we have a property, P. ***P is a dispositional property iff there are two sets of properties, A and B, such that if an object x were to possess A, it would possess B,*** where A and B are sets of categorical properties.[11] Or

---

[8] I assume here that all categorical properties are actual, but this might be controversial. On the one hand, there doesn't seem to be anything wrong with defining categorical properties as world-relative: each possible world has a set of properties 'actually' obtaining in that world, but on the other hand, the very idea of any property obtaining in a possible world other than the actual one is defined counterfactually: property P obtains in world **w** rather than in the actual world means that if **w** were actual, property P (instead of some actually obtained property) would obtain. This seems to imply that an object can only have P in the actual world counterfactually; therefore, P itself is dispositional.

[9] In quantitative sciences, an object's properties and their values are usually distinguished as if the latter are not properties, but in fact that is not the case: they are all properties. For instance, we have a flag pole which is five feet long. Having length is one of the pole's properties and so is being five feet long. Therefore, the distinction is a pragmatic one for quantitative sciences which should not cause any philosophical confusion.

[10] For philosophers, these are called categorists (cf. Armstrong 1983; Lewis 1981). Physicists usually do not state their commitment to categoricism explicitly, but it is taken for granted most of the time. For instance, anyone who regards the world as nothing more than the total collection of the worldlines in a 4-dimensional spacetime has to be a categorist. The values that uniquely determines a worldline are certainly values of categorical properties.

[11] Under most circumstances some laws are known to govern the behavior of an object, x, so that B is derivable from x's having A, but that may not always be the case. What is necessary is that the counterfactual statement, 'if x were to have A then it has B', be true. Whether a counterfactual statement can be true without being supported by some laws is another issue of which I cannot pursue here. Moreover, there may be cases where B consists of properties that are not of x's. For instance, the properties that display the 'penetrability' of a bullet (which is a common disposition) may belong to a sheet of glass or a block of wood, not at all to the bullet. Hence, a more inclusive definition of a dispositional property may be this: *∀X, X is a dispositional property of x iff ∃A, B, such that A and B are categorical properties, and if x were to have A then B would necessarily be instantiated by some set of objects, where the 'necessity' is physical necessity.*

similarly for propensity: *P is a propensity iff there are two sets of properties, A and B, such that if an object x were to possess A, it would possess B* ***with a probability*** *ρ*. B is usually known as the ***display*** of P. Now, an categoricist is someone who believes in the priority or the fundamentality of categorical properties.

The relations between these two kinds of properties are complex and sometimes controversial. Categorical properties are usually taken as primary properties and dispositional ones as secondary, but such a view loses its obviousness once it is scrutinized. It is possible, and even rigorously argued by some, that all properties are dispositional. The idea is to regard the obtaining of any property the result (and hence the display) of some procedure by which its values can be measured, and to use that or the law that governs it to define the property. For instance, a triangle's having three angles, one of the most categorical of all categorical properties, can be seen as a dispositional property (cf. Mellor 1982) if having three angles is the result of properly counting the angles of the triangle. Or the positions of a cannon ball are determined and defined by the mechanical law that governs its movement through space (cf. Ellis & Lierse 1994). Of course, the categoricists disagree; they regard the occurrent properties as prior or fundamental, while laws are contingent relations (or connections) among sets of those properties.

I have no intention to try to settle this debate, but the following fact needs to be acknowledged: although all properties can be consistently (and at least formally) rendered dispositional, the opposite is not true.[12] ***No real dispositional property can be consistently construed as categorical.*** It may be argued that this is not true because each disposition is definable in terms of two sets of categorical properties (see the A and B in the above definition) which are related by a counterfactual. But such a 'definition' has little consequence on the issues of categoricity: it does not imply that the object possesses those categorical properties when it has the disposition they define. Further, the very fact that a disposition is only definable by a counterfactual shows that it is something over and above the categorical properties in the definien (please ignore the use-mention distinction here): it is some sort of lawlike connection. The ***fragility*** of a vase or the ***fairness*** of a coin is a property that determines a lawlike behavior (rather than the

[12] I thank Kirk Ludwig who clarified for me matters discussed in this and the next paragraph.

states) of the object such that knowing it one knows not what the vase or the coin is but how it would behave were it subject to certain conditions.

From another angle, it may also be argued that dispositions (and propensities) are ***grounded*** in categorical properties because all dispositions either are ***reducible*** to or ***supervene*** on categorical properties of the same object. For example, suppose that D is a set of mechanico-chemical properties which make a vase fragile. Then, the vase breaks whenever it is thrown to a hard surface (*ceteris paribus*) because it has D. To generalize, it is plausible that for any disposition of an object, there is a set of categorical properties of the object that ***makes*** it, or ***causes*** it to, possess that disposition. In other words, x would exhibit B were it to have A because it has D, where A, B and D are sets of categorical properties which may not share any members. The relation between a disposition (or a propensity) and its categorical base (e.g. D) can be that of reduction or of supervenience, the discussion of whose differences and similarities has its own literature into which I have no time nor space to enter in this paper. The question that should be asked is whether the general thesis is true, namely, whether there is always a set of categorical properties to ground every disposition or propensity, whatever the nature of the grounding is. Some think that this is an empirical question that can only be answered via scientific investigations; and we shall see in a moment if it is so with our case.

Now, classical objects, such as fair coins and dice, have both categorical and propensional properties. It may be argued that the latter are grounded in the former, for example, a coin's propensity of landing heads or tails on a surface seems to be a groundable propensity. First of all, it seems safe to assume that the coin has definite categorical properties at every instant of its existence, i.e. where it is, how fast it moves or how its faces are oriented. Then, it seems equally obvious that the propensity in question supervenes on some of these properties in that no change of the probability of the coin's landing heads or tails can occur without a change of any of these properties. One may be able to generalize this result to all dispositional properties (probabilistic or not) of classical objects.

Let us now analyze the case of quantum objects. To cast the wave-particle duality into the categorico-dispositional scheme, we may say the following. The particle picture of an electron presents it as a ***propensional object*** (an object defined primarily by its propensional properties), i.e. what it would appear to be were it to be prepared and measured in a certain way. And the picture does not work in a categorical mode: not only can we not always infer from the known position of an electron after a measurement where it has been before that, but also it is inconsistent to assume, thanks to Bell's theorem, that it is definitely somewhere though we know not where.[13] The wave picture may be understood as presenting the electron as a ***categorical object*** (an object defined primarily by its categorical properties), except that the properties, such as the shape, movement and frequencies, etc., of the wave packet are of a peculiar type: they are not something one can directly measure with any apparatus.

Now, we can see why electrons and all quantum objects seem puzzling but are not: they do not have any classical (or observable) categorical properties that ground their propensities. But it does not mean that they do not have categorical properties at all which can do the grounding (in fact, if that is true, the states of a quantum object would be indeterminable, which is certainly not the case). If one denies that the wave function of an electron denotes a set of its categorical properties (e.g. its shape, movement and frequency) because of which it exhibits the propensities displayed in the measurement results, one must consider the electron as an ***irreducibly propensional*** object. Otherwise, one may opt to believe in the reality of electrons as wave packets and take the propensities to be grounded in their properties. Now, one must note that just because the dispositions of electrons are probabilistic (hence they are propensities), it does not mean that the grounding of them in the categorical properties is so as well. The propensities are uniquely determined (up to a set of phase transformations) by the wave functions. So, even the believers of the strong reduction thesis can have their case made if they can find ways to convince themselves that wave functions denote real categorical properties of quantum objects.[14]

---

[13] This is not true in one particular interpretation of quantum theory, namely, Bohm's interpretation (or Bohmian mechanics). But there is good reason to believe that Bohm's interpretation is not really an interpretation of quantum theory but rather a new theory for quantum objects (cf. note 5). It postulates new dynamical laws which are not present in the original quantum theory. Hence, my discussion in this paper should be understood as being confined in quantum theory and its interpretations.

[14] To be more precise, the idea can be rendered as follows. Quantum objects are (ontologically, which also means categorically) identical to their wave packets which have categorical and genuinely wavelike properties, such as amplitudes, frequencies, etc. and which also uniquely (up to a phase factor) determine their physical states (this is also why

Even if one has difficulty accepting the reification of wave functions, one does not necessarily have to give up categoricism (i.e. there are always categorical properties in which the propensities of quantum objects are grounded). The advance of physics may one day discover (by empirical means or not) some new categorical properties which may have a better ontological profile than wave functions and which may ground the propensities in a more satisfactory way.

To summarize, whether an object is irreducibly dispositional (or propensional) depends on whether there exits a set of its categorical properties which grounds all of its dispositions and/or propensities. It is if there doesn't. Some believe that this is an empirical question, but our analysis of quantum objects so far suggests otherwise. It is an empirical question only if the grounding categorical properties must be observables, but I do not see why this must be so. Had observability been a necessary condition for introducing new categorical properties, none of the hidden variables theories would even have gotten off the ground.[15] Therefore, this is one of those places in science where metaphysical speculations seem inevitable. One may adopt an agnostic (or a positivist) view on this matter: I don't care unless a set of observable grounding categorical properties is found. This is itself a sort of metaphysical commitment.

I therefore conclude: both coins and electrons possess categorical as well as propensional properties. The latter may be grounded in the former, but the categorical properties that ***ground*** a propensity may not be the ones that ***display*** it. Those who infer from the positions of an electron after measurements to its having definite positions between measurements have confused these two sets of categorical properties: the positions after measurements display one of the electron's propensities but do not in this case ground it. While it is likely the case that all dispositions and propensities of classical objects are grounded in observable categorical properties, it is

---

they are categorical and intrinsic properties of the objects in question). Wave functions are mathematical representations of those properties; in other words, the extensions of wave functions are exactly these wavelike properties.

[15] For a brief but clear account of the hidden variables theories, see Redhead 1987, 45-48. One of the essential constraints of a hidden variables theory is that from it no observational result be deducible that differs from the ones in quantum theory. If any hidden variable has independently observable values, the above constraint is violated. Therefore, no hidden variable can be an observable.

an open question whether the propensities of quantum objects are also so grounded. It is mostly likely that they are grounded but not in ***observable*** properties.

References

Armstrong, D. M. (1983). What Is a Law of Nature? Cambridge, Cambridge University Press.

Armstrong, D. (1969) 'Dispositions are Causes', Analysis, 30: 23-6.

Cushing, J. T., A. Fine, et al., Eds. (1996). Bohmian Mechanics and Quantum Theory: An Appraisal. Boston Studies in the Philosophy of Science. Dordrecht, Kluwer Academic Publishers.

Cushing, J. T. and E. McMullin, Eds. (1989). Philosophical Consequences of Quantum Theory. Nortre Dame, University of Nortre Dame Press.

Earman, J. (1986). A Primer on Determinism. Dordrecht, D. Reidel.

Ellis, B. and C. Lierse, (1994). "Dispositional Essentialism." Australasian Journal of Philosophy **72**: 27-45.

Hawson, C. (1995). "Theories of Probability." British Journal for the Philosophy of Science **46**: 1-32.

Lewis, D. (1981). A Subjectivist's Guide to Objective Chance. Ifs Ed. W. Harper et al. Dordrecht, D. Reidel.

Martin-Löf, P. (1966). "The Definition of Random Sequences." Information and Control **9**: 602-619.

Martin-Löf, P. (1969). "The Literature on von MIses' Kollectivs Revisited." Theoria **35**: 12-37.

Martin-Löf, P. (1971). "Complexity Oscillations in Infinite Binary Sequences." Z. Wahrscheinlichkeitstheorie verw. Geb. **19**: 225-230.

Mellor, D. H. (1971). The Matter of Chance. Cambridge, Cambridge University Press.

Mellor, D. H. (1982). "Counting Corners Correctly." Analysis **42**: 96-97.

Mellor, D. H. (1974) 'In Defense of Dispositions', The Philosophical Review, 83: 157-81.

Pargetter, R. J. and Prior, E. W. (1982) 'The Dispositional and the Categorical', Pacific Philosophical Review, 63: 366-70.

Particularism and Holism Revisited

The present essay aims at broadening the recent discussion on the issue of holism vs. particularism in quantum physics. I begin with a clarification of the relation between the holism/particularism debate and the discussion of supervenience relation. I then defend particularism in physics (including quantum physics) by considering a new classification of properties of physical systems. With such a classification, the results in the Bell theorem are shown to have violated spatial separability but not physical particularism.

I

The debate over holism and individualism (or particularism), which until recently only exists among philosophers of biology or the social sciences, has spread over to the heart-land of physics. In a series of articles probing the philosophical significance of Bell's theorem, authors such as Paul Teller and Don Howard [1, 2] argue that quantum systems exhibit holistic behavior. The effort culminated recently in an article by Richard Healey [3], in which Healey attempted to clarify various senses of holism and particularism as well as the real philosophical implication of the Bell theorem on the debate.

The present essay aims at broadening this line of inquiry. I shall begin with a clarification of the relation between the holism/particularism debate and the discussion of supervenience relation. Then I defend particularism in physics (including quantum physics) by considering a new classification of properties of physical systems. With such a classification, the results in the Bell theorem are shown to violate spatial separability but not physical particularism.

First, we need a clear notion of supervenience and distinguish it from that of reduction. It is well-known that certain type of properties *supervene* on but are not *reducible* to other properties. For instance, it is argued that moral properties supervene on but are not reducible to physical properties.[i] The same may be true between mental and physical properties. A precise definition of supervenience is still a matter of debate; but the following statement by David Lewis [4] captures the gist:

> A supervenience thesis is a denial of independent variation. ... To say that so-and-so supervenes on such-and-such is to say that there can be no difference in respect of so-and-so without difference in respect of such-and-such. ([4], 358.)

Or, more precisely, Jaegwon Kim [5, 6] claims:

> A *strongly supervenes* on B just in case, necessarily, for each x and each property F in A, if x has F, then there is a property G in B such that x has G, and *necessarily* if any y has G, it has F. ([5], 165.)

But if the relation between so-and-so and such-and-such is explanatory or definitional, it becomes reduction.

II

With this much preamble, let us examine the notion of particularism and of holism. For particularism, Teller claims:

> [T] *Particularism* states that the world is composed of individuals, that the individuals have nonrelational properties, and that all relations between individuals supervene on the nonrelational properties of the relata. This means that if two objects, 1 and 2, bear a relation *R* to each other, then, necessarily, if two further objects, 1' and 2', have respectively the same nonrelational properties which 1 and 2 have, then 1' and 2' will also bear the relation *R* to each other. I call such a relation a *Supervening Relation.* ([1], 213.)

The opposite of [T] is what is called Relational Holism, where either the system of interest has non-supervenient relations or it has relations that supervene on relational properties of its parts.

I see two difficulties of this formulation. First, the notion of supervenience does not seem to apply in this case. For a supervenient relation to work in a domain of objects, the Supervenience Basis (SB) — the set of properties on which other properties supervene — has to be complete in that a complete description of any object or system in the domain is feasible with the properties in the basis alone. Otherwise, the first condition in Kim's definition of supervenience may not be true since there may not always be a G in B for each F in A. Or to put it differently, the notion of supervenience has the connotation that the supervenient properties are in some sense dispensable in respect of the supervening properties in the SB. But it won't be the case if the SB is not complete in that the supervenient properties may be needed to complete a description. The following example might help drive this point home. In the thermodynamic theory of ideal gas, P (pressure), V (volume) and T (temperature) form a complete set of properties in describing the thermal phenomena of ideal gases. We know that they are related by Boyle's law, PV = RT, where R is a constant. It won't make any sense to say that T *supervenes* on P and V, even though T is completely determined by them.[ii] P and V are not a complete set of variables for describing the thermal behavior of ideal gases. Similar cases abound in physical sciences. The set of all physical properties (relational and non-relational) is clearly a good candidate for a SB, but that of biological properties might not be so. Nor, I believe, is the set of non-relational properties.

Second, it is not clear that relational and non-relational properties are independently definable. In the physical sciences, many seemingly non-relational properties are in fact relational by nature, e.g., the velocity of an object. Not only is a body's speed relative to some other body, the direction of its motion is also dependent on the body's relation with other bodies in space. The same is true with spatial and temporal properties. Is it not rather intuitively true that the spatial location of an object is determined by its spatial relation with other objects? In order for Teller's relational particularism to hold among spatial relations, one must believe in a very implausible version of absolute space. One must take the absolute space as a container wherein every point is a namable individual, so that for any, say, two-place spatial relation, xRy, the relation can be reduced to a set of one-place properties, $F^a x, F^b x$, ..., where $F^a x = xRa$, ..., and a, b, ... are individual absolute spatial positions. This is very problematic to say the least. It is not clear which kind of properties is more basic — relational properties among individuals or properties pertaining to single individuals.

R. Healey has a more general definition for what he calls 'Pure Physical Particularism'.

> [H] *Pure physical particularism:* Every qualitative, intrinsic physical property and relation of a set of physical objects from any domain D subject only to processes of type P is supervenient upon the qualitative, intrinsic physical properties and relations of their basic physical parts (relative to D and P). ([3], 402)

And the pure physical holism is the negation of this thesis by asserting that not all qualitative, intrinsic physical properties of the set (relative to D and P) are supervenient on those of their parts. Healey's formulation has several merits, one of which is that it acknowledges the relativeness of the division between the 'whole' of a system and its parts. Unless one is talking about the whole universe, every system is necessarily a part of some other system(s). And every part is necessarily the whole of some other parts, *mutadis mutantis.*

More importantly, it includes the 'qualitative, intrinsic relations of basic physical parts' in the SB. At the same time, it is also the Achilles heel of this formulation. What are qualitative, intrinsic relations of the parts of a physical system? They are, according to Healey,

> the ... relations that these (basic parts) bear in and of themselves, without regard to any other objects, and irrespective of any further consequences of their bearing these properties for the properties of any whole they might compose. ([3], 401.)

Healey acknowledges immediately after this characterization that it 'resists precise formulation.'(*ibid.*) And it is 'notoriously difficult' how the qualitative, intrinsic relations can be precisely defined. To illustrate this point, and also for a better formulation of particularism later, let us look at a simple example in chemistry.

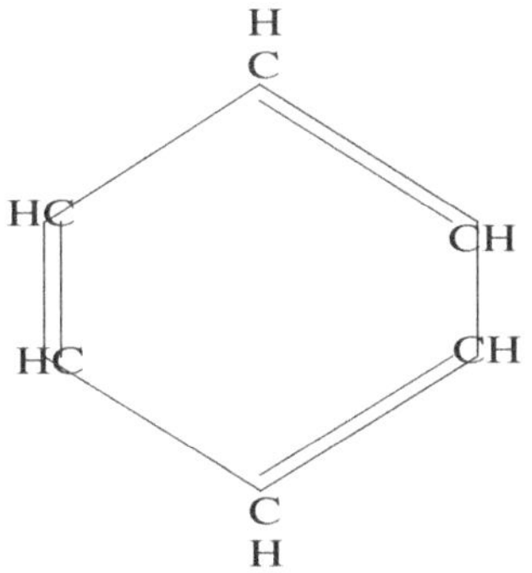

Fig.1 The Benzene Ring

In figure 1 the structure of a benzene molecule, C_6H_6, is drawn (keep in mind that the lines only represent physico-chemical relations, not spatial ones). Healey's physical particularism ought to apply to this simple example. The properties of the benzene molecule are determined by the physico-chemical properties of each hydrogen (H) atom and carbon (C) atom and the structure that these atoms support. However, the number of H's and C's is crucial to this structure, but this number has nothing to do with the qualitative, intrinsic properties or relations of each H and C in the benzene ring. Furthermore, the relations among the H's and C's depend crucially on which H or C is hooked up with which other H or C; they also depend on where the 'ports' for potential substitution are located so that other atoms can be attached to form the derivatives of benzene. I do not see how these structural relations can be viewed as 'qualitative, intrinsic relations' among the H's and C's upon which benzene's chemical properties supervene. Hence, either Healey's physical particularism is flawed or this is a case of holism. The former is more likely, since without the structural properties among the atoms, the SB is not complete, and almost no molecular properties supervene on atomic ones.[iii]

III

Physical sciences, especially classical physics, are considered the safe haven for particularism. One of the frequently cited examples of particularism in classical physics refers to the temperature of, say, a tank of gas. It is said that temperature is nothing more than the average kinetic energy of the gas molecules. The property of the whole system — the temperature — is thus reducible (methodologically) to and/or determined (metaphysically) by the properties of each non-overlapping parts — the molecules. Is the average kinetic energy or, in other words, the mean squared speed, $\overline{v^2}$, a 'qualitative, intrinsic property' of the molecules? But no molecule in the gas moves with that speed (some may in fact do so, but that is beside the point). Therefore, the temperature of a gas is determined by (or reducible to) a property of single molecules (parts of the gas) that strictly belongs to the gas as a whole. The same molecules moving with the same speed may have different 'temperatures' in different tanks of gas. In other words, the second condition in Kim's definition is violated.

There is a more problematic case in the same area of physics — the thermal transport of gases. The phenomenon refers to well-known cases such as the transportation of gas from one chamber to another when a dividing wall is removed. The rate (or speed) of transport in this case is a global property of the gas. The question now is whether this rate is reducible to and/or determined by the velocity of each individual molecule in the gas. The answer is negative, because what each molecule is doing in that process is nothing like gradually shifting from one place to another in a defined direction and speed. In fact, their movement is completely random in respect of its speed and direction. Neither Teller nor Healey's definition of particularism can survive in this case. From the properties and relations of each gas molecule, it is not determined whether the gas is stationary, symmetrically expanding or transporting gradually from one region to another. Thus, I conclude that particularism as construed by either Teller or Healey does not even apply to some of the well-known cases in *classical* physics.

IV

However, I believe that particularism is true. The question is how to formulate a precise definition of it, so that it would not fail in those cases where intuitively particularism appears to hold. The challenge seems to be one of how to adequately classify the properties of a system into different sets so that the supervenience relation applies among them. We have seen that it is not correct to separate the relational from the non-relational physical properties into two different sets and claiming that the former supervened on the latter. We also saw, in the case of the benzene ring, that the particular way in which a definite number of atoms connect to one another is crucial to the chemical properties of the molecule. This leads me into putting the physical properties of each individual part of any compound system into the following two categories: the *free properties* of the parts and their *confined properties*.

The free-property set is perhaps coextensive with Healey's SB, but more simply defined as follows. The properties are called free just in case they are the properties and dispositions of a system when it is not a part of any larger systems. In other words, free properties are *intrinsic* to the system that possesses them. In the case of an atom, for instance, they include the atom's rest mass as well as its dispositions (i.e. chemical properties) of forming various relations with other atoms to form chemical compounds.

The confined-property set includes mostly the relational (or structural) properties that each part of a particular system must bear to form the system. The properties belong to the parts, but are *not intrinsic* to them. They may or may not be highly dependent on the nature of the system in question and the realization of some of the free dispositions. In our case of the benzene ring, being one of the 12 atoms which is connected on the one side to a hydrogen and on the other a carbon atom is such a property of the carbon atom in the benzene molecule.

It is well-known that a set of free properties of atoms or molecules is compatible with different global properties of the compounds the atoms or molecules compose (e.g. many different molecules are made of the same atoms), so the globel properties cannot be said to *supervene* on the free ones if Lewis and Kim's definitions of supervenience are used (cf. p. 2). It is, therefore, too simplistic to think that whether a global property of a system supervenes on the intrinsic (or free) properties of its none-overlapping parts should decide if particularism (or holism) holds. Many simple global properties, such as the observable chemical properties of the benzene ring, are determined by (thus supervene on) the free as well as the confined properties of a system's parts, and they are good examples of particularism. Hence, I propose that we understand particularism as follows:

> [L] Physical Particularism holds just in case every qualitative, intrinsic property and disposition of a set of physical objects (from any domain D subject only to processes of type P) is supervenient upon the free and the confined properties and dispositions of their parts (relative to D and P).

I purposely followed Healey's definition closely because I wanted to show where the difference exactly lies between his and mine.

One may think that the confined properties of a system (or object) might be hard to precisely define, so [L] is not any better off than [H] or [T]. On the contrary, confined properties are easy to define because they are no more than the structural or mechanistic relations that each part in a system bears with others. One can hardly say that a compound system is completely determined by its parts without knowing the structure or mechanism of the system. All the previously discussed examples indicate that by making the division of the free and the confined properties (and dispositions), the ambiguity or vagueness of the previous definitions of particularism (or holism) disappears. As long as the parts of a compound system are known to be capable of free existence and the precise structure or mechanism of the system is determinable, the two sets of properties (and dispositions) for its parts can be well-defined. In the case of gas transport discussed earlier, if one includes how each gas molecule is related to every other molecule in the gas at every moment (i.e. the confined properties of the molecules), the transport rate is determined and particularism holds. One may think it odd to regard these as structural properties of the gas, but they in fact are, even though the structure is constantly changing.

The opposite of [L] is physical holism, which also better suits our intuition about holism. Take the presonality of a person for example. Even if all the free and confined properties and dispositions of the molecules that compose the person's body are known to be determinable, her presonality may still fail to supervene on them. Furthermore, there is no danger of anyone mistaken a trait of personality for a confined property of the molecules.[iv]

V

As a consequence of Heisenberg's uncertainty principle, the basic structure of quantum physics becomes fundamentally different from its classical counterpart. The principle rules out the traditional way of describing the state of a system by the simultaneous values of its observable properties, such as its position and momentum. Instead, one has to settle for what is known as the state function (or wave function). From the function, one only knows the possible values of the observable of interest and the weights (or probabilities) associated with the values. This means that if system S is in state ψ (ψ being the function that describes the state), then if one measures any observable, O, on S in ψ, one will get one of the possible values of O with the appropriate probability given by $|\psi|^2$.

Quantum objects, such as electrons, behave partly like classical waves (that is why ψ is a wave function), and one of the most important principles about waves is the *superposition principle*. We use the principle to explain novel quantum mechanical phenomena, such as the result of the well-known two-slit experiment on electrons. Suppose that we represent the state of an electron coming from one slit by $\psi 1$ and the state of it coming from the other slit by $\psi 2$. According to quantum theory, the probability that we find the electron in $\psi 1$ is $|\psi 1|^2$ and that in $\psi 2$ $|\psi 2|^2$. According to the classical theory of probability, the probability that the electron is in $\psi 1$ and $\psi 2$ should be $|\psi 1|^2 + |\psi 2|^2$. But this disagrees with the result of the experiment, which shows the effect of interference. This effect is explainable only by the superposition principle, according to which the sum, $\psi = \psi 1 + \psi 2$, represents the real joint state of the electron from both slits. This means that the probability of finding the electron in ψ is $|\psi|^2 = |\psi 1 + \psi 2|^2$, which, as it turns out, is not the sum of $|\psi 1|^2$ and $|\psi 2|^2$, as in the classical case, but $|\psi 1|^2 + |\psi 2|^2 + \psi 1^* \psi 2 + \psi 2^* \psi 1$.[v] The term, $\psi 1^* \psi 2 + \psi 2^* \psi 1$, accounts for the effect of interference.

Even though the effect of interference has produced profound metaphysical problems in quantum mechanics (e.g. the wave-particle duality of quantum objects), I fail to see how it violates physical particularism, as Teller has argued.[vi] First, the issue of holism vs. particularism does not seem to arise in this case since only one quantum object is involved. One may think that $\psi 1$ and $\psi 2$ describe two different parts of the object, but such a view is known to cause grave theoretical difficulties. Second, even if the object can be so described, I still do not see how particularism is violated. Is not the superimposed state, $|\psi|^2$, supervenient on $|\psi 1|^2$ and $|\psi 2|^2$ in the sense that the former cannot assume a different value without some change in the latter two? What stronger relation of supervenience can one expect than $|\psi|^2 = |\psi 1 + \psi 2|^2$? One should not equate particularism with what I call the classical particularism; i.e., if $\psi = \psi 1 + \psi 2$, then $|\psi|^2 = |\psi 1|^2 + |\psi 2|^2$. This is clearly too strong.

However, more problematic cases arise in quantum physics (cf. [1], 214ff), in which it does seem impossible to hold on to particularism. The most prominent among them is the Bell theorem.[vii] The current consensus on the implication of the Bell theorem is that one of its two premises has to be given up: (a) *locality condition*: there does not exist superluminal transmission of any information through space; (b) *separability condition*: the outcome of the spin value at one wing is independent of that at the other.

More precisely, let us assume that $p^{AB}(s^A, s^B \mid \sigma^A, \sigma^B)$ represents the probability of electron A having spin value s^A when the measuring device at wing A is set at direction σ^A and electron B having spin value s^B when the measuring device at wing B is set at direction σ^B.[viii] If the two conditions are satisfied, the two-electron system has the property of *factorizability,*

$$p^{AB}(s^A, s^B \mid \sigma^A, \sigma^B) = p^A(s^A \mid \sigma^A)p^B(s^B \mid \sigma^B) \tag{1}$$

(i.e., the states of A and B are independent). But this yields results which disagree with quantum mechanics and the experiments.

By simple logic, *modus tollens,* one must give up either the locality or the separability condition. Some believe that the locality condition must go, so quantum phenomena permit superluminal signaling. One may argue for this view more effectively as follows. It is well-known that quantum mechanics (more accurately referred to as non-relativistic quantum mechanics in contrast to relativistic quantum mechanics or quantum field theory) does not satisfy the principles of special relativity. From the non-Lorentz-invariance of the Schrödinger equation — the law that governs the evolution of the state function, ψ — it can be inferred that the effect of ψ can be transmitted instantaneously through space. We see essentially the same fact in Newtonian mechanics, in which that the Newtonian force acts instantaneously over distances is a consequence of the non-Lorentz-invariance of Newton's law of motion. We, therefore, do not need the Bell theorem to know that predictions derivable from quantum theory violate special relativity. If quantum mechanics (assuming the negation of locality condition) predicts one thing, the Bell inequalities (assuming the locality condition) predict the negation of it, and the experiment confirms the prediction of quantum mechanics, isn't it obvious that the locality condition should be given up?

The problem with such an argument is that it holds too simplistic a picture of quantum mechanical theory. The real nature of ψ is not known in quantum mechanics: we do not yet know whether it represents a field or wave that is locally definable (as the classical field or wave) or something which is intrinsically non-local.[ix] The only thing we know so far is that we have not yet seen anything in relativistic quantum mechanics which contradicts results in quantum mechanics. This suggests that quantum mechanics is not incompatible with special relativity. The apparent discrepancy may be the results of some other conditions, such as the non-separability condition.

Ever since the distinction between the locality and separability condition was made,[x] it has been vigorously argued that the separability condition should be the one to give up. The violation of this condition implies, according to Teller and Howard, a violation of physical particularism, and thus, quantum mechanical processes exemplify physical holism. To evaluate this point, one must clearly distinguish separability from particularism. For Teller, particularism implies separability (or outcome independence) (cf. [1], 220-1), so nonseparability means holism. However, Teller's particularism requires that all relational properties supervene on non-relational properties, which, as argued earlier, can not hold. What Teller has demonstrated is in fact the failure of classical particularism, which I have shown above as very different from the failure of particularism.

Because Healey allowed relational properties in the SB of his version of physical particularism, the implication of nonseparability becomes obscure. For instance, in an effort to link physical particularism with the separability condition, Healey formulated the notion of spatial separability: If one confines the relations among individual components to pure spatial relations, physical particularism reduces to,

> *Spatial separability*: The qualitative, intrinsic physical properties of a compound system are supervenient on the qualitative, intrinsic physical properties of its spatially separated component systems *together with the spatial relations among these component systems.* ([3], 410, my italics.)

But because the SB here includes 'the spatial relations' which are clearly not *intrinsic* to the 'component systems', spatial separability so defined is no longer a special case of Healey's pure physical particularism (cf. p. 5), as is intended. Further, even if we can make the spatial relations intrinsic so that it becomes a special case of physical particularism, the violation of it may still fail to warrant physical holism, because other 'intrinsic relations' of the component systems in the SB may make particularism hold. Hence, quantum phenomena which are said to violate spatial separability do not give evidence to holism (or physical holism).

Moreover, I want to argue positively that particularism (or physical particularism) is indeed not violated in the case of the Bell theorem. I shall here use the version of physical particularism given in section IV. The free-properties of the two electrons in the Bell theorem are those if the electrons are not in a singlet spin state and if the two measurement devices do not exist at the same time. It is obvious that if with only such free properties in the SB, particularism is violated. But so are many cases in classical physics as shown earlier. The set of confined-properties in this case contains among other things the correlation of measurement outcomes at both wings given by the theory of quantum mechanics. In other words, the property of non-factorizability of the probability function, which represents nothing more than a result of the interference between the two electron states, should also be included in the SB. When this is done, particularism survives. To compare this with Healey's notion of spatial separability, we see that there *are* non-spatial relations between the two electrons upon which the anti-correlation property of the whole system supervene, and these relations are not intrinsic to either electron. Therefore, if the question is whether the anti-correlation of the spin values supervenes on the spin states of the two electrons, as a question about physical particularism should be asked, the answer is affirmative since the superposition principle indicates exactly how that correlation is *determined*. But if the question is whether the anti-correlation supervenes on the *spatially separated(or free)* spin state of each electron, then the answer is negative, as is well known through the Bell theorem. *Thus, quantum mechanics violates spatial separability but not physical particularism.*

If I am allowed to speculate on this inclination of seeing holism in quantum mechanics, I would say that it is a confusion between classical particularism (cf. 12) and particularism. The classical particularists view quantum objects as primarily particles with locally definable classical properties, such as positions and spins. It is true that in the case of a singlet two-electron system, the anti-correlations of their spin values is not determinable by the two spatially separated electrons if they are treated as particles whose states are determined locally and independently of each other. However, it is well-known that quantum objects are partly wave-like (especially in respect of the behavior between experimental measurements) and their states represented by wave (or state) functions, which obey the superposition principle. If we accept, as I think we should, that wave functions describe genuine physical properties of quantum objects and the superposition principle is a law of interactions among them, I do not see why physical particularism should fail in quantum mechanics.

Let me conclude this essay by noting that holism is a doctrine which, however formulated, implies that there are properties of a system which can be better understood by investigating them independently of the properties of its parts. Whenever one has a case in which a property of a system is explicitly related to some properties of its parts, even though the relation is out of the ordinary, one is not in a position to question particularism. The cases in quantum mechanics are clearly of this kind.

References

[1] Teller, P. 1989. Relativity, Relational Holism, and the Bell Inequalities, in [8], 208-23.

[2] Howard, D. 1989. Holism, Separability, and the Metaphysical Implications of the Bell Experiments, in [8], 224-53.

[3] Healey, R. 1991. Holism and Nonseparability, *Journal of Philosophy* 88: 393-421.

[4] Lewis, D. 1983. New Work for a Theory of Universals, *Australasian Journal of Philosophy* 61: 343-77.

[5] Kim, J. 1984. Concepts of Supervenience, *Philosophy and Phenomenological Research* 45: 153-76.

[6] Horgan, T. 1982. Supervenience and Microphysics, *Pacific Philosophical Quarterly* 63: 29-42.

[7] Jarret, D. 1984. On the Physical Significance of the Locality Conditions in the Bell Arguments, *Noûs* 18: 569-89.

[8] Cushing, J. T. & E. McMullin, (eds.) 1989. *Philosophical Consequences of Quantum Theory* Nortre Dame: University of Nortre Dame Press

[i] The notion of supervenience as a technical term of this kind is first introduced by R. M. Hare. Cf. Hare, R. M., *The Language of Morals.* (London, 1952) p. 145.

[ii] This example may even serve as a counter-example of Kim's notion of strong supervenience relation, since the two conditions in Kim's definition seem to be satisfied in this case, and yet the supervenience of T upon P and V does not hold. Whether this is the case must be examined in detail before I can be certain, and that goes beyond the scope of this paper.

[iii] One may include, by fiat, those structural properties in the set of qualitative, intrinsic properties of atoms. I, however, do not see the benefit of doing so. It is counter-intuitive and threatens our notion of atoms as individual objects.

[iv] I am not committed, in this example, to whether or not a person's personality should be considered as a physical property. The possibility I alluded to here is that even if the personality is a physical property, it can still be a holistic one since it may not be supervenient on the physical properties of the molecules.

[v] Here we have, as a simple mathematical theorem, $|\psi_1 + \psi_2|^2 = (\psi_1 + \psi_2)^*(\psi_1 + \psi_2) = |\psi_1|^2 + |\psi_2|^2 + \psi_1^*\psi_2 + \psi_2^*\psi_1$, where ψ_1^* and ψ_2^* are the complex conjugates of ψ_1 and ψ_2.

[vi] Cf. [1], 214, especially the following passage.

> Indeed, we should question particularism because quantum mechanics itself gives every strong reasons for rejecting it. Quantum mechanics employs something called the Superposition Principle. A quantum mechanical state can attribute a property, p, to a particle. A second state can attribute p′. One can then always consider a third state, formed, literally, by adding the first two states. This superimposed state attributes neither p nor p′ but some new property, q, which can be considered to be a property somehow resulting by superimposing the properties p and p′.

[vii] I shall not describe the details of the Bell theorem, for there are many excellent accounts of it in the recent literature of philosophy of physics. See the articles in [8].

[viii] Here I omit the hidden variable, λ, for I will not be concerned with whether or not the hidden-variable theory is viable.

[ix] It goes beyond the scope of this paper to explain this point. Suffice it to say that there is an interpretation of quantum mechanics which identifies quantum objects with the wave packets (represented by the ψ functions), so that entities like electrons or photons are not point particles but rather extended objects of the size of their associated wave packets. In the case of the two-electron system in the Bell theorem, the wave packet may occupy the

whole space between the two measuring devices.

[x] Jarrett [1] is the first to distinguish the two conditions and argue for relinquishing the separability rather than locality condition. Also see [1, 2].

CHUANG LIU

POTENTIAL, PROPENSITY, AND CATEGORICAL REALISM

ABSTRACT. I argue that categorical realism, contrary to what most believe today, holds for quantum (and indeed for all) objects and substances. The main argument consists of two steps: (i) the recent experimental verification of the AB effect gives strong empirical evidence for taking quantum potentials as physically real (or substantival), which suggests a change of the data upon which any viable interpretation of quantum theory must rely, and (ii) quantum potentials may be consistently taken as the categorical properties of quantum objects so that categorical realism can be restored.

1. INTRODUCTION: NOT ANOTHER INTERPRETATION OF QUANTUM THEORY

Physical objects are commonly thought to have two kinds of properties, the categorical (or occurrent) and the dispositional. The former consists of the usual primary qualities of physical objects and substances, qualities whose values an object may possess intrinsically or independently of any other objects. Dispositional properties are often suspect in philosophy, but the quantifiable ones among them are widely used in physics and known as *potentials*. For classical (or macroscopic) objects and substances, their dispositions always seem to be reducible to categorical ones (by virtue of physical laws) unless it should turn out that all properties are dispositional, while quantum (or microscopic) objects appear to have propensities (i.e. probabilistic dispositional properties) that are *sui generis*. An electron's propensity, for instance, of jumping to a lower energy level which causes radioactive decay does not seem to be reducible to, or supervenient on, any of its categorical properties, such as its position and momentum before the decay, because either the electron does not in fact possess any definite position and momentum before the decay or even if it does, its position and momentum at that time do not determine if and when it decays.

Categorical realists (cf. Armstrong 1968: 85–8, 1969; Prior 1982, 1985; Pargetter and Prior 1982; Prior, Pargetter and Jackson 1982; also see, Ellis and Lierse 1994) believe that states and trajectories of physical objects are determined solely by their categorical properties. All dispositional properties of an object, if there are any, supervene on the categorical ones through laws (deterministic or probabilistic) that govern the object's behavior.

Erkenntnis **45:** 45–68, 1996.

The above seems to indicate that categorical realism cannot hold for quantum objects. That is not true, or so shall I argue in this essay. When genuinely probabilistic laws are incorporated, categorical realism is true in general. The argument consists of two main steps.

i. The recent experimental verification of the Aharonov-Bohm effect (AB effect, henceforth) gives strong empirical evidence for taking the quantum potentials (or wave packets) as *physically real*, which suggests a change of the data for any viable interpretation of quantum theory.
ii. I argue that quantum potentials are the categorical properties of quantum objects, and since they completely determine the objects' states, categorical realism holds even though the relationship between the categorical and the dispositional is no longer deterministically law-like.

The AB effect was first conjectured in 1959 via a thought experiment and its existence frequently debated until recently (cf. Aharonov and Bohm 1959; Peshkin and Tonomura 1989; Imry and Webb 1989; Liu 1994), the concept of a quantity being physically real is articulated in Cartwright (1987) and Liu (1994), and the substantival interpretation of wave packets was proposed in the early days of quantum theory and has been defended by a minority ever since (cf. de Broglie 1926; Schrödinger 1927; Bohm 1952).

Lest I be misunderstood, let me emphasize that I'm not trying to offer a new interpretation of quantum theory. The substantival (or realist) interpretation of wave packets is as old as quantum mechanics itself (cf. Schrödinger 1927), but since the empirical data upon which any adequate interpretation must rely have remained more or less the same over the years and militated against such interpretation, theories that postulate new entities or variables which merely save those data are naturally met with scepticism (hence, the attitude that such interpretations, even if correct, are pointless). What we need is some new data which weight in favor of such interpretation and which, I argue, are provided by the AB effect. From now on, any interpretation of quantum theory must square with the AB effect that requires the quantum potentials to play a causal role (just as it requires electromagnetic potentials to play such a role so that a merely instrumental interpretation of them is no longer adequate (cf. Wu and Yang 1975)).

Before we investigate the quantum potentials, let us see why potentials in classical physics are taken to be secondary, or even non-, properties.

2. POTENTIALS IN CLASSICAL PHYSICS

2.1. *Particles and Potentials*

(See Landau and Lifshitz 1976, chs. 1, 7) Most fundamental to classical physics are mechanics and electrodynamics: one is a theory for particles and the other for fields. According to the mechanical theory of point masses, the world consists of point particles moving in empty spacetime. Suppose that we have a system, S, of n particles. The states of S are completely determined by n pairs of quantities: $\langle \mathbf{q}_j(t), \mathbf{p}_j(t) \rangle$, where, $j = 1, 2, \ldots, n$; and $\mathbf{q}_j(t)$, $\mathbf{p}_j(t)$ are the 3-dimensional position and momentum vector respectively for the jth particle. The law that governs the motion of S (i.e. the law of motion) is a set of differential equations known as the Hamiltonian equations:

$$(1) \qquad d\mathbf{q}_j(t)/dt = \partial H/\partial \mathbf{p}_j(t), \quad d\mathbf{p}_j(t)/dt = -\partial H/\partial \mathbf{q}_j(t)$$

where H is the Hamiltonian function, which has the following general form:

$$(2) \qquad H = K + U(\mathbf{q}_1, \mathbf{q}_2, \ldots) = (1/2) \sum m_j (\mathbf{p}_j \cdot \mathbf{p}_j) + U$$

where K (and the sum $\sum(\ldots)$ runs from 1 to n) is the *kinetic energy*, and U the generic function for the *potential energy*, of the particles. Once a set of values at some time t_0 of $\langle \mathbf{q}_j(t), \mathbf{p}_j(t) \rangle$ is determined and the form of function $U(\mathbf{q}_1, \mathbf{q}_2, \ldots)$ known, the state of S at any time $t > t_0$ is also determined by Equation (1).

Now, from Equation (2) and the second equation in Equation (1), it is easy to see that we have

$$(3) \qquad d\mathbf{p}_j/dt = -\partial U/\partial \mathbf{q}_j$$

which is none other than $m_j \mathbf{a}_j = \mathbf{F}_j$, Newton's law of motion, where m_j is the mass of, $\mathbf{a}_j$ the acceleration of, and $\mathbf{F}_j$ the force which acts upon, the jth particle. The form of U, and hence that of $\mathbf{F}_j$, is determined by the forces acting among the particles or from outside S.

A different approach to the above is to characterize the states of S by a set of dispositional functions (cf. Hughes 1989; and Jauch 1968). Suppose O is an observable (e.g. $\mathbf{q}_j(t)$ or $\mathbf{p}_j(t)$, etc.) of the jth particle in S, and o a subset of its possible values. We can then define a function, $P(O, o)$, such that, $P(O, o) = 1$ if and only if when the particle is measured for O, O's values fall in o, and $P(O, o) = 0$ if and only if O's values fall outside of o. Since it has a definite answer (1 or 0) at any moment and represents how the particle would behave if measured, $P(O, o)$ characterizes its states completely and dispositionally. Nevertheless, this function cannot be used to replace $\langle \mathbf{q}_j(t), \mathbf{p}_j(t) \rangle$, partly because it does not appear in the laws of motion and partly because the use of it is either superfluous or problematic: superfluous if the notion of measurement is purely formal, and problematic if it is not, namely, that it is problematic if we adopt a strong operationalist view towards S's properties, which thereby commits us to all the problems that come with operationalism. More on this type of dispositionalism later in Sections 3 and 4.

The non-superfluous dispositional properties of particles pertain to their movement and force. They are known as 'potentials': a particle with certain kinetic energy has the *potential*, when colliding with another particle, to alter each other's movement; and every particle, by its nature, exerts gravitational force on each other so that its potential energy measures its *potential* to produce motion under that influence. This is elegantly represented in the general form of the Hamiltonian (cf. Equation (2)). If we think of H as S's disposition or capacity of producing various effects, the explicit separation of the kinetic and the potential term is significant. The kinetic term represents an intrinsic disposition of S, while the potential term an extrinsic one, because the kinetic term of a particle is non-relational and depends only on its own categorical properties ($K = f(m, v^2)$), while the potential term depends not only on the positions of other particles, but also on forces they exert. That is why it is not surprising that the potential (energy) is usually considered a non-reducible disposition of classical systems, and such familiar dispositional terns as 'fragility' and 'solubility' refer to the potential of a system that has a certain particular internal structure. However, the existence of such dispositions does not refute categorical realism, for the potential of S is determined if S is isolated and the positions and momenta of all the particles inside S are determined.

2.2. *Fields and Potentials*

(See Landau and Lifshitz 1970, ch. 3) The field theory of electrodynamics is a result (historically at least) of our refusal to accept the notion of action-at-a-distance. From the experimental works of Faraday, Maxwell discovered the equations for electromagnetic fields which bear his name. Just as any particle is subject to the action of a gravitational force field, a *charged* particle is subject, in addition, to that of an electromagnetic field. Since the action of gravitation appears in the Hamiltonian (cf. Equation (2)) by way of a potential term, U, the action a charged particle receives when it is placed in an electromagnetic field is also expressed in a potential term, U_e, which has the form:

$$(4) \qquad U_e = e(\mathbf{A} \cdot \mathbf{v} - \varphi)$$

where e is the charge, $\mathbf{A}$ the vector potential, φ the scalar potential, and $\mathbf{v}$ the velocity of the charged particle. (The unit is so chosen that $c = 1$.) And by solving Equation (1) with U_e, we have,

$$(5) \qquad m(d\mathbf{v}/dt) = e(\mathbf{E} + \mathbf{v} \times \mathbf{B})$$

where,

$$(6) \qquad \mathbf{E} = -\nabla\varphi - \frac{\partial \mathbf{A}}{\partial t}, \quad \mathbf{B} = \nabla \times \mathbf{A},$$

where $\mathbf{E}$ and $\mathbf{B}$ are what is known respectively as the strength of electric and magnetic field. Since ($e\mathbf{E}$) and ($e(\mathbf{v} \times \mathbf{B})$) are, respectively, the electric and magnetic force acting on the charged mass in question, $\mathbf{E}(\mathbf{r}, t)$ and $\mathbf{B}(\mathbf{r}, t)$ fully represent how the fields at $(\mathbf{r}, t)$ act on it. ($\mathbf{F} = e(\mathbf{E} + \mathbf{v} \times \mathbf{B})$ is known as the Lorentz force.) Field strengths are intrinsic to the fields in that they are their primary attributes upon which all other properties depend; for instance, the energy and momentum of a train of electromagnetic waves are defined in terms of $\mathbf{E}$ and $\mathbf{B}$. The states of the fields are uniquely determined by the values of these attributes, which do not supervene on any other quantities.

Are they then categorical properties? It depends. They are categorical in that they define the states of the fields and therefore describe what they are. But they are also dispositional in that they represent the field's capacity of producing certain effects upon charged particles. However, as I mentioned earlier, most properties, such as kinetic energy, of classical particles are also like that: the kinetic energy of a particle defines in part its states and also represents its capacity of producing motion in other particles when they collide. What is important is the following observation. Forces without a field-theoretic account are dispositional properties of particle systems that are not reducible to, or do not supervene on, the particles' positions and momenta. That is because force propagates with finite speed, and if it acts at a distance between two particles, its speed has to be a property of the two-particle system which is neither reducible nor supervenient on the particles' categorical properties (i.e. the values of their positions and momenta). Hence, categorical realism cannot survive in such an impoverished system. The field-theoretic account of forces in fact saved it in classical physics.

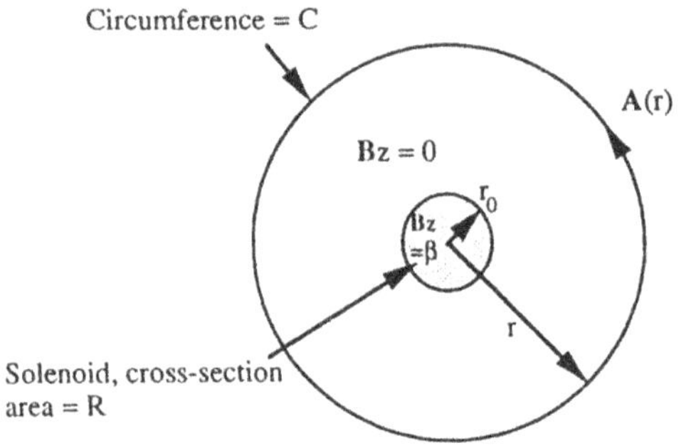

Figure 1. A solenoid perpendicular to the page confines the **B**-field within. The **A**-potential outside is nevertheless nonvanishing.

Now the question is which set of functions, the field-potentials (**A**, φ) or the field-strengths (**E**, **B**), better represents the categorical properties of the electromagnetic fields, their relationship being given by Equation (6). That which represents an electromagnetic field well must be capable of yielding values that uniquely determine all of the field's states and its contributions to every interaction it has with other fields and/or particles. In this respect, the potentials have a less robust status than the field-strengths. The following examples should suffice.

(a) Suppose we have the following linear electric field along axis-x:

$$E_x \begin{cases} = \alpha \neq 0 & x_1 < x \leq x_2 \\ = 0 & -\infty < x \leq x_1, x_2 < x < \infty \end{cases}$$

Since $E_x = \partial U / \partial x$, we have

$$U(x) = \int_{-\infty}^{x} E_x dx = \int_{x_1}^{x} E_x dx + \int_{-\infty}^{x_1} E_x dx = \alpha(x - x_1)$$

and, for this, we also have $U(x) = U(x_2)$, for all $x \geq x_2$. Since $E_x = 0$ at all $x > x_2$, how can we say that those points possess a disposition of affecting any particle (with value $U(x)$)?

(b) Suppose we have the following magnetic field along axis-z (see Figure 1):

$$(7) \qquad B_z \begin{cases} = \beta \neq 0, & r \leq r_0 \\ = 0 & r > r_0 \end{cases}$$

and from Equation (6) and Stokes' theorem, we have

$$(8) \qquad AC = \oint_C \mathbf{A} \cdot d\mathbf{l} = \int\int_R \mathbf{B} \cdot d\mathbf{s} = \beta R,$$

hence, we have $A(r) = \beta(R/C) = \beta(r/2)$, when $r \leq r_0$. Similar to the previous case, we have $A(r) = A(r_0) \neq 0$, when $r > r_0$, which produces a similar unpalatable result as in the previous case.

We say that there is no field in a region unless the strength there is nonvanishing because particles of the appropriate kind will not be affected in such regions. In case (**a**), an otherwise free particle will remain so by being put at a point outside the range of strength E_x. Even though its potential energy is non-zero, it possesses no 'disposition' to produce kinetic energy when unobstructed. The same, *mutatis mutandis*, can be said of the potentials in case (**b**). This phenomenon is known as the *gauge freedom* of electrodynamic potentials, namely that $(\mathbf{A}, \varphi)$ is physically equivalent[1] to $(\mathbf{A}', \varphi')$ if

$$\text{(9)} \qquad \mathbf{A}' = \mathbf{A} + \nabla f, \quad \varphi' = \varphi - \partial f/\partial t.$$

where, f is an arbitrary function. It is easy to see that by substituting $(\mathbf{A}', \varphi')$ for $(\mathbf{A}, \varphi)$ in Equation (6), the **E** and **B** remain unchanged, and hence, they, as well as the Maxwell equations, are gauge invariant with respect to the potentials. [One can easily see that Newton's law of motion is also gauge invariant with respect to the mechanical potential, $U(\mathbf{q})$, since $\mathbf{F} = \nabla U$ is invariant under a transformation of $U' = U + a$, where a is an arbitrary function of t.]

Moreover, whether an **A**-potential is nonvanishing at a point where $\mathbf{B} = \mathbf{0}$ depends on whether or not the loop of integration (in Equation (8)) which goes through the point in question encloses any region in which $\mathbf{B} \neq \mathbf{0}$. **A** is nonvanishing if the loop does so, but **A** is not if it does not. And since the loop is otherwise arbitrary, it follows that the values of **A** at that point are not determined independently of the **B**-field outside the point's immediate neighborhood.

3. POTENTIALS IN QUANTUM PHYSICS

3.1. *Wave Functions* (Ψ) *vs. Quantum Probabilities* ($|\Psi|^2$)

At the quantum level[2], states of a system, say S, can no longer be defined by definite values of its observables, for such values are not always obtainable simultaneously or, in other words, quantum states are not always dispersion-free. Because of this complication, a Hilbert vector space is used to support quantum theory. In this space, an observable is taken to be a linear operator, $\hat{O}$, and a state of S, a wave function, Ψ. Associated with each operator is a set of projection operators, P_i, each i corresponding to a possible value of $\hat{O}$ obtainable in Ψ. $P_i\Psi = |\phi_i\rangle$, where $\{|\phi_i\rangle\}$, for all i, is a set of eigenfunctions in which definite values of $\hat{O}$ obtain. The states of S are completely determined by the wave function, and the time evolution of S satisfies the Schrödinger equation of Ψ:

$$(10) \quad i\hbar\partial\Psi(t)/\partial t = \hat{H}\Psi(t)$$

where $\hat{H}$ is the Hamiltonian operator (the total energy) for S. An exact solution of Equation (10) is only possible for a few very simple Hamiltonians, but for whatever situation, the solution has the following generic form:

$$(11) \quad \Psi(t) = I(t)e^{-iS/\hbar},$$

which consists of two parts, one a real function – I (the amplitude), and the other an imaginary one – $e^{-iS/\hbar}$ (the phase function), in which S is the phase factor. This phase function disappears in $|\Psi|^2$, since $|\Psi|^2 = \langle\Psi|\Psi\rangle = |I|^2 e^{i(S-S)/\hbar} = |I|^2$ (= 1 for normalized Ψ), where $\langle\Psi|$ is the complex conjugate of $|\Psi\rangle$.

According to the Born probabilistic interpretation of $|\Psi|^2$, $|I|^2$ is a probability distribution of the following sort. Given $\hat{O}$, $\{|\phi_i\rangle | i = 1, 2, \ldots, n\}$, and a set of eigenvalues, $\{\lambda_i\}$, such that $\hat{O}|\phi_i\rangle = \lambda_i|\phi_i\rangle$, we can express Ψ by its Fourier expansion:

$$(12) \quad \Psi = \sum_{i=1}^{n} c_i|\phi_i\rangle,$$

where $c_i = \langle\phi_i|\Psi\rangle$, $i = 1, 2, \ldots, n$, is the expansion coefficient, and we also have

$$(13) \quad |\Psi|^2 = \langle\Psi|\Psi\rangle = \sum_i \langle\Psi|\phi_i\rangle\langle\phi_i|\Psi\rangle = \sum_i |c_i|^2.$$

Born takes $|I|^2 = \sum_i |c_i|^2$ to be the probability distribution in which each $|c_i|^2$ gives the probability that S in state Ψ ends up in the eigenstate $|\phi_i\rangle$ if it is measured for $\hat{O}$. And the average value of $\hat{O}$ is obtained by

$$(14) \quad \langle\Psi|\hat{O}|\Psi\rangle = \sum_i \langle\Psi|\phi_i\rangle\langle\phi_i|\hat{O}|\Psi\rangle = \sum_i \lambda_i|c_i|^2$$

which gives the weighed average of its eigenvalues.

Can categorical realism survive in a world thus described? What are the occurrent properties of an electron, for instance? Its states will be determined by its Ψ-function. But as a complex function, Ψ has no occurrent values obtainable by measurements (unlike $\langle \mathbf{q}_j(t), \mathbf{p}_j(t)\rangle$ in the classical case). The electron's observables (i.e. observable properties), such as its position and momentum, are not occurrent properties either. The non-dispersion-free nature of quantum states prohibits an attribution of definite values of those observables to the electron at all times. From what is said on pp. 3–4 regarding the alternative way of describing a physical system, it seems that quantum objects can only be described dispositionally if they are to be described at all by the observables, and only probabilistically so at best. Instead of $P(O, o) = 1$ or 0 whether the value of O when measured falls in or out of o, we have here, $P_\Psi(\hat{O}, o) = |c_i|^2 = |\langle \phi_i|\Psi\rangle|^2$, for $\lambda_i \in o$, which says that the probability of getting a value λ_i by measuring $\hat{O}$ in Ψ is $|c_i|^2$. Because quantum states are never dispersion-free, the answer to the experimental question as to whether or not $\hat{O}$ has a value in o is not either 'yes' or 'no', as in the classical case, but 'maybe' with a specific probability. It now seems that categorical realism is doomed because what works (i.e. the observables) in the classical theory becomes irreducibly dispositional in the quantum theory, whereas that which describes quantum states fully has no acceptable occurrent values (because the wave functions admit complex values).

To recover categorical realism, one may adopt either of the following approaches: search for a new set of variables (usually hidden ones) that do have occurrent values which uniquely determine a quantum object's states; or give up completely the corpuscularian view of quantum objects and make a substantival interpretation of what the wave functions represent. What the hidden-variable theorists have been trying to do belongs to the former approach, while the latter approach has always been a view of a small minority. Here I explain the pros and cons of interpreting the wave function, Ψ, substantivally before I discuss the effect that changes the data.

The difference between Ψ and $|\Psi|^2$ is the phase factor (cf. Equation (11)). $|\Psi|^2$ of course suffices if we are only concerned with the possible values of observables and their probabilistic distributions, but that is not all. In many situations, the role of the phase function in Ψ, $e^{-iS/\hbar}$, is crucial. In a two-slit experiment, for instance, when a single quantum object passes through the two openings, its wave function splits into two, Ψ_1 and Ψ_2, which determine, when the object is later intersected by a receptive screen, where the impact occurs with what probability. However, $|\Psi|^2 \neq |\Psi_1|^2 + |\Psi_2|^2 = |I_1|^2 + |I_2|^2$, as it should be if the object goes through either opening 1 or opening 2. Instead, Ψ_1 and Ψ_2 seem to interact with each other in the process which results in,

$$(15) \quad |\Psi|^2 = |I_1|^2 + |I_2|^2 + 2|I_1||I_2|e^{-i(S_2-S_1)/\hbar},$$

where the last term explains the interference pattern (the fringes) on the screen. While this phenomenon seems to suggest a causal interpretation of the interference, with the Ψs as functions representing categorical properties involved in the interaction, it does not necessitate it. The dispositional view of quantum states together with a corpuscularian ontology for quantum objects can still be defended and the theory that supports it is known as quantum logic (Hughes 1989, ch. 7). Instead of taking interference literally (as if it is a causal interaction of fields), quantum logic provides a non-Boolean algebra to account for the non-classical nature of quantum probabilities (i.e. the answers to the experimental questions). More on this later in sections 3.2 and 4.

Further, Ψ-functions also enjoy a kind of gauge freedom. The eigenvalues and their probabilistic distributions are what is observable for a system, but for each set of such values and distributions, the system's state is determined by Ψ up only to an arbitrary phase function (i.e. $e^{-iS/\hbar}$), where S is an arbitrary scalar function (cf. Equation (11)). In this sense, Ψ is similar to the electromagnetic potentials ($\mathbf{A}$, φ), and hence is often referred to as the *quantum potential.* One must note that I use the term 'quantum potential' to mean what is mathematically represented by the wave function, which is different from Bohm's, and perhaps the standard use of that term. The standard usage only includes the real part of what my term refers to. We shall see in the context of the AB effect that it is better to use my definition.

To highlight the differences between Ψ and $|\Psi|^2$ (cf. Liu 1994), let me note that Ψ is a function that represents the states of a quantum object (or system) and, as such, it evolves deterministically through time, obeying Schrödinger's law (Equation (10)). Structurally, it consists of an amplitude and a phase function: the former determines the probabilities and the latter the interference patterns (Equation (15)). However, it does not represent anything experimentally obtainable, while $|\Psi|^2$ does because when a large number of quantum objects prepared in state Ψ are measured for any observable, the probability distribution of the eigenvalues can be observed. Moreover, Ψ is underdetermined because of the gauge freedom it enjoys. We now turn to an experiment that threatens to change this view and more.

3.2. *The Aharonov-Bohm Effect*

The AB effect was first predicted by Aharonov and Bohm in 1959 as the result of a thought experiment (Aharonov and Bohm 1959) and later experimentally confirmed (cf. Peshkin and Tonomura 1989). After a period of intense debate on whether the effect is genuine (cf. Peshkin and Tonomura 1989 for a summary of the debate and the relevant literature), it is now clear that the AB effect is real and the original prediction of Aharonov and Bohm correct, subtle differences notwithstanding. While theoretical physicists use it to fathom deeper meanings of electromagnetism (cf. Feynman et al. 1964, ch. 15, Wu and Yang 1975) and experimental physicists explore its practical applications (cf. Imry and Webb 1989), philosophers are yet to find out its implications for the philosophy of quantum theory (one exception is Zangari 1992, also see Liu 1994). First, let us see how it changes our conception about potentials – the classical and the quantum.

There are two types of AB effects: the electric and the magnetic. The former is theoretically simpler but more difficult to be experimentally tested, while the latter is the opposite. The effects show that a difference in the potentials without the presence of the corresponding force-fields can cause a fringe shift of the interference patterns in the two-slit experiments with quantum objects (e.g. electrons). Low-temperature superconductivity and homologous imaging techniques recently made it possible for researchers to completely separate (as completely as is physically possible) the **B**-field from the electron beams (cf. Peshkin and Tonomura 1989). As a result, the magnetic AB effect is experimentally verified. The leak of **E**-field is more difficult to eliminate in an actual experiment, therefore the electric AB effect is yet to be verified.

For the electric AB effect, we imagine that some quantum objects (e.g. electrons) are fired, one at a time, towards two parallel tubes, and then, after emerging from the tubes, they hit a measuring screen which is either photosensitive or composed of an array of Geiger counters. Suppose the electrons are all prepared to be in state Ψ when leaving the source and are in a superposition of Ψ_1 and Ψ_2 after emerging from the two tubes. The two tubes are charged such that in one tube the potential is φ_1 and in the other tube φ_2, while the **E**-field is completely shielded from the interior of the tubes. By repeating this otherwise identical experiment many times under two different settings, i.e. charged and uncharged tubes, we observe a shift of the interference fringes. To explain such a shift, we note that from Equation (10), and with the two Hamiltonians: $\hat{H}_1 = \hat{H}_0 - e\varphi_1$ and $\hat{H}_2 = \hat{H}_0 - e\varphi_2$, one can solve the two equations by first assuming that Ψ_1 and Ψ_2 have the form, $\Psi_i = \psi_i^0 e^{-iS/\hbar}$, $i = 1, 2$, where ψ_i^0 is the function when the tubes are uncharged. The final result is as follows:

$$(16) \quad |\Psi|^2 = |\psi_1^0|^2 + |\psi_2^0|^2 + 2|\psi_1^0||\psi_2^0| \cos(\Delta S),$$

where

$$\Delta S = \frac{e}{\hbar} \int_0^t [\varphi_2(t) - \varphi_1(t)]dt$$

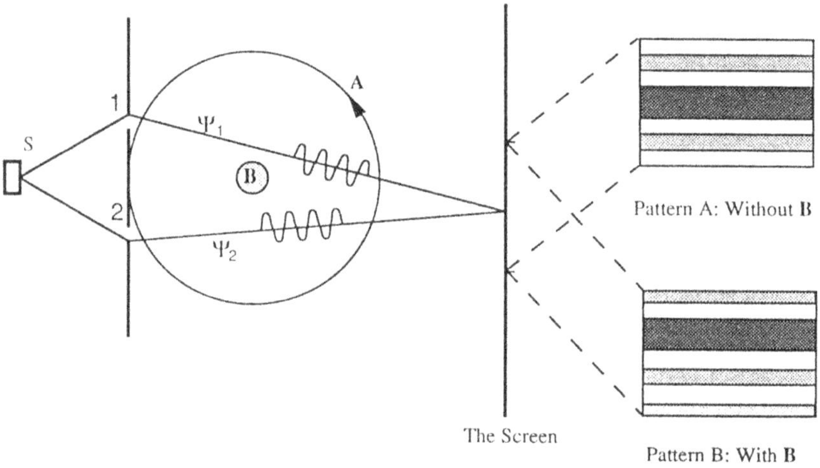

Figure 2. The magnetic AB experiment: Ψ_1 and Ψ_2 are the wavelets from the two openings. A small portion of the screen is shown with interference fringes. There is a distinct fringe shift in Pattern B when **B** is present.

is the phase difference. The last term in Equation (16) theoretically explains the fringe shift.

Figure 2 gives some details of the modified two-slit experiment which produces the magnetic AB effect. Comparing it with the **B**-field in Figure 1, we know (from Equations (7) and (8)) that $A(= \beta(r/2))$ has non-zero value outside the solenoid even though the **B**-field is completely confined within. Since the 'path'[3] of $\Psi_1 + \Psi_2$ encloses the solenoid, the magnetic potential is non-zero everywhere on this path whenever the **B**-field is present inside the solenoid. Repeating this experiment by changing, *ceteris paribus*, from having to not having **B**, one finds a shift of the interference fringes, as shown in Figure 2, patterns A and B. The theoretical explanation of this is similar to that of the electric AB effect. The electrons begin with a state, Ψ, and upon emerging from the two openings, they evolve into a superposition of Ψ_1 and Ψ_2. By some simple calculation, we derive the same equation as Equation (16) except that the phase difference is:

$$(17) \quad \Delta S = -\frac{e}{\hbar} \oint_C \mathbf{A} \cdot d\mathbf{l} = -\frac{e}{\hbar} \beta R.$$

Since the **E**- or **B**-field in either case and the wave packet of any electron are mutually excluded from each other in the experiment, the interaction between them, if there is any, has to be action at a distance. Barring such an action (a natural thing to do), the fringe shifts recorded in the experiments are best construed as the result of interactions between the electrons and the potentials. This has significant ramifications for electrodynamics and the quantum theory.

The AB effect first shows that electromagnetic potentials represent genuine physical properties of the fields. This is indeed the conclusion many physicists have begun to embrace. For instance, in a paper in 1975, Wu and Yang argued that because of the AB effect, "*[e]lectromagnetism is thus the gauge-invariant manifestation of a nonintegrable phase factor*" (Wu and Yang 1975, 3846). (The difference of such a phase factor is given in eq. 17.) Then, it shows that the gauge freedom embodied in Equation (9) is *broken* in that some $(\mathbf{A}', \varphi')$ may induce a different effect from what $(\mathbf{A}, \varphi)$ does, even though they yield the same values for **B** and **E** (Equation (6)). This is especially interesting if we note that a gauge freedom is viewed in physics as fundamentally different from a case of degeneracy (i.e. conditions under which distinct values of some property are not differentiable); the former being an equivalence of different representations and the latter a failure (usually under special conditions) of differentiating physically distinct values (for a detailed discussion of these two principles, see Liu 1995). The case of the AB effect seems to have shown that it might be impossible to know or show whether a certain principle is a principle of gauge freedom or one of degeneracy. Although we thought that the invariance of the **A**-potentials is a gauge invariance, it turned out to be a case of degeneracy.

A further interesting question about the AB effect is whether it is compatible with the notion of causality or locality. On this question, physicists are just as interested as philosophers. According to Peshkin, 'the AB effect does not violate Einstein causality; it sends no signal faster than light' (Peshkin and Tonomura 1989, 18–20). He invites us to imagine the following scenario: in Figure 2, we switch on the current in the solenoid at time t (where a **B**-field is almost instantly established) and ask whether the electron which is just about to hit the measuring screen (presumably at some distance from the solenoid) will be affected by such a change. Peshkin's answer is no, because calculation shows that the propagation of the potential, **A**, upon a sudden change of **B**, has a speed less than that of light. So, the Einstein causality is preserved for the propagation of the **A**-potentials. However, the theoretical result is not experimentally tested and the argument open to objection.

Let us suppose that Peshkin's argument is right, i.e. no instantaneous transmission of a modification of the **A**-field (or the **A**-potential field) through space. Still, it does not follow that the electron which is just about to hit the measuring screen is not affected at the moment when the current is switched on. This is because the electron in question is still around the solenoid when the current is switched on since its wave packet, no matter how localized, spreads out in space and so overlaps with the solenoid's neighborhood. Even though the modification of the **A**-potential takes time to reach the most probable location of the electron, the electron's probabilistic behavior should be modified instantaneously because the moment the **A**-field is modified at the solenoid's neighborhood, the wave packet as a whole is affected, and so is the probabilistic behavior of the electron. [Theoretically, this is clearly shown by the coupling term's being non-zero, i.e. $A_i\Psi \neq 0$, (where A_i: $i = 1, 2, 3$, are the coordinate components of **A**).]

3.3. *Bell's Theorem and Non-separability*

Two objections may be raised against what is said in the last paragraph. (i) Although an overlap between Ψ and **A** (the coupling $\neq 0$) does indicate a possible *instantaneous* modification of Ψ, it does not show that there actually is one because nothing in $A_i\Psi$ says whether or not the modification of the electron's behavior is instantaneous. To reach that conclusion, I will have to assume that the electron is identical to its wave packet, which is just what I am arguing for in this paper. (ii) It seems that I have to conclude that locality (or causality) *is* and *is not* preserved in the AB effect, a clear contradiction.

To rebut these objections, let me make a small, but I hope illuminating, detour into another related and well-known subject, namely, the discussion of the Bell theorem (cf. Bell 1967, Cushing 1989). The surface resemblance of the Bell theorem to the case at hand is that quantum objects seem to be *non-separable,* or entangled, systems in which the values of some parameters at one location seem to depend (instantaneously) on their values at other remote locations. In the case of EPR experiments with, say, pairs of electrons, whose quantum mechanical description violates the Bell inequalities, the value of a spin component at one wing is determined the moment the value of the same component at the other obtains. In terms of the electrons' probabilistic behavior, the probability of the electron at one wing assuming certain spin value changes from 0.5 to 1 the moment the electron at the other wing assumes a definite value of the same spin component; and most importantly, this is true regardless of which spin component is chosen.

This phenomenon for a system of two or more quantum objects whose state is in a superposition of those of its parts was initially taken to necessitate a violation of the so-called locality condition, a condition that was thought to follow directly from Einstein's causality condition, namely, no superluminal signaling (cf. Bell 1967). It was then realized that this strong locality condition is in fact the sum of two different conditions: the locality and the separability condition (cf. Jarrett 1984, 1989; Shimony 1984, 1989; Howard 1989).[4] The former is implied by the no-superluminal-signaling principle, but the latter spells out the entangled nature of superimposed wave packets. Roughly speaking (i.e. without using the precise technical terms in the Bell theorem discussion), Jarrett's (and Howard's) locality condition says that the outcome of measuring a spin component at one wing of the EPR experiment on electron-pairs is independent of the device orientation and the measuring at the other wing. The separability condition holds if and only if the states of two spatially separated quantum systems are determined independently of each other. Many have since argued that the negative results from the experiments should be interpreted as showing that quantum systems (in pure states) are *non-separable* systems which, nonetheless, do not violate the no-action-at-a-distance principle.

It is not difficult to see that both potentials, $(\mathbf{A}, \varphi)$ and Ψ, are creatures of the same kind. The values of either at a certain spacetime region are *inseparably* related to, or dependent on, certain conditions outside of that region. For the **A**-potential, we have seen earlier (cf. 3.1.) that whether it is zero at a certain point depends on whether the integration loop which goes through that point encloses any **B**-field. And in order to be affected by the **A**-potential, Ψ_1 and Ψ_2 have to form (according to Equation (8)) a closed loop which encloses the non-vanishing magnetic flux. The situation is illustrated in Figure 3 (a) and (b). In Figure 3 (a), the loop encloses the source and the fringe-shift occurs, whereas in Figure 3 (b), the loop does not enclose the source and no such shift is predicted to occur. This means that whether the potential affects quantum objects depends not only on its values at the neighborhoods of those objects, but also on their relations to the source and on how the objects go through different parts of the **A** region. In other words, the effect is *non–separable* because what transpires between the **A**-potential and the Ψs at a certain point x_0 depends on values of **A** and Ψ at points other than x_0 and its neighborhood. Therefore, locality is preserved for the **A**-field, as Peshkin has presumably proved, but separability is not, in the case of the AB effect.

Can we say that what is affected by the **A**-potentials is corpuscular? We know that electrons as macroscopic corpuscles (e.g. when seen in a cloud chamber) are not affected by the potentials $(\mathbf{A}, \varphi)$. The electrons' trajectories must be in contact with the non-vanishing **B**- or **E**-field for them to be bent in a cloud chamber or some such devices (Equation (5)). Nor are the probabilities of the electrons affected by the potentials.[5] This can be seen from the following consideration. Suppose that in Figure 2,

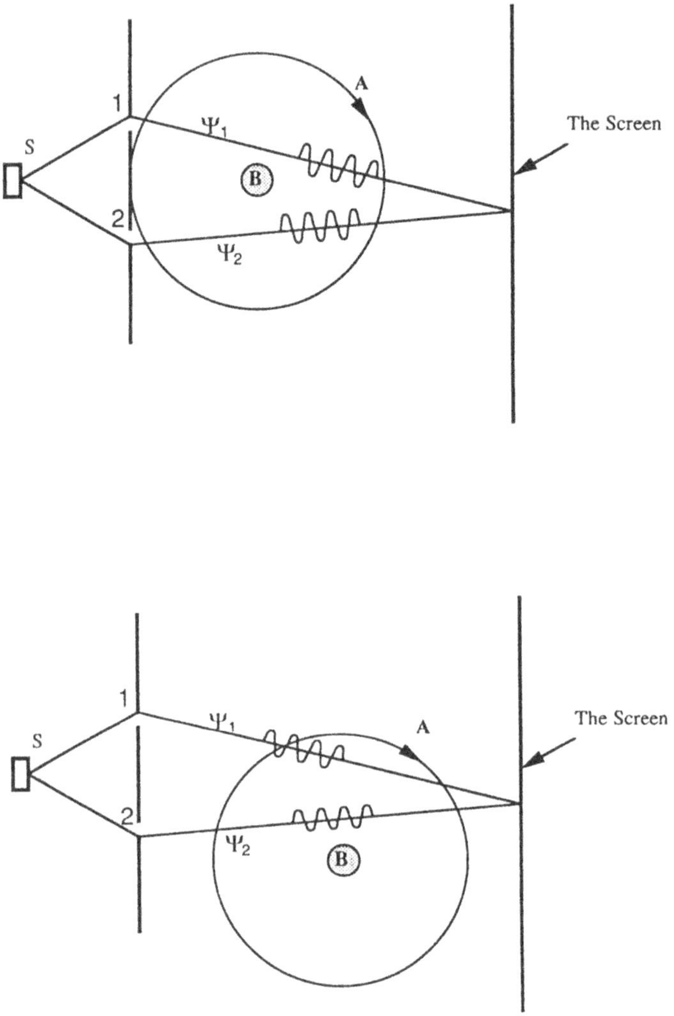

Figure 3. (a) The magnetic AB experiment where fringe-shift occurs. (b) The magnetic AB experiment where fringe-shift does not occur.

we put two detectors at the two openings so that the electrons emerging from them are described by two *incoherent* wave functions, Ψ_1 and Ψ_2. In other words, the detectors break the initial pure state into a mixture of two independent quantum states. According to quantum theory, for a mixed state of Ψ_1 and Ψ_2, we have, $|\Psi|^2 = |\Psi_1|^2 + |\Psi_2|^2$. From Equation (16), it follows that no fringe-shift will be caused by the presence of $(\mathbf{A}, \varphi)$. Thus, the AB effect is best seen as the result of an interaction between the two potentials. (NB: *the quantum potential and the wave packet are the same thing.*)[6]

I take the verification of the magnetic AB effect to be the *coup de grâce* for any corpuscularian conception of quantum objects, especially the quantum logic approach. Orthomodular lattices may be able to accommodate effects of interference of coherent quantum states by giving quantum probabilities a non-classical calculus, but it is incapable of calculating the exact amount of the fringe-shifts in the AB effect. The reason is that in quantum logic, no place is given to Ψ, for it only considers $|\Psi|^2$ and its Fourier expansions as elements of the theory. But for the AB effect, one cannot get away from not using Ψ, as I have explained in detail. The exactness of using Ψ to explain the fringe-shift and $|\Psi|^2$'s apparent lack of such capacity strongly suggests that the effect can only be explained as a result of the *interaction* between the electromagnetic potentials and the wave packets. We conclude, therefore, that (i) potentials (electromagnetic or quantum) are genuine physical properties in quantum reality, and (ii) quantum objects are ontologically wave-like entities of which the wave functions represent real properties.

4. POTENTIALS AND PROPENSIONAL OBJECTS

Let us now return to the issue of categorical realism in quantum reality. We have seen earlier that categorical realism seems secure in the classical world, but there are others who argue for dispositionalism, i.e., that all physical properties are dispositional (cf. Popper 1957, p. 70; Mellor 1974, 1982), and so regard categorical realism pointless. Such dispositionalism is indeed possible from a formal point of view. As seen earlier, it is always possible to characterize any state, classical or quantum, dispositionally by the system's responses to the experimental questions (as given by $P(O, o)$ and $P_\Psi(\hat{O}, o)$, cf. 2.1. and 3.1.). Whether dispositionalism is true, a dispute that does not concern us here, the following is not disputable. In classical physics, it makes little difference whether an object's basic properties are characterized dispositionally or categorically. For any property, O, and any subset of its possible values, o, $P(O, o) = 1$ if and only if the object in fact possesses a value of O in o, and $P(O, o) = 0$ if and only if it doesn't. This is no longer true for quantum objects: it is no longer tenable to assume that if a measurement results in an object's possessing value λ for an arbitrarily chosen observable (property) O, the object must have λ immediately before the measurement.

Because of this, it is tempting to believe that categorical realism fails in quantum reality, and therefore, quantum objects are exclusively dispositional, or more accurately propensional, objects (i.e. they are probabilistically dispositional). It is tempting to believe that

> [w]hat makes something an electron, . . . , is its causal power, capacities and propensities. . . . what an electron is disposed to do, . . . , is what makes it the kind of thing it is. (Ellis and Lierse 1994, 32–3)

If what I have argued so far in this essay is true, the opposite of the aforesaid should hold.

> (1) Wave packets (i.e. quantum objects such as electrons) are real objects; like all such objects, they have categorical properties such as actual sizes and shapes.

Of course, a wave packet is very different from some cloud or gaseous object in ordinary spacetime. Its size is determined by the range of possible values of the observables of which Ψ is a function, and its shape by the probability distribution of those values. These are the *possible* values that obtain when the packet is subject to measurement, and they determine its *actual* sizes and shapes. Further, the behavior of each packet — the change of its size and/or shape through time — obeys the Schrödinger equation.

Now, if the Ψ-function represents a property of a *particle*, the property must be relational, while if it represents that of a *wave packet,* the property is non-relational . A similar case in electromagnetism may illustrate this point. In the aether theory of electromagnetism, **E** or **B** describe a relational property (correlated vibratory motions) of the relevant aether particles. But in the field theory, it is a non-relational property of the fields. For the wave function of the two-electron system in the EPR experiment, treating the electrons as particles would make the property the function represents a relational one, while treating them as a superposition of wave packets would make it non-relational (see 2.2. for a similar point on classical fields).

> (2) Wave packets have propensional properties which supervene on their categorical ones.

So, a quantum object is a potential object (hence, quantum potential) in the same sense that a porcelain vase is one. Similar to a vase being fragile, an electron is 'locatable' in that it has the propensity of being found when measured at one of the possible locations within the extension of the packet. And just as the fragility of the vase supervenes on its categorical (i.e. structural) properties, the locatability of an electron, or some other propensity of it, also supervenes on the size and shape of its wave packet.

Less one thinks that the comparison is too far-fetched, let us compare the tossing of a fair coin with the two-slit experiment. The probability of a fair coin landing on heads when properly tossed is 0.5 and so is its landing on tails, where landing on heads and tails are the only two possible outcomes of the toss. Let us assume, for the time being, that the probabilistic behavior of a fair coin when properly tossed is fundamental, namely, that no further specification of the physical conditions before and during the toss can improve the odds of heads or tails.

We know that the measured distribution of a two-slit experiment on electrons is not $|\Psi|^2 = |\Psi_1|^2 + |\Psi_2|^2$ (cf. 3.1.). This indicates that the electrons do not have definite trajectories between their source and the screen, for otherwise, each electron would go through either the upper or the lower opening before reaching the recording screen, and so the above equation would be correct. That is why an electron is not characterizable by such categorical properties as its position and momentum. However, it does have categorical properties, such as the size and shape of its wave packet. One may say that such properties are not even relevant, for they are not among those whose values appear as possible outcomes of the experiment.

But what about the fair coin? It is true that we can calculate as well as see a fair coin's trajectory after it is tossed and before it hits the surface. However, we can similarly say that the tossed coin's having a definite trajectory is not relevant. The possible outcomes – heads up or tails up – are not possessed by the coin anywhere in the toss. To be more precise, the outcomes of the toss are not the positions and orientations of the coin (which make up its trajectory), but rather whether the heads or the tails of the coin face up when it settles on a hard surface. But nowhere in the toss is the coin lying on a surface, so, like the electron, it does not possess the values of that which is to be observed at the end of the experiment. Furthermore, it will not help the case if one lets a similar surface intersect the coin midway in the toss and finds out whether or not it lands on its head, because (i) if the experiment stops there, we still have the same problem; and (ii) if one lets the experiment continue by releasing the coin again and letting it fall on the final surface, the probability distribution will not be the same. Similarly, we may intersect an electron midway in the two-slit experiment with a Geiger counter and find out its definite position at that moment, but we either end the experiment there and face the same problem or let the electron go through to reach the recording screen on which only to see an altered probability distribution.

Therefore, if the odds of a fair coin's landing on heads or tails are fundamental and a coin is characterized by such properties alone, a fair coin is just as mysterious (or non-mysterious) as an electron. That an electron appears more so is because we are misled by looking at it in the wrong way: thinking that it *ought to* possess values of those properties whose values are only obtainable by measurement observations. If we apply the same assumption to a fair coin, we would, as we have seen above, get an equally puzzling object. The fact that a coin is characterized in some other way, for instance, by its actual trajectory in the air, is beside the point. We can choose to do the same for an electron, namely, to characterize it with the actual size and shape of its wave packet.

Therefore, I reach the following conclusion.

(3) Categorical realism is true in the quantum world. At least, it holds as well as in the classical world.

Categorical realism is true if and only if all dispositional/propensional properties of physical objects supervene on categorical properties and laws of nature.[7] The shift that takes place in the transition from the classical to the quantum with respect to categorical realism is one from *primarily deterministic* laws to *primarily probabilistic* ones. While in most cases, what a macroscopic object *is* is what it would be were it measured for that, what a microscopic object *is* (a wave packet) is not deterministically related to what would be were it measured for that (a corpuscle).

5. CONCLUDING REMARKS

If one asks what characterizes a coin primarily: its trajectory or its landing on heads or tails, the answer is that it is the former, but when one is confronted with a choice between characterizing an electron primarily with a wave packet and with possible locations or momenta, one seems to be forced to settle for the latter. That is why those who advocated the Copenhagen interpretation of quantum theories (Bohr and Heisenberg, among others) are right in saying that quantum theories are primarily about observables. Since quantum objects can only possess values of observables potentially between measurements, quantum objects are primarily potential or propensional objects. But the primacy of observables and classical theories is a result of our epistemic limitations; it is not anything intrinsic to microscopic objects. The fact that in the AB experiments we can manipulate the interactions between the two kinds of potentials should convince us of the substantivalness of both, even though the interactions can only be indirectly observed through their statistical effects.

Popper (1957, 1960, 1967) and Cartwright (1987), among others, proposed and defended a propensity interpretation of probabilities (classical or quantum). We now see that, contrary to Popper and Cartwright, the function that fully represents quantum propensity is not the probability (i.e. the $|\Psi|^2$-functions) but the wave function (i.e. the Ψ-function). Mellor (1971, pp. 76–82) is right in pointing out that a probability distribution is the display of some propensional property rather than the property itself. In the classical case, the probability of a coin's landing on heads is not a propensity of the coin; it is rather the manifestation of such a propensity. The same is true with quantum probabilities. For instance, that which interacts with the **A**-potential in the AB experiment is the quantum potential, not the interference pattern and its shift. The latter is a *display* of the former. Salmon (1979, 212–5, and 1984, 202–5) argues that propensities are causal structures carried by causal processes, and he identifies them in the quantum case with the amplitudes of the wave packets. We now see that amplitudes are still not enough; we must take the entire wave packet for the propensional object.

Further, as Mellor rightly points out again, objects possess genuine propensional properties only if those properties may affect and/or be affected by other physical objects independent of their display. Hence, the Ψ-function qualifies more than the quantum probability as a measure of propensity in that it not only determines (probabilistically) what happens when a measurement occurs but also determines (deterministically) how quantum potentials interact with one another (interference) and with other entities, such as the electromagnetic potentials (the AB effect). The mystery of quantum phenomena lies not so much in the probabilistic relation among observable results as in the *deterministic evolution and interaction* among wave packets and with electromagnetic (and presumably other) processes. Since a propensity, as any other dispositional properties, may never be displayed, an object must have it independently of any conditions that display it. This is exactly the kind of property that Ψ-functions represent and by taking it as physically real, the mystery disappears.

One further benefit can be seen as follows. Knowing states Ψ_1 and Ψ_2 as possible states of a quantum object, we know that the propensity of its being in Ψ_1 or Ψ_2 is $\Psi = \Psi_1 + \Psi_2$, which in turn determines its probability as given by Equation (15). With such a propensity interpretation of the quantum theory (NB: it is not a propensity interpretation of probability, classical or quantum), we understand why quantum probabilities do not behave like classical ones. Quantum logic only provides a description, but this interpretation gives an explanation, of this and perhaps other non-classical features of quantum probabilities.

NOTES

* I would like to thank Mark Zangari, Robert Rynasiewicz, Niall Shanks, Kirk Ludwig, and Greg Ray for helpful discussions for this paper. I am also grateful to the comments of the anonymous referees.

1 I shall tentatively define the concept of physical equivalence as follows. For any two magnitudes, A and B, in a physical theory, T, and a system, S, whose behavior is completely accounted for by T, A is *physically equivalent* to B if a transition from S's possessing A to its possessing B produces no recognizable change in T regarding S.

2 I shall use the non-relativistic quantum theory in this paper. Because the philosophical issues discussed in this paper and in most of the literature concerning quantum physics deal with non-relativistic phenomena, it will be inappropriate to introduce relativistic constraints, which are only necessary when the speed of the system in question is comparable to that of light.

3 By the path of an electron or Ψ_1, I only mean those regions through which it is most likely to pass. So, it should not be thought of as a trajectory.

4 Different terms are used for essentially the same set of conditions: Jarrett used 'locality' and 'completeness', Shimony 'outcome independence' and 'parameter independence', and the terms I used here are borrowed from Don Howard who used 'locality' and 'separability'. It is possible to quibble about the subtle differences among these pairs, but Bell's original locality condition is divided more or less at the same place with these pairs of conditions.

5 I consider this point because (i) according to the quantum logic approach, if the electrons are affected by the potentials, so must be the probabilities, for the states of the electrons (or other quantum objects) are determined by the probability distributions of the values of their observables (i.e. by the answers to the experimental questions); and (ii) if Nancy Cartwright (1987) is right about the reality of quantum probabilities, the potentials should affect the probabilities if any property of the electrons is affected (cf. Liu 1994). Both views are rendered inadequate by the AB effect, or so I have been arguing.

6 One may point out that Bohm's conception of quantum objects as waves plus particles (i.e. both the corpuscular electron and its guiding wave are real objects) is not threatened in this case. There is some truth in this. However, I am not sure how much of Bohm's conception remains, in light of the Bell theorem, that differs significantly from mine. From the Bell results we know that *local* hidden-variable theory is ruled out, namely, whatever the theory is, it cannot be one in which the state of an object is determined solely by the values of the hidden variable(s) in its neighborhood. This follows directly from the violation of separability condition mentioned earlier. Now, I am not sure if it makes much sense to insist that there is a particle inside every wave packet when its states are not always determinable by local values of its variables. In what sense is it still a corpuscle? The construction of non-local or contextual hidden-variable theories is in progress (cf. Shimony 1984), but such theories share more or less the same ontology as mine.

7 Depending on how laws of nature are understood, there may be two different positions for categorical realism: (i) if laws exist in reality over and above facts, dispositions/propensities supervene on categorical properties and laws of nature; and (ii) if laws are no more than some special kind of relations among facts, dispositions/propensities supervene on categorical properties alone (laws merely being the ways in which the supervenient is related to the supervening basis).

REFERENCES

Aharonov, Y. and D. Bohm: 1959, 'Significance of Electromagnetic Potentials in the Quantum Theory', *Physical Review* **115**, 485–91.

Armstrong, D. M.: 1968, *A Material Theory of Mind*, Routledge, London.

Armstrong, D. M.: 1969, 'Dispositions are Causes', *Analysis* **30**, 23–6.

Bell, J. S.: 1964, 'On the Einstein-Podolsky-Rosen Paradox', *Physics* **1**, 195–200.

Bohm, D.: 1952, 'A Suggested Interpretation of Quantum Theory in Terms of "Hidden Variables" ', Parts I & II, *Physical Review* **85**, 166–93.

Cartwright, N.: 1987, 'Max Born and the Reality of Quantum Probabilities', in L. Krüger, L. J. Daston and M. Heidelberger (eds.), *The Probabilistic Revolution* **2**, 409–16, The MIT Press, Cambridge, MA.

Cushing, J. T.: 1989, 'A Background Essay', in J. T. Cushing and E. M. McMullin (eds.), *Philosophical Consequences of Quantum Theory: Reflections on Bell's Theorem*, Univ. of Notre Dame Press, Notre Dame, 1–24.

de Broglie, L.: 1926, *La Physique Quantique Restera-t-elle Indeterministique*, Gauther-Villars, Paris 1953, 25–27.

Ellis, B. and C. Lierse: 1994, 'Dispositional Essentialism', *Australasian Journal of Philosophy* **72**, 27–45.

Feynman, R., R. Leighton, and M. Sands: 1964, *Feynman's Lectures on Physics*, 3 vols., Addison-Wesley, Reading, MA.

Howard, D.: 1989, 'Holism, Separability, and the Metaphysical Implications of the Bell Experiments', in J. T. Cushing and E. M. McMullin (eds.), *Philosophical Consequences of Quantum Theory: Reflections on Bell's Theorem*, Univ. of Notre Dame Press, Notre Dame, pp. 224–53.

Hughes, R. I. G.: 1989, *The Structure and Interpretation of Quantum Mechanics*, Harvard University Press, Cambridge, MA.

Imry, Y. and R. A. Webb: 1989, 'Quantum Interference and the Aharonov-Bohm Effect', *Scientific American* (April), pp. 56–62.

Jarrett, J.: 1984, 'On the Physical Significance of the Locality Conditions in the Bell Arguments', *Noûs* **18**, 569–89.

Jarrett, J.: 1989, 'Bell's Theorem: A Guide to the Implications', in J. T. Cushing and E. M. McMullin (eds.), *Philosophical Consequences of Quantum Theory: Reflections on Bell's Theorem*, Univ. of Notre Dame Press, Notre Dame, pp. 60–79.

Jauch, J. M.: 1968, *Foundations of Quantum Mechanics*, Addison-Wesley, London.

Landau, L. D. and E. M. Lifshitz: 1976, *Mechanics*, Pergamon Press, Oxford.

Landau, L. D. and E. M. Lifshitz: 1970, *The Classical Theory of Fields*, Pergamon Press, Oxford.

Liu, C.: 1994, 'The Aharonov-Bohm Effect and the Reality of Wave Packets', *British Journal for the Philosophy of Science* **45**, 977–1000.

Liu, C.: 1995, 'Gauge Invariance, Cauchy Problem, Indeterminism, and Symmetry Breaking', manuscript under review.

Mellor, H.: 1971, *The Matter of Chance*, Cambridge University Press, Cambridge.

Mellor, H.: 1974, 'In Defense of Dispositions', *The Philosophical Review* **83**, 157–81.

Mellor, H.: 1982, 'Counting Corners Correctly', *Analysis*, **42**, 96–7.

Pargetter, R. J. and E. W. Prior: 1982, 'The Dispositional and the Categorical', *Pacific Philosophical Review* **63**, 366–70.

Peshkin, M. and A. Tonomura (eds.): 1989, *The Aharonov-Bohm Effect*, Springer-Verlag, New York.

Popper, K.: 1957, 'The Propensity Interpretation of the Calculus of Probability, and the Quantum Theory', in S. Körner (ed.), *Observation and Interpretation*, Butterworths Scientific Publications, London, pp. 65–70.
Popper, K.: 1960, 'The Propensity Interpretation of Probability', *British Journal for the Philosophy of Science* **10**, 25–42.
Popper, K.: 1967, 'Quantum Mechanics Without "The Observer"', in M. Bunge (ed.), *Quantum Theory and Reality*, Springer-Verlag, New York, pp. 7–44.
Prior, E. W.: 1982, 'The Dispositional/Categorical Distinction', *Analysis* **42**, 93–6.
Prior, E. W., R. J. Pargetter, and F. C. Jackson: 1982, 'Three Theses about Dispositions', *American Philosophical Quarterly* **19**, 251–7.
Prior, E. W.: 1985, *Dispositions*, Aberdeen University Press, Aberdeen.
Salmon, W.: 1979, 'Propensities: A Discussion-Review', *Erkenntnis* **14**, 183–216.
Salmon, W.: 1984, *Scientific Explanation and the Causal Structure of the World*, Princeton University Press, Princeton, NJ.
Schrödinger, E.: 1927, 'Energieaustausch nach der Wellenmechanik', *Annalen der Physik* **83**, 956–68.
Shimony, A.: 1984, 'Contextual Hidden Variable Theories and Bell's Inequalities', *British Journal for the Philosophy of Science* **35**, 25–45.
Shimony, A.: 1989, 'Search for a Worldview Which Can Accommodate Our Knowledge of Microphysics', in J. T. Cushing and E. M. McMullin (eds.), *Philosophical Consequences of Quantum Theory: Reflections on Bell's Theorem*, Univ. of Notre Dame Press, Notre Dame, pp. 25–37.
Wu, T. T. and C. N. Yang: 1975, 'Concept of Nonintegrable Phase Factors and Global Formulation of Gauge Fields', *Physical Review D* **12**, 3845–57.
Zangari, M.: 1992, 'Adding Potential to a Physical Theory of Causation', *PSA 1992* vol. 1 261–273.

Manuscript submitted March 24, 1995
Final version received February 8, 1996

Brit. J. Phil. Sci. **45** (1994), 977–1000

The Aharonov–Bohm Effect and the Reality of Wave Packets

Chuang Liu

ABSTRACT

The objective of this paper is to show that, instead of quantum probabilities, wave packets are physically real. First, Cartwright's recent argument for the reality of quantum probabilities is criticized. Then, the notion of 'physically real' is precisely defined and the difference between wave functions and quantum probabilities clarified. Being thus prepared, some strong reasons are discussed for considering the wave packet to be physically real. Finding the reasons inconclusive, I explain how the Aharonov–Bohm effect delivers the final punch. I conclude that wave packets are the quantum objects that underlie the indeterministic quantum processes and have the propensity of displaying probabilistic (or indeterministic) behavior.

1 Introduction

In this essay I tackle the metaphysical question of what microscopic (or quantum) objects really are. It is well known that quantum objects have corpuscular and wavelike properties. But if you ask what the wavelike properties are, you get the answer that they are probabilities of obtaining possible values of the corpuscular properties when the objects are subject to measurement. Hence, many people, especially working physicists, find it natural to think that the denizens of the quantum world are particles which possess their corpuscular properties probabilistically or indeterministically. However, some philosophers of science (e.g. Popper [1957] and Cartwright [1987], among others) have argued that quantum probabilities are also *physically real.* Popper (also [1960] and [1967]) likened probability, when interpreted as propensity, to a force field, and

Cartwright wanted to expand the ontology to include particles as well as probabilities. In more precise terms, this controversy is about the interpretation of two theoretical entities in quantum theory: the wave (or state) function, $|\Psi\rangle$, which determines the states of any quantum object, and $|\Psi|^2$ which determines the quantum probabilities. Popper, Cartwright, and others view $|\Psi|^2$ as representing something physically real, something independent of the quantum particles, while the textbook tradition takes it to be a secondary property of the particles, something like the probability of a fair coin's turning up heads when properly tossed. Whether or not quantum probabilities are taken as physically real, it is generally agreed that the Ψ-function cannot be considered as such; it does not refer to any real property or substance. I argue in this paper that the opposite is true: the Ψ-function is physically real, namely, what it represents—the wave packet—is capable of playing a role in causal processes just as any other objects or substances are.

The aim of this paper is threefold. First, I shall clarify the issue about whether or not quantum probabilities are physically real. Thus, I begin with a criticism of Cartwright's recent argument (Cartwright [1987]) for the reality of quantum probabilities ($|\Psi|^2$). I will then examine the notion of 'physically real' and clarify the difference between the Ψ-function and the $|\Psi|^2$-function; the former notion is widely used but seldom precisely defined and the latter difference usually glossed over in the literature. Second, I discuss some strong reasons for considering the wave packet (instead of the quantum probability) as physically real. Finding the reasons inconclusive, I explain how the Aharonov–Bohm effect (AB effect, henceforth) delivers the final punch (Aharonov and Bohm [1959]; Wu and Yang [1975]; Peshkin and Tonomura [1989]; Imry and Webb [1989]; and Zangari [1992]). It is one thing, however, to show that wave packets are physically real, and it is another to find out what they are. So third and last, I explore the notion of propensity and see if a model can be constructed for wave packets. I see a model emerging in the criticisms of Popper's original propensity interpretation of probability (Kyburg [1974]; Mellor [1971]; Salmon [1979, 1984]; Maxwell [1982, 1988]; and Cartwright [1987, 1989]). I suggest that quantum objects are wave packets which have the propensity of exhibiting probabilistically distributed corpuscular properties. However, I am not proposing any new realist interpretation of quantum theory in this essay; rather I try to set some limits for the possible candidates for such an interpretation.

2 Cartwright on the reality of quantum probabilities

Suppose that we have a quantum system S which is in a state $|\Psi\rangle$.

According to quantum theory, it can be Fourier expanded by the eigenfunctions of any observable of which $|\Psi\rangle$ is a function. Suppose, again, that P and Q are two such observables and P's eigenfunctions are $\{|\phi_i\rangle\}$, where $i = 1, 2, \ldots, n$, while Q's eigenfunctions are $\{|\psi_j\rangle\}$, where $j = 1, 2, \ldots, m$. (For simplicity, I assume that P and Q have finite and discrete eigenvalues, p_i and q_j.) We can expand $|\Psi\rangle$ into a series of ϕ's or ψ's as follows.

$$|\Psi\rangle = \sum_{i=1}^{n} \langle\phi_i|\Psi\rangle|\phi_i\rangle, \tag{1}$$

or

$$|\Psi\rangle = \sum_{j=1}^{m} \langle\psi_j|\Psi\rangle|\psi_j\rangle. \tag{2}$$

From these, we have,

$$\begin{aligned}|\Psi|^2 &= \langle\Psi|\Psi\rangle = \Sigma_i\langle\Psi|\phi_i\rangle\langle\phi_i|\Psi\rangle = \Sigma_i|\langle\phi_i|\Psi\rangle|^2 \\ &= \Sigma_j\langle\Psi|\psi_j\rangle\langle\psi_j|\Psi\rangle = \Sigma_j|\langle\psi_j|\Psi\rangle|^2,\end{aligned} \tag{3}$$

where the Dirac notation is used for the algebraic operations of state functions.

Now, here is the core of Born's probabilistic interpretation: $|\langle\phi_i|\Psi\rangle|^2$ *and* $|\langle\psi_j|\Psi\rangle|^2$ *are the probabilities of obtaining the eigenvalues,* p_i and q_j, *when system S is measured for P or Q in state* $|\Psi\rangle$. In other words, $|\Psi|^2$ determines the probabilities. (For Born's original interpretation, see Born [1926].) This is the minimal (or partial) interpretation of quantum theory, for, on the one hand, any complete interpretation must be consistent with this and, on the other hand, other aspects of quantum theory are left uninterpreted, e.g. what is $|\Psi\rangle$?

Among the views concerning quantum objects which are consistent with Born's probabilistic interpretation is most famously the corpuscularian view. It holds that the denizens of the world are corpuscles, although they are not classical ones. Roughly put, quantum objects such as electrons and photons are small particles which possess classical properties indeterministically. The so-called wavelike properties of these particles are nothing more than the manifestation of this indeterminacy of their classical properties. To be more precise on this point, let us consider the following. When an electron or a photon is measured, it is always a single quantum of energy that is delivered instantaneously at a spatial point (or almost instantaneously delivered in a very small region). Even a quantity such as spin, which is non-classical, is none the less corpuscular. The difference between a classical particle and a quantum one is that, for the former, the values of their properties are always uniquely determinable by the appropriate laws of nature and initial/boundary conditions, while, for the latter, the properties are only determinable up to a set of possible values with given probabilities (or weights). This fact of indeterminacy and the probabilities, $|\Psi|^2$, are what makes the particles wavelike.

Such an ontological view is held by those who defend the Copenhagen interpretation of quantum physics (or some variation of it). Indeed, Heisenberg, one of the most influential Copenhagenists, even constructed a thought experiment with electrons and photons to show how the corpuscularity of light quantum made their classical properties not simultaneously determinable.[1] Max Born has always been considered to be among the pioneers of the Copenhagen interpretation. Was Born committed to the corpuscularian ontology?

The answer is 'No', according to Cartwright [1987]. Cartwright argues in her article that the 'pedestrian' story of Born's position is mistaken and that Born believes in the reality of quantum probabilities (for historical details of Born's original interpretation of the wave functions, see also Wessels [1980]). Let us now see how she argues for this conclusion. The pedestrian story of Born's position describes him as treating the wave functions as giving *epistemic* probabilities which measure the degree of our ignorance of the precise values of observables. In other words, $|\Psi\rangle$ determines $|\Psi|^2$ which measures the degree of our ignorance. This means that as the probability of a fair coin, neither the wave function nor its modulus squared are physically real quantities. Cartwright shows that a careful reading of Born's original papers reveals that the opposite is true: Born takes quantum probabilities as no less real than the particles. Here are her reasons:

1. Born, like all Copenhagenists, certainly believes in the reality of quantum particles, for singular measurements on quantum objects only produce values of corpuscular properties. But he also believes that the eigenstates of such particles (represented by eigenfunctions like the $|\phi_i\rangle$s and $|\psi_j\rangle$s in eqs. (1) and (2)) are physically real states, because these are the states in which singular measurements yield definite values of corpuscular properties.
2. In contrast, Born never considers a general wave function, $|\Psi\rangle$, as physically real, which are results of superpositions of eigenfunctions (as shown in eqs. (1) and (2)) because, as Cartwright argues, he never

[1] *Cf.* Heisenberg [1930], pp. 13ff. It should be noted that whether Heisenberg's thought experiment gives a sound argument for his indeterminacy relation is beside the point. The fact that such a thought experiment was constructed and taken seriously by most Copenhagenists shows their commitment to a corpuscularian ontological view.

believes in the reduction of wave packet but only the transition from one eigenstate to another.[2]

3. Born says that forces, such as the electromagnetic force, instead of acting directly on particles, act on the wave function, $|\Psi\rangle$. This should be understood as saying that the forces act on the wave functions of eigenstates. To make this point more precise, let us look at the case of a quantum collision, the study of which brought Born to his original interpretation. Suppose that an initially free electron which is in some eigenstate $|\phi_0\rangle$ approaches another object and gets into a superimposed state, $|\Psi\rangle$, under the influence of that object's force field. To simplify the discussion, let us assume that $|\Psi\rangle$ is a superposition of the $|\phi_i\rangle$s in the form of eq. (1). When Born talks about the force influencing the wave packet, what he really meant is that the eigenstates which constitute the superposition state are the ones that are affected by the force and the effect is shown in the coefficients of the eigenfunctions in eq. (1)—the $\langle\phi_i|\Psi\rangle$s.

Since the meaningful quantities of those coefficients in the Fourier expanson of $|\Psi\rangle$ are the $|\langle\phi_i|\Psi\rangle|^2$s and $|\langle\psi_j|\Psi\rangle|^2$s in eq. (3), it is natural to conclude that Born must have meant to take these probabilities to be physically real. We may view these probabilities as the *transitional probabilities*: $|\phi_0\rangle \mapsto \{|\phi_i\rangle\}$, where $i = 1, 2, \ldots, n$. In Cartwright's words:

> Jammer says, 'The Ψ-function must therefore be something physically real.' But for Born himself, since he explicitly denies that the superpositions are real physical states, it must be the probabilities themselves that are real. (Cartwright [1987], p. 414.)

Moreover,

> Probabilities, like particles, are a part of nature itself; they are objects to be operated on and evolve by nature's laws (*ibid.*, p. 415).

It is obvious that Cartwright is not merely doing exegesis of Born's original interpretation; she really believes that that is the right interpretation, in contrast to its Copenhagenized version. Cartwright's ontology for the microscopic world is richer than is usually acceptable: there are particles as well as transitional probabilities between eigenstates. As much as I agree with Cartwright that the simple corpuscularian ontology must be abandoned, I see some difficulties with her argument and the enriched ontology. First, if the reduction of the wave packet is a problem and should be denied, confining oneself to transitional probabilities will

[2] This means roughly the following. Suppose a system is initially in an eigenstate, $|\psi_1\rangle$, then evolves into a superposition state, $|\Psi\rangle$, and finally is measured to be in $|\psi_{10}\rangle$. Born does not believe that the measurement reduces the wave packet $|\Psi\rangle$ to $|\psi_{10}\rangle$ but rather there occurs a transition from $|\psi_1\rangle$ to $|\psi_{10}\rangle$.

not help to solve it. The transition from one eigenstate to another is no less mysterious than the transition (or reduction) from a superposition state to an eigenstate. Both are discontinuous and indeterministic. What in fact transpires in the former transition is not any better known than in the latter. Second and more importantly, there is no real distinction between a superposition state and an eigenstate. One observable's eigenstate is another observable's superposition, especially if the two observables are non-commuting. Without any loss of generality, let us suppose that the P and Q we have used before are two non-commuting observables. Then, any eigenfunction of P, $|\phi_i\rangle$, can be expressed, in the same way as in eq. (2), as a superposition of the $|\psi_j\rangle$s:

$$|\phi_i\rangle = \Sigma_j \langle\psi_j|\phi_i\rangle|\psi_j\rangle, \tag{4}$$

where, $|\langle\psi_j|\phi_i\rangle|^2$ will be the *transitional probability* (or is it the *reduction* probability?) from the i state of P to the j state of Q. Third, for any eigenfunction, $|\psi_j\rangle$, $\langle\psi_j|\psi_j\rangle$ is also a probability measure (always = 1). Now, which one is physically real according to Cartwright? If the quantum probability is real only because the wave function does not refer to a real state, we have to deny that $|\psi_j\rangle$ refers to a physically real state, which contradicts the original position.

Moreover, it is not entirely clear exactly in what sense Cartwright is using the notion of physical realness. From the conclusion she draws from her argument (the second quote above), Cartwright seems to think that something is physically real if it represents some kind of substance, something ontologically real. But it is not clear that the same sense is being used when she discusses Born's view of quantum states (either eigenstates or superimposed ones) (*cf.* the first quote above).

All this indicates that if wave packets are to be considered physically real, one has to take them all, in eigenstates or otherwise, as real. And it strongly suggests that if quantum probabilities are real, it is because they are real (dispositional) properties of the wave packet, which exists in the first place.

3 The reality of the wave packet

What does it mean to say that a quantity or function is physically real? Even though this notion is widely used in physics literature, it has never been explicitly defined or explained. Nor could we find any clarification of it in Cartwright's work we examined above. I shall begin by considering some examples from which the intuitive meaning of physical realness can be best extracted. Let us consider a fair coin. It has certain mass, size, shape, and velocity at some moment. We have no problem calling these quantities of the coin physically real quantities, for they are easily measurable and together they define the states of the coin. We also know that if tossed in a normal environment, the probability that the coin comes up heads is one-half. But we hesitate to call this quantity a physically real one, mainly because we think that it supervenes on quantities of the previous kind (mass, size, etc.): a change of the coin's mass, size, or environment, etc., may cause a change in its probability of landing on heads or tails, but one cannot manipulate the probability of the coin in order to change its mass or size, etc. It hardly makes sense to say that because of the interaction between the probabilities of two coins, the energy, say, of either or both of them is altered. It is for the same reason that probabilistic distributions of gas molecules are not usually taken to be physically real quantities, since they are reducible (stronger than supervening) to the kinetic energy of the gas molecules. Because of this, one cannot use the distribution functions to describe the interaction between two gases. Thus, it seems that only basic quantities, ones that do not supervene on other quantities, can be taken as physically real.

The case of electromagnetism gives more illustration, because it was not until the early years of this century that quantities like **E**s (the electric field strength) and **B**s (the magnetic field strength) acquired the status of physical realness. If the mechanistic interpretation of electromagnetism were true and luminiferous aether real, the **E**s and the **B**s would be in the category of quantities similar to the probability of a fair coin. Physically real would be those mechanical quantities which described the vibratory or undulatory motions of the aether particles; and the 'field' quantities would supervene on the mechanistic ones. The mechanistic interpretation is now known to be false and the mechanical aether a fiction, so the **E**s and the **B**s are recognized as physically real and represent a substance—the electromagnetic field. They not only are the fundamental quantities whose values are directly measurable and determine the states of electromagnetic fields, but also represent a physical substance that causally interacts with other objects. It seems to me that Cartwright is making a parallel move on behalf of the quantum probabilities. Quantities such as $|\langle\phi_i|\Psi\rangle|^2$ (the transitional probabilities), argues Cartwright, are mistakenly taken to be parasitic on corpuscular quantities of microscopic particles, but in fact they, like the electromagnetic field-strength functions, represent 'a part of nature itself' (*ibid.*). This example adds two more characteristics for the physically real: that they are measurable quantities and that they refer to something that is capable of causal interaction.

Thus, the following criterion seems to offer a precise notion of the physically real.

> (R) A quantity or function is *physically real* if (i) it is a basic quantity (i.e. not supervening on other quantities); (ii) it is experimentally measurable; and (iii) it refers to something that is capable of affecting, and/or being affected by, other physical objects or processes.[3]

Let us briefly look at one more clear example. In *classical* mechanics, for instance, **r** (position) and **p** (momentum) satisfy (R) and, hence, are physically real. However, the phase function of a particle or a body in the Hamilton-Jacobi equation—S(**r**, t)—does not satisfy (R) because it is not basic and experimentally measurable. Furthermore, what it refers, *i.e.* a phase front similar to a wave front, can never exert or receive causal influence in that capacity (the energy and momentum of the particle is not distributed over the surface).

The question now is how well (R) can survive at the quantum level. Let us look at the case in point: the physical realness of Ψ-functions and $|\Psi|^2$-functions. There is no question that Ψ-functions are basic quantities which do not supervene, or are not reducible to, other more fundamental quantities. This is even more so if the more fundamental quantities are corpuscular ones (e.g. position, energy, and spin, etc.). After the revelation from Bell's work on hidden-variables theory (the Bell theorem), it seems obvious that neither $|\Psi\rangle$ nor $|\Psi|^2$ are supervenient on the observables of quantum objects. Also, the Ψ-function is basic in that it gives a complete account of the state of quantum objects (whatever they are).

However, the Ψ-function is a complex function and not an observable; its values are thus not experimentally measurable. Besides, we do not have a good spacetime model for the *wave packet* which is what the Ψ-function refers to if it is to be taken literally. This seems to imply that the Ψ-function does not represent anything that is capable of causal interaction. If, as many believe, wave packets are entities in a fictitious mathematical space (the Hilbert vector space), the Ψ-function may only be real in the Platonic world.

To see a way around such objections to taking wave functions as physically real, let us now look at the details of a wave function. In most

[3] One may want to use 'if and only if' instead of 'if' in (R). But I think that will make it unnecessarily strong. In fact, to say something is physically real only if (i)..., (ii)..., and (iii)..., one runs into similar problems that a strong version of verificationism has. For instance, such a criterion would rule out empty spacetime regions as physically real, since they do not causally interact with any other physical objects or processes. (Here I take 'empty' to also include 'gravitation free', i.e. such regions are all flat.) Empty spacetime regions may in fact be unreal, but it cannot be so just for this reason alone.

cases, a wave function can be expressed in the following general form:

$$|\Psi\rangle = A\,e^{-iS}, \tag{5}$$

which consists of two parts, one real—A (the amplitude), and one imaginary—$|\Psi\rangle_{ph} = e^{-iS}$ (the phase function) (we assume that $\hbar = c = 1$). $|\Psi\rangle_{ph}$ disappears in $|\Psi|^2$ $(= |A|^2)$ and contains the phase factor, S, of the quantum object in question. $|A|^2$ gives the probability measure of, say, where an impact is going to occur when an electron in state $|\Psi\rangle$ interacts with some other objects, while S explains, *inter alia*, the interference effect we see in such familiar experiments as the two-slit experiment. Thus, even though we cannot imagine what kind of substance the wave packet is, we know exactly how it behaves if it exists. This suggests that it may be more fruitful if we examine condition (iii) in (R) first, and then, see if a modified condition (ii) can be satisfied. Let us ask, instead, whatever wave packets are, if they causally affect, and/or are capable of being affected by, other objects and processes. For the following reasons, the answer I think is a positive one, even though we may not be able to answer the question: how can anything be like that which $|\Psi\rangle$ represents?

(1) Wave functions represent things that interfere with one another, as shown most vividly in the case of the two-slit experiment. Very briefly, in a two-slit experiment, when a single quantum object passes through the two openings, the wave packet, $|\Psi\rangle$, splits into two, $|\Psi_1\rangle$ and $|\Psi_2\rangle$, which determine, when the object is later intersected by a receptive screen, where the impact occurs with what probability.[4] However, $|\Psi|^2 \neq |\Psi_1|^2 + |\Psi_2|^2$, as it should if the object went through *either* opening 1 *or* opening 2. The truth is that $|\Psi_1\rangle$ and $|\Psi_2\rangle$ interfere with each other, and thus

$$|\Psi|^2 = |\Psi_1|^2 + |\Psi_2|^2 + (\langle\Psi_2|\Psi_1\rangle + \langle\Psi_1|\Psi_2\rangle), \tag{6}$$

where the last term represents the interference effect.

(2) External forces, such as electromagnetic force, act on wave packets rather than on the particles. Recall that Born made this observation, even though Cartwright reinterpreted it to mean that forces act on quantum probabilities (*cf.* p. 981, and Cartwright [1987], pp. 410, 414–15).

(3) There is no other way of understanding the AB effect except by taking wave packets as physically real, as will be seen in the next section.

[4] Henceforth, $|\Psi\rangle$ will be used to refer to the wave function or the wave packet it represents. It should be clear which is meant in the context of a statement. Also, 'wave function', 'Ψ-function', and '$|\Psi\rangle$' are different names for the same item.

However, the problem with **(1)** is that it is difficult to prove that the interference effect shown in such cases as in eq. (6) is the result of a genuine interaction between $|\Psi_1\rangle$ and $|\Psi_2\rangle$. Whenever the object reaches the measuring screen, one gets either a single spot on a photo plate or a single click in a Geiger counter. To infer that interference has 'really' taken place, one must do the experiment on multiple objects (identically prepared) and look at the accumulated results (i.e. the interference pattern). It is well known that such results can be explained without evoking the existence of the wave packet: quantum objects are particles which behave probabilistically (i.e. obeying genuinely probabilistic laws) and the probabilities given by $|\Psi|^2$ are of a non-classical kind.[5] The talk of interference between $|\Psi_1\rangle$ and $|\Psi_2\rangle$ might have to be understood metaphorically as a description of two pure mathematical entities.

Similar to **(1)**, the problem with **(2)** is that one cannot prove that, as Born and Cartwright believe, forces act on the wave packet or the quantum probability rather than on the particles. That the forces seem to change the probabilities, rather than the velocities, of an electron may simply mean that its probabilistic behavior is the only means by which we know of its being influenced by the forces. A force field may change the odds of getting heads by flipping a fair coin in it, but it by no means proves that it is the probability that causally interacts with the force. Similarly, a tank of gas is affected by a uniform gravitational field in such a way that its distribution function, F, according to Boltzmann's theory, is

$$F \propto e^{-(\mu+E+mg)/kT}, \qquad (7)$$

rather than (when the field is absent),

$$F \propto e^{-(\mu+E)/kT},$$

where μ is the chemical potential, E the energy of a single molecule, k the Boltzmann constant, and T the temperature of the gas. The 'mg' is the gravitational potential. The probability distribution *is* the only means by which the influence of gravity on the gas can be known. The question now is: does the force act directly on the molecules or on the probability? Obviously, the right answer is: 'on the molecules!' The reason has a lot to do with what we think we know about gases and their molecules, but this much seems sufficient: gravitational force acts on singular particles similar to the gas molecules (if only larger in size). Why can we not then say the same in the case of **(2)**? We know that forces, whether gravitational or electromagnetic, act on charged particles (including electrons going

[5] This approach usually goes by the name of quantum logic (*cf.* Hughes [1989] and the references therein). The bottom line is that by developing a non-classical theory for quantum probabilities one can consistently treat quantum objects as corpuscles. It is the 'normal-ontology-but-queer-property' approach.

through a cloud chamber). This should put, it seems, sufficient doubt on Born's claim that those forces act on $|\Psi\rangle$ instead of on the particle in question. This leaves us with point (**3**).

Before we discuss the AB effect, let us see what becomes of (**R**) from our discussion above. It seems that we should give up condition (ii) in (**R**) and just take whatever that satisfies condition (i) and (iii) as physically real. In fact, there are independent reasons for rejecting condition (ii) at the quantum level. Taking (ii) too seriously may produce prima-facie absurd results. One may be forced to say that, for instance, the spin of an electron in the x-direction is a physically real quantity except when the spin of the same electron in the y- or z-direction is measured, because, according to Heisenberg's indeterminacy principle, the spin in the x-direction does not have a definite measurable value at those moments.

4 The Aharonov–Bohm effect

Very much like the EPR thought experiment, the AB effect was first predicted by Aharonov and Bohm in 1959 as the result of a thought experiment (Aharonov and Bohm [1959]) and later experimentally confirmed (*cf.* Peshkin and Tonomura [1989]). After a period of intense debate on whether the effect is genuine (*cf.* Peshkin and Tonomura [1989] for a summary of the debate and the relevant literature), it is now clear that the AB effect is real and the original prediction of Aharonov and Bohm correct, subtle differences notwithstanding. While theoretical physicists use it to fathom deeper meanings of electromagnetism (*cf.* Feynman *et al.* [1964], Ch. 15; Wu and Yang [1975]) and experimental physicists have been exploring its practical applications (*cf.* Imry and Webb [1989]), philosophers are yet to find out its implications for philosophy of quantum theory (one exception is Zangari [1992]). Part of my task in this section is to change that. Let me begin by giving a sketch of the AB effect.

There are two types of AB effect: the electric and the magnetic. The original theoretical concern of Aharonov and Bohm is whether electromagnetic potentials have effect on charged quantum objects, such as electrons. It is well known that according to the classical theory of electrodynamics, only the **E**s and **B**s are considered physically real in the sense that charged particles are affected in regions where **E** and/or **B** are non-vanishing. The potentials, the **A**s and φs, are pure mathematical artifacts. Electromagnetic potentials are different, as theoretical entities, from the fields, but they are parasitic on the fields and incapable of delivering any physical effect without their presence (at the same spacetime locale).

More precisely, the **A**s and φs are introduced as follows. At the core of classical electrodynamics are the four Maxwell equations which are differential equations of the **E**s and **B**s. To make the equations simpler and mathematically more elegant, the potentials are introduced by the following equations:

$$\mathbf{E} = -\partial \mathbf{A}/\partial t - \nabla\varphi \tag{8}$$

$$\mathbf{H} = \nabla \times \mathbf{A} \tag{9}$$

(where we assume that $c = 1$). We can also use the potentials to define a single 2nd-rank electromagnetic tensor, $F_{\mu\nu}$, and reduce the number of Maxwell equations from four to two. Further, the Lagrangian equation has to be construed in terms of various differentials of the potentials. But all equations in electrodynamics has the property that if two fields have the same values of **E**s and **B**s, they are physically identical. That means that no experimental results derivable from the equations can distinguish the two fields. Therefore, the potentials have what is called the gauge freedom. For any scalar function, f, a (gauge) transformation of the following kind does not produce any physically detectable difference:

$$\mathbf{A}' = \mathbf{A} + \nabla f, \quad \varphi' = \varphi - \partial \mathbf{f}/\partial t. \tag{10}$$

By substituting the **A** and φ with the **A**$'$ and φ' in eqs. (8) and (9), we get the same values of the **E** and **B**.

Indeed, if it is not for the integral equations of the following kind, one might think that the potentials are pure mathematical fluff. For the magnetic field, according to the well-established Stoke theorem, we have

$$\oint_L \mathbf{A} \cdot d\sigma = \oint\!\!\oint_S \mathbf{B} \cdot d\mathbf{s}, \tag{11}$$

where $d\sigma$ is the line element on the loop L which contains the area S of which $d\mathbf{s}$ is an element. (In words, eq. (11) says that the potential, **A**, is such that its integral along any closed loop, L, outside and enclosing area S equals the magnetic flux within, i.e. the rhs of the equation.) This means that if the **B**-field is confined in a region whose cross-section is S, there will be potential, **A**, outside of that region, since it is non-vanishing on any loop which encloses S (or any part of it). Hence, there are regions in space where the **B**-field is vanishing while the **A**-potential is not. As I said before, no physical effect is either theoretically predicted or experimentally discovered in such regions at the classical level.

Moreover, there is a theoretical reason not to take the potentials as physically real. The electromagnetic fields may exist independently of the source that produces them (in the form of electromagnetic waves). We are thus justified in speaking of **B** and **E** being magnitudes for something independent of charged particles. We cannot say the same for the potentials: they apparently do not represent anything independent of the **B**- or **E**-fields. This can be seen in the following peculiar characteristics of the potentials. Even though a magnetic potential can be non-vanishing at a point where $\mathbf{B} = \mathbf{0}$, (**0** being a vector with zero 'length'), it is only so when it is considered on a loop which encloses some non-vanishing **B**-field. In other words, whether the **A**-potential is non-vanishing at some spacetime point depends on whether or not the loop of integration, L, which goes through the point in question includes any **B**-field region. If the loop does include the **B**-field region, **A** is non-vanishing; if it does not, **A** is zero. And since the loop is otherwise arbitrary, it follows that the value of **A** at that point is not determined independently of the **B**-field. This is why it seems unquestionable in the classical context that the electromagnetic potentials should be treated as pure theoretical terms which are not physically real.

The AB effect changed all this and showed the contrary. Roughly put, the AB effect unequivocally shows that there exists physical interaction between the electromagnetic potential and charged microscopic objects. Low-temperature superconductivity and homologous imaging technique made it possible in recent experiments for researchers to isolate completely (as completely as is physically possible) the **B**-field and separate it from the electron beams. The fringe shifts recorded in the experiments cannot but be interpreted as the result of interactions between the electrons and the potentials that exist outside of the **B**-field regions.

Figure 1 gives some details of the modified two-slit experiment which produces the magnetic AB effect. **A** ($\propto$ distance from the **B**-field) is present outside the solenoid even though the **B**-field is completely confined within; and we have, according to eq. (11),

$$\oint_L \mathbf{A} \cdot d\sigma = \mathbf{B} \cdot \mathbf{S}.$$

Repeating this experiment as is usually done—one set with another set without **B**—one finds a shift of the interference fringes, as shown in Figure 1(A) and (B). The theoretical explanation of this is not at all difficult. It is well known that after passing the two openings, the wave packet, $|\Psi\rangle$, splits into $|\Psi_1\rangle$ and $|\Psi_2\rangle$ (*cf.* p. 990, Figure 1); and $|\Psi_i\rangle = |\Psi_i^0\rangle e^{-iS_i}$, where $i = 1, 2$, and $|\Psi_i^0\rangle$ is the wave packet without **B**. From eq. (6) and some simple calculation, we have

$$|\Psi|^2 = |\Psi_1^0|^2 + |\Psi_2^0|^2 + 2|\Psi_1^0||\Psi_2^0| \cos(\Delta S), \tag{12}$$

where ΔS ($= S_2 - S_1$) is the phase difference that causes the fringe shift. It

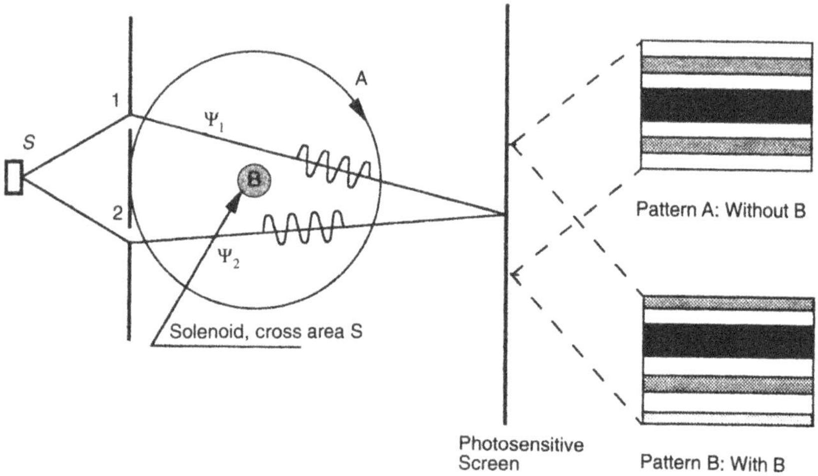

Fig. 1. The magnetic AB experiment: Ψ_1 and Ψ_2 are the wavelets from the two openings; **B**, the magnetic field, is confined inside the solenoid; **A** is the potential outside. A small portion of the screen is shown with interference fringes. There is a distinct fringes shift in Pattern B when **B** is present

is shown (*cf.* Tonomura, in Peshkin and Tonomura [1989], pp. 39–43) that

$$\Delta S = -\frac{e}{\hbar}\oint \mathbf{A}\cdot d\sigma = -\frac{e}{\hbar}\mathbf{B}\cdot\mathbf{S} \tag{13}$$

Since the wave packets can never enter the solenoid nor can the **B**-field ever get out, if there is interaction between **B** and the electrons, it has to be action at a distance. Barring such action (a natural thing to do), the fringe shift must be caused by the interaction between **A** and the electrons.

5 The reality of the wave packet again

The question, now, is how much the AB effect says about electromagnetism and microscopic objects. One must, first of all, note that the AB effect is a quantum effect. The effect is the induced shift of the interference fringes, which does not occur with macroscopic objects. Moreover, as will be shown later, there can be no such effect if the experiment is so conducted that electrons appear as macroscopic objects (e.g. when they pass through a Wilson cloud chamber).

Second, even though we know that microscopic objects are affected by the potentials, they are affected in an unusual manner if they are considered as corpuscles. One might think that wherever the **A**-potential is non-vanishing, there is a field, somewhat similar to the **B**-field, that interacts with the charged particles which pass through the region. That is not true. The reason is that what the potential represents, if anything, is *non-local* with respect to its effect on such particles. According to eq. (11), in order for the potential to have any effect, the electrons have to pass through the region in a closed loop which includes the non-vanishing magnetic flux. The situation is illustrated in Figure 2 (a) and (b). In Figure 2 (a), the loop includes the source and the fringe-shift occurs, whereas in Figure 2 (b), the loop does not include the source and no such shift occurs even though the electrons pass through the region with a non-vanishing **A**-potential. This means that whether the potential affects quantum objects as corpuscles depends not only on its values at the neighborhoods of those objects, but also on their relations to the source and on how the objects go through different parts of the region of **A**.

Third, this non-localness of the potential is, however, different from the classical concept of action-at-a-distance. What is acting at a distance in the magnetic AB experiment is the **B**-field which is confined in the solenoid. In fact, avoiding action-at-a-distance is the primary motivation for taking potentials as capable of affecting and being affected. Thus, there is no action-at-a-distance in the case of the AB effect when the **A**-potential is taken to represent the magnetic effect on electrons. But the effect is none the less not *local* because what transpires between the **A**-potential and $|\Psi\rangle$ at certain point x_0 depends on values of **A** and $|\Psi\rangle$ at points other than x_0 and its neighborhood.

Let us now turn our attention from the affecting to the affected. This is where the AB effect becomes relevant, and indeed essential, to the main topic of this paper. We want to know if wave packets are real. At the end of the previous section, we reach the conclusion that the quantum objects are affected by other objects or processes either as corpuscles or as wave packets. Now, I maintain that in the light of the AB effect, one has to conclude that what is affected in the experiment must be the wave packets. Thus, wave packets rather than the probabilities are physically real.

First, contrary to the case of the Boltzmann gas, we know that potentials do not influence charged particles which are large enough to be seen as macroscopic objects. What delivers causal influence on such particles is the Lorentz force, $\mathbf{F} = q(\mathbf{E} + \mathbf{v} \times \mathbf{B})$ (where q is the charge and **v** the particle's velocity), which vanishes wherever and whenver $\mathbf{E} = \mathbf{0}$ and $\mathbf{B} = \mathbf{0}$. Therefore, it does not seem plausible to infer that, for those charged particles whose size drops below the limit of macroscopic observability, what affects them turns out to be the potentials.

Second, because of the non-local feature of the interactions in the AB experiment I discussed earlier, the talk about interactions between corpuscles and potentials does not make much sense. It has the unpleasant

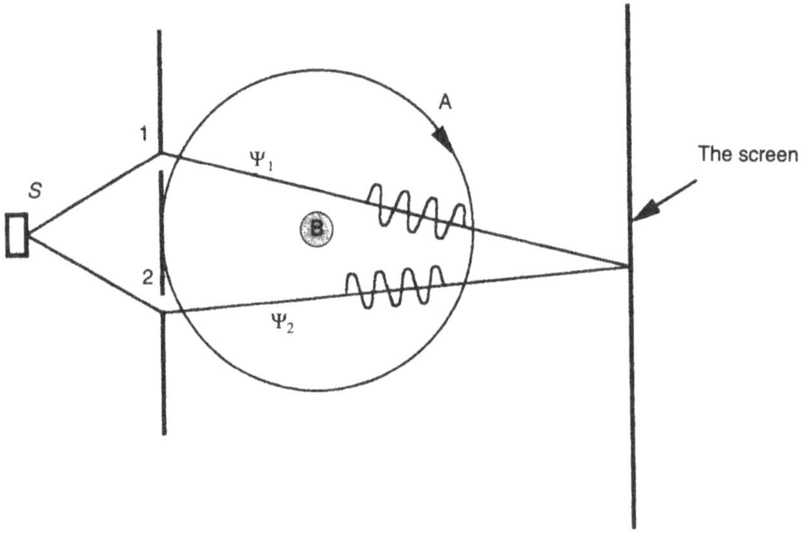

(a)

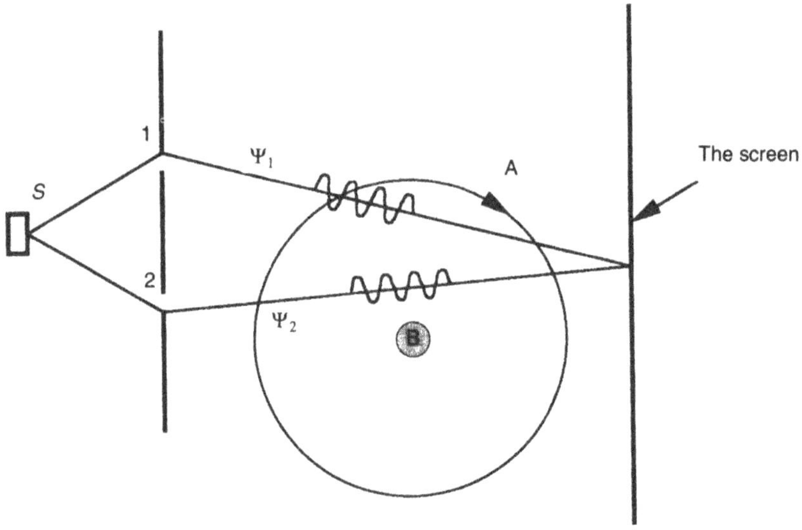

(b)

Fig. 2. (a). The magnetic AB experiment where fringe-shift occurs. (b) The magnetic AB experiment where fringe-shift does not occur.

implication that quantum particles are localized entities but they affect, and/or are affected by, other things through non-local means. For instance, in the AB experiment, where an electron lands after it emerges from the two openings and intersects with the screen depends not only on the values of the potentials at those points through which it passes, but also on the values at points through which it *may* move through. Therefore, it seems better if we construe the interaction as taking place between two non-localized entities—the electromagnetic potential and the wave packets. This implies that the wave packet exists and is the object that the **A**-potential affects.

Third, if the potentials causally influence the corpuscles rather than the wave packets, as in the case of a fair coin and the gas molecules, such influence must manifest in other ways than the shifts of the interference patterns. Such influence should especially manifest itself in those situations where quantum objects are indeed corpuscular. For instance, it seems natural that we should expect detectable effect on the trajectory of an electron in a cloud chamber when the electron goes through a region where the **A**-potential is non-vanishing. That is not true. Numerous experiments show that only by direct contact with an **E**- or **B**-field can an electron be deflected and its deflected trajectory observed.

It also seems natural that we should expect detectable effect that a non-vanishing **A** has on a beam of quantum objects when they are in a mixture. The idea is that if the potential affects quantum objects as corpuscles, then whether they are in a pure state or in a mixture should not make any difference. (The effect may be different in the two cases, but there should be effect in both.) To see why the answer is no, let us consider the AB experiment again. The AB experiments are all carried out with coherent electron beams (electrons that are prepared in a pure state). To create a mixture,[6] all one needs to do is to place detecting devices at the two openings (or the corresponding places whatever the actual experimental set-up may be) and break the coherency of the electron beams. Suppose that in Figure 1, we place two detectors at the upper and the lower opening, so that the electrons emerge from the two openings in a mixture state composed of $|\Psi_1\rangle$ and $|\Psi_2\rangle$, which are statistically independent. Such two states are called orthonormal, viz., $\langle\Psi_2|\Psi_1\rangle = \langle\Psi_1|\Psi_2\rangle = 0$. Therefore, in this mixed state, the interference terms in eq. (6) disappears and $|\Psi|^2 = |\Psi_1|^2 + |\Psi_2|^2$. Similarly, the interference term in eq. (12) also disappears in such a mixture, which means that $|\Psi|^2 = |\Psi_1^0|^2 + |\Psi_2^0|^2$. In

[6] Mixed states are represented in quantum theory by the density operators. Suppose that $|i\rangle$, $i = 1, 2, \ldots, n$, are statistically independent pure states from which a mixed state is formed. The density operator for this state is $\rho = \Sigma_i |i\rangle p_i \langle i|$ (the summation is from $i = 1$ to $i = n$), where p_i is the weight (or probability) that state $|i\rangle$ has in the mixture.

other words, the probability distribution of the electrons is not affected by the magnetic potential. The theoretical explanation for this is obvious: electromagnetic potentials only affect phase shifts in wave packets, but there is no phase coherence in a mixed beam of quantum objects, so there is no **AB** effect. Therefore, the potentials do not act directly on the objects (as small, individual corpuscles).

Fourth, does the potential have any effect on the expected values of a quantum object's observable properties? Does it have any effect on its average position, momentum, spin, and so on? Theoretically, the answer is again, no. Consider an electron in the **AB** experiment (Figure 1). Is any of its observables, P, affected by **A** when it passes through the region? The average value of P in the time interval t to t′, where t′ > t can be found as follows,

$$\overline{P_\psi} = \int_t^{t'} \sum_{i=1}^{2}\sum_{j=1}^{2} \langle \Psi_i | P | \Psi_j \rangle dt$$

$$= \int_t^{t'} \sum_{i=1}^{2}\sum_{j=1}^{2} \langle \Psi_i^0 | P | \Psi_j^0 \rangle dt = \overline{P_{\psi^0}} \qquad (14)$$

The second equation holds because $\Psi_i = \Psi_i^0 e^{-iS_i}$ and $P\Psi_i = e^{-iS_i}(P\Psi_i^0)$ (*cf.* p. 989); and this holds because in quantum theory (at least in the theory from which the AB effect is derived), all observables commute with the components of **A**-potential of which S_i is a function (e.g. $PA_x - A_xP = 0$). Hence, the expected (or average) value of P is not affected by the potentials. If one is not convinced by this argument since the commuting of the observables and the potentials depends on treating the field as unquantized, one can always settle the issue by designing experiments to measure the concerned averaged values of the observables with and without the presence of the **A**-potential. My guess is that there will be no difference because average values are usually insensitive to the relative change of a wave's phases.

For these reasons, I conclude that the wave packet is a real substance. I think that we can say that quantum objects, such as electrons, are wave packets, just as we can say that electromagnetism is the gauge-invariant manifestation of the potentials (the manifestation is through a non-integrable phase factor similar to eq. (13) (*cf.* Wu and Yang [1975], p. 3846).

6 Propensity and the wave packet

It is one thing to prove that the wave packet is physically real but quite another to say what it is; the latter is more difficult but less important. For should we not take it to be whatever the Ψ-function and the quantum theory say it is? Something analogous to this happens with the theory of electrodynamics: the **AB** effect prompted physicists to reconsider what electromagnetism *really* is, but it did not do much to expand the theory, for the potentials, **A** and φ, were well understood theoretical entities before the 1959 paper by Aharonov and Bohm. However, we all like models and wave packets seem to defy our imagination. We know that their mathematical representations are complex functions, so they cannot be waves in the ordinary sense. Here is where I see how the notion of propensity can help. Popper defended many years ago a propensity interpretation of probability which he thought was applicable to quantum probabilities, and so did many other philosophers. For instance, Cartwright argued in her article that 'there are two important aspects of the quantum probabilities ... that justify calling them propensities and ascribing a real existence to them' (Cartwright [1987], pp. 414–15). In what follows, I examine the place for propensity in the quantum world.[7]

For Popper [1957, 1960] probabilities are simply propensities. Instead of being the relative frequencies of (virtually) infinite sequences, probabilities are dispositional properties—propensities—of those generating conditions (or experimental set-ups) that produce, when infinitely repeated, the sequences (with the frequencies). Further, they are dispositional and relational properties. They are dispositional because they tell us what would happen if the object were to be prepared and measured in certain conditions (even if they are not so prepared and measured); and they are relational because they are determined by the relation between the object and the experimental set-up under which it is prepared and examined. Popper also compares propensity with the strength of an electric field and observes that 'just as we can consider the field as *physically real*, so we can consider the propensities as *physically real* (Popper [1957], p. 68, my italics). This strongly suggests that Popper, like Cartwright, is taking the quantum probabilities to be physically real. However, he also claims that the propensities are determined by the Ψ-functions, and since propensities are physically real, the Ψ-functions are so as well. Therefore, Popper wishes to conclude that quantum theory becomes no more mysterious than 'any [probabilistic] theory of any other game of chance, such as the bagatelle board [pin board]' (*ibid.*).

Even though Popper's conclusion sounds like the one I defended in this paper, it is in fact very different: what Popper meant by the Ψ-functions is different from mine; and his notion of physical realness is also different

[7] It is not the task of this paper to conduct a full investigation of the notion of propensity and its relation to interpretations of quantum theory. For that I will have to discuss in detail works by, among others, Mellor, Fetzer, and Salmon. Also, I will have to discuss Bohm's notion of potentiality (more precisely, the 'incompletely defined potentiality'). I shall pursue such an investigation in another paper, 'Potentiality and Propensity in Quantum Theory' (work in progress).

from mine. As is, Popper's interpretation cannot even be rendered consistently. First, Popper seems unaware of the importance of distinguishing the Ψ-function from the $|\Psi|^2$-function, which I have emphasized earlier. If the notion of propensity is used to interpret quantum probability, as obviously is Popper's intention, it is the latter function rather than the former that should be rendered as a measure of propensity. This is especially true when Popper claims that, by such a propensity interpretation of quantum theory, quantum probability is no more mysterious than that in a pin board game. However, when so rendered, Popper's propensity interpretation cannot work (Popper realized the difficulty himself, *cf.* Popper [1967], pp. 33–4; also *cf.* Salmon [1979], p. 193). For instance, in the two-slit experiment examined above, the propensity of an electron being in state $|\Psi_1\rangle$ is $|\Psi_1|^2$ and that in state $|\Psi_2\rangle$ is $|\Psi_2|^2$, and the electron, like the ball going through the pin board, is either in $|\Psi_1\rangle$ or in $|\Psi_2\rangle$. But propensity, being the proposed interpretation of probability, must satisfy the axioms of the calculus. However, by eq. (6), we know that the propensity of being in $|\Psi_1\rangle$ *or* $|\Psi_2\rangle$ (*i.e.* $|\Psi|^2$) is not the sum of the two propensities. This violates one of the axioms, and no interpretation of probability is acceptable if it is not consistent with the probability calculus.

Second, Popper has the right intuition when he says that the Ψ-function represents something physically real and it 'determines the propensities of the state of the electron' (*ibid.*), but then we do not have a propensity interpretation of quantum probability any more. This suggests that we give up the propensity interpretation of probability and take propensity to be something else. It seems to me (partly from Popper's intuition) that the wave packet is a good candidate for a propensity interpretation. I maintain that *quantum objects are wave packets that possess,* inter alia, *a dispositional property—a propensity—of exhibiting a probabilistic behavior in relation to how they are prepared and measured.* For those who are familiar with Bohm's idea of quantum properties as potentialities (*cf.* Bohm [1951], pp. 132–3, 138–9), the above may be rendered as follows. With respect to observable properties, wave packets are potential objects. So they only have dispositional properties whose display are the probabilistic distributions of possible values of corpuscular properties. Whenever they are actualized through measurement, they emerge as corpuscles.

Incorporating what I have argued before this section, I give the following reasons to support my propensity interpretation of wave packets.

1. In Popper's original interpretation, probability is taken to be the propensity of the generating conditions for producing a (virtually infinite) sequence whose relative frequency gives the measure. But in an experimental set-up, which are generating conditions and which are

not? It turns out that Popper's interpretation is incapable of giving a clear answer to this question (*cf.* Salmon [1979], pp. 192ff.). The same problem of course appears at the quantum level. Now, according to Mellor, propensity is a dispositional property of the object of interest (not of the experimental set-up) which is displayed not by the results of trials but by the probability distribution over the possible results of such trials (*cf.* Mellor [1971], pp. 68–71). This notion of propensity avoids the problem given above and is especially congenial to taking the wave packet as physically real and the Ψ-function as its representation. If the Ψ-function is taken to represent a property of a quantum *particle*, it, as just mentioned, can only be a relational property which describes a state of the particle under certain preparation conditions. But if it is about a *wave packet*, it represents a non-relational property. A similar case in electromagnetism may illustrate this point. In the mechanical theory of electromagnetism (aether), the **E**- and the **B**-functions describe relational properties (correlated vibratory motions) of the aether particles in the region of interest. But in the field theory, they are non-relational properties of the fields in the region. The dispositional property of the wave packet (described by the Ψ-function) is displayed by quantum probabilities—the $|\Psi|^2$-functions. The realization that propensity is a non-relational property of objects and its display not the individual results of trials but the probabilistic distributions of possible outcomes matches exactly the distinction between $|\Psi\rangle$ and $|\Psi|^2$, for the latter gives exactly the needed probabilistic distributions in various trials (the measuring of various observables) that display the dispositional property to which $|\Psi\rangle$ refers.

The Ψ-function qualifies more than quantum probability for a measure of propensity in that it not only determines (probabilistically) what happens when a measurement occurs but also determines (deterministically) how wave packets interact with one another (interference) and with other entities, such as the electromagnetic potentials (the AB effect). The mystery of quantum theory lies not in the probabilistic relation among observable results but in the *deterministic evolution and interaction* among wave packets and with electromagnetic (and presumably other) processes. According to Mellor, propensity, as any other dispositional properties, may never be displayed, so an object must have it independently of any conditions that display it. This is exactly the kind of properties that Ψ-functions represent and by taking it as physically real, the mystery disappears. Hence, quantum probabilities are not propensities, but rather a display of them: propensities determine quantum probabilities. Popper is right in introducing propensity to solve conceptual problems in quantum theory, but he wrongly identifies it with the quantum probability, and so does Cartwright ([1987], pp. 414–15).

3. If the wave function is taken to be a measure of propensity, the problem for Popper's propensity interpretation discussed above is naturally avoided. Knowing states $|\Psi_1\rangle$ and $|\Psi_2\rangle$, we know that the propensity of being in $|\Psi_1\rangle$ *or* $|\Psi_2\rangle$ is $|\Psi\rangle = |\Psi_1\rangle + |\Psi_2\rangle$, which in turn determines its probability as given by eq. (6). With this propensity interpretation of the wave packet, we understand why quantum probabilities do not behave like classical probabilities. Philosophers have developed a whole new logic—quantum logic—to account for the non-classical features of quantum probabilities, but it only provides a description of such features. This interpretation explains why quantum objects have such non-classical probabilities.

7 Conclusion

The view given in Section **6** has its origin in part in Mellor's view that propensity determines a chance distribution (Mellor [1971], pp. 63ff.) and Salmon's view on causal propensity (Salmon [1979], pp. 212–15; [1984], pp. 202–5). The influence by Maxwell's theory of 'smearon' or 'propensiton' (Maxwell [1982, 1988]) is also obvious. At the end of his 1979 article, Salmon suggests that we identify the wave amplitudes with propensities and predicts that if the details are worked out, we might gain a better insight in the quantum domain ('but I do *not* believe it could take the mystery out of quantum theory' (p. 214)). From eq. (5), we see that the amplitude is not robust enough for such a role, for it excludes the phase factor. It is now clear to me that one has not only to take the entire wave function as the measure of the propensity but also to grant the existence of wave packets.

Acknowledgements

I wish to thank Mark Zangari for getting me interested in the AB effect and its philosophical significance. I also want to thank Robert Rynasiewicz and Niall Shanks for e-mail exchanges on problems related to this paper. I also want to thank an anonymous referee for helpful comments.

Department of Philosophy
University of Florida
Gainsville, FL 32611
USA

References

Aharonov, Y. and Bohm, D. [1959]: 'Significance of Electromagnetic Potentials in the Quantum Theory', *Physical Review*, **115**, pp. 485–91.

Bohm, D. [1951]: *Quantum Theory*, Englewood Cliffs, NJ, Prentice-Hall.

Born, M. [1926]: 'Zur Quantenmechanik der Stossvorgänger', *Zeitschrift für Physik*, **37**, pp. 863–7.

Born, M. [1969]: *Physics in My Generation*, New York, Springer-Verlag.

Cartwright, N. [1987]: 'Max Born and the Reality of Quantum Probabilities', in L. Krüger, L. J. Daston, and M. Heidelberger (*eds*), *The Probabilistic Revolution*, Volume 2, Cambridge, MA, MIT Press, pp. 409–16.

Cartwright, N. [1989]: *Nature's Capacities and Their Measurement*, Oxford, Clarendon Press.

Feynman, R., Leighton, R. and Sands, M. [1964]: *Feynman's Lectures on Physics*, Volume II, Reading, MA, Addison-Wesley.

Heisenberg, W. [1930]: *The Physical Principles of the Quantum Theory*, Chicago, University of Chicago Press.

Hughes, R. I. G. [1989]: *The Structure and Interpretation of Quantum Mechanics*, Cambridge, MA, Harvard University Press.

Imry, Y. and Webb, R. A. [1989]: 'Quantum Interference and the Aharonov–Bohm Effect', *Scientific American* (April), pp. 56–62.

Kyburg, H. E. [1974]: 'Propensity and Probabilities', *British Journal for the Philosophy of Science*, **25**, pp. 358–75.

Maxwell, N. [1982]: 'Instead of Particles and Fields: A Micro Realistic Quantum "Smearon" Theory', *Foundations of Physics*, **12**, pp. 607–31.

Maxwell, N. [1988]: 'Quantum Propensiton Theory: A Testable Resolution of the Wave/Particle Dilemma', *British Journal for the Philosophy of Science*, **39**, pp. 1–50.

Mellor, D. H. [1971]: *The Matter of Chance*, Cambridge, Cambridge University Press.

Peshkin, M. and Tonomura, A. (*eds*) [1989]: *The Aharonov–Bohm Effect*, New York, Springer-Verlag.

Popper, K. [1957]: 'The Propensity Interpretation of the Calculus of Probability, and the Quantum Theory', in S. Körner (*ed.*), *Observation and Interpretation*, London, Butterworths Scientific Publications, pp. 65–70.

Popper, K. [1960]: 'The Propensity Interpretation of Probability', *British Journal for the Philosophy of Science*, **10**, pp. 25–42.

Popper K. [1967]: 'Quantum Mechanics Without "The Observer"', in M. Bunge (*ed.*), *Quantum Theory and Reality*, New York, Springer-Verlag, pp. 7–44.

Salmon, W. [1979]: 'Propensities: A Discussion-Review', *Erkenntnis*, **14**, pp. 183–216.

Salmon, W. [1984]: *Scientific Explanation and the Causal Structure of the World*, Princeton, Princeton University Press.

Wessels, L. [1980]: 'What Was Born's Statistical Interpretation?' *PSA 1980*, Volume 2, East Lansing, MI, Philosophy of Science Association, pp. 187–200.

Wu, T. T. and Yang, C. N. [1975]: 'Concept of Nonintegrable Phase Factors and Global Formulation of Gauge Fields', *Physical Review D*, **12**, pp. 3845–57.

Zangari, M. [1992]: 'Adding Potential to a Physical Theory of Causation', *PSA 1992*, Volume 1, East Lansing, MI, Philosophy of Science Association, pp. 261–73.

Part 7

Space-Time Theories and Realism

This part contains three previously published papers, which are loosely connected by the theme of history and philosophy of space-time theories.

The paper, "Is There a Relativistic Thermodynamics?" is a historico-philosophical study of a debate that took place in the 1960s and 1970s regarding the correct (special) relativistic formulation of thermodynamics. Unbeknownst to the physicists engaged in this debate, a mini-debate had taken place between Einstein and von Laue in their correspondence in 1952. Thermodynamics was given a relativistic formulation shortly after Einstein formulated the special theory of relativity in 1905. Planck and von Laue, among others, produced the standard formulation as a corollary of the relativistic theory of mechanics, and the formulation appears in von Laue's book on relativity theory. In 1952, upon receiving a copy of the new edition of von Laue's book, Einstein wrote to von Laue raising the objection to that formulation of relativistic thermodynamics. For a detailed study of this Einstein-von Laue correspondence, see another paper of mine, "Einstein and Relativistic Thermodynamics in 1952: A Historical and Critical Study of a Strange Episode in the History of Modern Physics," *The British Journal for the History of Science*, Vol. 25, No. 2 (Jun., 1992), pp. 185-206.

For more recent works in the physics literature of relativistic thermodynamics, I recommend the following. H. Callen and G. Horwitz, "Relativistic Thermodynamics," AJP **39**: 938 (1971); I-Shih Liu, I. Mueller, and T. Ruggeri, "Relativistic Thermodynamics of Gases," *Annals of Physics* **169**: 191-219 (1986). The first is a nice summary of the state of art towards the end of the debate on the subject and the second is a "field-theoretic" approach to relativistic thermodynamics. Thermodynamic quantities are treated in this approach as field intensities (there are 14 of them) and the special relativistic principles are applied on these fields to yield a formulation for relativistic thermodynamics. Relativistic thermodynamics becomes a standard subject in black hole thermodynamics; and for works in this connection, see S. A. Hayward, "Unified Law of Black-Hole Dynamics and Relativistic Thermodynamics," *Classical and Quantum Gravity* **15**: 3147 (1998).

The other two papers, "Gauge Invariance, Cauchy Problem, Indeterminism, and Symmetry Breaking," "Realism and Spacetime: Of Arguments Against Metaphysical Realism and Manifold Realism," are arguments made to deal with the "hole argument" in philosophy of spacetime. Despite its heavy reliance on the substantival conception of spacetime (in contrast to the relationist conception), general relativity is beset by an argument (first raised by Einstein himself) later known as "the hole argument," which appears to saddle the substantivalists with the dilemma of either giving up substantivalism or accepting a trivial form of indeterminism in physics. The two papers included here attempt to broaden the scope of the discussion by, first, to connect the hole argument discussion with that of gauge invariance and spontaneous symmetry breaking, and, secondly, to compare the hole argument with one of Putnam's argument against metaphysical realism. For more recent studies on the hole argument and the related issues concerning substantivalism vs. relationism in general relativity see the following selected bibliography (starting from the most recent).

1. J.D. Moulton, "Substantivalism, Haecceitism, and Representation," manuscript (2015)
2. M. Iftime and J. Stachel, "The Hole Argument for Covariant Theories," *General Relativity and Gravitation* **28**: 1241-1252 (2006).
3. L. Lusanna and M. Pauri, "Explaining Leibniz equivalence as difference of non-inertial appearances: Dis-solution of the Hole Argument and physical individuation of point-events" *Studies in History and Philosophy of Modern Physics* **37** (2006) 692-725.
4. D.P. Rickles, "A New Spin on the Hole Argument," *Studies in History and Philosophy of Modern Physics* **36** (2005) 415–434.
5. S. Saunders, "Indiscernibles, General Covariance, and Other Symmetries: The Case for Non-Reductive Relationalism," *in Revisiting*

the Foundations of Relativistic Physics: Festschrift in Honour of John Stachel, A. Ashtekar, D. Howard, J. Renn, S. Sarkar, and A. Shimony, (eds.), Kluwer (2003) 151-173.

Realism and Spacetime: Of Arguments against Metaphysical Realism and Manifold Realism[i]

In this essay I explore the relation between Putnam's model-theoretic argument against metaphysical realism and the hole argument against spacetime manifold realism. It is shown that the hole argument is really Putnam's argument restricted to spacetime predicates. This is because the gauge theorem is about inscrutability of reference, not about the indeterminacy of possible worlds. Thus, the gauge freedom in spacetime theories expresses a semantic fact rather than one about ontology; and the hole argument is really against the metaphysical determinacy of reference. Thus, the indeterminism crisis does not even arise. I then defend my construe of the hole argument by refuting a dualist view of the physical world and an incorrect version of the relationism-substantivalism debate.

I. Introduction

Metaphysical Realism is a position for those who believe that the world and its constituents exist independently of our knowledge, the truth of our theories is non-epistemic, and the terms in them genuinely refer. Over the years, Putnam has articulated a model-theoretic argument against this position, and the argument has become a highly controversial subject in philosophy (Plantinga, 1982; Lewis, 1984; Hansen, 1987; van Cleve, 1992; Anderson, 1993). Manifold Realism[ii] is held by those who believe that spacetime exists as a substance whose parts are spacetime points (or points and their neighborhoods)[iii], and terms in our spacetime theories genuinely refer to those points and their physical properties. Recently, Earman and Norton (1987) discovered that manifold realism was threatened by a dire consequence of radical indeterminism if one applies Einstein's 'hole argument' to the future domain of any physical system. The 'hole mess' (as is accurately referred to by Rynasiewicz) has since become a highly controversial subject in philosophy of spacetime (Butterfield, 1989; Maudlin, 1990; Mundy, 1992; Bartels, 1994; Rynasiewicz, 1994; and Healey 1993).

Rynasiewicz (1994) drew attention in a recent paper to the relation between the hole argument and the inscrutability of reference, which suggests that the issue is about semantics rather than ontology. Mundy (1992) also made a similar move when he argued that the hole argument shows nothing more than that a spacetime theory[iv] only determines its models up to an isomorphism. In this essay, I further explore this way of solving the hole-argument problem and its ramifications on other related issues by first argue that, technical difficulties aside, the hole argument is really Putnam's argument restricted to spacetime theories. The key to this move is to realize that the gauge theorem is about inscrutability of reference, not about the indeterminacy of possible worlds. Thus, the gauge freedom in spacetime theories expresses a semantic fact rather than one about ontology. And the hole argument is really against the metaphysical determinacy of reference. From this, I argue that the indeterminism crisis does not even arise. This conclusion is further supported by the well-known passive formulation of the gauge theorem. I then defend my construal of the hole argument by refuting a dualist view of the physical world (that it consists of spacetime and material processes as two distinct substances) and an incorrect version of the relationism-substantivalism debate. In the end, I give some thoughts on the ramifications of the modified hole argument.

II. Putnam's Argument

I do not pretend that the following discussion can do full justice to Putnam's model-theoretic argument against metaphysical realism. The aim is, however, to give it an accurate and useful account which sheds light on the analysis of the hole argument. What makes Putnam's argument powerful is its central theorem (the Putnam theorem hereafter) (Putnam, 1981, pp. 22-48, pp. 217-218; Putnam, 1983, pp. 1-25). Roughly put, the theorem claims that for any theory, a consistent set of sentences in some language, if it is satisfied by one relational structure, m, it is also satisfied by any other relational structure which is isomorphic to m. Therefore, the reference (or interpretation) of terms in a theory is radically underdetermined by what the theory says about the world. More precisely, it can be rendered as follows[v].

> *The Putnam Theorem* Suppose that we have a language, L, which consists of, inter alia, a set of n-place predicates, $P_1, \ldots, P_S$. For any theory,

> T, of L, if a first-order structure is a model of T under a (non-trivial) interpretation, **I**, there is always another first-order structure under another (non-trivial) interpretation, **K** (different from **I**), which is also a model of T that makes all statements in T true in exactly those possible worlds in which they are true under **I**.

The proof of this theorem centers on the availability of a permutation function which produces isomorphic models. Suppose that m = $\{D^{\alpha}; R^{\alpha}{}_{i} \mid i = 1, \ldots, s\}$ is a model of T in a possible world, W^{α}, under the interpretation, **I**, which has a domain, D^{α}, and an extension, $R^{\alpha}{}_{i}$, for each P_i in T. Suppose again that $R^{\alpha}{}_{i}$ is neither empty nor universal. Then, there always exists a permutation, $\hat{P}: D^{\alpha} \Rightarrow D^{\alpha}$, such that $\hat{P}*(R^{\alpha}{}_{i}) = R^{\alpha}{}_{j}$ (where there is at least one i, such that $i \neq j$ and $\hat{P}*$ is induced by $\hat{P}$), and m' = $\{D^{\alpha}; R^{\alpha}{}_{j} \mid j = 1, \ldots, s\}$ is different from m. Since $\hat{P}$ is a 1-1, onto mapping that, while exchanging individuals in D^{α}, exchanges $R^{\alpha}{}_{i}$'s accordingly, T is satisfied by m in W^{α} under a different interpretation, **K**, such that $\mathbf{K}(P_i)$ is $R^{\alpha}{}_{j}$ (or $\hat{P}*(R^{\alpha}{}_{i})$). Moreover, since W^{α} is chosen arbitrarily, m' satisfies T in any possible world just in case m does in it.

One may illustrate the theorem as follows (Putnam, 1981, pp. 32-38). Suppose that a theory, T, consists of two sentences: 'a cat is on a mat' and 'a cherry is on a tree'; and T is true in those possible worlds where a cat is on a mat and a cherry is on a tree. By permutating the extension of 'cat' with that of 'cherry' and the extension of 'mat' with that of 'tree', while leaving the extension of 'is on' unchanged, the new relational structure makes the two sentences true under a new interpretation of 'cat', 'cherry', 'mat', and 'tree' in exactly the same possible worlds as under the original interpretation. Therefore, it is inscrutable whether, say, 'cat' refers to cat or cherry in those possible worlds.

As a theorem in logic, the Putnam theorem's validity is hardly questionable; the question is how much philosophical ice it can cut. Putnam believes that, one can use it to do away with metaphysical realism (Putnam, 1979, p. 1981; Putnam, 1983). Putnam's argument can be summarized as follows.

Suppose that we have a theory, T.

1. Metaphysical realism is true.
2. The Putnam theorem is true, which means that no theoretical or

observational constraints can determine T's model beyond an isomorphism.

3. It follows from 1 that which term in T refers to what object or property in the world is uniquely and non-epistemically determined.
4. It follows from 2 that no constraints, either theoretical or observational, will allow one to pick out one reference relation (or model) among the isomorphic ones as the actual relation (or model)
5. From 3 and 4, a metaphysical realist is never in the position to pick out the actual reference relation as the intended one (e.g., whether 'cat' refers to cats or dogs).

The last statement is absurd, therefore, reductio ad absurdum, metaphysical realism is false.

III. The Hole Argument And The Radical Indeterminism

The Gauge Theorem (or the active General Covariance Principle (GCP)) and the Leibniz Equivalence Principle form the center piece of the hole argument against manifold realism (Earman/Norton, 1987, pp. 517-521):

> *The Gauge Theorem* For any model of a local spacetime theory, T — m = $<M, O_i \mid i = 1,\ldots, n>$ (M is the manifold of spacetime points and the O_i's are the tensors fields on those points) — there is always a C^∞ differentiable, 1-1, and onto mapping, h: M => M, such that it transforms m into a new relational structure, m' = $<M, h^*O_i \mid i = 1,\ldots, n>$ (where h^* is induced by h), so that m' is also a model of T.
>
> *The Leibniz Equivalence Principle* Diffeomorphic models are identical or represent the same state of affairs.

The proof of the gauge theorem (ibid., pp. 520) depends on the availability of a diffeomorphic mapping that produces carry-alongs that map one model to another while keeping all the theory specified structural properties invariant. By those properties I mean the ones that are determined by the equations in the theory. Since diffeomorphisms make those equations remain satisfied, so do the structural properties they determine remain invariant. More precisely, a model of a spacetime theory takes a collection, M, of spacetime points as its domain and various fields, O_i, as the properties of the points, which give extensions to various quantities in T. The diffeomorphic mapping, h, from M onto itself induces a 'carry-along', h*, which drag the fields (including the metric field) along with the points, such that the structure (on M) in m is preserved in m'. This diffeomorphic mapping can be a 'hole' diffeomorphism if it is an identity map for everything outside of, and a non-identity map for everything inside, a region (small or large) in M. Since diffeomorphic models are the same in their theory specified structure, they all satisfy T if one of them does.

As a theorem or principle in physics, the gauge theorem has stood the test of time. The question now is how much it implies metaphysically about the nature of the spacetime manifold. Earman and Norton believe that a radical indeterminism, clearly a dire consequence, follows from it if one wants to be a manifold realist (Earman/Norton, 1987, pp. 522-524; Earman, 1989, pp. 179-183). Suppose that our manifold, M, admits global time slices, and let's choose a Cauchy surface at a certain time, t_0. Because of the gauge theorem, for every model, m, there is a diffeomorphism, ϕ, such that $\phi^*m|_{t\leq t_0} = m|_{t\leq t_0}$, and $\phi^*m|_{t>t_0} \neq m|_{t>t_0}$, and if $m \in \mathfrak{M}_T$, then $m' = \phi^*m \in \mathfrak{M}_T$ ($\mathfrak{M}_T$ being the set of models for T). More explicitly, we have,

1. Manifold realism is true.
2. The gauge theorem is true, which means that T determines its models up to a diffeomorphism.
3. It follows from 1 that the Leibniz equivalence principle must be denied, so that two models differing in spacetime arrangement of events represent different possible states of affairs.
4. It follow from 2 and 3 that there are two possible states of affairs that are identical in their past and present, but differ in their future (the availability of ϕ).
5. Hence, from 4, determinism is violated, and it is violated in every possible model of any spacetime theory.

However, for whatever reason determinism may fail, it should not do so necessarily. Therefore, reductio ad absurdem, manifold realism is false. The radical indeterminism argument is a hole argument in that the entire future domain of a physical system is taken to be the hole.

IV. A Putnam-Type Argument Against Manifold Realism

Before modifying the hole argument, a few preliminary remarks are in order. First, the inscrutability of reference among isomorphic models is perhaps the weakest kind of inscrutability. According to the Löwenheim-Skolem theorems, a theory which has the expressive power of a decent scientific theory will be satisfiable by infinitely many non-isomorphic models of different cardinalities (Hansen, 1987, pp. 85-87). So, if metaphysical realism is true, the theoretical and observational constraints may lead one to pick out a model which has more (or less) individuals in its domain than in that of the actual one. Second, the model-theoretic characterization of the truth of a theory (or of whose statements) is different from a direct 'possible-world' characterization of models; the latter is clearly assumed in the indeterminism argument. A model of T in this model-theoretic sense gives one an interpretation of T that distinguishes those possible worlds in which every statement of T is true from the rest. It does not give one a particular possible world in which T is true. And, finally, how the absurd conclusion of Putnam's reductio argument should be construed is still a bit of controversy. For instance, Anderson (1993) states the conclusion thus.

> Our statements are semantically indeterminate. There is no fact of the matter concerning the reference of our terms. [The indeterminacy of reference.] (p. 313)

Then, Anderson puzzles over why Putnam takes this as a dire consequence. This construal seems wrong, since the indeterminacy of reference is the result of the Putnam theorem, not the absurd conclusion of the reductio argument. In fact, there is nothing wrong with the inscrutability of reference; what is absurd is that (because of this) metaphysical realist will never be in the position to pick out the actual reference relation. If the realist is right, we will have to accept reality as a transcendental realm which we have no way of latching onto.

I now argue, in this and the next section, that the hole argument is really a Putnam-type argument, which does not give rise to radical indeterminism. And then, I discuss why the hole argument was mistaken for being capable of producing radical indeterminism.

Suppose that we have a spacetime theory, T^{st}, which has terms mentioning and quantifying over points on M.

1. Manifold realism is true.
2. The gauge theorem is the Putnam theorem restricted to spacetime points and their physical properties.
3. It follows from 1 that terms in T^{st} refer to points and their properties in one of the diffeomorphic models and the reference is unique and a matter of fact.
4. It follows from 2 and 3 that realists are never in a position to determine which one among the diffeomorphic models is the actual one. However, 4 is absurd, ergo, manifold realism is false.

Now, the only thing that needs to be shown is the truth of (2).

> *The proof of 2* As a local spacetime theory, T^{st} typically consists of, among other things, some names, α, β, ..., and a set of variables, P_1, ..., P_s, which should be understood as referring to spacetime points and their properties; for instance, $P_i(\alpha)$ means that point α has property P_i; and for every x in region A, $f_i(x) = P_i$ means that every point in A has P_i. In other words, the quantifiers in a local spacetime theory only range over individual spacetime points.[vi] Suppose that T^{st} can be axiomatized and formulated independently of coordinate systems.[vii] It then, if consistent, can have first-order structures as its models. An interpretation of T^{st}, **I**, is given by mapping α, β, ... to specific points in M, and giving extensions, O_i, i = 1,..., s, to P_1, ..., P_s (tensors in T^{st}). $m = \langle M, O_i \mid i = 1, ..., s\rangle$ is a model of T^{st} under **I** if its statements are true in m. Now, according to the gauge theorem, there is a diffeomorphism, h: M => M, that maps m onto a new relational structure, $m' = \langle M, h{*}O_i \mid i = 1,..., s\rangle$, such that m' is a model of T^{st} under a different interpretation, **K**.
> The definition of diffeomorphic mappings from M onto itself shows that they are nothing but 'C^{∞} differentiable' permutations, clearly a subset of the permutations used in the Putnam theorem.

Hence, we do not need the radical indeterminism argument to undermine manifold realism, nor does the gauge theorem so understood leads to the indeterminism crisis. What is really at issue in the hole argument is the determinacy of a unique reference relation, not the coexistence of two possible states of affairs.

V. Radical Indeterminism Again

In this section I argue that the radical indeterminism does not even arise. It has been taken for granted that manifold realism entails the denial of the Leibniz equivalence principle, so that the gauge theorem is thought to yield the possibility of two states which violates determinism. That this seems true is due to a misleading understanding of the Leibniz principle. What that principle really says is that diffeomorphic models (as model-theoretic entities) make the theory they interpret true in the same states of affairs. Now, if the theory happens to be true in only one possible world, then the principle says that diffeomorphic models 'represent' the same possible world or state of affairs. So understood, the principle is entirely compatible with manifold realism. We can certainly construct the two diffeomorphic models of T^{st}, m and m', where $m' = \phi^*m$, and $m'|_{t\leq t_0} = m|_{t\leq t_0}$, and $m'|_{t>t_0} \neq m|_{t>t_0}$. However, since they are different interpretations of T^{st}, the 'indeterminism' is really the underdetermination or inscrutability of the interpretations, not that of different possible worlds. Thus, the availability of the diffeomorphism ϕ and the two models does not violate the Leibniz equivalence principle. In other words, claims 3 and 4 in section III (the radical indeterminism argument) do not hold.

We now see that in the original hole argument which supposedly engenders the radical indeterminism crisis, models as interpretations are confused with models as possible worlds or states of affairs. In the gauge theorem, models are model-theoretic entities which specify interpretations (or reference relations) between a theory and the items they refer; but that is not the way they are understood in the hole argument debate. There, they represent possible worlds (or states of affairs). In that case, one must consider the interpretation of a spacetime theory as already fixed, and the problem then is whether, under that interpretation (what interpretation?), all those possible worlds in which every statement of the theory is true are compatible so that determinism holds. In other words, whether there could be two worlds, w and w', where both share the same history up to certain time and then are radically different after that, and every statement in the theory (as interpreted) is true in both of them. This indeed violates determinism, but this is hardly anything derivable from the gauge theorem, which states no such thing as that a theory when fully interpreted can be true in two such possible worlds.

Furthermore, as an epistemological thesis (the Laplacian determinism), a deterministic theory is one in which, once fully interpreted, the equations determine from one complete set of data at one time to another set at another time. Whether the interpretation is determinate (a matter of semantics) cannot affect such a property of the theory. As a metaphysical thesis, whether or not the world is deterministic should have nothing to do with how the terms in a theory are related to objects and relations in the world. The diffeomorphic mappings which affect such relations but keep the truth-value distribution invariant in the theory clearly have no effect on what happens in the world, not to mention rendering an otherwise deterministic world indeterministic. This also explains why the radical indeterminism appears true a priori. It is because the gauge theorem expresses a semantic property, not a property of spacetime.

To sum up, if manifold realism is false, it is so not because it yields radical indeterminism, but rather because it is an inconsistent position in the light of the gauge theorem. A manifold realist, as a realist, must demand the determinacy of a unique intended model among the diffeomorphic ones, but that goes against the truth that a theory only determines its models up to a diffeomorphism. Unless some independent way of fixing a model can be argued as plausible, manifold realism is not defensible.

VI. The Passive General Covariance Principle

One might argue that I misconstrued the hole argument, which is not about the inscrutability of reference but rather about the indeterminacy of different states of affairs. In that sense manifold realist is indeed incompatible with Leibniz equivalence principle. To better understand this objection, let us see a well-known argument from Leibniz. Leibniz argues, for a relationist position, that God would have no reason,[viii] had absolute space and time existed, not to have created the universe, say, three miles to the east from where, or a year sooner than when, it was created (Alexander, 1956, pp. 26-27). This assumes that if empty space is a substance, then model A (where the world of matter is in one place) and model B (where it is three miles to the east of A) must represent different states of affairs. Hence, manifold realism does imply that the Leibniz equivalence principle is false and, so, which one of A and B obtains is indeterminate.

A rejoinder can be provided from several different angles. First, one may, and should, flatly refuse to take material processes as ontologically different from spacetime manifold. Modern spacetime theories are all formulated under a monist view of reality, viz. spacetime and the properties of its constituents. It seems completely out-of-date to follow Leibniz in speaking of 'placing' matter in empty space or of some bare points being 'occupied' by bodies. We would now rather speak of certain regions of space (or spacetime) having such-and-such properties. Under the monist view, the two models A and B above represent the same state satisfying the theory. More about this later.

Secondly, one may think that a body occupying a spatial point is similar to an ice cube occupying a cell of an ice tray. However, that is the wrong analogy. The latter refers to a relation between two objects, which is not the same as a body occupying a bare spacetime point. In principle, there can be no physical effect that distinguishes a body's coincidence with one bare point from its coincidence with another. For instance, a ball rolling on a pool table successively occupies different parts of the table at different times and results in a trajectory. But to map this trajectory to a different but diffeomorphic one, to 'move' it on the bare spacetime manifold, does not produce another trajectory. A dynamic mapping that carries the ball from one place to another is one among different states of affairs, whereas the diffeomorphic mapping we see in the gauge theorem is one among semantic relations; it is theoretically incapable of producing any physical effect. Hence, it is inappropriate to speak of diffeomorphic models as representing different states of affairs; and it would be better to abandon, except in everyday language, the talk of things or events occupying spacetime points. (One must note here that the metric of spacetime is also a dynamic field, so a coincidence with different portions of it may produce physical effect. Thus, diffeomorphisms do not apply to transformations on the metric.)

Thirdly, a different but equivalent formulation of the gauge theorem makes it almost impossible to interpret diffeomorphic mappings as anything other than among different reference relations. The gauge theorem actually is, or directly follows from, one of the formulations — the active one — of the General Covariance Principle (GCP) in the general theory of relativity. The other formulation — the passive one — is given in terms of coordinate transformations.

The passive GCP All equations of physical laws concerning spacetime are form-invariant under any smooth and differentiable coordinate transformations.

Suppose that we have a set of tensor fields, e.g. $(O_1, ..., O_n)$, which obeys certain physical laws. These fields are represented in a spacetime theory, T, in terms of coordinate (or reference) components. For instance, if $\{x^i \mid i = 0,1,2,3\}$ is a coordinate system, the fields are represented by a set of components, $(O_1|x^i, ..., O_n|x^i)$, and the laws by equations in the form of $S[O_1|x^i, ..., O_n|x^i] = 0$. With a choice of different coordinate systems, the fields will have different components, e.g. $(O_1|y^i, ..., O_n|y^i)$ for $\{y^i \mid i = 0,1,2,3\}$ (a different coordinate system). The passive GCP requires that $S[O_1|x^i, ..., O_n|x^i] = 0$ iff $S[O_1|y^i, ..., O_n|y^i] = 0$, provided that $(O_1|x^i, ..., O_n|x^i)$ and $(O_1|y^i, ..., O_n|y^i)$ are related by smooth and differentiable transformation functions. Since physical laws determine which process is physically possible and which not, satisfying the same equation(s) means that the solutions are physically indistinguishable; and $(O_1|x^i, ..., O_n|x^i)$ and $(O_1|y^i, ..., O_n|y^i)$ are simply two different ways of representing the same states of the tensor fields.

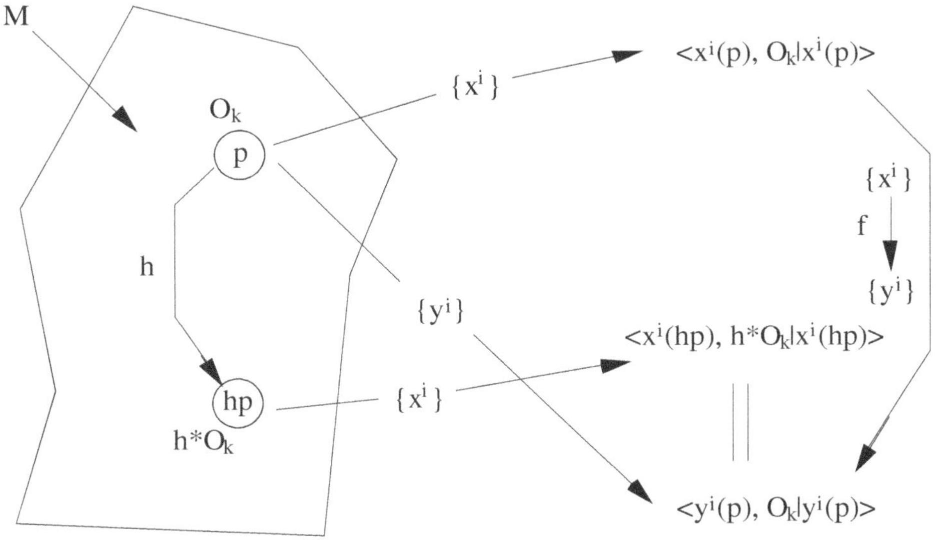

Figure 1 The equivalence of the passive and the active GCP: for any region, p, in M with tensor field O_k, a coordinate transformation, f, can always be matched by a diffeomorphic map, h, which produces the same effect.

It is easy to see that the two formulations are equivalent, and such equivalence further supports my claim that the inscrutability is about references and not about states.

Corresponding to any coordinate transformation, f: $\{x^i\}$ a $\{y^i\}$, which induces a f*: $O_k|x^i$ a $O_k|y^i$, there is a diffeomorphism, h: p a p', where p, p' $\in$ M (regions in M), such that $x^i(hp) = y^i(p)$ and $O_k|y^i\,(p) = h{*}O_k|x^i(hp)$ (see Fig. 1). This means that it is inscrutable whether, say, $O_k|x^i$ refers to O_k in region p or $h{*}O_k$ in region hp, etc. (the same, mutatis mutandis, is true for $O_k|y^i$). [ix] It is obvious that a coordinate transformation maps one reference relation between the coordinate components and the fields to another. It then seems inconceivable that an equivalent form of transformation would accomplish something totally different, namely, mapping one state of affairs to a different one; or that the passive principle is about a semantic property of a theory and the active principle about the ontology of reality. In this important equivalence of the two formulations lies the real nature of diffeomorphic mappings.

Finally, the conclusion that the gauge theorem, being derivable from the GCP, expresses a semantic rather than a physical fact is also supported by the following understanding of spacetime (or relativistic) theories. Even though it has seldom been realized, fundamental theories of physics carry their own metalanguage, or at least, some metalinguistic constraints. Principles such as the gauge theorem of spacetime are in fact such metalinguistic constraints. Einstein already understood this point in the early years of his special theory of relativity (Einstein, n.d., pp. 228-229). He called his theory a theory of principle rather than a constructive theory.

The difference and relation between these two types of theory are rather interesting, and spelling them out in detail would help us to see the semantic nature of the former. There is no pure theory of principle in physics. A relativistic (or spacetime) theory must be a theory of something more than relativity (or spacetime); thus, we have relativistic theory of mechanics or relativistic theory of electrodynamics, etc. For these theories of physics, the word 'relativistic' refers to the metalanguage part and the rest is the name for a theory in the object language about the phenomena (i.e. a constructive theory). For instance, a theory of mechanics assumes the existence of massive bodies whose movement obeys some mechanical laws (e.g. the conservation law of mechanical energy and momentum); but a relativistic theory of such massive bodies is different from a non-relativistic theory of the same bodies in that the laws obey a specific kind of transformation rules under which the truth of their solutions are preserved. The same is true for all physical theories of the form 'relativistic theory of ...', where the blank is filled out by the constructive part of the theory.

VII. Relationism vs. Substantivalism

The fact that diffeomorphic models represent the same state of affairs and that they are so because the spatial-temporal relations among events are invariant seems to suggest that relationism has won this round of the debate. However, such a conclusion is unwarranted. The issue is about whether the variables which range over spacetime points on M genuinely refer. Relationism in its original sense must be understood as denying such reference relations. A relationist must take anything pertaining to bare spacetime points as non-referring terms or expressions. The Putnam-type argument given above does not necessarily support such a position.

If relationism follows from the refutation of manifold realism by the Putnam-type argument, one might with similar reason say that logical positivism (the sense-data version) follows from Putnam's model-theoretic argument against metaphysical realism. One might argue that if metaphysical realism is false, our use of the 'thing-language' consists of non-referring expressions which are only convenient ways of talking about our sense impressions. It is, however, well known that logical positivism, at least this version of it, is not a necessary conclusion of Putnam's argument; it is not even the preferred one. Then, neither is relationism a necessary conclusion of the modified hole argument. In fact, I think that none of the reasons which revived the realist view of spacetime in the past decades has been refuted by the hole argument.[x]

At this juncture, one may raise another objection to my view by questioning whether there is still a place for the relationism-substantivalism debate in my theory: the traditional view of relationism and substantivalism does not seem to apply any more. On the one hand, a refutation of manifold realism (if the modified hole argument is sound) does not lead to the victory of relationism, whereas, one the other hand, it is not clear in what sense the bare spacetime manifold is a substance independent of the matter content it contains if diffeomorphic models represent the same state of affairs. A short response to this objection is: so what? A position that excludes the traditional debate between the two '-isms' is not necessarily a bad position. In fact, any version of this debate which crucially depends on a dualist view of material processes and spacetime should, in my view, be abandoned. Because, construed in that way, the debate is not affected in any way by the hole argument.[xi] The argument has nothing to say about whether or not literally placing the whole material world in two different regions of empty space should be counted as creating two different possible worlds. What it is really capable of refuting, if anything, is the manifold realism which takes a monist view of reality.

Our understanding of the nature of spacetime has undergone dramatic changes since the time of Leibniz and Newton, even though the philosophical controversy remained more or less the same. In the general framework of Newtonian mechanics (in which the original debate was understood), material bodies and empty space belonged to two radically different categories. It was in that context, what Leibniz said about God placing the whole material world into the space and time made sense. As the concept of physical fields became accepted in nineteenth century, the matter-space distinction became blurred. Today, it seems quite pointless to debate on whether the denizen of the world should include bare spacetime as a substance. It seems natural to consider that the physical reality consists of points on a manifold and their properties (which are represented by geometric objects such as tensors). For instance, instead of speaking of spacetime points being 'occupied' by, say, gravitational fields, we speak of points with vanishing or nonvanishing field strength.

In fact, since spacetime is likely to be completely filled with fields of some kind, there is no point of carrying on the relationism-substantivalism debate as an ontological debate, viz. whose ontology is the right one. Einstein once said:

> There is no such thing as an empty space, i.e., a space without field. Space-time does not claim existence on its own, but only as a structural quality of the field. (Einstein, 1961, p. 155.)

Despite of Einstein's favor of relationism, this remark can also be used for substantivalism. For this point, Friedman puts it very clearly:

> In fact, as far as we can tell, the real universe appears to be completely filled with electromagnetic background radiation. Hence, all space-time points turn out to be occupied. This means that the relationist's ontology is just as rich as the absolutist's: ... Hence, the traditional debate threatens to dissolve completely (Friedman, 1983, pp. 222).

Conclusive or not, these views show that it is no longer appropriate to take the debate as about ontological priorities. Since all field theories, like any theory in physics, are also only capable of determining their models up to a diffeomorphism, the best view seems to be that the denizen of reality are spacetime points, but the reference relations between terms in a spacetime theory and those points are inscrutable, which mathematically creates an abstract entity called the manifold, M. That M is structurally richer than the real spacetime and physical systems (including fields) can be 'moved' around (by diffeomorphisms) in M creates the illusion that there exits an entity that corresponds exactly to M. As I have argued so far, this is the wrong way of looking at this picture. The richness of the manifold expresses a semantic property of the spacetime theory, not some extra bit of reality.

VIII. Conclusion And Further Observations

By a way of concluding the essay, I offer the following tangential observations on the subject.

1. The relationism-substantivalism debate, when viewed as the material world versus the empty spacetime, is obsolete. From my perspective, the real debate is between a minimal realist (or substantivalist) and a full-blooded realist position on spacetime. The former holds that spacetime exists as a substance and is endowed with various physical properties, but does not make any commitment to the ways in which our spacetime theories are connected to that substance; whereas the latter does make that commitment. The hole argument is essentially an argument against any non-epistemic way of making that commitment. So, it has no bearing on the minimal position. Similarly, some philosophers correctly pointed out that Putnam's argument has no effect on a minimal position of realism (Hansen, 1987), which merely asserts the independent existence of the external world.

2. The discussion can hardly be considered complete without addressing the possibility of giving a realist rejoinder to the Putnam-type argument, which is along the same line as the realist rejoinder of Putnam's model-theoretic argument. The best and mostly defended rejoinder to Putnam's argument is what is generally known as the causal theory of reference. The theory may be roughly taken as saying that the reference relation, which is inscrutable under theoretical and observational constraints, is uniquely determined by a causal relation (whether or not we have epistemic access to it) between the speaker and the object to which her terms refer. The question now is whether manifold realists can make a similar rejoinder. Can one plausibly argue that we have causal access to one of the diffeomorphic models of a spacetime theory and, in that sense, render the reference relation fixed?

In the case of ordinary objects, such as cats and cherries, the causal theory is plausible enough, or so it seems to those who defend it. Our uttering the word 'cat' may be, under suitable circumstances, causally connected to cats or cherries and, thus, refer to cats or cherries as a matter of fact, even though either way, the sentence we utter, 'a cat is on a mat', is true. Can we say the same thing in the spacetime case? If we rule out the need for referring to bare spacetime points (assuming that reality is a plenum) I do not see why referring to points is any more problematic than referring to cats and cherries. After all, in a fundamental theory of physics, cats and cherries are nothing more than patches of points with properties of cathood and cherryhood.

To take a closer look at this point, let us examine the following comparison. In Putnam's example of cats and cherries, the permutation function, π, may be such that π(cat) = cherry (and π(cherry) = cat) and π(mat) = tree (and π(tree) = mat), which leaves sentences such as 'a cat is on a mat' true as in the unpermutated model. However, a π(cat) is physically different (distinguishable) from a cat. We do not hesitate to say that the term 'cat' referring to cats is a causally different relation, if it is indeed a causal relation, from 'cat' referring to cherries. So it makes a difference which one is the right interpretation. Can the same thing be said in the case of a spacetime model, $<M, O_i>$, and a diffeomorphism, ϕ, such that $\phi*(O_i) = O_i'$ and leaves all the solutions in the theory true? Could we plausibly argue that $\phi*(O_i)$ is physically distinguishable from O_i simply because they differ in bare spacetime points. A realist of spacetime theories will not see why the two can not be so distinguished. The fact that the two models, $<M, O_i>$ and $<M, O_i'>$, represent the same possible states of affairs by no means renders the two extensions of a tensor, i.e. O_i' and O_i, physically indistinguishable. In fact, if O_i is mapped to O_i' for a particular value of i (leaving the rest of the tensors untransformed), the resulting model may represent a different possible world.[xii] Therefore, it seems possible to defend manifold realism by a causal theory of reference.

3. Since the controversy over the hole argument has been based on the misunderstanding of the gauge theorem and the believe that manifold realism results in radical indeterminism, the available proposals for fixing the problem (on behalf of the realists) mostly missed the target. However, Maudlin's essentialist solution is different, even though it is taken to have resolved the indeterminism crisis which according to me, does not exist. In that solution, Maudlin (1990) proposes that we get rid of the unwanted diffeomorphic models by only recognizing the one with the right metric (the model in which points have their actual metric properties essentially). Even though I agree with Mundy that Maudlin, like the other participants in this controversy, mistook manifold realism to imply the denial of the Leibniz equivalence principle, his essentialist solution, when properly interpreted, may still provide us with a reference fixing apparatus. For instance, in the case of the two models, m and ϕ*m, where $\phi * m|_{t \le t_0} = m|_{t \le t_0}$, and $\phi * m|_{t > t_0} \neq m|_{t > t_0}$ (cf. p. 6), instead of saying, as Maudlin did, that only one of these are physically possible, we should say that only one of these give the right reference relation. In other words, we may say that 'cat' refers only to cats, not to cherries, because only the former gets the essential properties of cats right.

Of course, a rejoinder to this solution is already well known to philosophers who deal with Putnam's argument. Is not the determination of essential properties itself very much theory-laden? Like Putnam's attack on the causal theory of reference, one can say similar things about Maudlin's essentialist solution. Furthermore, among the objections that Maudlin himself anticipated and responded in his paper (Maudlin, 1990, p. 546), I would endorse the first objection which says that it is artificial, if not arbitrary, to put the metric field in the bag with spacetime points, and all the other fields in the bag with cats and dogs. Why can't we put all the fields in one bag and the rest in the other, and talk only about interactions between fields and bodies, which are all physical interactions and capable of producing different states of affairs? As Earman pointed out, 'it seems unfortunate to try to put metrical properties on the essential side of the essential/accidental cut in the context of GTR [general theory of relativity], where the space-time metric is a dynamic field (1989, p. 201).

4. As I have mentioned throughout this essay, Mundy's paper (1992) has much in common with the present essay. However, the conclusion I reached was different from Mundy's, namely, I concluded that the hole argument, when properly understood (as a restricted case of the Putnam argument), was a forceful argument against manifold realism, whereas, Mundy's conclusion was that 'neither the revived hole argument nor the original indiscernibility argument [Leibniz's] bear against manifold realism (1992, p. 519)'. For Mundy, what was real were the structural properties of spacetime models, and 'a possible world is simply an isomorphism-class of models (1992, p. 523)'. Furthermore, '[p]oints, like numbers, are structural roles in isomorphism-classes of models of certain theories, which roles may be played by different objects in different models or representations (ibid.)'. It appears that by this Mundy denied the uniqueness and the determinacy of a model among the isomorphic models (by whatever means). This seems to me to be more of a realist relationism (or structuralism) position which concedes only to the reality of spatial-temporal relations than a realist view of spacetime points.

Different philosophers draw different conclusions from the fact that the reference function of our language is inscrutable. Quine makes the most radical move by inferring from the inscrutability of reference to the relativity of ontology (Quine, 1969, pp. 27-68). According to Quine, the only acceptable notion of truth and reference is that within a model (an interpretation).

> It is thus meaningless within a theory to say which of the various possible models of our theory form is our real or intended model. Yet even here we can make sense still of there being many models. For we might be able to show that for each of the models, however unspecifiable, there is bound to be another which is a permutation or perhaps a diminution of the first (p. 54).

Mundy seems to have reached a similar conclusion with respect to terms concerning spacetime points on M. Since their reference to spacetime points is inscrutable, there is no intended model for them. If this is right, I do not see why this is a realist position on spacetime points.

References

Alexander, H. G., 1956, The Leibniz-Clarke Correspondence, Manchester.

Anderson, D. L., 1993, "What Is the Model-theoretic Argument?" Journal of Philosophy, 90, 311-322.

Bartels, A., 1994, "What is Spacetime, if not a Substance?" in: U. Majer and H-J. Schmitt, (Eds.), Semantical Aspects of Spacetime Theories, Mannheim, 41-51.

Butterfield, H., 1989, "The Hole Truth", British Journal for the Philosophy of Science, 40, 1-28.

Earman, J., 1989, World Enough and Space-Time, Cambridge, MA.

Earman, J. and J. Norton, 1987, "What Price Space-Time Substantivalism? The Hole Story", British Journal for the Philosophy of Science, 38, 515-525.

Einstein, A., 1961, Relativity: The Special and the General Theory, New York.

--------------. n.d., Ideas and Opinions, New York.

Friedman, M., 1983, Foundations of Space-Time Theories, Princeton, NJ.

Hansen, C., 1987, "Putnam's Indeterminacy Argument: The Skolemization of Absolutely Everything", Philosophical Studies, 51, 77-99.

Healey, R., 1993, "Substance, Modality and Spacetime", (manuscript.)

Lewis, D., 1984, "Putnam's Paradox", Australalasian Journal of Philosophy, 62, 221-236.

Maudlin, T., 1990, "Substances and Space-Time: What Aristotle Would Have Said to Einstein", Studies in History and Philosophy of Science, 21, 531-561.

Mundy, B., 1992, "Spacetime and Isomorphism", PSA 1992, vol. 1, 515-527.

Plantinga, A., 1982, "How to Be an Anti-Realist", Proceedings and Addresses of APA, 56, 47-77.

Putnam, H., 1979, Meaning and the Moral Sciences, London.

------------- 1981, Reason, Truth and History, Cambridge.

------------- 1983, Realism and Reason: Philosophical Papers, vol. 3, Cambridge.

Quine, W.V., 1969, Ontological Relativity and Other Essays, New York.

Rynasiewicz, R., (1994), "The Lessons of The Hole Argument", British Journal for the Philosophy of Science 45, 407-436.

van Cleve, J., 1992, "Semantic Supervenience and Referential Indeterminacy", Journal of Philosophy, 89, 334-361.

Wald, R., 1984, General Relativity, Chicago.

Footnotes

[i] A short version of this paper was read at the 1995 Eastern Division Meeting of the American Philosophical Society in Boston. Andreas Bartels, Rob Rynasiewicz and John Norton gave me helpful comments on an earlier draft. I am also grateful to Cory Juhl, Kirk Ludwig, Greg Ray and Niall Shanks with whom I discussed the paper either in person or through e-mail.

[ii] A more popular term for this is 'manifold substantivalism'. Here I adopt Mundy's term, manifold realism, for I think it is more accurate. One may believe in substantivalism without being a realist about spacetime and so avoid most of the arguments against it. I will discuss this point later in this essay.

[iii] Henceforth, the word 'point' means 'a point and its neighborhood'. This move avoids the problem of having to deal with dimensionless entities and is widely used in physics literature.

[iv] The term 'spacetime theory' is used in the literature interchangeably with 'relativistic theory'. Even though I think the latter is more accurate, I shall follow the convention in using the former. It refers to a class of theories currently used and/or investigated by physicists in general relativity.

[v] Here I adopt a simplified version of the theorem taken from the Appendix of Putnam 1981. I thereby avoid another version where Putnam uses the Löwenheim-Skolem (L-S) theorems. In the discussion of Putnam's argument, many considered the use of L-S theorems superfluous. After the work of, among others, Lewis (1984) and Hansen (1987), it is clear that Putnam's model-theoretic argument does not need the L-S theorems.

[vi] In common-sense language, we say that something is at some place at a certain time. In the language in which T^{st} is construed, we must say, instead, that a certain place and time (a point on the manifold) 'is occupied by event A', a one-place predicate in $\{P_i\}$ (where A is not a variable, but a description of the event in terms of the appropriate values of some variables). Or better still, one does not have to and should not use the phrase 'being occupied by such-and-such', but should just specify the various tensor fields which represent whatever is 'occupying' the point and give their values. One must note that our best formulation of a local spacetime theory is indeed so construed (Wald, 1984).

[vii] One of the objections to my view may be that it is not possible to axiomatize a spacetime theory. For a positive argument for the axiomatizability, see Mundy (1992). In that article, Mundy shows that it is possible to construct a coordinate-free spacetime theory which is axiomatizable. Even if Mundy's result proves to be controversial, one can still argue the case as follows. We have already proven that there exists diffeomorphic mappings that map models of a spacetime theory to their isomorphic counterparts which are different interpretations of the theory. Whether or not the theory is axiomatizable, what follows in this proof should hold. And that is enough to show that the gauge theorem is a restricted case of the Putnam theorem.

[viii] Here, Leibniz's principle of sufficient reason is obviously assumed.

[ix] For a more elaborate discussion of these two equivalent versions of GCP, see Friedman

1983, pp. 50-55. It is not clear, however, what Friedman takes to be the semantic relations, and in what sense Friedman uses the notion of models.

[x] According to Earman,

> The absolutist can point to three reasons for accepting a substratum of space-time points: the need to support the structures that define absolute motion, the need to support fields, and the need to ground the right/left distinction when parity conservation fails. (Earman, 1989, p. 173.)

None of these reasons is threatened by the inscrutability of diffeomorphic models of a spacetime theory.

[xi] On this point, I reached the same conclusion as Mundy's (1992, p. 519). However, I will later argue that the hole argument, if sound, does refute manifold realism in its proper sense.

[xii] For instance, suppose that we have a one-dimensional world in which there are three segments of a field with nonvanishing strength that are related as follows: A ---> B ---> C. A different model as the result of a mapping, ϕ, that reverses the arrows and maps C to A, A to B, and B to C represents the same possible world as before. But $\phi(A)$ is physically different from A, for were that the only change in the mapping, we would have a model: B ---> C, which would make the theory false in that world.

GAUGE INVARIANCE, CAUCHY PROBLEM, INDETERMINISM, AND SYMMETRY BREAKING

CHUANG LIU†‡

University of Florida

The concepts in the title refer to properties of physical theories (which are given, in this paper, a model-theoretic formulation and appropriate idealizations) and this paper investigates their nature and relations. The first three concepts, especially gauge invariance and indeterminism, have been widely discussed in connection to spacetime theories and the hole argument. Since the gauge invariance principle is at the crux of the issue, this paper aims at clarifying the nature of gauge invariance (either in general or as general covariance). I first explore the following chain of relations: gauge invariance ⇒ the conservation laws ⇒ the Cauchy problem ⇒ indeterminism. Then I discuss gauge invariance in light of our understanding of the above relations and the possibility of spontaneous symmetry breaking.

1. The Gauge Invariance Principles. Among the laws of physics is a category whose typical members are such items as *the general covariance principle* of general relativity and *the gauge invariance principles* of various matter fields. They all share a feature best described (first by Einstein) as non-constructive. These principles, unlike the usual physical laws, do not directly say anything about the properties of any specific physical system, rather they put constraints on them, so that lawlike statements won't qualify as physical laws unless they satisfy such constraints. In this sense, the principles appear to be 'meta-laws'. Moreover, since the constraints only concern the independence of physical processes from certain reference scales or systems relative to which they are measured or described (i.e. the gauge freedom of some sort), they are called 'gauge invariance principles'.

First, the general covariance principle in the general theory of relativity (Earman 1989, 47):[1]

[G] Given that a theory, T, is generally covariant, its solutions are only determinable up to a diffeomorphism or a coordinate transformation; namely, if $m = \langle M, g_{ab}, O^i_{ab}\rangle$ is a model of T, then for any diffeomorphism (a C^∞ differentiable, $1 - 1$, and onto mapping), $h: M =\rangle M$, $m' = \langle M, h{*}g_{ab}, h{*}O^i_{ab}\rangle$ is also a model of T.

Then, we have the gauge invariance principle in electrodynamics (in special relativity) (Landau & Lifshitz 1975; Wald 1984, 252–254):

[E] Given that a theory, T, is gauge invariant under S, its solutions are only determinable up to a gauge transformation in S; namely, if $m = \langle M, \eta_{ab}, A_a\rangle$

†I would like to thank John Norton for discussions on the nature of gauge principles. I am also indebted to Richard Healey and J. B. Kennedy for discussion on matters related to this paper.

‡Department of Philosophy, 330 Griffin-Floyd Hall, University of Florida, Gainesville, FL 32611.

[1] I here assume the usual tensor notion for the general theory of relativity: M is the spacetime manifold, g_{ab} the metric, and O^i_{ab} the ith arbitrary 2nd-rank tensor on M.

Philosophy of Science, 63 (Proceedings) pp. S71–S79. 0031-8248/96/63supp-0009$2.00

is a model of T, so is $m' = \langle M, \eta_{ab}, sA_a \rangle$, where $s \in S$ and $sA_a = A_a + \partial_a f(x)$ and $f(x)$ is an arbitrary function of x^a;

(where, $A_a = (\varphi, -A)$ is the 4 dimensional electromagnetic potential and η_{ab} the metric of a flat spacetime)

If one has qualms in believing that [G] and [E] are similar gauge invariance principles, one should note that g_{ab} is the potential tensor of gravitational fields and A_a that of electromagnetic fields. And despite their formal differences, [G] is remarkably similar to [E] when the field is weak, namely, when $g_{ab} = \eta_{ab} + d_{ab}$, where d_{ab} is a small deviation from η_{ab}, for any coordinate transformation, $x^a \Rightarrow x'^a = x^a + \varepsilon^a(x)$, we have,

$$d_{ab} \Rightarrow d'_{ab} = d_{ab} - \partial_a \varepsilon^b(x) - \partial_b \varepsilon^a(x) \tag{1}$$

which is the 2nd-rank analog of

$$A_a \Rightarrow A'_a = A_a + \partial_a f(x) \tag{2}$$

(Weinberg 1972, 252–5).

Moreover, the two principles, despite their different names, are principles of the same kind because of a remarkable theorem proven by Emmy Noether (Kastrup 1987). Noether proved two theorems concerning a general connection between the invariance of the action function under a group of transformations and the system's conservation laws, one of which concerns the invariance of the action function (characterized in the most general terms) under a gauge group (of infinite dimension) whose elements depend on some arbitrary smooth functions of x^a. Instantiating it in general relativity, the theorem says that [G] implies the famous (contracted) Bianchi identifies and the law of local energy-momentum conservation, while instantiating it in electrodynamics, it says that [E] implies the law of charge conservation. More on the details of this later.

Again, why are [G] and [E] meta-laws which differ fundamentally from, say, Newton's law of motion? As formulated above, [G] and [E] seem to say nothing more than a model closure under specific transformation groups.[2] But the law of motion is also a principle of model closure, namely, that the law gives a group of one parameter (i.e. time) transformations, $N(t): t \mapsto t'$ such that for any model, $m(t, p)$, $m'(t', p') = N(t' - t)m(t, p)$ is also a model. What is distinct about principles like [G] is that under their transformations the transformed models are not physically different. This is certainly not true between $m(t, p)$ and $m'(t', p')$. In the Earman-Norton original hole argument against substantivalism (Earman & Norton 1987), the above distinction is given by another principle—The Leibniz Equivalence Principle, of which the following is a natural generalization:

[LE] Two models that are related by a gauge transformation (including a diffeomorphism) represent the same state of affairs (or possible world).

I find this construal unnatural. It seems to suggest that we should treat all physical laws as model closures of various kinds and then specify whether two models represent the same state of affairs by an [LE] or anti-[LE]. I would prefer a *thick* interpretation so that whether or not different models represent different worlds is implied. But nothing is inconsistent about this construal.

2. The Cauchy Problem. Without much loss of rigor, we may formulate the Cauchy problem of a physical theory as the following question:

[2]More precisely, for a physical theory T and a group of transformations H, the set of all models M of T is closed under H if and only if for any $h \in H$ and any m, $h(m) \in M$ if $m \in M$.

[CP] Given a set of laws, does the state of a system on a hypersurface at a given moment uniquely determines its states on hypersurfaces of any later moments?

Here, the states of the system are determined by the values of the set of physical quantities appearing in the laws. When the answer is positive, we say there is a solution or the laws admit (or possess) an initial value formulation.

2.1 The Cauchy problem for electrodynamics. Classical electrodynamics is usually taken to consist of the Maxwell equations. However, there are two different formulations.[3] The more traditional formulation uses the electromagnetic field-strength functions, i.e. $\boldsymbol{E}$ and $\boldsymbol{B}$, and the well-known Maxwell equations. The other more profound and elegant formulation is the Lagrangian formulation, in which one begins with the Lagrangian density, L, of a field, ϕ, whose integral forms the 'action', $S(\phi) = \int L(\phi)dv$, where the integration is taken in some region on M. The least action principle says that for some one parameter family of the fields, ϕ_λ, which represent alternative field configurations with the same initial and boundary values, the variation of S with respect to ϕ_λ must be zero, i.e.,

$$\delta S = \int L(\phi_\lambda)\delta\phi_\lambda dv = 0. \tag{3}$$

In the case of electrodynamics, the field quantity is A_a, and $L = L_F + L_I$, i.e. the sum of the field Lagrangian and the interaction Lagrangian. By eq. (3), we have $\delta S = \int (L_F(A_a) + L_I(A_a))\delta A_a dv = 0$. Solving it gives us the other formulation of Maxwell equations as:

$$\partial^a(\partial_a A_b - \partial_b A_a) = -4\pi J_b, \tag{4}$$

where $J_a = (\rho, -\boldsymbol{J})$, the charge density ($\rho$) and current density ($\boldsymbol{J}$), and A_a is the potential which is functionally related to $\boldsymbol{E}$ and $\boldsymbol{B}$ and made indispensable by the Lagrangian formulation.

All models, $m = \langle M, \eta_{ab}, A_a\rangle$, should be uniquely determined at all times if it is determined at a given time, if the theory has a solution for the Cauchy problem. It looks as if eq. (4) is up to the job since it has four independent differential equations for the four unknown quantities A_a, $a = 0, 1, 2, 3$. This is however not true because all electromagnetic processes obey the law of charge conservation, which is expressed by the following equation:

$$\partial^b J_b = -\partial^b\partial^a(\partial_a A_b - \partial_b A_a) = 0, \tag{5}$$

which reduces the degrees of freedom to three; hence, the theory does not admit an initial value formulation. And since it is the lack of sufficient degrees of freedom that denies the theory its initial value formulation, no amount of strengthening the initial and/or boundary conditions can improve the situation.

To see a way out, one note that eq. (5), the law of charge conservation, is a direct consequence of [E]. Given [E] and that $L_I = -J_a A^a$, (the interaction Lagrangian), the action principle must hold under the gauge transformation: $A_a \rightarrow A_a + \partial_a f(x)$. So we have, $\delta A_a = \partial_a f(x)$, and,

$$0 = \delta \int L_I dv = -\int (J_a \partial^a f)dv = -\int \partial^a (f J_a)dv + \int f(\partial^a J_a)dv,$$

but $\int \partial^a (f J_a)dv = 0$ because its integrand is a 4-dimensional divergence; hence, $\int f(\partial^a J_a)dv = 0$ and $f(\partial^a J_a) = 0$. Since, $f(x)$ is arbitrary, we have $\partial^a J_a = 0$, which is just eq. (5). This is one application of Noether's theorem I mentioned earlier.

[3]They are equivalent in classical electrodynamics but not when quantum systems are involved (cf. the AB effect in Section 5.2).

Therefore, the apparent failure of eq. (4) to have a solution for the Cauchy problem is a consequence of the fact that, for any set of initial data on a hypersurface at a given time x_0, if A_a is a model at $x'_0(x'_0 > x_0)$, so is $A'_a = A_a + \partial_a f(x)$ at x'_0. Note that the traditional version of Maxwell equations (in terms of ***E*** and ***B***) does not suffer from this problem.

2.2 The Cauchy problem for general relativity. This case is remarkably similar to the above (Wald 1984, 244–267; Adler et al. 1965, 210–214; Hawking and Ellis 1973, 226–233).

Einstein's field equation is:

$$G_{ab} \equiv R_{ab} - (1/2)Rg_{ab} = 8\pi T_{ab} \tag{6}$$

which, because of $G_{ab} = G_{ba}$, gives 10 independent covariant differential equations for 10 components of the metric tensor, $g_{ba} = -g_{ba}$. However, because of the contracted Bianchi identity equation:

$$\nabla^a G_{ab} = 0 \tag{7}$$

(where ∇^a is the contravariant counterpart of ∂^a), there are only $10 - 4 = 6$ independent equations in eq. (6). Hence, similar to the previous case, the general theory of relativity also does not admit an initial value formulation.

Also similar to the pervious case, the Bianchi identity equation is implied by the general covariance principle, [G], in the Lagrangian formulation of general relativity. Without going through the derivation we note that by applying the variation generated by an arbitrary diffeomorphism, h, such as $\delta g_{ab} = h{*}g_{ab} - g_{ab}$, to the action in terms of R_{ab}, we can derive the Bianchi identity equation, eq. (7). we can also derive the local conservation law of energy-momentum from the action in terms of T_{ab}:

$$\nabla^a T_{ab} = 0. \tag{8}$$

To quote Wald at the end of his derivation of these relations:

> Thus, for a diffeomorphism invariant action, T_{ab} always is conserved by virtue of the matter field equation. . . . Note also that by applying the above argument to S_G [the action in terms of R_{ab}], it follows that . . . we have
>
> $$\nabla^a G_{ab} = 0$$
>
> Thus, in the Lagrangian formulation of general relativity, the contracted Bianchi identity may be viewed as a consequence of the invariance of the Hilbert action under diffeomorphism. (Wald 1984, 456)

This is another application of Noether's theorem I mentioned earlier.

Therefore, the apparent failure for general relativity to have a solution for its Cauchy problem is a consequence of the fact that, for any set of initial data on a hypersurface at a given time x_0, if $\langle M, g_{ab}\rangle$ is a model at $x'_0(x'_0 > x_0)$, so is $\langle M, h{*}g_{ab}\rangle$ at x'_0, where h is an arbitrary diffeomorphism.

One must note that unlike the case in electrodynamics, there does not seem to be another set of tensors that can be used to represent the gravitational fields. Hence, even though one could avoid the apparent lack of solution of the Cauchy problem by not using the electromagnetic potentials, one could not do the same in general relativity.

Here we see (though without complete details) how the lack of initial value formulation is a consequence of gauge invariance and, as I mentioned earlier, Noether's general theorem provides the crucial link.

2.3 The solution. For physicists, the problem is easily solved. In Wald's words, 'this difficulty is not of physical character' (Wald 1984, 253). Two electromagnetic potentials related by a gauge transformation 'represent the same physical electromagnetic field' (ibid.); and two spacetime models related by an arbitrary coordinate transformation or diffeomorphism 'describe the same geometrico-physical situation' (Adler et al. 1965, 214; see also Wald 1984, 259–260, Hawking and Ellis 1973, 227–228). Moreover, since all such models are physically equivalent, one may choose any one of them to describe the physical situation. To do that, one chooses a specific (e.g. the most convenient) gauge or coordinate convention. This is exactly what the physicists did: for A_a, they choose the Lorentz gauge and for g_{ab}, the harmonic coordinates.

Two observations seem to be in order. First, the physicists no doubt interpret the two gauge principles not as merely stating a model closure of their corresponding theories, but rather as saying that those models are physically equivalent. In other words, [LE], or at least a version of it, is thought to be implied by [G] and [E].

Second, the fact that choosing a convenient gauge is sufficient for a solution of the Cauchy problem seems to suggest that even though it looks as if each theory *underdescribes* the physically possible worlds, so that different models may evolve from the same initial data, it is in fact not so because the mathematical modeling of the electromagnetic field or spacetime *overdescribes* the substance. The underdescription can be corrected by eliminating the overdescription if they match, which is achieved by choosing a gauge or coordinate convention. For instance, the law in terms of A_a underdescribes the physics of electromagnetism to the same degree that the space of A_a overdescribes it. But if one chooses the Lorentz gauge, $\partial^a A_a = 0$, the overdescription is reduced, and so the law determines its solutions uniquely.

3. The Cauchy Problem and Determinism. Whether a theory is deterministic is closely related to whether it has a solution for the Cauchy problem (Earman 1986, 179–183). In the present context the concept of Laplacian determinism seems most appropriate. Following Earman (but weaker than what he has), we have:

[D^1] For the set of all physically possible worlds, W, the world $w \in W$ is Laplacian deterministic (L-deterministic) if and only if for any $w' \in W$, if w and w' agree at any time, they agree for all later times.

Translating [D^1] into a definition for deterministic theories, we have:

[D^2] Given [D^1], a theory is L-deterministic if and only if it determines a set of possible worlds in which no two worlds can agree at any time without agreeing at all later times.

Given [CP] and [D^1], a theory which is indeterministic will never possess an initial value formulation. But the reverse is not necessarily true since the initial value formulation may fail, for instance, in cases where hypersurfaces do not exist. Hence, we have IVF $\rightarrow$ LD, but not always $\sim$IVF $\rightarrow$ $\sim$LD. Since general relativity has an IVF, it is L-deterministic.

4. The Hole Argument Revisited. The connection between the discussion in the previous section and the recent debate over the hole argument is now obvious (Earman and Norton 1987, Earman 1989, Norton 1992). I take manifold substantivalism to be a doctrine of *spacetime realism,* which consists of two theses:

(a) spacetime exists independently of its matter content; (b) the general theory of relativity, when interpreted literally, describes spacetime. A realist must deny [LE] in general relativity. One can always find a hole diffeomorphism which transforms a model of T to a distinct diffeomorph of it such that they agree at a given time but differ after it. By [G], such a diffeomorph is also a model of T but according to the realists, the model and its diffeomorph represent different worlds. Hence T is indeterministic and so a priori since it is arbitrarily chosen. The conclusion is absurd; ergo, spacetime realism is false.

The challenge has always been to come up with a defense of spacetime realism that avoids the dire consequence of a priori indeterminism, and the focus on whether, or in what sense, a realist has to hold thesis (b) or deny [LE] (Butterfield 1989, Maudlin 1990, Rynasiewicz 1994, Bartels 1994, Brighouse 1994, Liu 1996).

However, in light of our discussion in Sections 2 & 3, a priori indeterminism is no longer a problem. But does this mean that none of the physicists who offer the solution for the Cauchy problem is a spacetime realist? From what they did to resolve the difficulty, it is clear that they think the manifold overdescribes the real spacetime, otherwise, one cannot arbitrarily choose a convenient gauge to recover the solution (and hence determinism). But it is not impossible, though unlikely, that they take distinct diffeomorphs to be physically equivalent but none the less to represent different states of affairs, different in some non-physical and yet real properties of spacetime. Note that both Wald and Adler et al. explicitly stated that the mathematical formalism of general relativity contains redundancy in describing the 'geometry', or 'geometrico-physical' properties, of spacetime. The problem remains.

5. The Nature of Gauge Invariance and Its Breaking.

5.1 Is it a physical principle? John Norton (1989, 1992, 1993, 1994), since his contribution to the hole argument, has exerted considerable effort in addressing this very question. His conclusion is that gauge principles (together with [LE]) are physical principles rather than purely mathematical ones, i.e. that they make factual statements about the physical world, which gives them empirical content. Rynasiewicz (1994) argued for the opposite view, namely, the difference among diffeomorphs is not physical but semantic, namely, the gauge freedom that general relativity enjoys is really a kind of inscrutability of reference. Liu (1995) further argued that the hole argument is a special case of the Putnam model-theoretic argument against metaphysical realism.

Reading Norton's work closely, one can see that in fact these two views, despite their surface differences, have a great deal in common. According to Norton (1992, 283–284), [G] says that the mathematical representation of spacetime distinguishes distinct diffeomorphs as different models, and [LE] that they represent the same spacetime, so the two together state the fact that those diffeomorphs have distinct properties but such properties have no physical significance or that only invariants under arbitrary diffeomorphisms are physically real.

Let us look at this more closely. We now have at least three conceptually distinct possibilities: the difference of the diffeomorphs is (i) purely mathematical, or (ii) purely spatio-temporal, or (iii) physical. If (i), it means roughly the following: Mathematical quantities and equations constitute the language by which we refer and represent the physical world. If the difference is purely mathematical, it must be a difference in the representation rather than one in the represented, in which case it would be one of (ii) or (iii).

However, even if (i) is true, it does not follow that the principle is not stating a fact of the world. And Norton seems to be saying exactly this: that the principle *does* state such a fact, namely, it is a fact of the physical world and stated by [G] that diffeomorphs have distinct properties (of reference relations) but such properties are not about properties of physical objects. If this is right, I do not see why Norton's view has to differ from the view of Rynasiewicz and Liu. There is no reason to believe that the inscrutability of reference is not stating a fact of the world even though its content concerns the word-world relation.

Norton might not be happy with such a construal of his view. It might be better construed as saying that non-identical diffeomorphs, though distinct in sense (ii), is nonetheless equivalent in sense (iii). This then implies that all *possible operations,* such as shifting, rotating and boosting a physical system in the spacetime manifold (as literally given by the mathematical object with the same name) produces no physical effect. This makes [G] not merely a statement of a fact but also one *about* some physical properties of systems.

This reading is, however, obviously too strong for [G]. It applies to relativity principles such as the principle of special relativity which gives us the equivalence of inertial reference frames. It is now well known that general covariance is a gauge invariance principle not a symmetry principle to which relativity principles belong (Earman 1989, Norton 1992). The proper reading of general covariance is that it guarantees that a theory gives 'an independent characterization of intrinsic reality', and 'a violation of general covariance involves a nonintrinsic reality' (Earman 1989, 47); it does not concern the symmetries of a physical system in spacetime.

Thus, gauge invariance principles are about representations while symmetry principles of spacetime are not. It is in fact questionable if it makes sense today to distinguish sense (ii) and (iii) in light of general relativity. Are there pure properties of spacetime that have no physical capacities (as there are of Newtonian or Minkowskian spacetime)?

5.2 Can it be broken? What may, and sometimes does, happen is that the apparent overdescription of some physical phenomena turns out to be no fluff at all but the description of some hitherto undiscovered physical properties. When and if that happens, new dynamical conditions must be introduced in respect of the Cauchy problem. Does this signify a breaking of the gauge invariance?

To answer this question we need to distinguish again a symmetry from a gauge invariance.[4] A symmetry is any physically possible operation which preserves certain physical structure, while a gauge invariance means the independence of a set of models from the unit or scale of measurement or representation. A symmetry can be broken by one of two ways: either when external influence is introduced or when some hitherto undiscovered cause is discovered (i.e., the spontaneous symmetry breaking). It is not however clear to me if it makes sense to speak of the breaking of a gauge invariance. If it breaks, the physical system the theory describes has no intrinsic nature, or its nature depends on how it is represented or measured.

It seems much better to reserve the notion of gauge invariance for those genuine cases of semantic redundancy of representation, and to take each breaking of a gauge as indicating a breaking of a symmetry, namely, that the breaking signifies that the invariance previously thought of as a gauge invariance is in fact a symmetry which harbors a degeneracy that is now revealed (spontaneously or not).

One of the best examples of the breaking of a gauge invariance (which is in fact a spontaneous symmetry breaking) is the breaking of [E] in what is now known

[4]In the standard literature of mathematical physics, gauge invariance groups form a proper subset of the set of all symmetry groups. Here, the notion of symmetry is a much more restricted one.

as the *Aharonov-Bohm effect* (the AB effect) (Aharonov and Bohm 1959, Peshkin and Tonomura 1989, Liu 1994). Recent experiments in verifying the magnetic AB effect made it no longer possible to doubt the existence of the effect.

Before the AB effect was predicted in 1959 (and even long after that, since its existence was debated until recently), the law that governs the interaction between quantum objects and electromagnetic fields are said to be gauge invariant under the following two gauge transformations: the e-m gauge: $A_a \Rightarrow A_a + \partial_a f(x)$ and the wave-function gauge: $\psi \Rightarrow \psi e^{ief(x)}$, where the first equation is simply eq. (2), and the second a transformation by adding an arbitrary gauge factor to the wave-function of the quantum object in question (and $c = \hbar = 1$). This gauge invariance principle applies to the nonrelativistic (the Schrödinger) equation as well as the relativistic (the Klein-Gordon) equation. One of the consequences of this gauge invariance is that only the field strengths (i.e., ***E*** and ***B***) of electromagnetism represent fields that affect quantum processes (Schiff 1968, 177, 398ff; Weinberg 1972, 111–113; Berestetskii et al. 1971, 98–99). This is no longer true because of the AB effect (for a discussion of the AB effect, see Liu 1994).

The AB effect shows that the joint gauge invariance of the potential gauge and the wave-function gauge transformations no longer holds in general. This means that the 'gauge invariance' given in the classical theory of electrodynamics is in fact no gauge invariance at all, but a degeneracy (or symmetry) whose fine structure is only revealed when the field interacts with quantum objects (as extended wave packets). The question now is whether this could also happen to [G]? I have shown its possibility, but how [G] might break is no longer a question for philosophers of science.

REFERENCES

Adler, R., Bazin, M. and Schiffer, M. (1965), *Introduction to General Relativity*. New York: McGraw-Hill.

Aharonov, Y. and Bohm, D. (1959), "Significance of Electromagnetic Potentials in the Quantum Theory", *Physical Review* 115: 485–491.

Bartels, A. (1994), "What is Spacetime, if not a Substance?" in U. Majer and H-J. Schmitt (eds.), *Semantical Aspects of Spacetime Theories*. Mannheim: BI-Wissenschaftsverlag, pp. 41–51.

Berestetskii, V. B., Lifshitz, E. M. and Pitaevskii, L. P. (1971), *Relativistic Quantum Theory*. Part I, London: Pergamon Press.

Brighouse, C. (1994), *Spacetime and Holes*. in D. Hull, M. Forbes and R. M. Burian (eds.), *PSA 1994*, Vol. 1, East Lansing, MI: Philosophy of Science Association, pp. 117–125.

Butterfield, J. (1989), "The Hole Truth", *British Journal for the Philosophy of Science* 40: 1–28.

Earman, J. (1986), *A Primer on Determinism*. Dordrecht: D. Reidel.

———. (1989), *World Enough and Space-Time*. Cambridge, MA: MIT Press.

Earman, J. and Norton, J. (1987), "What Price Space-Time Substantivalism? The Hole Story", *British Journal for the Philosophy of Science* 38: 515–525.

Kastrup, H. A. (1987), "The Contributions of Emmy Noether, Felix Klein and Sophus Lie to the Modern Concept of Symmetries in Physical Systems", in M. G. Doncel et al. (eds.), *Symmetries in Physics (1600–1980)*. Servei de Publicacions, UAB.

Hawking, S. W. and Ellis, G. F. R. (1973), *The Large Scale Structure of Space-Time*. Cambridge: Cambridge University Press.

Landau & Lifshitz, (1975) *The Classical Theory of Fields*. (Translated from the 6th rev. ed.) London: Pergamon Press.

Liu, C. (1994), "The Aharonov-Bohm Effect and the Reality of Wave Packets", *British Journal for the Philosophy of Science* 45: 977–1000.

———. (1996), "Realism and Spacetime: Of Arguments against Metaphysical Realism and Manifold Realism", (forthcoming in *Philosophia Naturalis* 33).

Maudlin, T. (1990), "Substances and Space-Time: What Aristotle Would Have Said to Einstein", *Studies in History and Philosophy of Science* 21: 531–561.

Norton, J. (1989), "Coordinates and Covariance: Einstein's View of Space-Time and the Modern View", *Foundation of Physics* 19: 1215–1263.

———. (1992), "The Physical Content of General Covariance", in J. Eisenstaedt & A. J. Kox (eds.), *Studies in the History of General Relativity*. Boston: Birkhäuser, pp. 281–315.

———. (1993), "General Covariance and the Foundations of General Relativity", *Reports on Progress in Physics* 56: 791–858.

———. (1994), "Why Geometry is not Conventional: The Verdict of Covariance Principles", in U. Majer and H-J. Schmitt (eds.), *Semantical Aspects of Spacetime Theories*. Mannheim: BI-Wissenschaftsverlag, pp. 159–167.

Peshkin, M. and Tonomura, A. (eds.), (1989), *The Aharonov-Bohm Effect*. New York: Springer-Verlag.

Rynasiewicz, R. (1994), "The Lessons of The Hole Arguments", *British Journal for the Philosophy of Science* 45: 407–436.

Schiff, L. (1968), *Quantum Mechanics*. 3rd ed., New York: McGrill-Hill.

Wald, R. (1984), *General Relativity*. Chicago: Chicago University Press.

Weinberg, S. (1972), *Gravitation and Cosmology*. New York: John Wiley & Sons.

0039-3681(94)00048-4

Is There a Relativistic Thermodynamics? A Case Study of the Meaning of Special Relativity

*Chuang Liu**

1. Introduction

The obscurity of the topic of this essay—Relativistic Thermodynamics (RTD)—calls for some explanation as to why anyone (except the experts) should need to know about it, especially about its history. First, it is precisely its history that is interesting and important, for it is full of twists and turns and its foundations run deep, touching the very meaning of the special principle of relativity. Thus, secondly, the history serves as a case study of the meaning of Einstein's special relativity, which has long become a permanent part of the landscape of modern physics but whose profound meaning still sometimes eludes us.

Let it be noted at the beginning that it is not my intention to provide any new solution to the problem of RTD, for which theoretical physicists have made and are still making substantial contributions (for the most recent works, see note 10). Rather, I aim to single out the key moves made in the history of RTD and to critically assess their moral (Sections 2 and 3). In 1952, Einstein suggested an interesting solution which was later rediscovered and articulated independently by Anderson. Section 4 is devoted to the examination of Anderson's theory. I do not think Anderson solved the problem, but his approach to it revealed the real difficulty that especially beset the heated debate about RTD in the 1960s and 70s. In other words, it revealed a deeper meaning of special relativity without which one could not even see what the problem was with RTD. I shall round out the essay (Section 5) with a brief account of some of the promising directions along which RTD is presently pursued, and of why they are so pursued.

*Department of Philosophy, 330 Griffin-Floyd Hall, University of Florida, Gainesville, FL 32611, U.S.A.

Received 13 *April* 1993; *in final form* 25 *April* 1994.

Pergamon

Stud. Hist. Phil. Sci., Vol. 25, No. 6, pp. 983–1004, 1994.

0039–3681/94 $7.00 + 0.00

2. Planck, Einstein, Pauli and the Early History of RTD

The reception of Einstein's special theory of relativity (STR) in the early years of this century was marked by two simultaneous processes, one for testing the theory with experiments (e.g., Kaufmann's β-ray experiments which initially seemed to have refuted STR (cf. Einstein, 1907, 436–9)), and the other for applying the principles of the theory to more physical phenomena. Max Planck was one of the first eminent physicists to have embraced and further developed Einstein's STR from the very beginning.[1] When STR was threatened by the negative results of Kaufmann's β-ray experiments, Planck's attitude was that one should have further attempts at 'finding out all the consequences it [STR] has' (Planck, 1906, 137). For 'if it is not correct, it should lead to absurdity' (*ibid.*). Thus Planck proceeded: RTD was one of the consequences Planck derived from STR—and it bore a simple and elegant form.

According to Einstein, STR is a 'theory of principle', namely, it consists mainly of principles that laws in all disciplines of physics must obey. That simply meant for Planck and others that laws of physics must be form invariant under the Lorentz-Transformations (L-Ts) for the physical magnitudes involved.[2] Planck's contribution to RTD was the result of his passion for principles of the greatest generality and his involvement with the work of blackbody radiation. Planck was then dissatisfied with the lack of an intrinsic connection between different branches of theoretical physics and could not see a way of unifying them with the only available general principle—the principle of the least action. Planck saw Einstein's STR as providing a second principle of equal generality which, together with the principle of least action, could unify physics. RTD was part of Planck's ambitious project for an 'Allgemeinen Dynamik', a theory in which laws for all physical phenomena, be they mechanical, electrodynamic or thermodynamic, are derivable from the two principles (cf. Planck, 1907, 564). At the same time, Planck supervised a dissertation by one of his students, Kurt von Mosengeil, in which the energy density and the pressure of the radiation inside a moving cavity were obtained. With the help of von Mosengeil's results and the special principle of relativity, Planck was able to derive a set of L-Ts for the thermodynamic magnitudes (see the next paragraph for the derivation and explanation).

The results are as follows:

$$dQ = dQ^0 \, (1 - u^2)^{1/2}, \tag{1}$$

$$T = T^0 \, (1 - u^2)^{1/2}, \tag{2}$$

[1]Among Planck's contributions to the reception of STR, Planck (1906) and (1907) are most important. It was in the latter, 'Zur Dynamik der bewegter Systeme', that Planck formulated his theory of RTD. For a detailed study of Planck's contributions to STR, see Liu (1991), ch. 2; also see Goldberg (1976).

[2]This view has been so natural and prevalent, one may find it in virtually every exposition of STR, especially works before, say, 1950. A relativistic theory is represented by the set of L-Ts for its main physical magnitudes.

$$S = S^0, \tag{3}$$

$$p = p^0, \tag{4}$$

where we assume there are two inertial reference frames, K and K^0, moving relative to each other with a speed, u; the thermodynamic magnitudes on the lhs are those measured in K, and on the rhs those in K^0; and dQ is the transferred heat, T temperature, S entropy and p pressure. (I shall also assume, throughout this paper, that $c = \hbar = 1$.) It is important to note, among other things, that the invariance of entropy, S, is supposedly proven by Planck with a simple argument.[3] From eqn (2) we see that the temperature of a body while moving appears cooler than when it is at rest.

Planck's derivation of these L-Ts was idiosyncratic and, thus, thoroughly revised by Einstein and von Laue (cf. Liu, 1991, ch. 1; Einstein, 1907; von Laue, 1911). Einstein's main contributions were: (i) he gave Planck's L-Ts a completely different derivation; (ii) he modified an equation given by von Mosengeil (cf. Mosengeil, 1907) and used it to show how the temperature of a moving body with respect to a stationary observer could be calculated from measurable quantities in the observer's frame [see eqn (9) below].[4] After Einstein, RTD was considered a natural extension of the relativistic mechanics of continua, and was given a four dimensional formulation by von Laue (1911). It was W. Pauli who put everything together and finalized the formulation in his 1921 encyclopedia article on the theory of relativity (Pauli, 1921). I shall henceforth call this the *Planck–Einstein formulation* of RTD; let us see how it is derived (cf. Pauli, *ibid.*, 134–5).

Consider the two major laws of thermodynamics:

$$dU = dQ - dW \tag{5}$$

and

$$dS \geq dQ/T, \tag{6}$$

[3]Planck gave two arguments for the invariance of entropy (cf. Planck, 1907, 522). The second of the two, which depends on nothing beyond the scope of pure thermodynamics, can be paraphrased as follows. Suppose we have a system which in its initial state has S^0_1 in K^0 (which is at rest with respect to the system). Then we accelerate it reversibly and adiabatically to a second state where its entropy becomes S^0_2 and it is at rest with respect to K, while K^0 is moving relative to K with a certain speed. Also if we look at the same process from the point of view of observers in K, we will have the system having S_1 in its initial state, and S_2 in its final state. Now, from the reversibility and adiabaticity of the process, we know that (i) $S^0_1 = S^0_2$ and (ii) $S_1 = S_2$; namely, no matter from which observer's point of view, the entropy values of the two states are equal. If $S_1 \neq S^0_1$, say, $S_1 > S^0_1$; this means that, with respect to observers in K^0, the 'moving' entropy is greater than the 'rest' entropy. Then, by the same token, we must have $S^0_2 > S_2$, with respect to observers in K, since the system is at rest in K in state 2 (the former is the 'moving' entropy and the latter the 'rest' one from K's point of view). From (i) we conclude that $S_1 > S_2$, which contradicts (ii) in the premise. The same contradictions will arise if $S_1 < S^0_1$. Therefore, it must be true in general that $S^0 = S$, namely, the entropy of a body is an invariant. Two assumptions are used in this argument: one is the equivalence of all inertial reference frames and the other is that reversible and adiabatic processes do not change a system's entropy.

[4]For a discussion of Einstein's derivation, see Liu (1992), 191–4.

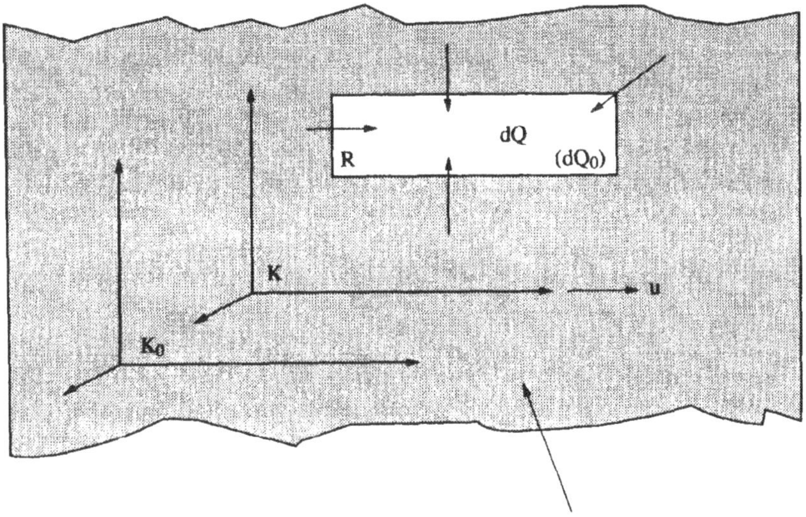

Fig. 1. Heat exchange with respect to different inertial reference frames, K and K_0. For the Planck–Einstein L-Ts, the environment is always at rest with the frame, but for the Einstein–Ott L-Ts, it is always at rest with R.

where dU is the change of internal energy of a system R, dQ the heat that R absorbs, and dW mechanical work that R produces on its environment (also suppose, for convenience, that the system consists of homogeneous fluid with volume V and constant pressure p). Since eqns (5) and (6) express fundamental laws in thermodynamics, they must be form invariant in all inertial reference frames (see above).

For the first law, when considered in the rest frame, K^0, the mechanical work is,

$$dW^0 = p^0 dV^0, \tag{7}$$

but in any other reference frame ($u \neq 0$), the work is (cf. Fig. 1),

$$dW = p dV - (\mathbf{u} \cdot d\mathbf{G}), \tag{8}$$

where the first term, pdV, is called the *compressional work*, which is a measure of the mechanical work due to a compression/expansion process, and the second term, $(\mathbf{u} \cdot d\mathbf{G})$, is the *translational work*, which is a compensation for the change of momentum, $d\mathbf{G}$, due to Rs radiation (or absorption) of heat energy. Since $(\mathbf{u} \cdot d\mathbf{G}) = 0$ when $u = 0$, the first law is form invariant under this modification. Also note that the inclusion of the translational work is due to the relativistic energy–mass relation, $E = mc^2$. In Newtonian physics, radiation does not carry inertial mass. Since the entropy is proven to be invariant (as Planck thought he had, see note 3), the form invariance of the second law in equilibrium yields,

$$T = (\mathrm{d}Q/\mathrm{d}Q^0)T^0. \tag{9}$$

This is the equation that, according to Einstein, one can use to obtain (indirectly) the values of the temperature of a moving body.

By 1907, Einstein had obtained the following L-Ts from electrodynamics of continua:

$$\mathrm{d}V = \mathrm{d}V^0/\gamma$$
$$p = p^0$$
$$\mathrm{d}U + p\mathrm{d}V = \gamma(\mathrm{d}U^0 + p^0\mathrm{d}V^0)$$

and

$$\mathbf{dG} = \mathbf{u}\gamma(\mathrm{d}U^0 + p^0\mathrm{d}V^0),$$

where $\gamma = (1 - u^2)^{-1/2}$. From eqn (5) and (8), we have

$$\mathrm{d}Q = \mathrm{d}U + p\mathrm{d}V - (\mathbf{u}\cdot\mathbf{dG}),$$

and from eqns (5) and (7), we have

$$\mathrm{d}Q^0 = \mathrm{d}U^0 + p^0\mathrm{d}V^0.$$

One can now easily derive from these equations the following,

$$\mathrm{d}Q/\mathrm{d}Q^0 = 1/\gamma = (1 - u^2)^{1/2}.$$

Now, this is eqn (1), and from this and eqn (9), we get eqn (2).

Fifty or so years had to pass before the Planck–Einstein formulation of RTD was first challenged in the literature.[5] The first criticism, however, came from Einstein himself in 1952 when he put his criticism and an alternative formulation to von Laue in a letter. The alternative excluded the translational work, ($\mathbf{u}\cdot\mathbf{dG}$), and thus gave a different set of L-Ts for thermodynamic quantities [see eqns (10) and (11) below]. Therefore, Einstein discovered first one of the most viable formulations in RTD, which was later rediscovered and vigorously defended in the 1960s. Von Laue did not seem to understand Einstein's point, for what he did in response was to point out the obvious, namely, the original derivation of the Planck–Einstein formulation. Einstein said that he could not see exactly what was wrong with the original formulation, but his alternative 'is so simple, that I can hardly imagine that it contains any mistake' (Einstein, 1952, quoted with translation in Liu, 1992, p. 195). Then, Einstein realized that neither the original nor his new alternative formulation contained the truth. He suggested yet another theory of RTD which took all

[5]The main development of RTD during those years was due to Tolman (1934). Tolman not only made some applications of the Planck–Einstein formulation, he also extended thermodynamics to general relativity. Some very significant work had also been done by Danzig (D. van Danzig, *Physica* **6** (1939), 637–704; *Proceedings of the Koninklijke Nederlandse Akademie van Wetenschappen* **42** (1939), 601–7), which had substantial influence on the later development of relativistic fluid mechanics. For discussion of these works, see Liu (1991).

thermodynamic magnitudes to be scalars; his reason being that in RTD one only needed the 'eigen'-values for those magnitudes, namely, values obtainable in the co-moving reference frame of the system in question. Since von Laue was not persuaded by this theory either, Einstein stopped pursuing the subject further, and his ideas were not known to the public, even though Einstein anticipated nearly every main position developed later.[6]

3. The Debate on RTD

The first published challenge of the Planck–Einstein formulation was given by H. Ott, which in part rediscovered Einstein's alternative, and which I shall call the *Einstein–Ott formulation* (Ott, 1963). In his criticism of the old formulation, Ott put his finger right on the problem. The question is whether the translational work as given in eqn (8) is needed when a system is moving relative to the observer. Planck, Einstein and others believed that it is needed because otherwise the system will accelerate (or decelerate) when losing (or gaining) radiation energy. Ott argued that whether this is so depends entirely on whether or not the environment is co-moving with the system (cf. Fig. 1). It has nothing to do with which inertial frame the system is observed in.

If the environment is not co-moving, the system will accelerate (or decelerate) regardless of which inertial frame it is observed in, because inertial mass is added to the system with a different velocity. In this case, however, there will be an impact and a loss or gain of kinetic energy, as happens in any inelastic collision. Even in classical mechanics, translational work is needed in this situation to prevent the system from accelerating (or decelerating). But since there is irreversible gain (or loss) of energy, the heat (i.e., mass) exchange is no longer a process that can be handled in the Planck–Einstein formulation of RTD (which only deals with equilibrium thermodynamics).

But if the environment is co-moving, the translational work is no longer needed, as can be easily shown in the following. In K, when dQ is transferred from the environment to R, it carries with it the amount of momentum $d\mathbf{G}$ which exactly compensates the energy (i.e., mass) increase in R, so no extra translational work is needed to keep R from decelerating. Without this term, everything else being the same, it is easily shown that the L-Ts become

$$dQ = dQ^0/(1-u^2)^{1/2} \tag{10}$$

$$T = T^0/(1-u^2)^{1/2}. \tag{11}$$

$$S = S^0,$$

$$p = p^0.$$

[6]For a detailed study of this episode of the Einstein–von Laue correspondence and how Einstein had anticipated positions in the later debate, see Liu (1992). It is very unlikely, though not impossible, that Einstein's ideas were known to anyone else besides von Laue. At least, they were not mentioned anywhere in the later publications concerning the subject.

Now the moving system appears to be warmer than it is at rest.

Ott's criticism of the Planck–Einstein formulation is sound in so far as the translational work term is concerned. However, it has its own problems, the most serious of which is that the notion of the temperature of a moving system becomes completely mysterious. If one does not consider heat exchanges between systems moving relative to each other, in what way can one obtain the values of that temperature? In Fig. 1, if the environment and R are at rest with respect to each other all the time (and they are both at rest with respect to K_0), the observer in K cannot 'look' at the exchanged heat or temperature in the two systems and obtain their values. There has to be some way of measuring these values, which brings us back to confront the situation of thermal interaction.

Even though Ott's posthumous article was the first published challenge to the old formulation, the debate did not begin until two years later when H. Arzeliés published a paper in *Nuovo Cimento* (Arzeliés, 1965) and apparently rediscovered the Einstein–Ott formulation (through a different argument from Ott's) (see Liu, 1991, pp. 99–103, for an analysis of the derivation). Several papers followed in the same journal debating the right way to Lorentz transform mechanical work.[7] One year later, P. T. Landsberg started a different line of argument in *Nature*, mainly investigating (using some relativistic kinetic model of thermal systems) whether the temperature of a moving body should appear cooler or warmer than when it is at rest.[8] These papers saw the initial form of Landsberg's argument for the invariance of temperature. At the same time a third approach began to emerge as a major contender, which could be traced back to Ott. It was the covariant formulation of RTD: the hope was that once one put thermodynamic laws into a four dimensional covariant form, one automatically took care of the problem of its relativization (cf. Ott, 1963; Arzeliés, 1965; Rohrlich, 1966; Kampen, 1968; Yuen, 1970).

The Planck–Einstein formulation also had its defenders. The support mainly came from works in relativistic statistical mechanics (or kinetic theory of gas).[9] At the end of the 1960s, the research turned towards doubts and reflection. Prominent figures in the debate, such as Landsberg and Møller published long articles assessing the pros and cons of each position (Landsberg, 1970a; Møller, 1968). A general consensus or attitude began to form among the participants that one should take RTD as a conventional theory. Which formulation was right, they contended, should depend

[7] *Nuovo Cimento* remained the major battleground for the debate throughout the years [see A. Gamba, *Nuovo Cimento* **37** (1965), 1792; T. W. B. Kibble, *Nuovo Cimento* **41B** (1966), 72; R. Penny, *Nuovo Cimento* **43A** (1966), 911; Rohrlich (1966), and as late as in 1984, we have P. Ammiraju, *Lettere al Nuovo Cimento* **39** (1984), 424–9].

[8] P. T. Landsberg, *Nature* **212** (1966), 571; **214** (1967), 903; **220** (1967), 1120; see also J. H. Fremlin, *Nature* **213** (1967), 277; I. P. Williams, *Nature* **213** (1967), 1118; **214** (1967), 1105. P. T. Landsberg became one of the key figures in the debate, charting out new directions at almost every turn. Cf. Landsberg and Johns (1967) and (1968); Landsberg (1970b) and (1978).

[9] cf. Pathria (1966) and (1967), Balescu (1968), Taff (1968), Treder (1977). Also see, J. H. Eberly and A. Kujawski, *Physical Review* **155** (1967), 10–19, *Acta Physica Polonica* **A41** (1972), 259–273.

on how one defined the basic thermodynamic quantities with respect to a moving observer (cf. Landsberg, 1970a, Yuen, 1970). Thus, the debate practically ended around the end of the 1970s without a clear verdict. However, the search for a better formulation and the solution of this puzzle never really stopped.[10] Below is a brief discussion of the important developments in the debate (for a more detailed discussion, see Liu, 1991, ch. 5).

Landsberg's investigation yielded some interesting results, one of which was the argument for the invariance of temperature (Landsberg, 1967 and 1970a). Suppose that two thermodynamically identical bodies are in thermal contact and moving relative to each other, and their temperature, T, in K is not equal to their temperature, T^0, in K^0. The observer attached to each of the two bodies will see the other body's temperature as higher (or lower) than her own. Since a temperature difference should cause a heat flow across the bodies, there will be such a flow in opposite directions with respect to the two observers, which seems absurd. Therefore, (by *reductio*) the temperature of the two bodies must be the same.[11] This, however, forces one of the two laws of thermodynamics to assume a different form in a moving reference frame, seemingly violating the special principle of relativity. But here comes the revelation (which was made by Landsberg but is logically independent of his argument). The form invariance of the two thermodynamic laws does not require that the laws be invariant under the classical form (Landsberg, 1970a). The classical laws may be written in the general form: $X = Y$, where X and Y are functions of thermodynamic quantities and/or their differentials. Since the laws in classical thermodynamics are only defined in the co-moving frame of the system in question, their relativized form may become: $X = f(u)Y + g(u)$, where u is the relative velocity, $f(u)$ and $g(u)$ are two arbitrary functions of u, such that $f(u) = 1$, $g(u) = 0$, when $u = 0$.[12] This fact not only provides justification for Landsberg's invariant temperature, it also renders obsolete Einstein's method of determining the temperature of a moving system [eqn (9)].

The support for the Einstein–Ott formulation formed the strongest camp in the debate. In addition to arguments similar to Ott's, the covariant theory of RTD was considered to provide the most promising support. By the 1960s, the majority of

[10]See R. G. Newburgh, *Nuovo Cimento* **52B**(2) (1979), 219–28; P. L. Kellerman, *Foundations of Physics* **10** (1980), 163–75; P. T. Landsberg, *Physical Review Letters* **45** (1980), 149–50, and *European Journal of Physics* **2** (1981), 203–7; Ammiraju, *op. cit.*, note 7; I. Ojima, *Letters in Mathematical Physics* **11** (1986), 73–80; V. J. Menon and D. C. Agrawal, *American Journal of Physics* **59** (1991), 258–60; P. T. Landsberg, *American Journal of Physics* **60** (1992), 561; V. J. Menon and D. C. Agrawal, *Journal of Physics* **25A** (1992), 4517–20.

[11]The problem with this argument is that heat flow considered only as energy transmission is not described by a tensor. Thus, it is not clear whether its direction is independent of observers. In relativistic physics, only a four dimensional tensor has frame-independent meaning; and a component of that is usually frame-dependent, e.g., temporal or spatial intervals. The problem with Landsberg's argument is that temperature difference only warrants an energy flow but not an energy-momentum flow; the latter would indeed be frame-independent. (This was brought to my attention by Andreas Bartels.)

[12]The translational work, ($\mathbf{u}\cdot d\mathbf{G}$), used in the Planck–Einstein formulation is in fact one form of the $g(u)$. Therefore, the arbitrariness was already there in the Planck–Einstein formulation when the translational work was added (cf. Section II). But its meaning was not realized until Landsberg's work.

physicists had become convinced that to fulfil the requirement of STR one only needed to put the laws in a theory into *covariant* form.[13] That is to say that once all the quantities in an equation are four dimensional tensors of the same kind, the equation (and so the law it represents) automatically satisfies the special principle of relativity. The two laws in this formulation are given as follows (cf. Arzeliès, 1965, 800):

$$dQ^i = dU^i + dW^i, \tag{12}$$

and

$$dQ^i \leq T^i dS, \tag{13}$$

where $dQ^i = u^i dQ^0$, $T^i = u^i T^0$, $dU^i = u^i dU^0$, and $dW^i = [(\gamma F dt), dW^0]$ are all four dimensional vectors on the Minkowski space-time. One could easily derive the L-Ts for T or dQ from these equations and they agree with the Einstein–Ott formulation. The problem with this formulation is that it is difficult to find physical interpretations for many of the spatial components of these 4-vectors. What could $\mathbf{T} = \mathbf{u}T^0$ possibly mean: a temperature current? But temperature is a state parameter. How could there be a temperature current?

The support for the Planck–Einstein formulation came from the following well-known fact: the metabolism of an organism is positively correlated to the intensity of the motion of its molecular components. According to the time dilation effect, the metabolism of a human body appears slower in a moving frame than in the rest one. This seems to entail that the body temperature is also lower in a moving frame than in the rest one, confirming the Planck–Einstein L-T for temperature (cf. Fremlin, 1967; Williams, 1967). A more sophisticated treatment along this line is roughly as follows. One starts with a relativistic distribution function for classical or quantum gases (including radiation as photon gas).[14] The distribution function, F, is taken to be a scalar for it is simply a probability function. In Pathria's work (1966 and 1967), the distribution function of quantum gases in a moving frame is given as follows,

$$F = \{\exp[(-\mathbf{u}\cdot\mathbf{p} - \mu)/kT] + a\}^{-1}, \tag{14}$$

where e is the energy of a molecule, $\mathbf{u}$ the velocity, $\mathbf{p}$ the momentum, μ the chemical

[13]We now know that this is a misinterpretation of relativistic theories, but it is harmless if proper constraints are introduced. The reason that this interpretation survived for so long was that in most cases the constraints were automatically in place. We will come back to this point in Section IV, but for a detailed study of the covariance principle and relativistic theories, see Earman (1974), Friedman (1983) and Norton (1989).

[14]The relativistic distribution function was obtained by F. von Jüttner in 1911. See, F. von Jüttner, *Annalen der Physik* **34** (1911), 856–82, and **35** (1911), 145–61, for the relativistic distribution function for classical gases; also F. von Jüttner, *Zeitschrift für Physik* **47** (1928), 542–66, for the same function for quantum gases.

potential, k the Boltzmann constant and T the temperature. When $a = -1$ the gas is Bosonic and when $a = 1$ the gas is Fermionic. Since F is Lorentz-invariant, we have,

$$(e - \mathbf{u}\cdot\mathbf{p} - \mu)/\mathrm{k}T = (e^0 - \mu^0)/\mathrm{k}T^0,$$

where the rhs contains quantities in a rest frame. We know the L-Ts for energy and momentum for a single particle, hence it can be derived that $T = (1 - u^2)^{1/2}T^0$, matching the Planck–Einstein formula (and, *mutatis mutandis*, for other thermodynamic quantities as well). Balescu, in his paper (1968), went through similar considerations to Pathria's, but with a final important twist. Noticing Landsberg's work in pure thermodynamics, Balescu realized that the same kind of freedom in formulating relativistic laws also applied to the kinetic theory of gases. The distribution function in a moving frame could be

$$F = \{\exp[\alpha(\mathbf{u})(e - \mathbf{u}\cdot\mathbf{p} - \mu/\beta(\mathbf{u}))/\mathrm{k}T] + a\}^{-1}, \qquad (15)$$

where $\alpha(\mathbf{u})$ and $\beta(\mathbf{u})$, satisfying $\alpha(\mathbf{0}) = \beta(\mathbf{0}) = 1$, are two arbitrary functions of the system's velocity. One can see that these are the same kind of functions as the $f(u)$ and $g(u)$ in Landsberg's case. Hence, if Balescu is right, resorting to the molecular components of a system does not help in solving the problem of RTD.

The meaning of the 'moving' values of thermodynamic quantities was a sensitive issue in the debate, for it was not clear whether they should be defined directly (with explicit operations) or indirectly. Although many participants voiced the need for an operational definition (e.g. Taff, 1968; Rohrlich, 1966; Schmid, 1970; Møller, 1968; Yuen, 1970, and others), Landsberg and his colleague, K. A. Johns, were probably the only people who seriously considered the possibilities (see Landsberg, 1970a, 1968; Landsberg and Johns, 1968 and 1972). Landsberg in his 1970 (1970a) paper listed five possible ways of measuring the quantity, four of which used radiation in one way or another. The most plausible method was perhaps the following (cf. also Landsberg and Johns, 1968). Suppose that a blackbody, S, is moving within another blackbody, L (a cavity), which is completely sealed. In order to ensure quasi-equilibrium, Landsberg employed what he called the *basic trick*, namely, sealing off the radiation of S inside for most of the time, only letting it be released into L from time to time instantaneously. However, even with such an artificial arrangement, the result is not encouraging. The L-T for temperature so obtained turns out to depend on the shape of S. Such results suggest that thermal equilibrium is impossible among bodies with relative motion. The fact seems to be that *temperature measurement requires genuine thermal interaction and the state of equilibrium, but when relative macroscopic motion is present, such interaction always disrupts the state of equilibrium and thus renders temperature measurement impossible*. There were even theoretical proofs that thermal equilibrium excluded relative motion among interacting bodies.[15]

However, if one gives up genuine thermal interaction and measures the temperature indirectly, one may find it very simply, if not trivially.[16] For many materials, their temperature is a linear function of their volume (that is how thermometers are made). If we assume that the same relation holds in the relativistic case, the temperature L-T will then be the same as the volume L-T, and we know that the latter is simply $V = (1 - u^2)^{1/2} V^0$. It is not surprising that the L-T obtained in this way agrees with the Planck–Einstein formula, for it is the same kind of argument as that of the metabolism of organisms (see above). But such consideration is no good precisely because it has nothing to do with the thermal interaction.

Through these cases I hope that I have shown why the debate ended in a mood of resignation. Such feeling was perhaps further strengthened when in a paper in *Physica* (Paiva-Veretennicoff, 1974) it was shown that the Galilean transformations of thermodynamic magnitudes were equally arbitrary. It seemed pointless indeed, if Paiva-Veretennicoff was right, to look for a unique set of L-Ts for thermodynamic magnitudes.[17]

4. The Special Principle of Relativity and the Nature of Relativization

Unknown to the participants of the debate on RTD was J. L. Anderson's (1964) simple solution for the problem. There are several reasons for examining in detail Anderson's solution, none of which hinge on whether it ultimately solves the problem in the debate (it does not; See section 5 below). First, it resembles one of Einstein's suggestions for this problem (cf. Section 2); and like that suggestion, it has never been noticed in the debate. Second, it is a solution of a different sort, namely, it questions the very foundation of RTD; in other words, it questions not how but whether one should search for the thermodynamic L-Ts. Third, Anderson's principles of the theory of relativity and of space-time set the foundation for the current philosophical investigation of relativistic theories (cf. Friedman, 1983; Earman, 1989). Therefore, this discussion will shed light on both the foundations of RTD and the nature of relativization.

Let us first examine the basic assumption that almost everyone in the business of RTD took for granted. *To obtain a relativistic theory, one must find a unique set of L-Ts for the theory's magnitudes.* In the case of thermodynamics, this necessitates

[15]Within the covariant theory of RTD, Kampen (1968, 173) showed that two bodies moving relative to each other cannot be in the state of equilibrium unless they move with the same velocity, given that the state of equilibrium was defined as zero entropy change. Landsberg obtained the same result with his theory (Landsberg, 1978, 378–9), and Eimerl even elevated it to the status of a postulate of RTD (D. Eimerl, *Physica* **91** (1975), 481–98).

[16]This was suggested to me by Clark Glymour.

[17]It is difficult to assess the actual influence of Paiva-Veretennicoff's paper on the whole debate of RTD. It was widely cited by subsequent works of the participants, but never really carefully examined or further developed.

the search for a set of L-Ts for thermodynamic magnitudes. I said 'almost everyone', for Landsberg did question the assumption in one of his seminal papers (Landsberg, 1970a). In the section 'Why transform thermodynamic variables at all?' Landsberg answers:

> To refuse Lorentz transformations for temperature (and possibly entropy and heat) is of course inelegant because, by accepting relativistic mechanics, many thermodynamic quantities are already subject to accepted transformations: pressure and volume are notable examples. This leads to the transformation of force and hence work and energy. By dealing with work and energy, relativistic mechanics takes us to the very topics of interest in thermodynamics. (Landsberg 1970a, p. 260.)

Therefore, Landsberg concludes that even though there is no practical need for transforming thermodynamic variables ('nobody in his senses will do a thermodynamic calculation in anything but the rest frame of the system!' *ibid.*), it is nevertheless required in principle. This is precisely the point that Anderson argues against. By a different interpretation of the special principle of relativity, Anderson is able to show that there is a sense in which a theory is relativized without the need for a set of L-Ts.[18]

Anderson argues that the special principle of relativity embodies two distinct interpretations—the active and the passive, which are traditionally taken as merely different expressions (Anderson, 1964, pp. 175–8). The active version expresses the non-distinguishability of inertial reference frames by physical experiments, whereas the passive version demands the form invariance of laws under the appropriate L-Ts.

The active principle (Fig. 2) can be seen as a faithful interpretation of the statement of the principle by Einstein (1907): '*The Laws of Nature are independent of the state of motion of the reference system, at least if the latter is without acceleration*' (my translation, Einstein, 1907, p. 416). The boost serves in this principle a similar, but not equivalent, function as the L-Ts in the passive one. Anderson adds emphatically,

> Note that observer A [Σ] never looks at system B [Σ'] and vice versa; they merely compare results of what they see their own system doing. (Anderson, 1964, pp. 176–7.)

The principle, so interpreted, implies that space-time is such that no uniform motion of the whole system in question produces any effect on processes therein.[19]

The passive principle (Fig. 3) says essentially that by using the appropriate L-Ts, results obtained from different inertial reference frames (or by different inertial

[18]Anderson's objection to the relativization of thermodynamics first appeared in his 1964 paper. In addition to the objection (it is a rather small part of the whole paper), Anderson proposed his idea of the relationship between the notion of covariance and that of symmetry. For that he made the distinction between a dynamical geometric object and an absolute one, which became an important issue in the foundations of space-time theories. But in the first section of the paper, where the objection was made, Anderson gave his new interpretation of STR, which in turn provided a solution for the problem of RTD (even though Anderson was not aware of it as a problem). See also Anderson (1967), 215–7, 313–7.

[19]For historical interest, I have given Einstein and Anderson's formulation of the active version of the principle. For a precise formulation of this in the current literature of philosophy of space-time, see Friedman (1983), 129ff; also, pp. 16–17 below.

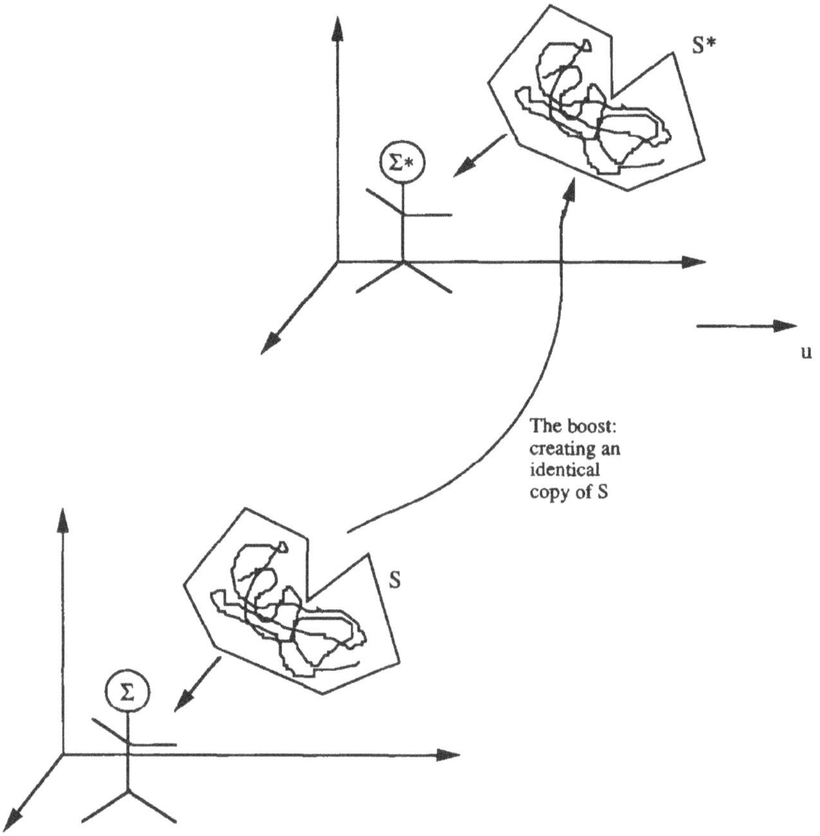

Fig. 2. The active verion of the special principle of relativity. No results of any experiment can tell whether the experiment is done by Σ on S or by Σ on S*. No experiment can tell S's (uniform) state of motion.*

observers) satisfy the same equations (i.e., laws). This principle clearly says more than the indistinguishability of inertial reference frames.

Anderson (1964, pp. 179–81) contends that the active interpretation captures the whole content of the principle with which every theory in physical sciences must comply; yet only to those theories which deal directly with 'geometric objects' on the space-time manifold, is the passive version of the principle applicable. Most theories in physics, such as mechanics and electrodynamics (of particles or of continuous matter), belong to this category. Thermodynamics, however, is different. Anderson argues that *only the active principle applies to thermodynamics, and it is already relativistic in its classical form*. According to Anderson, geometric objects are physical magnitudes defined on space-time regions that could be *directly* measured by different inertial observers. Magnitudes in the theory of equilibrium thermodynamics are not objects of this sort, because they are defined and measured

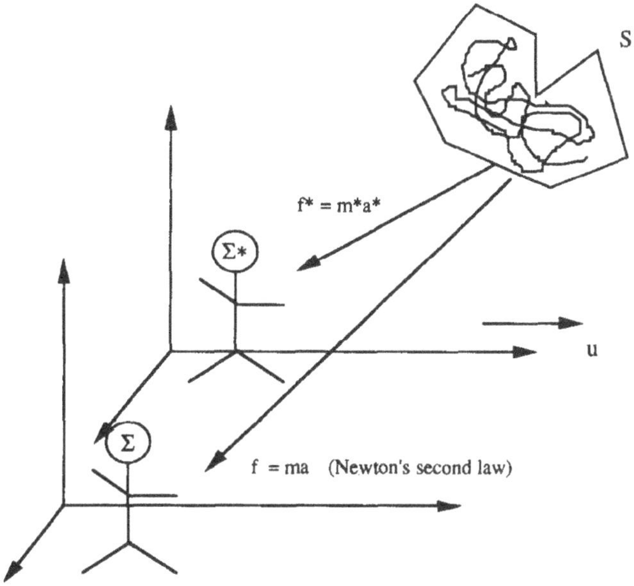

Fig. 3. The passive verion of the special principle of relativity. S is observed by two different inertial observers and the resulting physical laws, such as f = ma, have the same form, e.g., when (f, m*, a*) is related to (f, m, a) by the appropriate L-Ts.*

only in the co-moving frame of the system in question. As long as the thermal processes in a system are not affected by its uniform motion, thermodynamic theory satisfies the active principle; and since no magnitudes are measured by observers other than the co-moving ones, the question for L-Ts does not even arise. One may impose some L-Ts on thermodynamic quantities, but they can only be purely formal ones: neither essential nor unique. By this, it seems, Anderson resolves the problem of RTD. We will evaluate this argument later; but let us now consider its implication for the foundations of relativistic theories.

As I alluded earlier, Anderson's view on relativistic theories initiated a gradual shift in our understanding of the theoretical and philosophical aspects of space-time (see also point (ii) below).[20] A detailed account of this shift is beyond the scope of this paper; it suffices to mention a few crucial points.

It was a long-held belief among physicists, as manifested in the covariance approach towards RTD (see Section 3), that relativization of a physical theory entailed no more than putting its laws into a covariant form (i.e. four-dimensional tensor equations of the appropriate kind), and this simply meant that all reference frames, inertial or not, were equivalent. All this was taken to be the consequence of

[20]The transition is the result of many people's work. cf. Anderson (1967), ch. 4; Havas (1964); Trautman (1964); Earman (1974) and (1989); Friedman (1983); Wald (1984); Hiskes (1984); and Norton (1989).

the principle of general relativity. However, gradually, two problems surfaced. First, the general relativity principle did not mean the equivalence of all reference frames, if it was to be understood as in the same way as the equivalence of all inertial frames. Second, it turned out that all theories, relativistic or not, could be put into a covariant form without ever changing their physical content.[21] Thus, in principle, the formal requirement of covariance had no necessary connection to the relativization of a physical theory. Anderson's active interpretation of relativity principles pointed to the direction in which these two difficulties could be dealt with satisfactorily.

Through Anderson's work, we now realize that relativistic theories are about the structure of space-time, rather than about transformational properties of laws. Hence, these theories are simply taken as theories about space-time and the geometric objects on it. As an example, let us look at the current formulation of the special principle of relativity (cf. Earman, 1974 and 1989 and Friedman, 1989). Instead of looking at physical theory as mainly consisting of a set of laws, we represent the theory, T, with the set of models, $\mathfrak{M}_T$, which satisfy those laws. The special principle of relativity is then a principle of closure. Thus, T satisfies the principle if and only if for appropriate f, $\forall m \in \mathfrak{M}_T, f(m) \in \mathfrak{M}_T$. More precisely, the models in $\mathfrak{M}_T$ can be seen as $<M, \eta, O_1, \ldots O_n>$, where M is the manifold, η the Minkowski metric, and $O_1, \ldots O_n$ are geometric objects defined on regions of M. An automorphism, $\phi: M \rightarrow M$ is a continuous, one-to-one, onto mapping function which maps points onto points in M and leaves η unchanged, and which induces a function, ϕ^*, that carries along the geometrical objects in M. So in the brief formulation above, f represents the automorphism and the carry-alongs. (A more explicit formulation of the principle is: T satisfies the principle if and only if for any model $<M, g, O_1, \ldots O_n> \in \mathfrak{M}_T$, the model $<M, \eta, \phi^*O_1, \ldots \phi^*O_n>$ is also in $\mathfrak{M}_T$, where $\phi^*(\eta) = \eta$.) It can be shown that merely putting an equation into a covariant form of four-tensors is by no means sufficient to satisfy this principle. Hence, the attempt in the debate to construct RTD by giving it a four-dimensional covariant formulation is not the right approach (cf. Section III).

Several profound changes took place with such a shift, of which I will discuss three.

(i) The presence of a space-time manifold and its appropriate structural elements become prerequisites for a relativistic theory. Physically meaningful magnitudes must be geometric objects (or object fields) defined on the manifold, which are independent of any coordinate descriptions (including inertial frames). With this formulation, one is able to clearly distinguish the notion of *symmetry* from that of *covariance*. The equivalence of inertial reference frames is captured by a symmetry group of automorphisms defined on the space-time manifold, whereas the admissibility of all reference frames (though they are by no means equivalent) is captured by a covariance group of transformations among the coordinate representations of geometric objects.

[21]For example, we can have covariant theories of Newtonian mechanics (cf. Havas, 1964 and Friedman, 1983).

From this, it is realized that different relativity principles are about different symmetry properties of the space-time manifold, not about different transformational behaviors of physical laws among reference frames.

(ii) This construal of the relativistic theories lends strong support to the realist (or substantivalist) view of space-time. It comes as no surprise that the resurgence of the realist view of space-time went hand in hand with the development of 'space-time' theories under the active interpretation of the relativity principles.[22] Through this movement, a conceptual transition took place in our understanding of relativistic theories, which moved away from Einstein's original idea. Whereas Einstein talked about standard rulers and ideal clocks as giving measures of spatio-temporal relations among events and about how laws of physics were affected by the changed meaning of such measurements, the new formulation takes space-time seriously as a four dimensional continuum and requires that all physical theories be about geometric object fields on this continuum. Thus, we find out the structure of space-time by looking at the effects of moving those objects around in the manifold. Taking STR as an example, one can clearly see that with the active interpretation, the special principle of relativity is no longer about the equivalence of different views (from different inertial observers) of the same system, as implied in the passive interpretation, but rather it is about the flatness, the homogeneity and the isotropy of the gravity-free space-time. It asserts that one can have the 'freedom to reposition, reorient, or boost physical systems in the space-time container', instead of having 'the freedom to describe the same system in many ways' (Earman, 1989, p. 56).

(iii) The conceptual shift can also be seen as one from the syntactic conception of scientific theories to a semantic one.[23] With the passive interpretation of the principle, a relativistic theory is taken to be a set of lawlike propositions (mostly equations), some of which are primitive (axioms) and some derivative (theorems). The principle of relativity is given as a meta-theoretical constraint on the transformational properties of these propositions. This picture squares well with the syntactic conception. But now, instead of looking at the laws and conditions, we focus our attention on the relational structures that satisfy them. In other words, we take a theory to be a set of models rather than a collection of interpreted linguistic entities. Rather than finding out how equations behave under coordinate transformations of their

[22]I am here speaking of a movement which H. Field referred to as the ' "new wave" of space-time theorists' (Field, 1980, 35). Field mentioned J. Earman and M. Friedman; one may also include G. Nerlich.

[23]The syntactic conception emerged together with the 'received view' of the structure of scientific theories and still remains an integral part of it today, even though most of the other components of that view have long been given up (cf. F. Suppe, *The Structure of Scientific Theories* 2nd ed., Urbana, Ill: Univ. of Illinois Press, 1977). It is not entirely clear who first conceived the semantic conception, but philosophers who defended and developed the conception in its early days include P. Suppes ('What is A Scientific Theory?' in S. Morgenbesser (ed.), *Philosophy of Science Today*, New York: Basic Books, 1967) and B. van Fraassen (*Philosophy of Science* **37** (1970), 325–34). For the most recent works, see F. Suppe, *The Semantic Conception of Theories and Scientific Realism*, Urbana, Ill: Univ. of Illinois Press, 1989, and B. van Fraassen, *Quantum Mechanics*, Oxford: Clarendon Press, 1992.

variables, we mathematically manipulate elements in a model to see what properties of the space-time manifold and the geometric objects can be found. One can hardly expect a more appropriate application of the semantic conception of scientific theories.

Despite its success, this approach may still be untenable. Besides the problem that arises from Anderson's analysis of RTD, which I will presently discuss, there are other internal problems, such as the lack of a general coordinate-free definition of geometrical objects and of a clean separation between the absolute and the dynamic objects (cf. Anderson, 1967, pp. 14–16; Friedman, 1983, pp. 46–62; Earman, 1989, pp. 38–40 and 55–60). For the former, the most general definition of geometric objects still requires the use of coordinate transformations, even though we can define tensors (the present majority among geometric objects) independently of that. And for the latter, the intuitive idea of absolute geometric objects—objects which pertain to the immutable properties of the space-time background—has so far eluded every attempt at a precise definition.

Moreover, if Anderson is right in his argument against RTD as given above, part of the meaning of the special principle of relativity is missing in this 'space-time theory' approach to relativistic theories. Thermodynamic theory does not contain geometric objects and yet its laws satisfy the principle, and thus it is a relativistic theory. The truth seems to be the following. Principles of relativity are about space-time. Even when applied to thermodynamics, the special principle says that space-time is uniform in that it has no effect on thermodynamic processes whether or not the whole system in question is in uniform motion. However, this fact about space-time is not obtainable, in this case, from the 'manipulations of geometric objects' on the space-time manifold, as the space-time theory approach demands. Here, Einstein's original conception of the relativity principle—a meta-theoretical constraint on laws of nature—seems closer to the truth. A similar case is found in relativistic quantum mechanics (or quantum field theories), where the state vectors have L-Ts but are not geometric objects.[24] This is, however, a far more complex case, for the state vectors which represent quantum fields are nevertheless vectors in Hilbert space (a complex mathematical space with infinite dimensions), not in real space-time. Together, these two cases raise serious doubts about the tenability of the space-time theory approach to relativity principles.

5. Beyond the Formalistic Debate of RTD

Anderson's objection to RTD may have brought enlightening insight into the meaning of the special principle of relativity, yet the validity of his argument depends on thermodynamic quantities not being geometric objects. For this, Anderson has two arguments. First, he points to the fact that classical thermodynamics of equilibrium

[24] I thank an anonymous referee for pointing this out to me. See, also Sypel and Brown (1992).

only deals with processes in their co-moving frame. This is probably true (he has the backing of Einstein of 1952 and Landsberg). But one should not take what *is not* the case for what *cannot be* the case. Secondly, Anderson maintains that there is no direct way of measuring a thermodynamic quantity, such as temperature, if the instrument is not at rest with the measured system. This second point, although putting the finger on the real problem, is far from obviously true. The discouraging results from the search for an operational definition of the temperature of moving body (see Section 3) may have lent strong support to Anderson's view; they do not, however, prove it.[25]

It is difficult to imagine that the thermodynamic properties of the 3 K background cosmic radiation are somehow not accessible to us; but if Anderson's argument holds, they must be so, for we cannot eliminate the relative motion between the radiation and the earth upon which it is measured. Besides, there are a few areas in which an RTD with L-Ts may not be dispensable: for example, for a general and exact kinetic theory of systems consisting of a large number of particles and of irreversible phenomena, for a pure thermodynamic theory for irreversible phenomena, and for a field theory of thermodynamics (just to mention the most noteworthy ones). Without going into details, I shall briefly discuss some recent developments of RTD where certain L-Ts do seem justified.

RTD originated, as discussed earlier, from the investigation of blackbody radiation of a moving cavity by Planck and his student von Mosengeil. During the years of the big debate, there were also derivations of the thermodynamic L-Ts with respect to radiation, which appeared to be strikingly simple and non-arbitrary (cf. Treder, 1977). Suppose that there are two bodies, A and B; A moving slowly inside B and producing blackbody radiation which is completely absorbed by B. With blackbody radiation, its energy distribution is known to be proportional to the famous Planck function,

$$F(\nu) = [\exp(\nu/kT) - 1]^{-1} \tag{16}$$

where ν is the frequency of the radiation. F is invariant [cf. eqn (14)]; so for an observer at rest with A, the function is $F^0(\nu) = [\exp(\nu^0/kT^0) - 1]^{-1}$, while for an observer attached to B, it is $F(\nu) = [\exp(\nu/kT) - 1]^{-1}$. Since, $F(\nu) = F^0(\nu)$ (invariant),

$$\nu^0/T^0 = \nu/T. \tag{17}$$

From the well-known Doppler-effect of a moving light (radiation) ray, we have, $\nu = (1 - u^2)^{1/2}\nu^0$, the L-T for the frequency. Hence, we obtain the Planck–Einstein L-T for temperature: $T = T^0(1 - u^2)^{1/2}$. It is true that if the radiation is weak and rare enough (e.g. in the case of the 3 K background radiation), its distribution is very close to the Planck distribution. This result is indeed used in the cosmological study of such

[25]Anderson might be wrong about thermodynamic magnitudes not being geometric objects, for what follows seems to be contrary to his view. But his observation that a theory such as classical thermodynamics (containing no geometric objects) can be relativistic without L-Ts is still true. The very fact that there is at least one theory of thermodynamics which is relativistic only by means of the active principle is sufficient to reject the idea that all relativistic theories are about geometric object fields.

radiation (cf. Demiänski, 1985, pp. 35–8 and 274–9). Here, the following transformation formula is used to measure the temperature of the background cosmic radiation (with its direction oblique to the motion of the earth with angle θ):

$$T(\theta) = T^0[(1-u^2)^{1/2}/(1-u\cos\theta)]. \quad (18)$$

When $\theta = \pi/2$, namely, when the radiation is perpendicular to the motion of the earth, we have, $T(\pi/2) = T^0(1-u^2)^{1/2}$.[26]

Synge's book, *The Relativistic Gas* (1957), was the first attempt to construct a general and rigorous relativistic kinetic theory of gas. It was general and rigorous because Synge attempted constructing the stress-energy tensor for the gas system, from such tensors of individual gas molecules. In so doing, Synge was able to begin discussing energy (heat) or matter transfer among systems of gas, setting the foundation for a relativistic theory of thermal transport (including irreversible processes). And the study of shock waves as consequences of violent explosion in a gas was also one of Synge's main contributions to relativistic gas theory. Later, in the 1970s, Ehlers greatly improved Synge's theory, not only by putting it in a more rigorous mathematical formulation but also by extending it to general relativity (Ehlers, 1971).[27] Even though the L-Ts are not explicitly mentioned and studied in the theory, it is not difficult to derive them from it; for instance, the temperature turns out to be a scalar, matching Landsberg's result (for details, see Liu, 1991, pp. 180–2).

C. Eckart and D. Landau (cf. Eckart, 1940, Landau and Lifshitz, 1959) were considered the pioneers of the field theory of thermodynamic phenomena. Works in this area essentially involved studies of heat conduction in dissipative fluids, namely, processes involving viscosity and thermal conduction. The task for Eckart and Landau was to upgrade Fourier's law of heat conduction from classical theory to STR. The theories of both men required that temperature and entropy be scalars (cf. Eckart, 1940, p. 922). Among those who further developed the field theory of thermodynamics, I. Müller was the most instrumental.[28] At the beginning of his book, *Thermodynamics*, Müller introduces the field theory of thermodynamics as follows:

> This chapter defines thermodynamics of a single body as a field theory of density, motion and temperature. It shows how equations of balance and constitutive equations combine to give the field equations of thermodynamics and describes the need for restrictive conditions on the constitutive functions. Such conditions are material objectivity, the entropy principle and thermodynamic stability which are formulated here and exploited for viscous, heat-conducting fluids (Müller, 1985, p. 1).

Like all field quantities, density (scalar), temperature (scalar), and velocity (vector)

[26]For a study of RTD of blackbody radiation, see C. Liu, 'The History of Relativistic Thermodynamics as Originated and Developed in the Study of Blackbody Radiation', in W. Schröder (ed.), *The Earth and the Universe: Festschrift in Honor of Hans-Jürgen Treder*, Bremen-Rönnebeck: IAGA, 1993, 273–288.

[27]J. Ehlers made many contributions to relativistic kinetic theory and statistical mechanics, among them Ehlers (1971) is the most comprehensive. Also see W. Israel and J. M. Stewart, *Annals of Physics* **118** (1979), 341, for a more recent treatment.

[28]I. Müller, *Zeitschrift für Physik* **198** (1967), 329.

are geometric objects. The conception of a field theory for thermodynamical phenomena may sound peculiar since such phenomena are average statistical effects of molecules. But is it not the case that the electromagnetic radiation is nothing but the average statistical effect of photons? If electrodynamics can be a field theory, so can thermodynamics. Recent development in this direction includes works by Liu, Müller, and Ruggeri and by Geroch and Lindblom.[29] RTD may just have acquired a new lease of life in the study of irreversible thermal phenomena.

6. Concluding Remarks

Let me repeat that it is obvious that this essay is not written with the intention of solving the problem of RTD as a physical problem. Otherwise, I should have started from where the most recent research in this area has left off. In the historical account of RTD I have examined the major approaches and their theoretical, as well as methodological, difficulties. It is interesting to see that a theory initially as simple and innocent as RTD has turned out to be one of the most recalcitrant in resisting the efforts of relativization. In the study of the meaning of the special principle of relativity, I have shown that RTD of equilibrium thermal phenomena is the theory that truly distinguishes the active from the passive interpretation of the principle. Anderson's interpretation of the principle does explain why, in the purely formalistic search for the L-Ts of thermodynamic quantities, the results are arbitrary or conventional. But Anderson fails to prove that thermodynamic quantities *cannot* be made into genuine geometric objects. Radiation appears to be by far the simplest phenomenon in which the study of RTD can be conducted without too many assumptions and too much concern with microscopic details. RTD began with blackbody radiation: that may be where it will stay.

Acknowledgements—I wish to thank John Earman, John Norton, Clark Glymour and Al Janis for helping in the research which led to this paper. Special thanks go to Peter Landsberg who gave me constructive criticism and invaluable advice during the writing of previous drafts of this paper. I thank an anonymous referee for the comments. I also thank Mingmin Zhu for suggestions on grammar and style.

References

Anderson, J. L. (1964), 'Relativity Principles and the Role of Coordinates in Physics', in H.-Y. Chin and W. F. Hoffman (eds), *Gravitation and Relativity*. New York: W. A. Benjamin.

Anderson, J. L. (1967), *Principles of Relativity Physics*. New York: Academic Press.

Arzeliés, H. (1965), 'Transformation Relativiste de la Temperature et de Quelques Autres Grandeurs Thermodynamique', *Nuovo Cimento* **35,** 792–804.

Balescu, R. (1968), 'Relativistic Statistical Thermodynamics', *Physica* **40,** 309–38.

[29] I. S. Liu, I. Müller and T. Ruggeri, *Annals of Physics* **169** (1986), 191–219; R. Geroch and L. Lindblom, (1989), preprint, to appear in *Physical Review*.

Demiänski, M. (1985), *Relativistic Astrophysics*. London: Pergamon Press.
Earman, J. (1974), 'Covariance, Invariance, and the Equivalence of Frames', *Foundations of Physics* **4,** 267–89.
Earman, J. (1989), *World Enough and Space-Time*. Cambridge, Mass: The MIT Press.
Earman, J. and Norton, J. (1987), 'What Price Space-Time Substantivalism? The Hole Story', *British Journal for Philosophy of Science* **38,** 515–25.
Eckart, C. 'Thermodynamics of Irreversible Processes', *Physical Review* **58,** 267, 919.
Ehlers, C. (1971), 'General Relativity and Kinetic Theory', in R. K. Sachs, (ed.), *General Relativity and Cosmology*. New York: Academic Press, pp. 1–70.
Einstein, A. (1907) 'Über das Relativitäts prinzip und die aus demselben gezogenen Folgerungen', *Jahrbuch der Radioaktivität und Elektron* **4,** 411–62.
Einstein, A. and von Laue, M. (1952–1953) *The Einstein–von Laue Correspondence: 1952–1953*, in The Einstein Papers in Boston University.
Field, H. (1980), *Science without Numbers*. Princeton, NJ: Princeton University Press.
Friedman, M. (1983), *Foundations of Space-Time Theories*. Princeton, NJ: Princeton University Press.
Goldberg, S. (1969), Ph.D. Thesis. Harvard University.
Havas, P. (1964), 'Four-Dimensional Formulations of Newtonian Mechanics and Their Relations to the Special and the General Theory of Relativity', *Physical Review* **36,** 938–65.
Hiskes, A. L. D. (1984), 'Space-Time Theories and Symmetry Groups', *Foundations of Physics* **14,** 307–22.
van Kampen, N. G. (1968), 'Relativistic Thermodynamics of Moving Systems', *Physical Review* **173,** 296–301.
Landau, L. D. and Lifshitz, E. M. (1959), *Fluid Mechanics*. Oxford: Pergamon Press.
Landsberg, P. T. (1970a), 'Special Relativistic Thermodynamics—A Review', in E. B. Stuart, B. Gal-Or and A. T. Brainard, (eds), *Critical Review of Thermodynamics*. Baltimore: Mono Book Corp., pp. 253–72.
Landsberg, P. T. (1970b), 'Concepts in Special Relativistic Statistical Thermodynamics', in G. T. Conn and G. N. Fowler (eds), *Essays in Physics* vol. 2. New York: Academic Press, pp. 93–129.
Landsberg, P. T. (1978), *Thermodynamics and Statistical Mechanics*. Oxford: Clarendon Press.
Landsberg, P. T. and Johns, K. A. (1967), 'A Relativistic Generalization of Thermodynamics', *Nuovo Cimento* **52B,** 28.
Landsberg, P. T. and Johns, K. A. (1968), 'The Problem of Moving Thermometers', *Proceedings of the Royal Society* (London) **A306,** 477–86.
Landsberg, P. T. and Johns, K. A. (1972), 'Carnot Efficiencies for Moving Reservoirs', *Journal of Physics* **5A,** 1433–37.
Liu, C. (1991), *Relativistic Thermodynamics: Its History and Foundations*, Ph.D. Thesis, The University of Pittsburgh.
Liu, C. (1992), 'Einstein and Relativistic Thermodynamics in 1952: a Historical and Critical Study of a Strange Episode in the History of Modern Physics', *British Journal for Philosophy of Science* **25,** 185–206.
Møller, C. (1968), 'Thermodynamics in the Special and the General Theory of Relativity', in G. Puppe, (ed), *Old and New Problems in Elementary Particles*. New York: Academic Press.
Müller, I. (1985), *Thermodynamics*. Boston: Pitman Advanced Publishing Program.
Norton, J. (1988), 'Coordinates and Covariance: Einstein's View of Spacetime and the Modern View', *Foundations of Physics* **19,** 1215–64.
Ott, H. (1963), 'Lorentz-Transformation der Wärme und der Temperatur', *Zeitschrift für Physik* **175,** 70–104.
Paiva-Veretennicoff, I. (1974), 'Is Galilean Relativistic Thermodynamics Well Defined?' *Physica* **75,** 194–7.
Pathria, R. K. (1966), 'Lorentz Transformation of Thermodynamic Quantities: Part I', *Proceedings of the Physical Society* (London) **88,** 791–9.

Pathria, R. K. (1967), 'Lorentz Transformation of Thermodynamic Quantities: Part II', *Proceedings of the Physical Society* (London) **91,** 1–7.
Pauli, W. (1921), 'Relativitätstheorie', in *Encyklopädie der matematischen Wissenschaften* **V19,** Leipzig: B. G. Teubner; or *Theory of Relativity*. New York: Dover.
Planck, M. (1906), 'Das Prinzip der relativität und die Grundgleichungen der Mechanik', *Verandlungen der Deutschen Physikalischen Gesellschaft* **8,** 136–41.
Planck, M. (1907), 'Zur Dynamik bewegter Systeme', *Sitzungsberichte der Preussischen Akademie der Wissenschaften* (Berlin) **13,** 542–70.
Rohrlich, F. (1966), 'True and Apparent Transformations, Classical Electrons, and Relativistic Thermodynamics', *Nuovo Cimento* **45B,** 76.
Schmid, L. A. (1970), 'Effects of Heat Exchange on Relativistic Fluid Flow', in E. B. Stuart, B. Gal-Or and A. T. Brainard, (eds), *Critical Review of Thermodynamics*. Baltimore: Mono Book Corp., pp. 161–200.
Synge, J. L. (1957), *Relativistic Gas*. Amsterdam: North–Holland.
Sypel, R. and Brown, H. R. (1992), 'When is a Physical Theory Relativistic?', *PSA 1992* vol. 1, pp. 507–14.
Taff, L. G. (1968), 'A Relativistic Transformation for Temperature', *Physics Letters* **27A,** 605–6.
Tolman, R. C. (1934), *Relativity, Thermodynamics and Cosmology*. Oxford: Oxford University Press.
Trautman, A. (1964), 'Foundations and Current Problems of General Relativity', in *Lectures on General Relativity*. Englewood Cliffs, NJ: Prentice Hall.
von Treder, H. J. (1977), 'Die Strahlungs-Temperatur bewegter Korper', *Annalen der Physik* **34,** 23–9.
von Laue, M. (1911), *Die Relativitätsprincip*. Braunschweig: F. Vieweg.
von Laue, M. (1952), *Die Relativitätstheorie*. Braunschweig: F. Vieweg.
von Mosengail, K. (1907), 'Theorie der stationaren Strahlung in einem gleichformig bewegten Hohlraum', *Annalen der Physik* **22,** 867–904.
Wald, R. (1984), *General Relativity*. Chicago: Chicago University Press.
Yuen, C. K. (1970), 'Lorentz Transformations of Thermodynamic Quantities', *American Journal of Physics* **38,** 246–52.

Chuang Liu is Professor of Philosophy at University of Florida, and has been Beiren Project Professor at Shanxi University and Yangtze River Scholar at Tsinghua University, China.

www.ingramcontent.com/pod-product-compliance
Ingram Content Group UK Ltd.
Pitfield, Milton Keynes, MK11 3LW, UK
UKHW052114170926
8624IPUK00026B/392

9 789814 632188